D0912978

ULTRASTRUCTURAL PLANT CYTOLOGY

ULTRASTRUCTURAL PLANT CYTOLOGY

with an Introduction to Molecular Biology

BY

A. FREY-WYSSLING
Professor of General Botany

AND

K. MÜHLETHALER
Professor of Molecular Biology

AT THE SWISS FEDERAL INSTITUTE OF TECHNOLOGY
Zurich, Switzerland

ELSEVIER PUBLISHING COMPANY
AMSTERDAM - LONDON - NEW YORK
1965

ELSEVIER PUBLISHING COMPANY
335 JAN VAN GALENSTRAAT, P.O. BOX 211, AMSTERDAM

AMERICAN ELSEVIER PUBLISHING COMPANY, INC.
52 VANDERBILT AVENUE, NEW YORK, N.Y. 10017

ELSEVIER PUBLISHING COMPANY LIMITED
RIPPLESIDE COMMERCIAL ESTATE, BARKING, ESSEX

LIBRARY OF CONGRESS CATALOG CARD NUMBER 65-13236

WITH 223 ILLUSTRATIONS AND 44 TABLES

PRINTED IN THE NETHERLANDS

Foreword

This monograph on 'Ultrastructural Plant Cytology' is an extension of a part of the second edition of 'Submicroscopic Morphology of Protoplasm', edited by the senior author in 1953. At that time, electron microscopy was still in its infancy, whereas it has now developed into a decisive tool in cytological research, giving us an insight into the world of ultrastructures. The methods then in use for observing the fine structure of cells, such as ultra- and polarisation microscopy, have now receded into the background, and whereas formerly structures invisible under ordinary light microscopes had to be inferred by indirect means, today they may be represented directly. As a result, a large volume of data has been collected in the course of a few years, so that the whole field of ultrastructural morphology as a branch of general cytology can no longer be encompassed in a manageable text-book. For this reason, we have decided to restrict the description of the ultrastructures and the molecular biology aspects of cytology to the plant cells, commencing with a brief description of the general morphology of biogenic molecules. Despite this limitation, a full account of current methods on the new direction of research has had to be omitted; the basic principles will be found in the 1953 edition, which is still available.

We hope that this monograph of plant cytology brought up to date in the light of the present state of our knowledge will, besides being of assistance to biologists, physiologists, and biochemists, also reveal something of the close liaison existing between the spectacular achievements of organic chemistry and the classical realisations of microscopy to the more technically interested students of agronomy and forestry.

We wish to express our thanks to Dr. Elsa Häusermann for her help with the manuscript, to Ruth Rickenbacher for managing the bibliography, and to Mr. H. Eggmann, Dipl. Sc. Nat., for preparing the drawings.

Swiss Federal Institute of Technology
Department of General Botany
Laboratory of Electron Microscopy

A. FREY-WYSSLING
K. MÜHLETHALER

Zurich, January 1, 1964

Contents

PART II

Introduction

Ultrastructural Morphology

Ultrastructural cytology is based on molecular biology (Astbury, 1961). Its morphological branch encompasses the spatial prerequisites for the co- and interpenetration of the biochemical processes occurring in the cell, and thus provides the basis for cellular physiology. Just as human physiology was only able to develop to its present level after exhaustive clarification of the histological structure of the organs of the body, so cellular physiology will expand to an unprecedented extent once the cellular ultrastructures have been finally resolved.

The morphological sciences describe the spatial relationships of the structural elements. As shown in Table I they form a hierarchical system, since its components may be of very different dimensions. According to the nature of the units being studied, different instruments are required for their resolution. It will thus be seen that ultrastructural cytology lies between classical microscopic cytology and structural chemistry. While structural chemistry describes the interrelationships of certain radicals (methyl, hydroxyl, and aldehyde groups, for example) in organic molecules, and macromolecular chemistry the linking of such micromolecules to high-polymer macromolecules, ultrastructural morphology is concerned with the arrangements of these molecules forming particles, elementary fibrils, helices, or lamellae invisible under the light microscope, and with the associations of such structures in all kinds of tissues. In the chemistry of proteins, the order of succession of the micromolecular monomers along the high polymer molecular chain is designated as a primary structure, the shape of the chains as a secondary structure, the formation of globular or fibrillar particles as a tertiary structure, and the

TABLE I

MORPHOLOGY

Morphological hierarchy		Instruments of research	Scales	Order of magnitude
organs	organography	eye, magnifying glass	millimetre scale	> 0.1 mm
tissues	histology	light microscope	micron scale	> 1 μ
cells	cytology	immersion, phase contrast, and ultra-violet microscope	wavelengths of light	> 0.1 μ
ultrastructures	ultrastructural morphology	electron microscope	millimicron scale	< 100 Å
molecular structures	structural chemistry	X-rays	wavelengths of X-rays	> 1 Å

incorporation of such particles into associations as a quaternary structure (see p. 57). Whilst the primary and secondary arrangements can be taken as belonging to the sphere of structural chemistry, the tertiary and quaternary structures belong to the field of ultrastructures.

We can thus see how molecular biology joins the former 'inimical brothers' of histological morphology and physiological chemistry in fruitful cooperation. There was a time when the two sciences hardly understood each other: the biochemist spoke sarcastically of the histologist as a stamp collector because his purpose was to prepare and assemble flawless mounts, whilst the histologist regarded the biochemist as a poor biologist making a homogenate of all the various cell constituents and drawing whimsical conclusions from the analysis of this awful mixture. Morphology now knows that, in the last resort, its structural units are macromolecules, and the biochemist is aware that the cycles discovered by him from numerous reaction equilibria can function only when the enzymes are placed in series of well-ordered structures. Thus the old animosity between the morphological and chemical sciences is surmounted and forgotten by the discoveries in the field of ultrastructures.

The size of the structural units considered in morphology is distinguished according to their dimensions as macroscopic, microscopic, sublight-microscopic and amicroscopic. The limitation of these domains is provided by the resolving power of the eye (about 0.09 mm), of the light microscope (about 0.3 μ) and of the electron microscope (about 10 Å). If the measurements are expressed in nm, a proportion of 90 000:300:1 is obtained, showing that light microscopy contributed a threehundred-fold extension of what was observable with the naked eye, and that a similar extension was again achieved with the discovery of the electron microscope.

The measurements of length normally used in morphology are given logarithmically in Table II. As presented here, the microscopic and sublight-microscopic (ultrastructural) fields are equally large. It is recognised that organographical objects are macroscopic and that both histological and cytological structural units are microscopic. Ultrastructures and macromolecules are invisible under the light microscope and are consequently sublight-microscopic, but they are capable of resolution and portrayal under the electron microscope. Micromolecules (molecular weight < 500) and atoms cannot be seen at present even with the electron microscope. They are amicroscopic, but, unlike the case of the light microscope, the ultimate limits of resolution for the electron microscope have not as yet been reached. In 1940 the limit was situated at about 50 Å, and has since been improved to less than 10 Å. As a result, the ultrastructural domain has been considerably extended beyond the region of the smallest colloidal particles, which are what the gold particles with a diameter of about 60 Å detected under the ultramicroscope by Zsigmondy (1925) are held to be. At the beginning of electron microscopy, the sublight-microscopic region was almost congruent with the region of the colloidal particles, whereas today hemicolloid particles can also be shown. Whilst, therefore, the upper limit of the ultrastructural field is fixed by the resolution limit of the light microscope, the lower limit is variable. For this reason, and because ultrastructures have become portrayable under the electron microscope, the original term 'submicroscopic' morphology has been criticised and replaced by the new

TABLE II

THE DOMAINS OF MORPHOLOGY

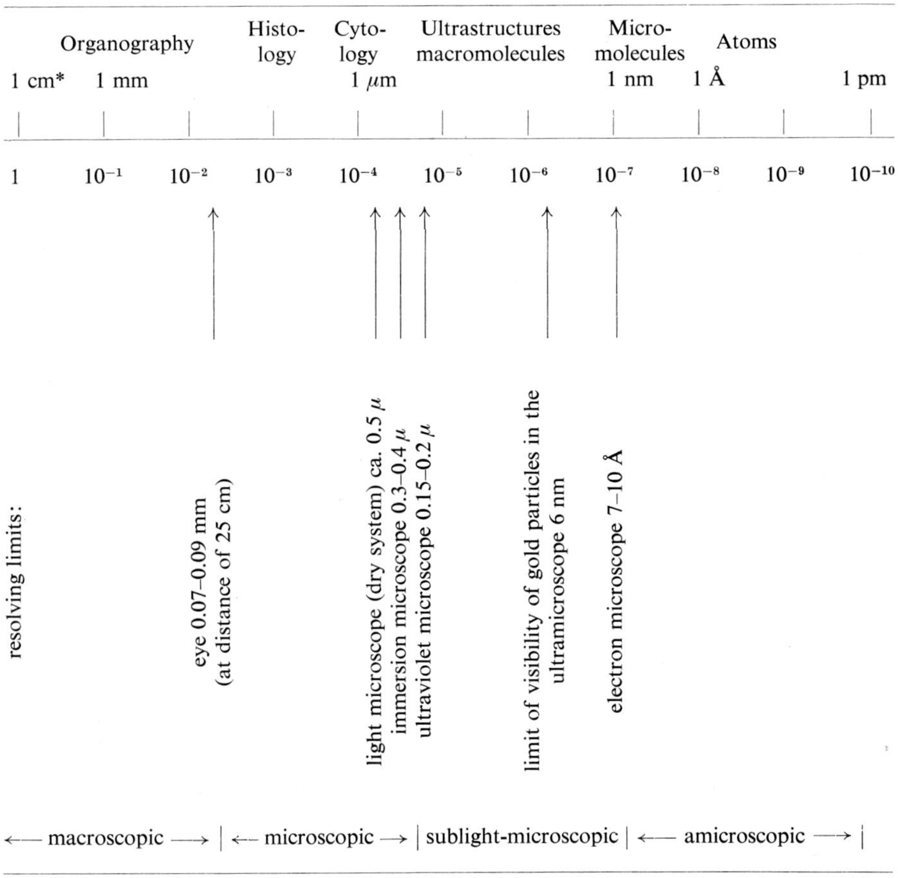

Organography			Histo-logy	Cyto-logy	Ultrastructures macromolecules		Micro-molecules	Atoms		
1 cm*	1 mm			1 μm			1 nm	1 Å		1 pm
1	10^{-1}	10^{-2}	10^{-3}	10^{-4}	10^{-5}	10^{-6}	10^{-7}	10^{-8}	10^{-9}	10^{-10}
resolving limits:		eye 0.07–0.09 mm (at distance of 25 cm)		light microscope (dry system) ca. 0.5 μ; immersion microscope 0.3–0.4 μ; ultraviolet microscope 0.15–0.2 μ		limit of visibility of gold particles in the ultramicroscope 6 nm	electron microscope 7–10 Å			

← macroscopic → | ← microscopic → | sublight-microscopic | ← amicroscopic → |

* In recent times, by international convention, the scale of lengths is no longer based on the centimetre (cm) but on the metre (m). 10^{-3} m = 1 millimetre (mm), 10^{-6} m = 1 micrometre (μm; abbrev. micron, μ), 10^{-9} m = 1 nanometre (nm; formerly millimicron mμ) and 10^{-12} m = 1 picometre (pm). The Ångström unit (Å) = 10^{-8} cm or 10^{-10} m, which was introduced by crystallographers and is generally used in electron microscopy, has now been dropped. Since ultrastructure research has not as yet endorsed this innovation, in the present book we thought it best to adhere to the traditional scale.

term 'ultrastructural' morphology. However, this term is not necessarily an improvement, because it is taken from the terminology of colloidal chemistry (e.g. ultracentrifuge, ultrafiltration, ultramicroscope) which uses the prefix 'ultra' to designate particle sizes which are beyond the resolving power of the light microscope. As a matter of fact the term 'submicroscopic' meant much the same thing. The objections of the electron microscopist can be met by defining the term more precisely by the use of the word 'sublight-microscopic'.

It is of great importance that we should start with a correct idea of the dimensions in the ultrastructural domain. The electron microscope, capable of more than 100,000-fold magnification, is nowadays able of producing images that frequently leave us unaware of the excessively small units observed, or of how far we have already advanced into the realm of molecular dimensions. The transition, therefore, from objects visible under the light microscope to ultrastructural objects is presented in Table III in a sliding scale of particle sizes. On this scale, however, only particles up to the size of the T2 coli phages can be given; red blood cells, which are a hundred times larger, would show a diameter of 3.75 m.

In earlier times, the invisible sublight-microscopic particles were characterised by their particle weight as determined in the ultracentrifuge, and these data are given under the heading of 'molecular weight' in Table III. It is known that the so-called large viruses possess molecular weights of more than 2 billion, with particle diameters (210–320 nm) which are larger than the resolution capacity of the ultraviolet microscope (150 nm). The small viruses with particle weights of the order of 10 million and diameters of about 10–20 nm are no larger than the largest globular protein and carbohydrate macromolecules. Both viruses and haemocyanin

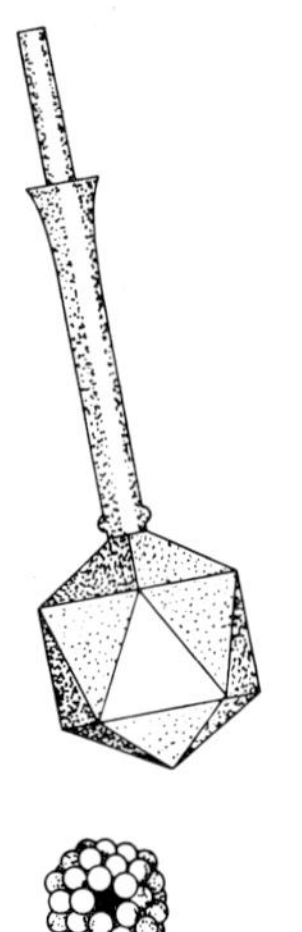

TABLE III

PARTICLE SIZES

Object	Molecular weight	Diameter in nm	Authors
1. Red blood cell		7500	
2. *Escherichia coli*		3000×6000	
3. Rickettsia		300	
4. Smallpox virus	$> 2 \times 10^9$	230 — 320	Peters (1960)
5. Coli-phage T2	210×10^6	65×95	Taylor, Epstein and Lauffer (1955) Kellenberger (1961)
6. Coli-phage T7	38×10^6	59 — 65	Davison and Freifelder (1962)
7. Tomato bushy stunt virus	8.9×10^6	30	Hersh and Schachmann (1958)
8. Polio virus	6.8×10^6	28	Schaffer and Schwerdt (1959)
9. Haemocyanin (*Helix pomatia*)	6.6×10^6	30×33	Van Bruggen *et al.* (1962)
10. Ribosomes (50 S)	1.85×10^6	14×16	Tissière and Watson (1958) Huxley and Zubay (1960)
11. Smallest ultramicroscopically visible gold particle	2.7×10^9	6	Zsigmondy (1925)
12. Haemoglobin	67×10^3	$5.5 \times 6.4 \times 5.0$	Perutz *et al.* (1960)
13. Myoglobin	17×10^3	$4.3 \times 3.5 \times 2.3$	Kendrew *et al.* (1958)
14. Saccharose	342	0.5×1.0	
15. Hydrogen molecule	2	0.2	

or haemoglobin are composed of sub-units (see p. 72), from which arises a further analogy of these objects. The protein macromolecules referred to are sometimes larger and sometimes smaller than the smallest colloidal gold particles seen under Zsigmondy's ultramicroscope. Cane sugar and hydrogen are given as examples of amicroscopic micromolecules; they are so small that they cannot be indicated on the scale we have used.

The smaller the particles become at this level, the less we are permitted to think of their internal morphology as something fixed, for the sub-units, radicals, atomic nuclei, and electrons of which they are composed are constantly in a state of oscillation and rotation. The same thing applies to the structural units themselves, when they are associated in larger groups. The molecular fine-structure of the cell organelles which is considered in this book should not therefore be taken as inflexible; we must be aware that ultrastructures, due to metabolism and growth, are in a state of constant rearrangement.

Growth includes the problem of the ontogenetic development of the cell organelles. It is a great advance that cytological ontogenesis can today be followed under the electron microscope, so that ultrastructural research has been brought from the realm of biophysical science back to pure biology. Whilst previously the only method available for the understanding of an ultrastructure consisted of a static description, starting from its macromolecular building blocks through their associations and ultrastructural arrangements up to the structure visible in the light microscope, today the main task is the dynamic elucidation of their origin from micromolecules engaged in metabolic processes. For this reason a knowledge of molecular morphology must precede the undertaking of this task, and therefore, in the first part of this monograph, the forms of macromolecules and their interactions will be described. This is followed by a detailed description of investigations on the fine structure and development of the cell organelles, as a contribution of cytology to the modern science of molecular biology.

PART I

Molecular Morphology

A. Principles of Molecular Structure

Study of the positions of atoms within molecules is the main object of structural chemistry, which in this respect appears as a morphological science. For example, the familiar tetravalent representation of the carbon atom, or the hexagonal representation of benzene, are morphological illustrations (Fig. A–1). The exact spatial orientation of the bonds and the interatomic distances remained unknown for a long time, and the directions and lengths of valencies were represented in a rather arbitrary manner (see Fig. B–2b). At the present time a large volume of data required for an exact morphological representation has become available, and at least the simpler chemical formulae can be drawn to represent actual three-dimensional molecular models projected on the plane of the paper. The exact knowledge of distances and directions is largely due to X-ray analysis, which allows the measurement of distances of the order of an X-ray wavelength (e.g. 1.54 Å for copper radiation), provided that the distances in question are repeated systematically and behave as a lattice. Such lattices cause interference of the incident X-radiation, and give rise to macroscopic effects which can be recorded photographically. It is therefore this principle of repetition which enables us to explore the morphology of molecular structure; the more regularly the given distances are arranged, the greater the accuracy with which the absolute lengths and directions can be determined. This means that X-rays cannot help us in the study of the morphology of molecules in liquids and gases, although the solutions of certain very large molecules, whose construction itself shows a certain periodicity (e.g. carbon chains), constitute an exception. In such cases the measurements are however associated with a certain degree of uncertainty, since the molecules are not oriented in fixed directions. The most reliable values of interatomic distances, frequently reaching an almost unbelievable precision (up to 0.01 Å), have therefore been determined in crystal lattices.

The usefulness of X-ray analysis is unfortunately rather limited in cytology. Although we must attribute a certain structure to the protoplasm, this is not governed by the principle of repetition to an extent sufficient for X-ray study. Periodicity does play an important part in all living matter, but more with respect to time than to the arrangement in space. A rigidly periodic order in space would presuppose an equilibrium of forces, whilst life is based on movement and on the maintenance of non-equilibria. However, as soon as a chemical substance is withdrawn from the metabolic process, the ordering forces can intervene and form

Fig. A–1. Projection of simple molecules on to a plane.

periodic structures, such as the skeleton substances cellulose, chitin, collagen, keratin, etc. Structure of the protoplasm must therefore be studied by other methods, which are however based in part on the investigations of crystal structure. For this reason a brief description will now be given of this important branch of morphology.

1. CRYSTAL STRUCTURE

(*a*) *Lattices*

The essential characteristic of lattices is that certain locations of points, which in the simpler cases correspond to the centre of gravity of the atoms, repeat themselves periodically in space in three directions coinciding with the crystallographic axes. The distances from any one point to the next identical one are known as identity periods or lattice spacings. Such spacings may be equal in all three directions (cubic system), in only two directions (tetragonal, hexagonal, and rhombohedral systems), or they may be different in all three directions (as in the rhombic, monoclinic, and triclinic systems). The regularly repeated points thus form rows, which may be displaced by constant amounts in directions either perpendicular or oblique to their own direction to give lattice planes, which may in turn be displaced to yield the crystal lattice. If a certain point in such a lattice is moved in the three principal directions, each time covering the identity period concerned, and if the resulting three vectors are formed into a three-dimensional parallelepiped, we obtain the so-called elementary or unit cell characteristic of the crystal lattice. By analogy with a molecule of gas which represents the smallest unit possessing all the chemical properties of the macroscopic gaseous phase, the elementary cell is the smallest complete unit of the pattern, showing all physical and symmetry properties of the macroscopic crystal. The elementary cell can consist of one or more molecules or, in the case of high polymers, even parts of molecules. We are therefore dealing with a geometrical and not with a chemical concept, as the crystalline properties disappear when the unit cell is decomposed into its elements. Since the unit cell is characteristic of the lattice as a whole, structural analysis is concerned with the determination of the dimensions and the symmetry of such unit cells. The shape is defined by the three identity periods a, b, and c (in Å) to which, in the monoclinic and triclinic systems, we must add the angle β or the angles α, β, γ formed by the axes of the cell. The macroscopically determined ratios of the crystal axes are in good agreement with the ratios measured in the unit cells, provided that analogous planes are considered.

The unit cell must be such as to permit complete filling of space when the cell is translated along the crystal axes. Only two-, three-, four-, and sixfold rotation axes are therefore possible in crystal lattices (see p. 122). There are 14 such space lattices (Bravais lattices), containing one, two (body-centred, base-centred), or four (face-centred) non-identical points. If these points are not ideally centred they are multiplied by the relevant symmetry elements (mirror planes, axes of rotation, helical axes), giving rise to 230 space groups which fulfil the requirement of complete space filling (see point groups, p. 122).

X-ray analysis allows us to measure the interplanar distances in crystal

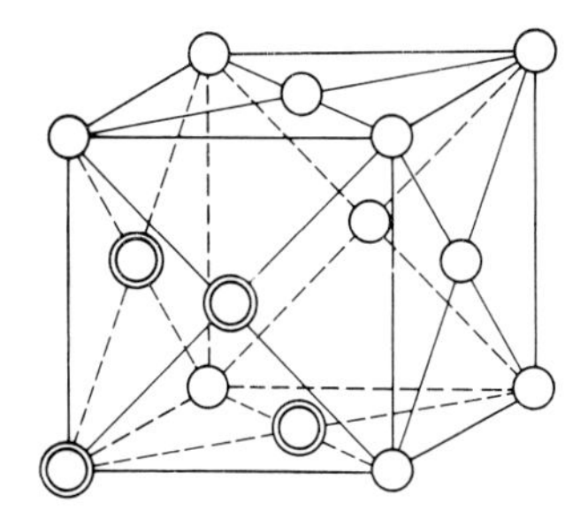

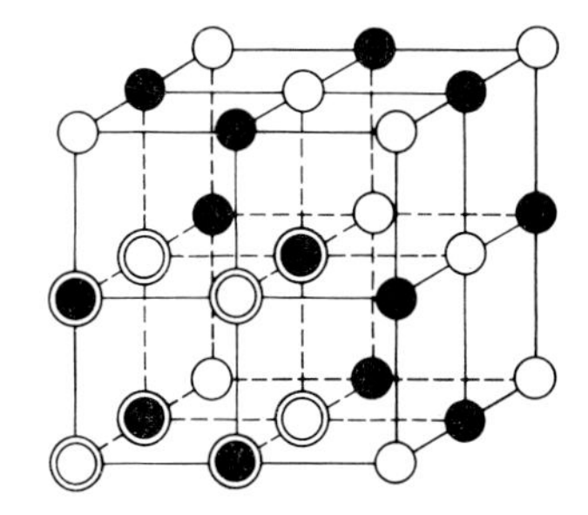

Figs. A–2 and A–3. Crystal lattices. The encircled points belong to the unit cell. Fig. A–2. Gold; a = 4.07 Å, ○ = Au. Fig. A–3. Sodium chloride; a = 5.60 Å, ● = Na, ○ = Cl.

lattices. In crystals showing a high degree of symmetry, such as those belonging to the cubic system, the lattice points are identical with the points of intersection of the symmetry planes, and lattice spacings can therefore be calculated from the distances in the X-ray patterns. The situation becomes more complex if the crystal has a lower degree of symmetry, as the points in the lattice planes possess a certain degree of freedom, and their positions are not defined unambiguously. Determination of such structures therefore requires measurement of the intensities of the diffraction lines, as well as of their spacings on the X-ray diagram. Frequently however the positions of the lattice points within the unit cell can only be obtained approximately (see Figs. B–7, B–8 and B–18b).

Figures A–2 and A–3 represent two of the best known crystal lattices, of metallic gold and of sodium chloride. Both lattices are cubic, so that the size and shape of each unit cell is defined by a single lattice spacing a, which is the same in three mutually perpendicular directions. Determination of a by X-ray diffraction, yields the volume a^3, and, if the density of the crystalline substance is known, the weight of the unit cell. The number of atoms or molecules in the unit cell (see p. 74) may then be calculated by dividing this weight by the absolute weight of the atoms or molecules concerned (atomic or molecular weight divided by Avogadro's number, 6.02×10^{23}).

Such calculations show that the unit cell of gold contains 4 atoms of gold, whilst the unit cell of sodium chloride consists of 4 sodium and 4 chloride ions; these points are shown circled in Figs. A–2 and A–3. The other uncircled points should be considered as originating from the circled ones by simple translation, and thus belong to the neighbouring unit cells. The gold lattice is designated as face-centred because the points at which the diagonals on the cube faces intersect are all occupied by atoms. Many elements, e.g. Ag, Cu, Al, Pb, etc. crystallise in this type of lattice, being distinguished crystallographically only by their lattice spacings a. Two such face-centred lattices overlap in the NaCl lattice, which is found in several binary compounds (NaF, KCl, PbS, etc.), again with different values of a.

(*b*) *Coordination*

In the above-described examples the lattice forces (see p. 19) are of different nature, although the two are alike. In Fig. A–2 the gold atoms are held together by metallic bonds, whilst in Fig. A–3 the bonding is ionic.

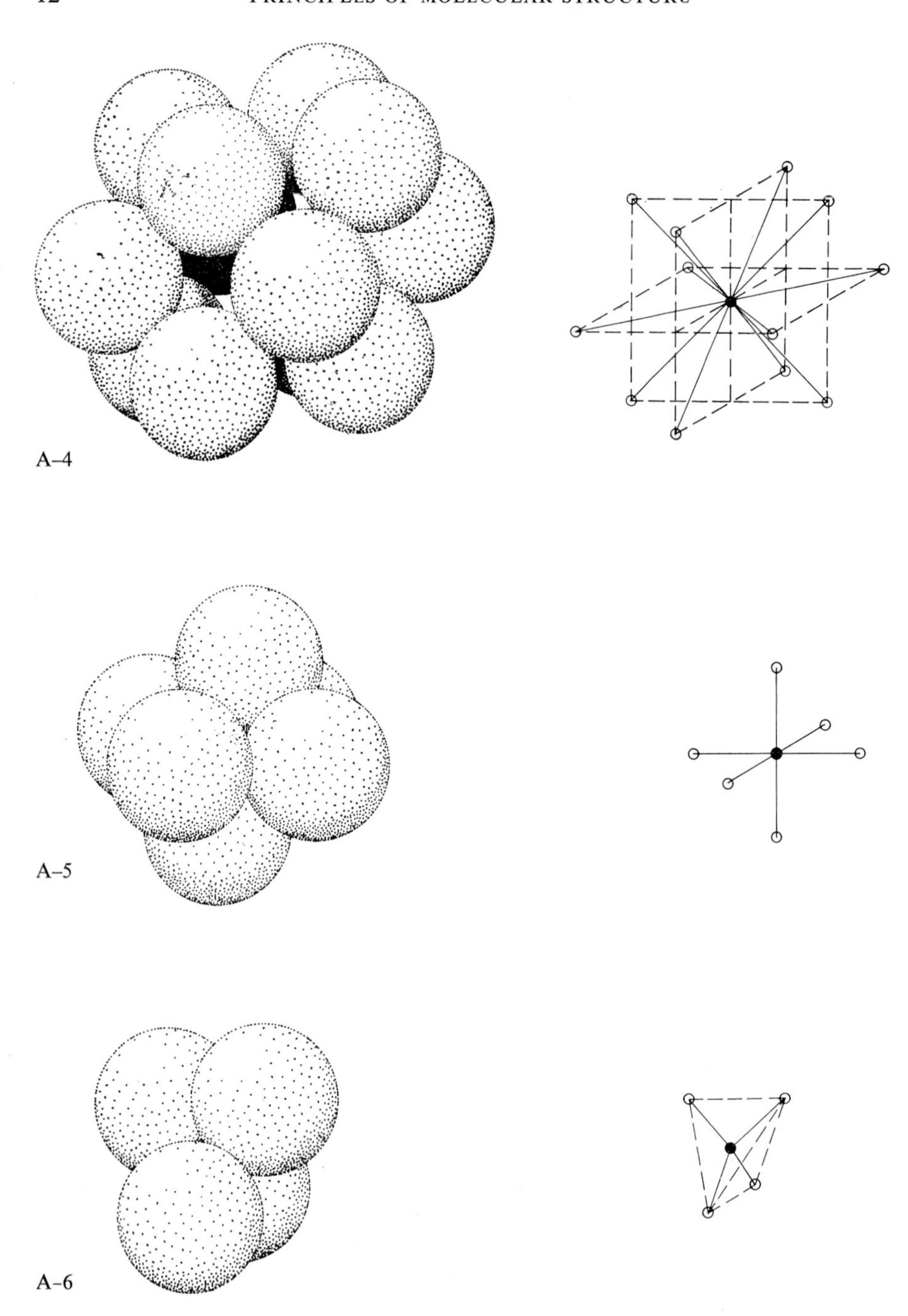

Figs. A–4 through A–6. Coordination numbers (Magnus, 1922).
Left side: space packing models; right side: steric frame models.
Fig. A–4. Number 12; e.g., Au $(Au)_{12}$ in crystallised gold.
Fig. A–5. Number 6; e.g., $Na(Cl)_6$ in sodium chloride; $Fe(CN)_6$ as ion.
Fig. A–6. Number 4; e.g., CCl_4, $C(C)_4$ in diamond.

This may be explained by the fact that in both cases construction of the lattice is governed by rules described in the theory of coordination. According to Werner's chemistry of complexes, every atom is surrounded by 4, 6, 8 or 12 neighbouring particles, the actual number depending on volume considerations. This theory, which was originally based on the composition of salts containing water of crystallisation [e.g. $Ca(H_2O)_6Cl_2$], and complex salts, has also proved useful in elucidating the crystal structures of other compounds and of the elements. Thus it may be seen that every gold atom in Fig. A–2 is surrounded by 12 neighbours, whilst every Na ion in Fig. A–3 is surrounded by 6 Cl ions and *vice versa.*

The theory of coordination has led to another fundamental discovery, which has become of great importance in the ultrastructural morphology of organic compounds. The lattice points of Figs. A–2 and A–3 represent only the centres

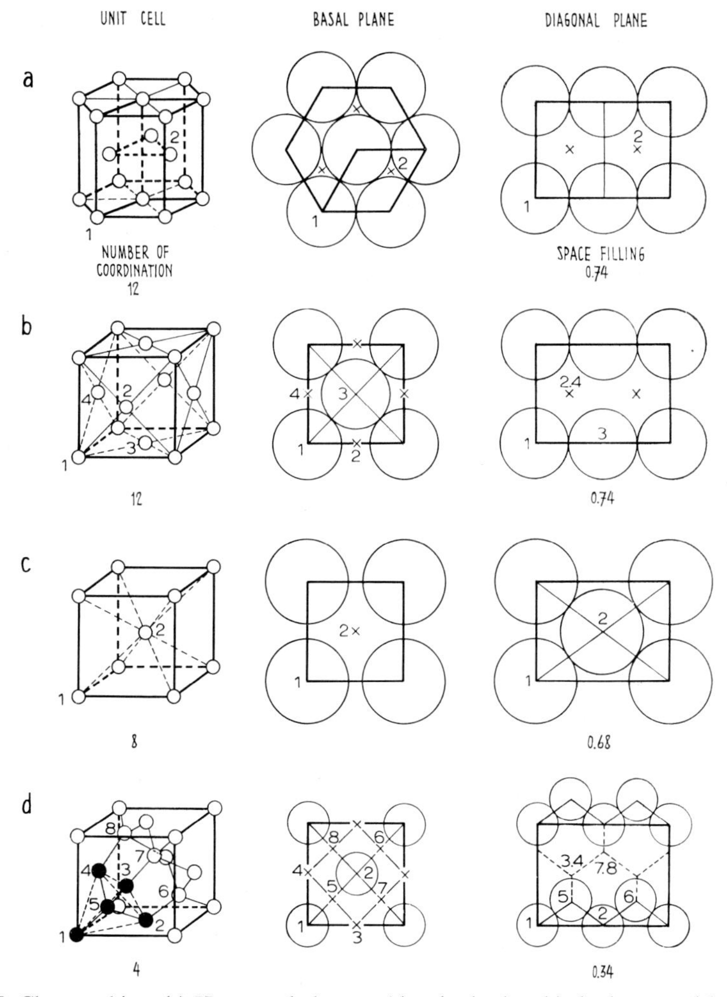

Fig. A–7. Close packing. (a) Hexagonal close packing (orthorhombic body-centred lattice with $b : a = \sqrt{3}$). (b) Cubic close packing (cubic face-centred lattice). (c) Cubic body-centred lattice. (d) Diamond lattice.

of the atoms. In actual fact the electron shells extend over relatively large distances around each such point, so that crystal lattices are better represented as regular collections of finite spheres in contact with each other (Figs. A–4 to A–6). Crystal lattices, held together by chemical bonds, are therefore much more closely packed than is suggested by the common representations, although the latter possess the great advantage of clarity. In a single plane, the use of spheres rather than points can be very successful (see Fig. A–8). The interatomic distances in the lattices of elements thus correspond to atomic diameters, and in binary compounds they represent the sum of the two radii of the constituent ions. In this way, it has been found possible to determine the volumes occupied by various atoms, and to account for the observed coordination numbers. For example, four chlorine atoms arranged at the corners of a tetrahedron enclose a space which is just filled by one silicon atom, accounting for the coordination number of 4 in $SiCl_4$, whilst as many as six of the smaller fluorine atoms are needed to enclose a corresponding volume, explaining the coordination number of 6 in SiF_6.

It is easy to see that the coordination number has a decisive influence on the density of the resulting crystal lattice. The closest possible packing (and the greatest density) is given by 12-fold coordination (Fig. A–4). Under these considerations the spheres occupy 74% of the lattice space, leaving only 26% of the volume for interstices. With a coordination number of 8, present in the body-centred cubic lattice (Fig. A–7c), the spheres fill only 68% of the available space, leading to a lower density. Finally, the tetrahedral 4-coordination (Fig. A–6) gives a rather loose structure, in which only 34% of the available volume is occupied (Fig. A–7d). This explains why in spite of their extremely high hardness, diamonds, which posses the latter structure, have rather a low density.

(*c*) *Interatomic distances and bond energies*

If the lattice contains covalent bonds, the distances between constituent atoms or, in other words, atomic diameters, exhibit a surprising constancy, not only in simple compounds, but also in very complicated organic molecules. In ionic lattices a disturbing effect may occur because the oppositely charged ions exert a polarizing effect upon each other. This can lead to deformations of the electron orbits (Fajans, 1925), so that the ions can no longer be regarded as spherical; for this reason ionic lattices often possess a low symmetry, and the interionic distances between two given partners are subject to fluctuations.

This complication does not arise in the structures of non-ionic organic compounds, and interatomic distances determined in such compounds are fairly characteristic of a given pair of atoms. We can therefore speak of distances rules. A number of interatomic distances determined in organic crystals with the aid of X-rays is given in Table A–I. In these considerations the hydrogen atoms can be neglected, since they do not scatter X-rays. Table A–I, for example, shows that in single bonds the atomic radius of carbon, r_C, amounts to 0.77 Å and that of nitrogen, r_N, is 0.70 Å. Thus in spite of the larger atomic weight of nitrogen, its sphere of influence is smaller than that of carbon. It may further be observed that the atomic radius of carbon is decreased by double bonding.

A definite amount of energy can be associated with every bond in an organic molecule (Meyer and Mark, 1930). It is found, for instance, that the molar heats

TABLE A-I

INTERATOMIC DISTANCES AND BOND ENERGIES IN ORGANIC COMPOUNDS

Type of bond	Interatomic distance[a] Å	Bond energy[b] kcal/mol
aliphatic C—C	1.54	79.3
(diamond lattice)	1.54	
aromatic C=C	1.39	116.4
(graphite lattice)	1.42	
aliphatic C=C	1.34	140.5
C≡C	1.20	196.7
C—O (ethers, alcohols)	1.43	79.6
(acids, esters)	1.36	
(epoxy compounds)	1.47	
C=O	1.23	168.7
C—N (amines)	1.47	65.9
(amides)	1.32	
C—H (aliphatic)	1.10	98.7
(aromatic)	1.08	100.7
O—H (alcoholic)	0.97	104.7
N—H	1.04	92.0
hydrogen bond O—H---O		
O—H	0.99 } in ice	2.5
O—H---O	2.76 } in ice	

[a] Tables of Interatomic Distances and Configuration in Molecules and Ions. The Chemica Society, London, 1958.
[b] Moelwyn-Hughes, 1957.

of combustion of homologous paraffins increase by a definite increment (~ 70 kcal) for each additional C-atom introduced into the paraffin chain. The energy equivalents for the other bonds listed in Table A–I have been determined in a similar way. It will be apparent that the bond energy increases with decreasing distance between the carbon atoms.

(*d*) *Bond angles*

The discovery of stereoisomerism demonstrated the importance of distinguishing between the different positions of substituents on the carbon atom. At first the results (Werner, 1904; Freudenberg, 1933) were of a rather qualitative nature, and referred in a general way to the directions radiating from the C-atom, as quantitative evaluations of the distances along and the exact angles between these directions were not possible at the time. Modern crystallographic techniques allow us however to determine not only qualitatively but also quantitatively the relative positions of the atoms in space.

The starting point of this development of structural chemistry was the lattice of diamond, which crystallises in the cubic system. The unit cell of diamond is a cube containing 8 carbon atoms, 4 of which belong to a face-centred cube as in the case of gold, and the remaining 4 atoms being situated on the body diagonals half-way between the corners of the cube and its centre (Fig. A–7d). Thus the unit

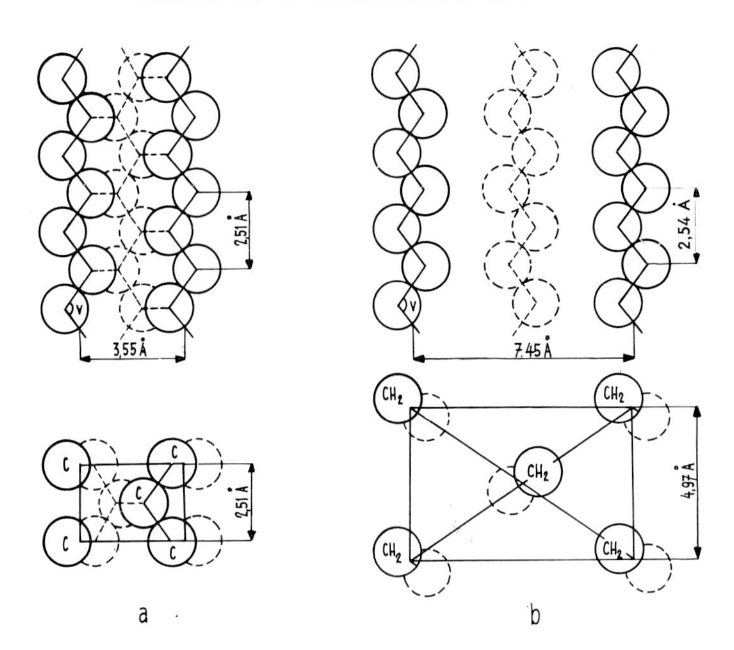

Fig. A–8. (a) Diamond lattice (covalent lattice) inclined at 45° to the orientation shown in Fig. A–7d. (b) Paraffin lattice (molecular lattice). Bond angle v = 109° 30′.

cell contains as it were 4 central atoms, surrounded tetrahedrally by 4 neighbouring atoms, in conformity with their coordination number (Fig. A–6). If this three-dimensional lattice is projected onto its base, a square is obtained, which shows the bond arrangement commonly employed in organic chemistry. The usual diagrammatic representation of tetravalent carbon (Fig. A–1) is therefore morphologically correct if it is considered as a plane projection of a tetrahedron.

According to X-ray analysis, the lattice spacing of diamond, i.e. the edge of the elementary cube, measures 3.55 Å. It follows that the distance between the lattice points on the face diagonal is $1/2 \times 3.55 \times \sqrt{2} = 2.51$ Å, and the shortest distance between two carbons on the body diagonal is $1/4 \times 3.55 \times \sqrt{3} = 1.54$ Å. The C—C distance corresponding to the sphere of action of a carbon atom in an aliphatic bond has been calculated in this simple way (Table A–1).

If a plane is drawn through the two body diagonals an arrangement of lattice points is obtained which is shown in Fig. A–8a. In this cross-section the carbon atoms are joined by a zig-zag line whose links enclose the so-called tetrahedral angle of 109° 30′. Further arrays of such zig-zag chains are found to lie in parallel planes, one of which has been represented by dotted lines. The planes are linked by covalent bonds.

The angle between the directions of two bonds radiating from an atom is called a bond angle. In the diamond lattice, this is the same as the tetrahedral angle, whilst in other carbon compounds the bond angle can be modified by the substituents around the carbon atom (see Fig. B–28). In general, the bond angle does not deviate considerably from 110°.

(*e*) *Aliphatic compounds*

The zig-zag ordering described above is fundamental to the morphology of saturated carbon compounds; it has been found that all aliphatic molecules form such kinked chains. Thus in paraffin molecules the increase in chain length is 1.27 Å for each additional C-atom, instead of the expected 1.54 Å. It can be readily calculated that this is in accordance with zig-zag chains showing the tetrahedral angle. In this way two adjacent carbon atoms have a spacing of 2.54 Å, which is the intramolecular period in paraffins (Hengstenberg, 1928; Müller, 1929).

Fig. A–8b shows how by parallel alignment such chains combine into the rhombic crystal lattice of the paraffins. At first sight it may seem paradoxical that the soft, plastic paraffin crystals should have a lattice structure so similar to the diamond model represented by Fig. A–8a. Despite the apparent analogy, however, fundamental differences exist between the two lattices, which explain the differences observed in the physical behaviour of the two substances. In particular, the paraffin crystals are built much more loosely, owing to the fact that the paraffin lattice is molecular not covalent. The paraffin chains are held only by the relatively weak Van der Waals' forces, since the CH_2 groups can bind only two neighbouring groups by covalent bonds. Thus in the paraffin lattice we find two orders of distances: molecular distances of the order of 5 Å, and atomic spacings of the order of 1.5 Å (Fig. A–8b). The fact that in the diamond lattice all carbon atoms are in contact explains the greater density and hardness of diamond. In contrast, the paraffin lattice has a smaller density, and the molecular chains can be readily shifted with respect to each other (Fig. A–9a); this accounts for the softness and plasticity of paraffin crystals.

As long as the paraffin chains are short, they crystallise easily into a molecular lattice, giving rise to flaky crystals exhibiting basal cleavage (Fig. A–9a). When

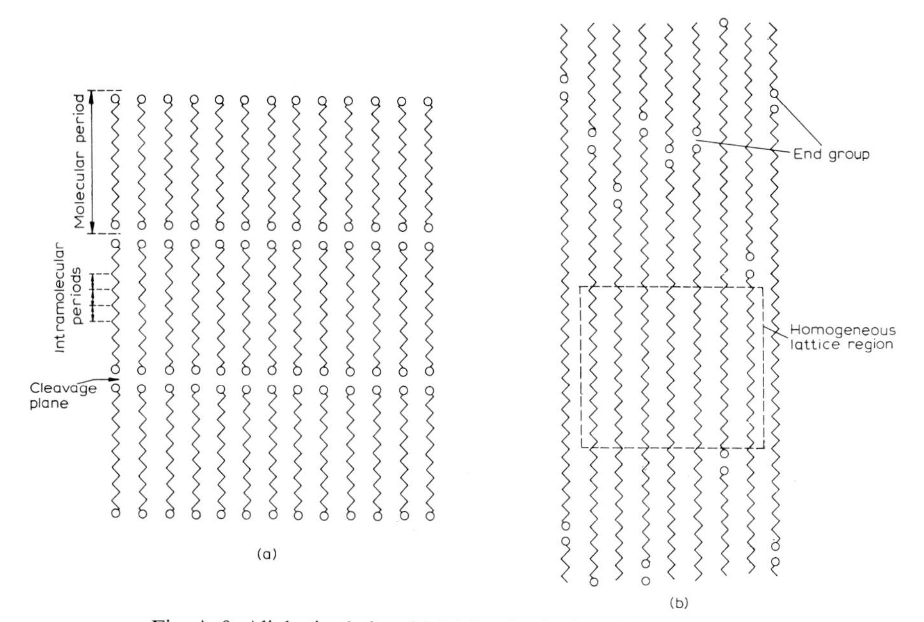

Fig. A–9. Aliphatic chains. (a) Molecular lattice. (b) Chain lattice.

however the chains become very long, it becomes increasingly difficult to arrange the terminal groups in fixed planes, and crystallisation takes place as in Fig. A–9b. No rigorous lattice order is obtained in this case, since longitudinal displacements of the chains with respect to each other over distances equal to an intramolecular spacing, that is through only a fraction of the chain length, do not affect the lattice structure. This is because the smaller intramolecular spacings (2.54 Å in the case of paraffins) become so important on account of their large number that the periodicity of the end groups becomes less significant. Such arrangements of long molecules are known as *chain lattices*. The chains cannot rotate around their longitudinal axis (since otherwise there would be no lattice order), so that the cross-section of the chain lattice is homogeneous. Longitudinal inhomogeneities do occur, which in Fig. A–9b are indicated by circles marking the end groups. It is clear, however, that longitudinally homogeneous lattice regions can also exist when the chains are sufficiently long.

(*f*) *Aromatic compounds*

Unlike the case of aliphatic compounds, aromatic structures cannot be derived from the structure of diamond. The aromatic structures are rather more similar to that of *graphite*. The latter modification of carbon crystallises in the hexagonal system, and possesses a lattice illustrated in Fig. A–10. The carbon atoms form hexagonal rings linked together in an uninterrupted plane, so that three bonds are engaged at each lattice point. The fourth valency is distributed

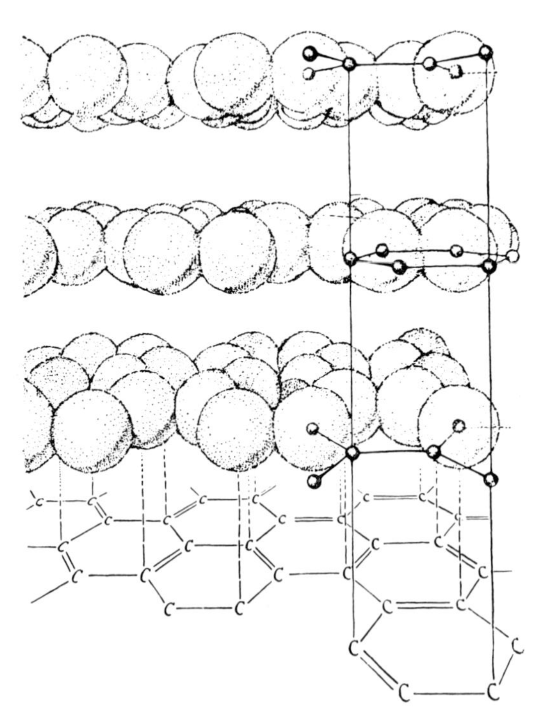

Fig. A–10. The graphite lattice (Pauling, 1953). C—C distances: within the ring 1.42 Å, between planes 3.41 Å.

between the neighbouring atoms in the same plane, as in the benzene ring (Fig. A–1). In consequence of the resulting higher bond energy, the C—C distance in graphite is reduced to 1.45 Å (see Fig. A–10). All bonds are thus engaged in planes, held by weaker forces to form a lattice. The layer spacing (3.41 Å) is therefore considerably larger than the C—C distance in the rings. A structure in which the lattice forces and spacings within a plane are so strikingly different from those in a direction perpendicular (or nearly perpendicular) to this plane is termed a layer lattice. Compounds of this type always crystallise in the form of flakes, and are in general easily cleavable along the base (mica, sericite). Many benzene derivatives and other aromatic compounds (naphthalene, anthracene, etc.) belong to this class. The division into aliphatic and aromatic substances is therefore based not only upon chemical behaviour, but also possesses a morphological background in that aliphatics tend to crystallise into chain lattices, whilst aromatics show a strong tendency towards the development of layer lattices.

2. BOND FORCES

As may be seen on the example of the graphite lattice, the shorter interatomic distances are caused by stronger bond forces. Thus in the paraffin lattice, the C—C distance of 1.54 Å is due to strong covalent bonding, whereas the 7.45 Å long spacing between the paraffin chains represents a weaker molecular bond due to cohesive forces, which are said to be physical. There is however no clear-cut difference between chemical and 'physical' types, because bonds of intermediate nature exist, which make such a classification illusory. Even the chemical bond is not an unambiguous concept, since its definition leaves open various possibilities. According to Pauling (1960), a chemical bond is realised, if the interatomic forces are such as to lead to the formation of an aggregate with sufficient stability to be considered as an independent molecular species. Three types of bonds are covered by this definition: covalent, ionic, and metallic.

(*a*) *Covalent bonds*

The most important bond in organic chemistry is the covalent or 'shared electron pair' bond. As these terms indicate, the origin of this bond lies in the electronic structure of atoms. Each atom consists of a positively charged nucleus consisting of protons and neutrons, surrounded by shells of negatively charged electrons. The outermost of these shells are responsible for the chemical behaviour and for the valencies of atoms, and the corresponding electrons are therefore known as valency electrons, represented by dots in formulae. Thus hydrogen with one such electron is written H·. This structure, with its lone electron, is unstable, but stability may be achieved if two electrons of opposite spin meet to form an electron pair. The hydrogen atom can therefore lose its electron for this purpose, leaving the atomic nucleus or a proton, H^+, also known as a hydrogen ion. Alternatively, H· can acquire a second electron to form a stable shell, e.g. by combining with another hydrogen atom to form the molecule H_2, written H:H. Since the valency electrons of two atoms are involved, the electron doublet is called a shared pair. It is easily seen that the colon representing such a pair corresponds to the conventional valency bond in organic chemistry, but since the valency electrons of two different atoms are involved, the bond is now said to be covalent.

The numbers of valency electrons in the first ten elements of the periodic system are as follows (Fig. A–11):

H·	He:	Li·	$\dot{\mathrm{Be}}\cdot$	$\dot{\underset{\cdot}{\mathrm{B}}}\cdot$	$\cdot\dot{\underset{\cdot}{\mathrm{C}}}\cdot$	$:\dot{\underset{\cdot}{\mathrm{N}}}\cdot$	$:\dot{\underset{\cdot\cdot}{\mathrm{O}}}\cdot$	$:\ddot{\underset{\cdot\cdot}{\mathrm{F}}}\cdot$	$:\ddot{\underset{\cdot\cdot}{\mathrm{Ne}}}:$

Fig. A–11. Valency electrons of the atoms in the first two periods of the periodic system.

The elements Na· through $:\ddot{\underset{\cdot\cdot}{\mathrm{Ar}}}:$, belonging to the next period, have the same electron configurations of their outermost shell as the corresponding elements Li· through $:\ddot{\underset{\cdot\cdot}{\mathrm{Ne}}}:$. For chemical purposes the electrons lying in inner shells are usually disregarded, so that homologous atoms can be written as Li·, Na·, K·, Rb·, Cs·, or $\dot{\mathrm{B}}$e·, $\dot{\mathrm{M}}$g·, $\dot{\mathrm{C}}$a·, $\dot{\mathrm{S}}$r·.

The two elements H_2 and He are exceptional in that they are stabilised with one electron pair. For this reason He: is a non-reactive noble gas. All the other elements tend to complete their valency shells up to four electron pairs, and thus achieve a stable state (Fig. A–12). The noble gases other than helium are chemically inert, because their outermost shells contain electron octets. Elements with 7, 6, or 5 valency electrons tend to take up sufficient electrons to complete the octets. Thus the unstable element $:\ddot{\underset{\cdot\cdot}{\mathrm{Cl}}}\cdot$ can complete its shell by taking up one electron, to become the stable ion Cl^-. This electron may be borrowed from H· or Na· etc., which readily part with their valency electron and become positive ions H^+ or Na^+. In chlorine molecules Cl_2 the covalent $:\ddot{\underset{\cdot\cdot}{\mathrm{Cl}}}:\ddot{\underset{\cdot\cdot}{\mathrm{Cl}}}:$ bond is formed. The same is true when oxygen, nitrogen, or carbon combine with hydrogen, to give H_2O, NH_3, or CH_4 (Fig. A–12).

$$\begin{array}{c} \mathrm{H} \\ :\ddot{\underset{\cdot\cdot}{\mathrm{O}}}:\mathrm{H} \end{array} \qquad \begin{array}{c} \mathrm{H} \\ :\ddot{\underset{\cdot\cdot}{\mathrm{N}}}:\mathrm{H} \\ \mathrm{H} \end{array} \qquad \begin{array}{c} \mathrm{H} \\ \mathrm{H}:\ddot{\underset{\cdot\cdot}{\mathrm{C}}}:\mathrm{H} \\ \mathrm{H} \end{array}$$

Fig. A–12. Simple compounds with shared electron pairs.

Electronic formulae also explain why the valency of certain elements is different depending on the partner with which the element reacts. Thus nitrogen is trivalent towards hydrogen, because it can take in 3 valency electrons, but pentavalent towards oxygen because it shares the 5 electrons of its exterior orbit. In the ion NO^-_3, for example, the three oxygen atoms require six valency electrons; since nitrogen can bring in only five, the sixth electron must be provided by hydrogen or a metallic atom, so that a negative charge is in excess and the ion NO^-_3 results.

Two electron pairs are shared in compounds with double bonds, and three electron pairs in compounds containing triple bonds. Thus ethylene, $CH_2{=}CH_2$, is written as in Fig. A–13. In acetylene, $CH{\equiv}CH$, three doublets have to be written between the two C symbols, and for the sake of simplicity these are not put in line, but side by side. In general this is also done with double bonds:

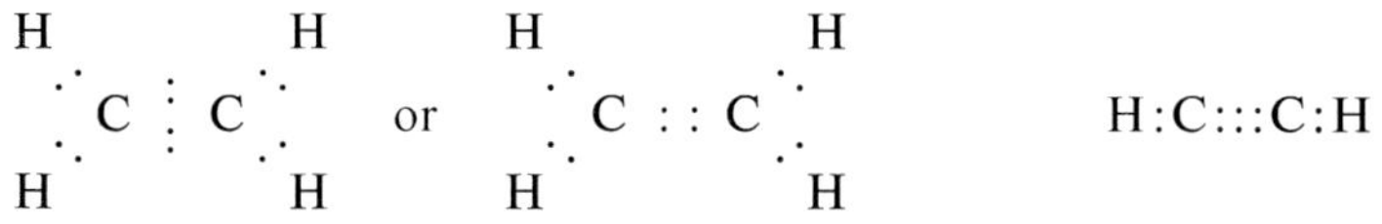

```
H        H      H         H
 ˙.    .˙        ˙.     .˙
   C ⋮ C    or     C :: C            H:C:::C:H
 .˙    ˙.        .˙     ˙.
H        H      H         H
   Ethylene                              Acetylene
```

Fig. A–13.

Since the above formulae correspond to the conventional representations, with the only difference that every valency line is replaced by a pair of dots, organic chemists continue to use the simple line for a covalent bond.

The bond energy of a shared electron pair varies depending on the nature of the atomic partners, and on their adjoining substituents. As a rule, bond energy and bond strength increase with decreasing bond distance (see Table A–I, p. 15).

(*b*) *Ionic bonds*

As has already been pointed out, elements having only one or two electrons on their outermost shell lose these electrons easily, forming ions such as Na^+, Ca^{2+}, etc. The elements tend to assume the electron structure of the noble gas preceding them in the periodic system with an atomic number lower by one or two units, e.g. $Ca^{2+} \sim Ne$, $K^+ \sim Ar$. On the other hand, elements having seven electrons on their outer orbit, such as the halogens, readily accept an additional electron to form a complete four-pair shell. The ions F^-, Cl^- etc. are produced in this way. These again display the electron structure of noble gases, *i.e.* that of the next higher element: $Cl^- \sim Ar$, $Br^- \sim Kr$. This is the reason why such ions are stable entities, whose reactivity is governed primarily by their electrical charge.

Although the saturated outer shells prevent these ions from forming covalent bonds, the oppositely charged ions attract each other. No fixed electrostatic bonds are possible in solution, because the charged particles are hydrated, and the surrounding water molecules prevent the ions from forming binary molecules. When the water is evaporated and the substance crystallises, definite combinations are formed by the so-called ionic bonds. The latter differ from covalent links in that no individual molecules are produced, but instead the charge of every ion is compensated by 4, 6, 8, or 12 oppositely charged neighbours, according to the coordination theory. Thus in Fig. A–3 every Na^+ ion is surrounded by 6 Cl^- ions, and every Cl^- ion by 6 Na^+ ions. The overall formula NaCl is due to the fact that the numbers of the two types of ions are equal.

Ionic bonds can be broken by polar liquids (see p. 23) such as water or liquid ammonia, and by hydrogen or hydroxyl ions. When this occurs, the ionic lattice is broken down and the solid compound is dissolved.

(*c*) *Metallic bonds*

In a metallic crystal the lattice is not formed by ions but by neutral atoms, which do not attract each other electrostatically but form covalent bonds according to their valency. For example, the following bonds are possible in the monovalent potassium between neighbouring atoms (Fig. A–14):

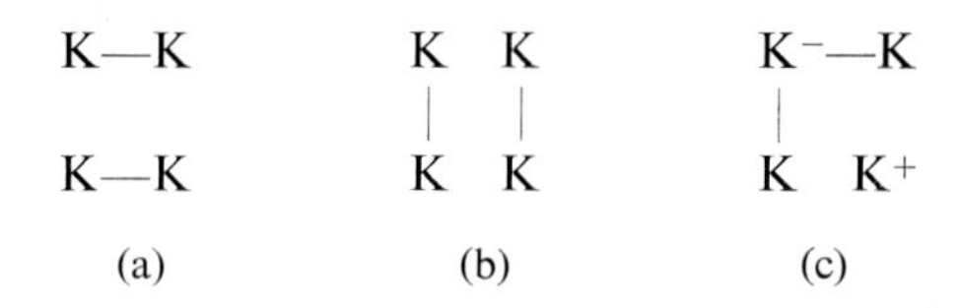

Fig. A–14. The metallic bond.

No binary molecules K_2 are formed in this case, because in metals the electrons readily change places by resonance. If this occurs simultaneously in two adjoining atoms, the bond distribution may change, e.g. from that in Fig. A–14a to that of Fig. A–14b. If only one such bond resonates, there are potassium atoms which have lost their valency electrons and temporarily become cations K^+, whilst others have an electron in excess and carry a transient negative charge, becoming anions K^- (Fig. A–14c). The metallic bond is thus distinguished from the stable covalent bond by its fluctuating resonance. This electronic mobility is the reason why metals conduct electricity.

(*d*) *Oxidation–reduction and bond formation*

Certain metals, such as iron or copper, can easily change their valency. Thus iron Fe can part with two or three electrons, to form the ferrous ion Fe^{2+} and the ferric ion Fe^{3+} respectively. In the same way, copper Cu can produce cuprous ions Cu^+ or cupric ions Cu^{2+}. In a solution of iron salts ferrous and ferric ions exist in equilibrium

$$Fe^{2+} \underset{+e}{\overset{-e}{\rightleftharpoons}} Fe^{3+}$$

which is known as an oxidation–reduction, or redox equilibrium. This demonstrates that oxygen or hydrogen are not necessarily involved in oxidation and reduction processes, which consist basically of an electron transfer. Reduction is defined as the addition and oxidation as the loss of valency electrons. Thus the anion Cl^- is the reduced form and the cation Na^+ the oxidised form of the neutral atoms Cl and Na.

In compounds such as H_2O, H_2S, NH_3, or CH_4, the non-metals whose outermost electron shells are filled up to the stable number of 8 (Fig. A–12) are considered as electron acceptors, in spite of the fact that in covalent bonds the electron pairs are shared. They are therefore reduced, whilst the hydrogen is oxidised.

The same is true in redox equilibria in which hydrogen transfer takes place. The formation of sulphur bridges between polypeptides, whereby macromolecules can aggregate or dissociate (Fig. A–15) may be mentioned as an example. The relevant equilibrium is

$$R:\ddot{\underset{\cdot\cdot}{S}}:H + H:\ddot{\underset{\cdot\cdot}{S}}:R \underset{+H:H}{\overset{-H:H}{\rightleftharpoons}} R:\ddot{\underset{\cdot\cdot}{S}}:\ddot{\underset{\cdot\cdot}{S}}:R$$

Cysteine Cystine

Fig. A–15. The cysteine–cystine equilibrium.

Since H_2 carries away two electrons, the compound on the right is oxidised and that on the left is reduced.

(e) Hydrogen bonds

The two covalent bonds in water are arranged asymmetrically with respect to the centre of the oxygen atom (Fig. A–16):

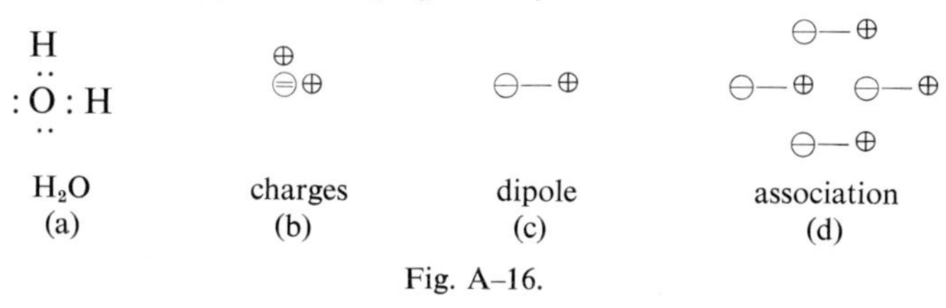

Fig. A–16.

As a result of this asymmetry, the distribution of electric charges within the molecule is such that the hydrogen nuclei appear to be positive with respect to the oxygen which has taken up two electrons. Such neutral molecules with a polar distribution of their internal charges are known as dipoles, and are symbolised as indicated in Fig. A–16c.

Dipolar molecules are attracted to each other to give associated clusters of definite structure, but of a size which remains undefined as long as the 'surfaces' of the associating entities display a polar distribution of positive and negative charges (Fig. A–16d). Such a diagram is however unsatisfactory morphologically, since the oxygen atoms which carry the negative charge are very much larger than the hydrogen atoms. A better representation of the true state of affairs, which takes into account the bond angle of 104°45′ in HOH (Pauling, 1960) is given in Fig. A–17a. It is seen that a hydrogen atom is situated between two oxygens, which is bonded covalently to one O atom and attracted electrostatically by the negative charge of the other O atom. Dissimilarity of these two types of bonding results in a slight elongation of the O—H distance and consequently in a weakening of the covalent bond; it is as if the single valency of the hydrogen atom were shared between the two oxygens. Such electrostatic linkage is known as a hydrogen bond. The fact that hydrogen bonds are considerably weaker than covalent bonds is conventionally denoted by O—H --- O.

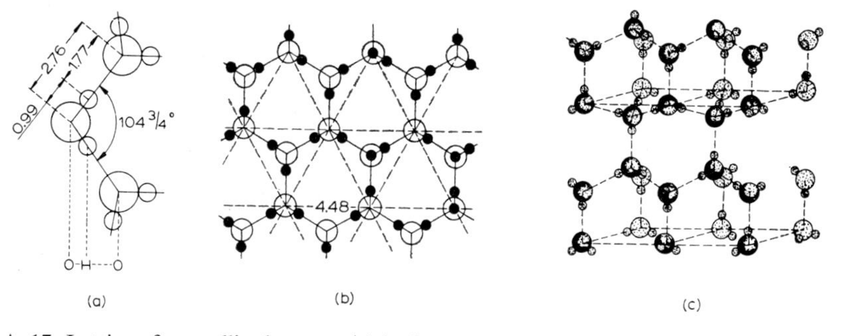

Fig. A–17. Lattice of crystallised water; (a) hydrogen bond in water; bond angle υ = 104° 45′; (b) projection of the lattice on to the base of the hexagonal ice crystal; (c) space model showing the coordination number 4.

If a fourth molecule of water would be added to the left in Fig. A–17a, an almost equilateral triangle would appear. If the H—O—H angle were equal to 120°, this could give rise to a monomolecular layer of H_2O molecules, possessing a hexagonal arrangement. The water angle of 104°45′ is however closer to the tetrahedral angle of 109°28′, favouring three-dimensional association. The oxygens assume a coordination number of 4, so that in ice each O atom is surrounded by four O atoms disposed tetrahedrally around it. A projection of a layer of such tetrahedra on to the base of the hexagonal ice lattice is illustrated in Fig. A–17b. The tetrahedral structure is fairly open, which accounts for the low density of ice; in the liquid state the H_2O molecules can fill the space more densely. The interatomic distance between the two oxygens in an O—H---O group is 2.76 Å, of which O—H accounts for 0.99 and H---O for 1.77 Å; the hydrogen bond covers the whole distance (see Table A–I). The proton is thus situated at a point approximately one third along the O—O distance, lying closer to the atom with which it shares its valency electron.

Hydrogen bonds are also observed with a few atoms other than oxygen. Of particular importance is the linkage between polypeptide chains, where hydrogen bonding occurs between >NH and O=C< groups to give a structure >N—H---O=C<. The N—O distance in N—H---O is 2.79 Å, *i.e.* almost identical with the O—O distance in ice.

Hydrogen bonding is distinct from ordinary 'chemical' bonds, nor can it be classified as a 'physical' Van der Waals' bond; its character places it between these extreme types. This is also shown by the bond energies, which are of the order of

100 kcal/mole for covalent bonds (Table A–I),
2.5 kcal/mole for hydrogen bonds (Table A–I),
1–2 kcal/mole for Van der Waals' bonds in paraffins (Table A–II).

(*f*) *Van der Waals' bonds*

Molecules possessing a dipolar character are of a special type. More frequently, and particularly in organic compounds, the charge distribution does not lead to the formation of positive and negative poles. This is clearly demonstrated by the structure of methane CH_4 (Fig. A–12). Such non-polar molecules are nevertheless attracted to each other, and give rise to crystalline lattices in the solid state. In lattices of this type the cohesive intermolecular forces are about a hundred times weaker than the covalent intramolecular bonds, and are called Van der Waals' forces after the Dutch physicist who was the first to study these interactions in full.

When liquids are evaporated or molecular lattices sublimed, the Van der Waals' forces which hold these phases together must be overcome. Evaporation and sublimation energies may therefore be used to calculate the bond energies of cohesion forces. Such bond energies are expressed in kcal/mole, in very much the same way as the bond energies of covalent linkages, which are derived from heats of combustion. In chainlike molecules, such as the paraffins, the cohesion energies increase with chain length, by constant increments for each —CH_2— group added to the chain. Such increments vary considerably for various organic groups (see Table A–II) and are additive, allowing an approximate calculation of the cohesion

TABLE A–II

COHESION FORCES BETWEEN ORGANIC GROUPS
(after Meyer and Mark, 1930)

Groups		Molar cohesion energy, kcal/mole
aliphatic C: methyl and	$—CH_3$ and $=CH_2$	1.78
methylene groups	$—CH_2—$, $=CH—$	0.99
ether bridge	—O—	1.63
amino group	$—NH_2$	3.53
carbonyl group	=CO	4.27
aldehyde group	—CHO	4.70
hydroxyl group	—OH	7.25
carboxyl group	—COOH	8.97

energy of simple molecules. Thus, for instance for pentane, $CH_3CH_2CH_2CH_2CH_3$, the cohesion energy is $1.8 + 1 + 1 + 1 + 1.8 = 6.6$ kcal/mole.

It may be seen from Table A–II that groups containing N or O give rise to considerably greater cohesion than radicals consisting only of C and H. This is because nitrogen and oxygen atoms are able to induce dipole effects.

Compounds consisting of non-polar groups such as $—CH_3$, $—CH_2—$, $=CH_2$ (e.g. the paraffins) which are insoluble in water but dissolve in benzene, ether and other fat solvents, are known as hydrophobic or lipophilic. Compounds which by virtue of groups such as —OH, —COOH, $—NH_2$, etc., possess an affinity towards the dipolar H_2O molecule, are known as hydrophilic.

Intermolecular Van der Waals' attraction is due to the interaction of electrons and nuclei between the molecules concerned. Although the positive and negative charges are balanced so that the molecule is neutral as a whole, the opposite charges are slightly separated, even if only by a fraction of an Ångström. From a certain direction, therefore, the negative electrical field of the electron shells will predominate over the positive field of the somewhat more remote nucleus. Since a neutral molecule would behave as positive in such a field, neutral molecules will thus be attracted. For argon, whose atoms behave as molecules, this attraction assumes maximum values at a distance of about 5 Å. At shorter distances, e.g. 4 Å, strong repulsion forces begin to operate due to the interpenetration of similarly charged outer electron shells (Pauling, 1953). Local positive fields operating in a similar manner may be postulated in a CH_4 molecule, in which the carbon and its electron shells are surrounded tetrahedrally by the four protons.

The Van der Waals' cohesive forces originate from the same electrostatic source as the dipole attraction. Whilst dipolar molecules with their considerably separated charges produce strong positive and negative fields at the opposite poles, in the so-called homopolar molecules, in which one of the charges has a slightly more peripheral position, the fields are very much weaker because the charge distribution is essentially symmetrical.

This explains why there is no sharp distinction between the bonding existing in homopolar and that in heteropolar organic molecules. As may be seen from

Table A–II, introduction of polar NH_2 or OH groups increases the cohesion forces considerably more than the introduction of a homopolar radical such as CH_2. The radicals listed in Table A–II are thus a series of increasing polarity.

(*g*) *Long-range forces*

The above-mentioned cohesion forces between micromolecules and the covalent bonds act only over short distances, of the order of 1–10 Å. Their effectiveness decreases very rapidly with increasing distance, and such short-range forces thus become inactive at greater ranges. Observations have however been reported (e.g. Wyckoff, 1947) that macromolecules in concentrated sols exhibit a certain regularity of arrangement, implying the existence of particle interactions over distances as large as 100 Å or even greater. The attractive forces responsible for this effect have therefore been termed long-range forces (Bernal, 1946).

Molecular long-range forces play a very important part in biology, since many cases are known in which macromolecules are mutually attracted or in which even microscopic particles combine with each other. The following mechanism may explain this behaviour.

In the crystallisation of globular proteins and viruses the formation of a three-dimensional lattice is always preceded by a para-crystalline phase in which the molecules assume a more or less uniform arrangement but are still able to rotate around their axes. This mobility gradually disappears, and in the final crystalline lattice the molecules are restricted solely to the moderate thermal oscillations. Para-crystalline ordering forces are particularly noticeable in the rodlike macromolecules, such as the tobacco mosaic virus (Oster, 1950). When this substance is properly crystallised, the centres of gravity of the particles are still separated by appreciable sublight-microscopic distances, so that tobacco mosaic virus crystals are capable of producing Bragg interferences in the visible region ($\lambda \sim 5000$ Å) of the spectrum (Wilkins *et al.*, 1950). Bearing in mind that such interferences are shown in inorganic crystals by X-rays ($\lambda \sim 1.5$ Å), this comparison is a remarkable illustration of the different orders of magnitude of the short-range and long-range forces.

The effect of long-range forces is also observed in immune reactions, when proteins are precipitated from solution or when bacteria are caused to aggregate and settle with the aid of agglutinins. In both these cases attraction forces are exercised between ultrastructural particles over appreciable distances.

Another puzzling phenomenon is syndesis during the fusion of chromosomes in meiosis, in which corresponding chromomeres of two unwound homologous chromosome threads are mutually attracted. In this case long-range forces must operate even over microscopic distances. The relations here appear to be analogous to those in the lateral aggregation of striped elementary fibrils of protein. In the latter case, it is observed that mass-rich and mass-poor bands attract each other over distances in excess of 500 Å, so that the striping shows no sublight-microscopic dislocations in the broad microfibrils. This conjoint alignment of the mutual banding in neighbouring fibrils is called the Beyersdorfer effect (Wohlfarth-Bottermann and Krüger, 1954).

Although the physical principle underlying the long-range forces has been the subject of heated arguments, it has not as yet been possible to demonstrate the existence of any special forces acting over macromolecular distances. It has in fact been shown that the familiar phenomena of uneven electron distribution and of dipole induction through electrical fields (Van der Waals' cohesion forces) which are known to act as short-range forces, are also significant in the attraction and repulsion of macromolecules.

It has been suggested (Oster, 1951) that the cohesion forces accumulate over certain regions of the surfaces of macromolecules, and can then bridge distances of the order of macroparticle diameters. The ordering observed in concentrated sols should rather be ascribed to the electrostatic repulsion of similarly charged molecules: such particles surround themselves with an electrical double layer, which can increase their hydrodynamic volumes to such an extent that the entire solvent is bound. If all particles then keep as far away as possible from their neighbours, a regular arrangement will necessarily result owing to particle morphology. Finally, a crystal lattice may result if the electrostatically bound water of hydration is removed. Thus essentially the same phenomena appear to take place as when simple molecules, e.g. cane sugar, crystallise or pass into solution. Either the repulsion (dissolution) or attraction (crystallisation) forces are preponderant, depending on the conditions.

The term 'long-range forces' does not therefore describe a specific physical concept, but is merely a convenient expression indicating that sublight-microscopic particles have a mutual ordering effect.

(*h*) *Theory of junctions*

A colloidal system may exist as a sol in a more fluid, or as a gel in a more solid state. The transition between these two states of aggregation may be reversible; this is illustrated by the transition of the cytoplasmic ground plasma from the active mobile into the inactive resting condition (Frey-Wyssling, 1949a):

$$\text{plasma sol} \rightleftharpoons \text{plasma gel.}$$

The gel–sol transformation may follow the general rule that as the temperature is increased, the system increases its volume, decreases its density, and eventually liquefies, as in the case of heated hydrated gelatin or agar.

The volume of a gel is not, however, invariably increased by liquefaction;

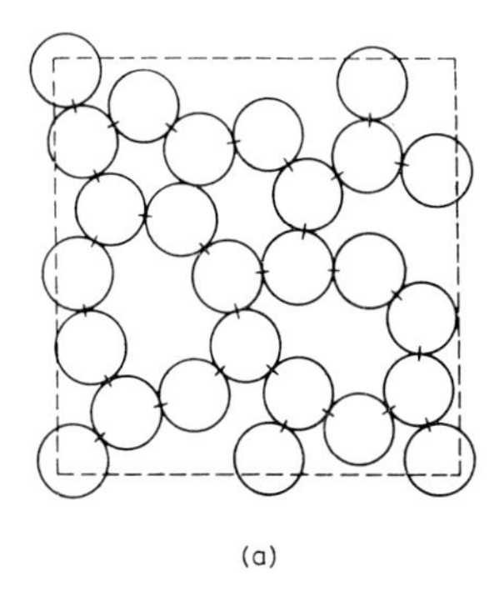

(a)

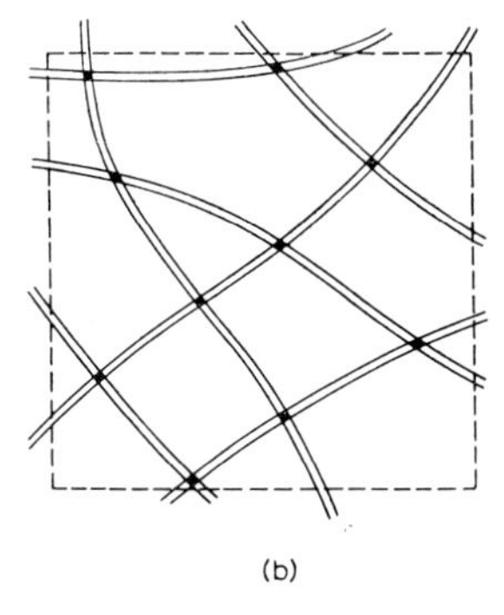

(b)

Fig. A–18. Formation of ultrastructures, (a) with spherical, (b) with filiform particles.

it may remain constant, as e.g. in the case of sodium oleate. Reversible gel–sol transformation may then be induced without any change of temperature or pressure, simply by stirring or shaking. A colloid of this type is termed a thixotropic gel. The volume may even decrease on forming the sol (e.g. methylcellulose in water), and in such cases the gel can be liquefied under isothermal conditions by increasing the pressure. Cytoplasm, which belongs to this category, is liquefied by high pressures, and this treatment (e.g. 300 atm – Brown, 1934; Marsland, 1942) which temporarily stops plasma flow, does no permanent harm to the cells.

Whenever a sol sets to a solidified gel, its dynamically more or less independent particles are linked together in an ultrastructure. As can be seen from Fig. A–18, it is considerably easier to form a gel from elongated filiform particles than from spheres, where many more points of contact between the particles are necessary for setting. The resulting linkages may have any of the characteristics described on the preceding pages in the classification of bond types (other than the metallic bond type), from loose Van der Waals' cohesion to firm covalent bonds. Saltlike linkages, hydrogen bonds, ester bridges, —O—, or —S—S— links may be formed. Hydrogen bonding may reduce the unlimited swelling capacity of hydrophilic gels by the intake of more and more water, to the limited swelling as found in cellulose. The solvation of the lipophilic gel polystyrene with its swelling agent benzene is similarly limited if the chain molecules are cross-linked here and there by divinylbenzene bridges (Staudinger, 1936).

Since in many cases it is not as yet known precisely which type or which combination of bonds participates in forming the structure, it is advisable not to

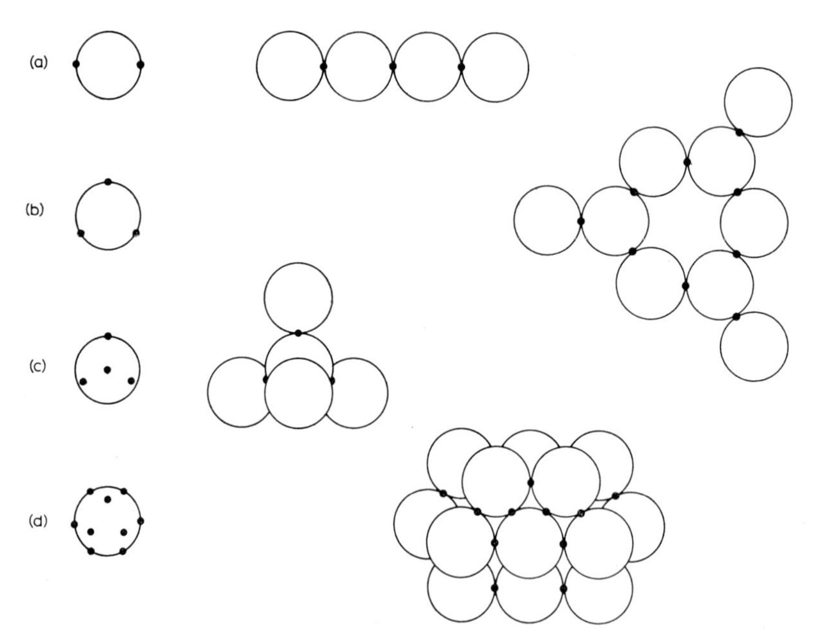

Fig. A–19. Aggregation of spherical macromolecules by junctions (from Frey-Wyssling, 1953). Number of coordination: (a) 2: beaded chains, (b) 3: sieve film, (c) 4: tetrahedral groups, capable of three dimensional structures (see Figs. A–7d, A–8a, diamond lattice and Fig. A–17c ice lattice), (d) 12: close packing (see Fig. A–7a, b).

employ terms which anticipate the character of bonds joining the dispersed particles together into a structural frame. A preferable terminology has therefore been created in the theory of junctions (Haftpunkt-Theorie) long before the modern elaborate bond classification was evolved (Frey-Wyssling, 1938), which defines any linkage between monomers, micromolecules, or macromolecules as a junction (Haftpunkt). The lability of possible junctions between chainlike molecules, which depends on temperature, hydration, pH, and redox potentials has already been discussed. Junction fields (Haftstellen) exist in macromolecules, in which the areas of contact may be of considerable size.

Globular macromolecular proteins may carry unevenly distributed charges, so that poles will occur on their surfaces which attract the opposite poles of neighbouring particles. Beaded chains (Rybak and Bricka, 1952; Seki, 1952) and elementary fibrils may originate in this way (Fig. A–19a). Fig. A–19b shows how three such active spots, emanating attractive forces, give rise to a two-dimensional layer representing a porous film. If six coplanar junctions operate, a dense film is produced, with only tiny interstices between the spheres. Four tetrahedrally oriented junctions (Fig. A–19c) would similarly bind the particles into a three-dimensional framework, and superposition of several dense films (Fig. A–19d) results in a close-packed crystal lattice. This suggests that junctions are induced wherever the globules are in contact. It is difficult to decide whether these 12 spotr are predeterminated by the internal structure of the macromolecule, or whethes cohesive forces are induced at arbitrary points of contact. In the first case, the coordination number of 12 of the macromolecule would correspond to 6 definite axes in its molecular structure (Fig. A–4), and all macromolecules would be incorporated into the lattice in an identical position. In the second case, the spheres would be free to rotate, and could occur in any orientation in the closely packed aggregate.

Such considerations are valuable in the description of observed structures as long as the attractive forces involved are insufficiently known. Sooner or later those dynamic aspects must however be cleared up, because a structure is only satisfactorily understood when not only its morphological features, but also the reasons why a certain and not a different structure originates, are thoroughly elucidated.

B. Macromolecules

1. CARBOHYDRATES

(*a*) *Glucose and its disaccharides*

When the bond angles between successive C—C bonds are taken into account, the structural formulae of aliphatic compounds bear a pronounced resemblance to molecular models. A chain such as that in hexane should therefore be drawn kinked rather than straight (Fig. B–1a, b). In molecules which do not form part of a crystal lattice but are free to move about in the gaseous state or in solution, the groups can rotate around the direction of the bonds. Such free rotation would not give rise to a new structure in straight chains as in Fig. B–1a. In zig-zag chains, however, the possibility of rotation means that e.g. group 1 in Fig. B–1b need not necessarily lie in the plane of the paper with 2 and 3, but may be located anywhere on the perimeter of a cone which has its top in group 2 and whose apex angle is equal to the bond angle. Among these various possibilities there is one special case, in which groups 4 and 5 are turned through 180°, resulting in a ring-shaped model. It is not difficult to see that this can easily lead to cyclic compounds. Fig. B–1c shows that rings of 5 or 6 atoms are favoured because the complement (70°30′)

Fig. B–1. Hexane; (a) conventional structural formula, (b) morphologically correct formula, (c) ring conformation; $v = 70°\ 30'$ (complement to the bond angle).

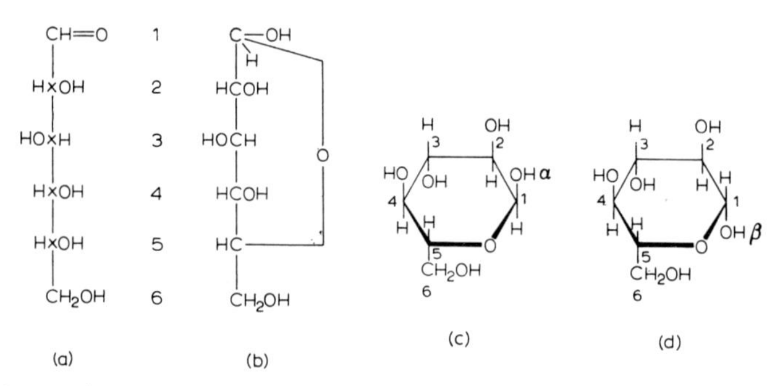

Fig. B–2. Glucose; (a) aliphatic, (b) heterocyclic structural formula; (c, d) α- and β-conformations (according to Haworth, 1925).

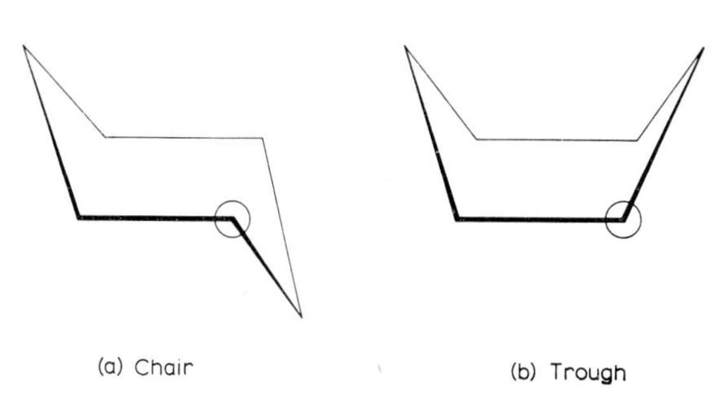

Fig. B–3. Glucose; (a) chair conformation, and (b) trough conformation.

of the bond angle is contained somewhat more than 5 and a bit less than 6 times in 360° ($5 \times 70°30' = 352°30'$, $6 \times 70°30' = 423°$). The different forms which a molecule can assume are called its conformations or configurations – thus Figs. B–1b and c represent two different conformations of hexane.

The monosaccharides, $C_6H_{12}O_6$, which were formerly believed to be open chains (Fig. B–2a) have now been shown to consist of heterocyclic rings with an oxygen bridge. In glucose this is usually a 1,5-bond, and is often represented in the manner of Fig. B–2b. This formula is not however a true representation, since the C—O distance is unduly large. For this reason, in 1925, Haworth proposed that sugars be represented by equilateral hexagons or pentagons, according to whether the oxygen bridge is situated between carbons 1 and 5 (derivatives of pyranose) or 1 and 4 (derivatives of furanose). Figs. B–2c and d illustrate the glucose pyranoses. The dimensions of a glucose molecule can be calculated with the aid of interatomic distances (Table A–1). For example, assuming that the ring is completely planar and represents an equilateral hexagon, an axis drawn through carbons 1 and 4 has a length of $2 \times 1.54 + 2 \times 1.43 = 5.94$ Å. This value is only approximate, because, to begin with, the hexagon is not completely equilateral on account of the somewhat smaller diameter of the oxygen atom, and because the carbons and the hydroxyls do not lie strictly in the plane in which the distances are measured. In other words, instead of the distances C—C and C—O, only their projections contribute to the length concerned. In this respect, two different conformations are possible, which are distinguished as chair and trough conformations (Figs. B–3a and b). The lengths of these two uneven rings are similar. The chair conformation occurs in cellulose (Fig. B–8b). If all these factors are taken into account, the 1,4-axis is calculated as 5.15 Å, rather than 5.94 Å. This is also the value derived from X-ray analysis. Fig. B–4 demonstrates the profound similarity between the modern structural formulae (Fig. B–2c) and the molecular models. The former are no longer arbitrary, but represent correctly proportioned two-dimensional projections of the true molecular structure.

According to the aliphatic formula (Fig. B–2a) glucose contains four asymmetric carbon atoms (marked ×), since only the $—CH_2—$ and the $>C{=}O$ groups possess a plane of symmetry. As a result of ring formation, however, the carbonyl carbon also becomes asymmetric, allowing the possibility of two different configurations of the heterocyclic ring; these are called α- and β-glucose (Figs. B–2c and d) and are distinguished by their optical rotation (β is characterised by the smaller

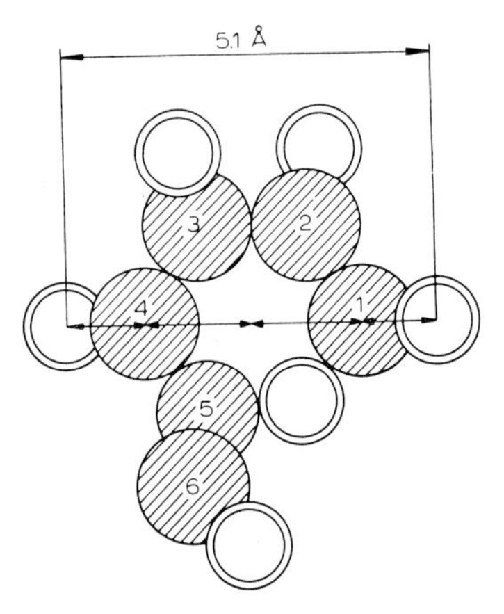

Fig. B–4. The molecular structure of glucose (from Meyer and Mark, 1930). C-atoms hatched, O-atoms surrounded.

rotation). It may be seen that in β-glucose the H- and OH-groups alternate regularly on both sides of the ring, whilst in α-glucose the hydroxyls on carbons 1 and 2 point in the same direction.

The fact that in β-glucose the OH on carbon 1 lies on the same side of the ring as the OH on carbon 6 may be demonstrated by closing a second oxygen bridge between these two carbons by dehydration (laevo-glucosan). This cannot be done with the α-conformation.

The α- and β-positions of the hydroxyls on carbon 1 are of fundamental importance for the understanding of the structures of disaccharides and polysaccharides. In disaccharides a 1,4-bridge between two glucose rings is formed by the loss of one molecule of water. In the case of the α-position, the two rings can be joined directly, whereas in the case of the β-position one of the rings must first rotate through an angle of 180° around its 1,4-axis in order to bring the required two OH-groups into a neighbouring position.

Both cases occur in nature; α-glucose gives rise to maltose, and β-glucose to cellobiose, the disaccharide unit of the cellulose chain (Fig. B–5). In maltose the two glucose rings can be made to coincide by a simple translation, whereas in cellobiose this requires a twofold screw axis. The cellobiose molecule possesses therefore a higher degree of symmetry, since the coincidence must be achieved by a combination of translation and rotation.

The bond pictured in Fig. B–5a is called α-glucosidic and the one in Fig. B–5b β-glucosidic. Other molecules containing hydroxyl groups can also combine with glucose according to these two schemes, giving α- and β-glucosides. This distinction

(a) (b)

Fig. B–5. Disaccharides from glucose; (a) maltose, (b) cellobiose.

is not only interesting and important from the point of view of molecular morphology, but is also of great physiological importance, since the α- and β-bridges are broken down by different enzymes. Thus for the hydrolysis of maltose we require an α-glucosidase which is incapable of splitting cellobiose, whilst β-glucosidases attack cellobiose but are inactive with respect to maltose. It seems that in plants the reserve substances, which must be quickly mobilised when required, are more often built according to the α-type (saccharose, starch); the glucosides, which cannot be utilised directly as reserves (e.g. amygdalin) and cellulose, are β-glucosides. This example shows that in the final analysis the problem of enzymes is also of a morphological nature, since the enzymes must possess quite a specific structure to be able to distinguish between α- and β-linkages. The well known comparison of lock and key is not only a symbol – the substrate and the enzyme must in fact fit morphologically together in the strict sense of the word.

(*b*) *Polysaccharides*

The formation of polysaccharides $(C_6H_{10}O_5)_n$, which are of outstanding importance in plant physiology, is governed by the same principles as those controlling the formation of disaccharides (Fig. B–5). The monoses can again interlink by 1,4-oxygen bridges with the elimination of water, and this polycondensation may involve a large number of units. In cellulose, the successive links of β-glucose are rotated through 180° with respect to each other, but in starch the

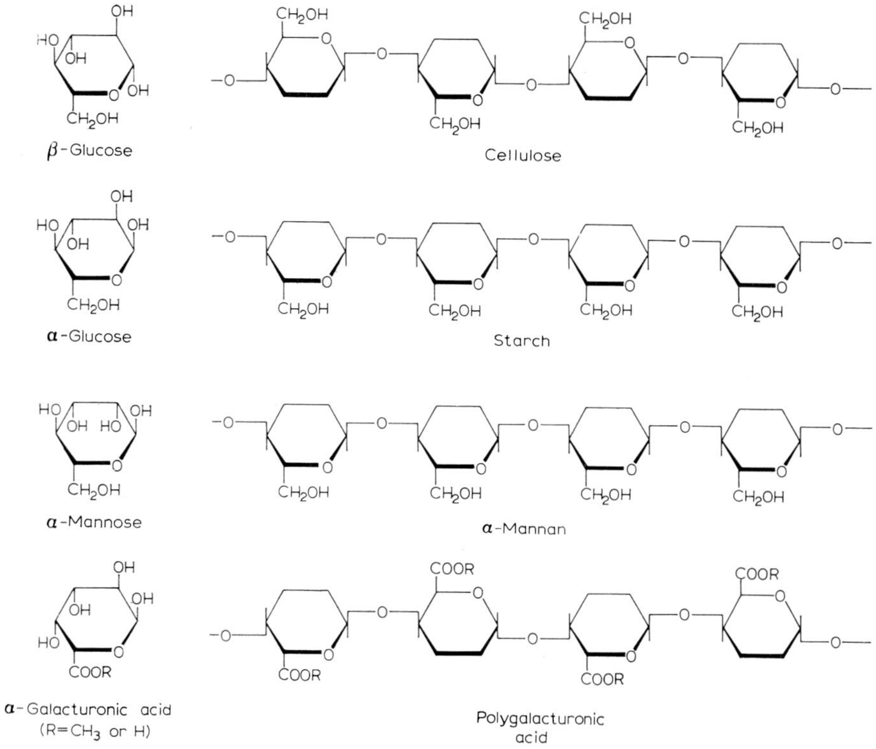

Fig. B–6. Polysaccharides.

α-glucose residues can react with each other without rotation (Fig. B–6). The cellulose chains have a twofold screw axis as an element of symmetry, in distinction to the starch chains which do not. The more symmetrical cellulose molecules are consequently more stable and straightened out, whereas the starch molecules tend to become more convolute. This morphological difference is undoubtedly one of the reasons for the different behaviour of starch and cellulose, and may also be responsible for the tendency of starch molecules towards branching (see Fig. B–11b).

Glucose can also polymerise with β-1,3-linkages. This is the case in laminarin, a polysaccharide of red algae, in callose (see p. 290), and in yeast glucan which is known both in amorphous and in crystalline states.

The mannans occurring in corozo nut and in the rhizomes of *Amorphophallus konjak* can be derived from mannose just as starch and cellulose are derived from glucose (Fig. B–6). The two monomers differ only in the positions of the H- and OH-groups on the second carbon atom. Glucomannans due to mixed polymerisation occur in the cell walls of soft woods (Table J–II).

The primary alcohol groups —CH_2OH in hexoses can be oxidised to the carboxyls —COOH, giving uronic acids. The oligomers or polymers of uronic acids are known as uronides. The pectic substances believed to be responsible for the coherence of plant tissues belong to this class of highly polymerised compounds. Owing to the presence of carboxyl groups, they are capable of forming salts. The polyacid itself is soluble in water, but the Ca salt is not, so that it can be precipitated by Ca ions. Some of the carboxyl groups may be esterified by methyl alcohol (e.g. Deuel, 1943).

The monomer of pectic acid is the α-galacturonic acid. As in α-galactose, the hydroxyl groups on the first and the fourth carbon atoms are situated on different sides of the pyranose ring (Fig. B–6); thus the α-glucosidic linkage causes a rotation of succeeding chain members. In the crystalline sodium pectate the screw axis is not twofold as in cellulose, but threefold (Palmer and Hartzog, 1945). The crystallising tendency of pectic substances is considerably smaller than that of cellulose, and in plants they occur only in the amorphous state (Wuhrmann and Pilnik, 1945).

Polymeric pentoses or pentosans fall partly under the heading of hemicelluloses which will be described when treating with the plant cell wall (p. 286). They have a structure similar to those of the polysaccharides already described, except for the absence of the side chains, that is the sixth carbon atom. Replacement of this group by H in cellulose or polygalacturonic acid yields the xylan chain or a polyaraban.

The polysaccharides are a striking demonstration of how slight morphological variations of the same structural principle may give rise to substances which behave quite differently from a physiological point of view.

(*c*) *Cellulose*

The chain length of cellulose can be determined by osmotic methods, taking into account that Van 't Hoff's law does not apply rigorously to macromolecules, and that corrections (similar to the constant *b* in Van der Waals' equation of state of gases) must therefore be introduced (Schulz, 1936). A method due to Staudinger is

TABLE B–I

THE HOMOLOGOUS SERIES OF CELLULOSE POLYMERS
(Staudinger, 1937)

	Degree of polymerisation	Chain length in Å	Mechanical properties and form	Formation of fibrils	Solubility in 10% NaOH
oligosaccharides cellodextrins	1–10	5–50	pulverisable, crystalline powder	none	true solution
hemicolloid 'β-cellulose'	10–100	50–500	pulverisable, short fibres	slight	formation of sol
mesocolloid regenerated cellulose	> 100	500–2500	fibres	large	formation of gel
native cellulose α-cellulose (fibre cellulose)	500–10,000	2500–50,000	long fibres	very large	limited swelling

based on the fact that the specific viscosity of chain molecules (the increase in viscosity imposed upon the solvent by the solute) is approximately a linear function of chain length within a certain range of molecular weights. Since cellulose is soluble in solutions of cuprammonium hydroxide (Schweizer's reagent) or copper ethylenediamine, viscosimetry of its copper solutions has become an important method of research in cellulose chemistry.

According to Staudinger, the experimental data lead to the following conclusions regarding the size of cellulose molecules: if some 10 glucose residues are linked together to a chain (Fig. B–6), one obtains easily soluble cellulose products which, owing to their particle length of 50 Å, already exhibit slight colloidal properties. Compounds of this kind are known as the degradation products of cellulose, termed cellodextrins (formerly γ-celluloses). If the number of units present in the chain increases to about 100, the β-celluloses are obtained, which dissolve in 10% sodium hydroxide without swelling to form viscous sols. When the degree of polymerisation exceeds 100 and approaches 800, we obtain regenerated celluloses which are no longer attacked by 1% sodium hydroxide and which find application in the cellulose industry (rayon, cellophane). These dissolve slowly and with swelling in 10% NaOH yielding viscous 'gel-solutions'. Native cellulose (the so-called α-cellulose) is distinguished by a still higher degree of polymerisation. Viscosity measurements carried out on native cellulose dissolved in fully deoxygenated Schweizer's solution yield a degree of polymerisation of over 2000 for the fibre cellulose of linen, hemp, ramie, and others. The values determined from the viscosity can be checked osmometrically up to lengths involving about 1000 units in the cellulose chain. Beyond this limit we must resort to extrapolation, according to the linear viscosity rule. Whether this may be rigorously applied to the entire range from 1000 to 10,000 linkages, the maximum value found viscometrically (Marx, 1955; Marx-Figini and Schulz, 1963) is still an open question. On the other hand, native fibres may contain still longer chains, which are degraded on disso-

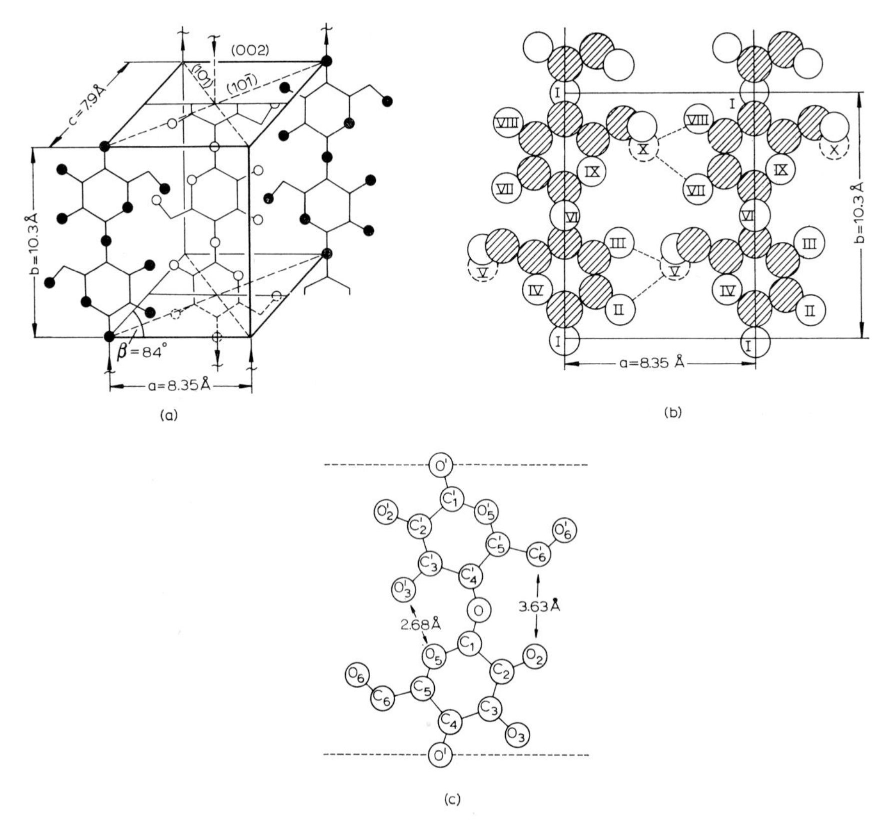

Fig. B–7. Crystal lattice of native cellulose; (a) unit cell (Meyer and Misch, 1937), τ = twofold helical axis; (b) plane (002) (Meyer and Misch, 1937), dotted lines are hydrogen bonds; (c) spatial arrangement of 'crooked' cellobiose unit (Liang and Marchessault, 1959).

lution in cuprammonium hydroxide. The number of 10,000 units for certain fibre celluloses is therefore not absolutely reliable, and is merely an experimentally determined value to which for the time being we must refer. Its magnitude is quite impressive, since a degree of polymerisation equal to 10,000 corresponds to a chain length of 5μ (!), each glucose residue measuring 5 Å. The molecules of cellulose are therefore of microscopic length and remain invisible only because their width is amicroscopic.

These cellulose chains do not occur as free molecules in nature, and in cell walls, even in primary walls, they always appear to be more or less crystallised into a chain lattice. In secondary walls these lattices yield excellent X-ray patterns, from which the unit cell of Fig. B–7 has been derived. The latter represents a monoclinic crystal lattice, containing four glucose residues of two antiparallel cellulose chains. One of these runs along the edge of the parallel-epiped and the other, in the reverse direction, along the line of the intersection of its crossed diagonal planes. Each of these lines corresponds to a twofold screw axis. According to Meyer and Misch (1937) the dimensions of the unit cell and the monoclinic angle β are

$$a:b:c = 8.35:10.3:7.9; \quad \beta = 84°.$$

The glucose rings lie in parallel lattice planes symbolised by (002); the indices of the shorter diagonal plane are (101) and those of the intersecting plane (10$\bar{1}$).

The unit cell of a chain lattice is peculiar in that it does not contain whole molecules but only the repeating unit of the molecular chain, which in the case of cellulose is the disaccharide cellobiose. The 10.3 Å long period along the *b*-axis, corresponding to two glucose residues (Fig. B–4), is called the fibre period, because it runs parallel to the cell axis of plant fibres with a so-called fibrous texture (s. p. 295).

The model of Fig. B–7a puts the plane of the glucosidic bond (bond angle 107°) into the (200) crystallographic plane of the lattice (projection in the direction of the *a*-axis; Fig. B–8b). As Carlström (1957) has shown for chitin, (Fig. B–18a), this is very unlikely. Therefore Liang and Marchessault (1959) place the glucosidic 107° angle of the β-1,4-bond in the (002) plane, resulting in a crooked chain (Fig. B–7c).

The cellulose lattice is held together by various types of bonds. Thus along the *b*-axis the bonds are covalent, whilst the thirty times weaker hydrogen bonds are exerted perpendicularly to this axis. This accounts for the pronounced optical and mechanical anisotropy observed in cellulose fibres: covalent bonds must be broken to overcome the tensile strength of such a fibre, whilst in lateral swelling only a weakening of the hydrogen bonds is involved.

Meyer and Misch (1937) located the hydrogen bonds in the (002) plane, as indicated in Fig. B–7b. This would mean that (002) is a cleavage plane of the lattice, and a plane of preferred growth. The crystallites of cellulose which appear in the form of elementary fibrils have a diameter of about 35 Å. The plane of preferred growth is however not (002) but (101) (see Fig. B–10). This can be explained by placing the hydrogen bonds in the (101) plane. Fig. B–8 and Table B–II show the distances for possible H-bonds between O-atoms in a cellulose lattice which is slightly modified as compared with Fig. B–7.

Since a primary hydroxyl group OH_V (Fig. B–7b) is free to rotate around the

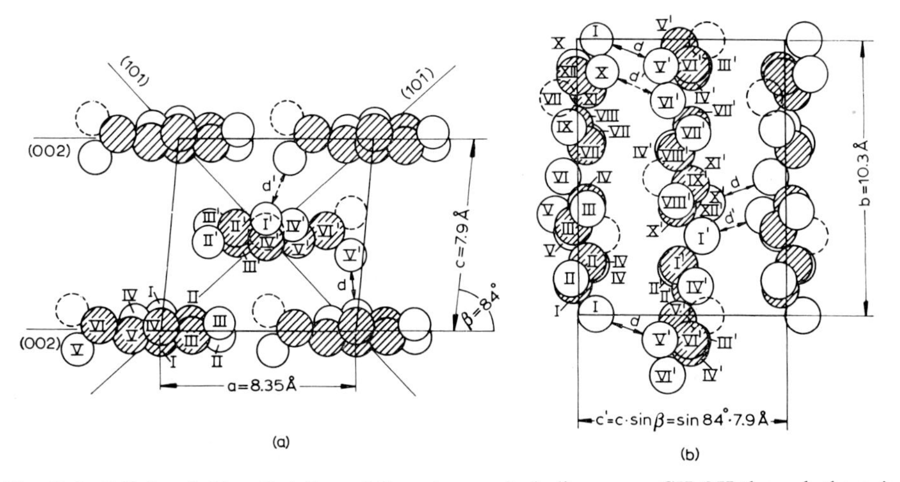

Fig. B–8. Cellulose lattice. Rotation of the primary alcoholic group $-CH_2OH$ through the axis C_V—C_{VI} establishes hydrogen bonds *d* and *d'* between chains of neighbouring layers. *d* binds chains of the (101) plane, *d'* binds chains of the (10$\bar{1}$) plane. – (a) Projection in direction of the *b* axis; (b) projection in direction of the α axis (Frey-Wyssling, 1955b).

axis of the covalent bond between carbons 5 and 6, it can be brought into a position in which it is at the minimum distance from the glucosidic oxygen of a neighbouring cellulose chain. Under these conditions the distances listed below are observed (Table B–II):

TABLE B–II

DISTANCES BETWEEN THE OXYGEN ATOMS IN THE MODIFIED CELLULOSE LATTICE
(Frey-Wyssling, 1955b; see Figs. B–7b, B–8a and b)

Possible H-bond	Distance in Å	Connects chains in plane	
O_V—H . . . O_I	2.54 = d	(101)	see Fig. B–8a and b
O_V—H . . . O_I'	2.80 = d′	$(10\bar{1})$	
O_V—H . . . O_{III}	2.65	(002)	see Fig. B–7b
O_V—H . . . O_{II}	3.42	(002)	

The Table shows that the smallest possible distance *d* for hydrogen bonds is in plane (101) (Fig. B–8a). This bond links a hydroxyl to a glucosidic oxygen, whereas in the model of Meyer and Misch (Fig. B–7b) bonds between two hydroxyl groups are proposed. In that case two hydrogen bridges must originate from each OH-group, as shown in Fig. B–7b, and an endless series of such shared protons would be needed, as e.g. in a crystal of ice (Fig. A–17b). It is true that such a bonding system zig-zagging along the entire length of two cellulose chains would provide an excellent explanation for the insolubility of this hydrophilic compound. Unfortunately, the available distances do not allow such an interpretation. The distance between oxygens V and II (Fig. B–7b) is already too large for a hydrogen bond, not to mention the distances between oxygens II and VIII or III and VII.

The firm linkage between the cellulose chains in the (101) plane (distance *d*)

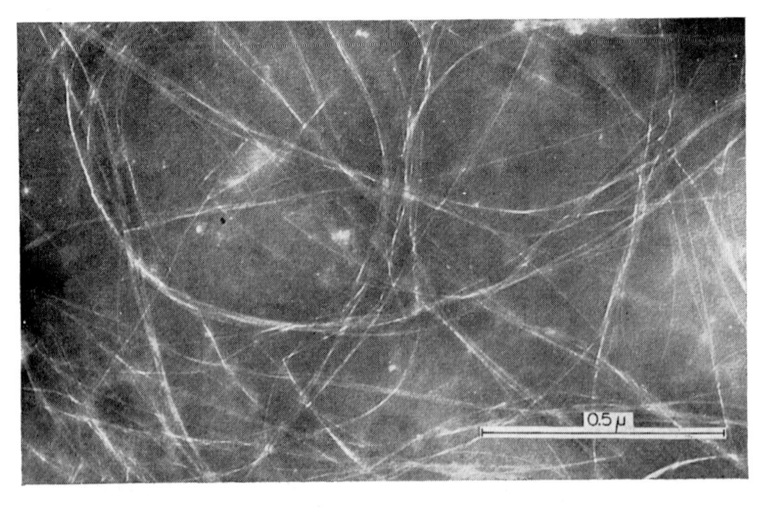

Fig. B–9. Elementary fibrils of cellulose (Mühlethaler, 1960a) from primary walls of a root tip *(Allium cepa)*.

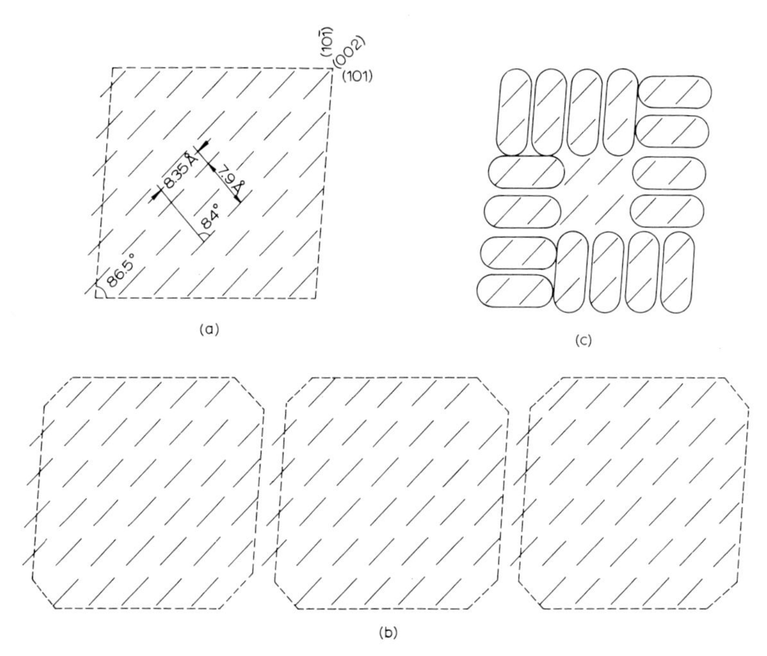

Fig. B–10. Cross-section of crystalline elementary fibrils showing the arrangement of the cellulose chains (Frey-Wyssling and Mühlethaler, 1963); (a) number of chains in fibrils of 35 Å diameter, (b) fasciation of elementary fibrils to microfibrils 35 × 100 Å, (c) antiparallel chains represent molecule pairs as a unit.

and the somewhat weaker bonding in the (10$\bar{1}$) plane (distance d') are the reason for the somewhat flattened cross-section of the elementary fibrils and for their diagonal orientation of the glucose rings. These crystalline fibrils are about 30 Å thick (Vogel, 1953) and 35 Å wide, and possess a pronounced tendency to aggregate laterally in the (101) plane by two, three, or more such units (Mühlethaler, 1960a). Fig. B–9 shows the elementary fibrils, and the structure of their cross-section is illustrated in Fig. B–10a.

From studies of infrared spectra, Liang and Marchessault (1959) confirmed that in their crooked chains cellulose model, hydrogen bonding is strongest in the (101) plane between the glucosidic oxygen and a hydroxyl of the neighbouring chain.

In the plant cell wall, the (101) plane is found to run tangentially and the (10$\bar{1}$) plane radially (Nicolai and Frey-Wyssling, 1938). This is a general rule from which only rare exceptions have been reported (Kreger, 1957). It is difficult to explain why the important plane (002) has a diagonal orientation. It should be mentioned that plane (101) has the densest packing of all possible lattice planes and is therefore the most hydrophilic plane. Furthermore, planes (101) and (10$\bar{1}$) intersect more nearly perpendicularly (86°30′) than planes (002) and (100) for which $\beta = 84°$, so that that angle meets better the orthogonal system requirement of the tangential and radial directions in the cell wall. Whether these arguments form a satisfying explanation is open to discussion.

The cross-section of an elementary fibril measures $30 \times 35 = 1050$ Å^2. Since the area occupied by a single cellulose chain is $1/2 \times 7.9 \times 8.35 \times \sin 84° = 32.8$ Å², an elementary fibril contains only about 32 cellulose molecules! It is true that there is a tendency for further growth by fasciation – lateral aggregation of fibrils (Fig. B–10b), to form microfibrils (Frey-Wyssling, 1951) – but why the crystalline chain lattice does not continue to grow in width appears at first sight to be an enigma. This behaviour is however due to the fact that in growing cell walls crystallised cellulose does not originate by the bundling of chain molecules, but by adding glucose rings through concomitant polymerisation and crystallisation to the ends of the fibrils. Once the formation of an elementary fibril has been started, its two ends will soon reach such a distance that growth will be restricted to a prolongation of the chain.

In this respect there is another riddle: according to structure analysis, the cellulose chains in the crystal lattice run anti-parallel with respect to each other, so that at the end of a fibril the glucose residues must be added in two different ways to form 1,4-bonds with half of the chains, and 4,1-bonds with the remaining chains. According to enzymology, there must be two different enzymes to activate the first and fourth carbon atoms in a glucose ring for its incorporation at the reducing or non-reducing chain end. A polar structure has been postulated for the elementary fibrils, with parallel chains needing but one enzyme (Roelofsen, 1959, p. 30) to avoid this complication. In this case the elementary fibril would grow only in one direction (like a growing cotton seed hair!) and its features should be those of a hemimorphic crystal. Morphologically and electrically (piezo- and pyro-electricity), such a crystal lattice displays completely different behaviour at the two ends, e.g. the top may be pointed whilst the base is flat; on stress or a change in temperature the ends carry opposite charges, and any enzymatic disintegration or attack ought to proceed in a strictly polar manner. No such effects have ever been observed in cellulose. On the contrary, Colvin (1963) found that both ends of fibril segments exhibited the same reducing power against silver nitrate. An antiparallel orientation of cellulose molecules in the chain lattice seems therefore more probable than a parallel one. If, as seems unlikely, the structure is in fact parallel, the polar fibrils must be arranged antiparallel in cell walls with a fibre texture, because no hemimorphic effects are known from plant fibres (see corrosive figures obtained by enzymatic hydrolysis of wood-destroying fungi – Bailey and Vestal, 1937).

Since polymerisation and crystallisation of cellulose occur simultaneously in growing elementary fibrils, for every glucose residue incorporated in the chain lattice a molecule of water or phosphoric acid must be removed from the growing elementary fibril, as otherwise no homogeneous lattice would result. If it is admitted that pairs of antiparallel glucose molecules are added to the end of the fibril, almost all of them are situated in its surface: in Fig. B–10c, for instance, 16 out of 18 such pairs. This principle might also be a factor limiting the growth of the elementary fibrils in thickness (Frey-Wyssling and Mühlethaler, 1963).

(*d*) *Starch*

The reserve carbohydrates saccharose, maltose, and starch are all α-glucosides (see p. 31), in distinction from the skeleton carbohydrates cellulose, xylan, etc., which possess the β-glucosidic structure. In contrast to the straight cellulose chains, the glucosan chains with α-1,4-bonds are helical (Fig. B–11a).

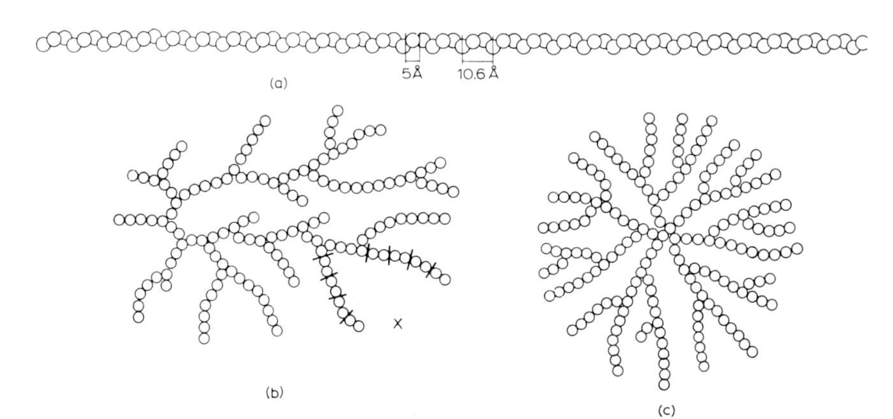

Fig. B–11. Diagram of molecular shapes of α-1-4-polyglucosan molecules; (a) amylose, (b) amylopectin, (c) glycogen.

The chemistry of starch is complicated by the presence of two chemically distinct substances, amylose and amylopectin. Amylose is soluble in hot water and is stained blue by iodine, whilst amylopectin swells in boiling water and gives a violet colour with iodine. Thus when the starch granules are made into a paste, amylose goes into solution, and amylopectin becomes a swollen, insoluble jelly. This behaviour is a result of their structures, for amyloses consist of unbranched chains whereas amylopectin is made up of branched chains which together form a gel framework. Neither component reduces Fehling's solution, showing that both contain but one free aldehyde group. Their other properties are summarised in Table B–IIIa.

TABLE B–IIIa

COMPARISON BETWEEN AMYLOSE AND AMYLOPECTIN

	Amylose	Amylopectin
configuration	unbranched chain	branched molecule
molecular weight (detd. osmotically)	up to $1/3 \times 10^6$	up to $80–100 \times 10^6$
degree of polymerisation	up to 2100	up to 620,000
length of molecule	up to 7400 Å	up to 1000 Å
β-amylase produces	complete hydrolysis	maltose + dextrin
pasting	forms no paste	forms paste
tetramethylglucose	0.05%	3–5%
dimethylglucose	—	3–5%

Amyloses represent a homologous family of polymers with degrees of polymerisation up to 2100 (Aspinall and Greenwood, 1962). Since stretched amylose helices contain three glucose residues, with 10.6 Å per turn, their maximum length reaches 7400 Å. In solution, the chain contracts forming a helix containing about six glucose residues per turn. Such an amylose helix encloses longitudinally a free space of a diameter which allows iodine molecules to penetrate, and it is conse-

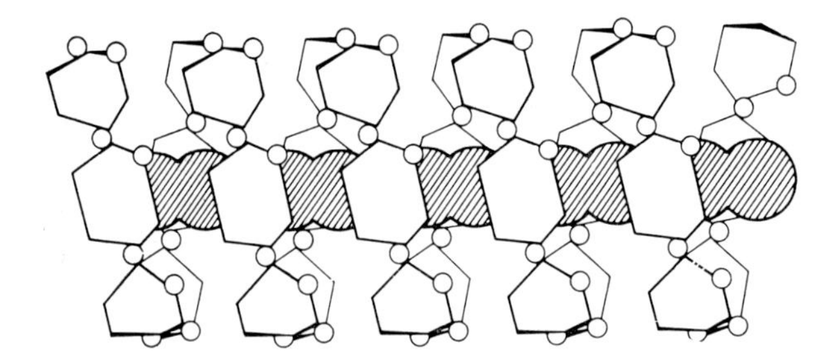

Fig. B–12. Amylose helix encircling iodine (Rundle *et al.*, 1944).

quently believed that the blue colour given by solutions of amylose with iodine is due to such an adsorption of I_2 inside the helical starch molecules (Fig. B–12).

The branching of amylopectin has been established by complete methylation, followed by hydrolysis of the chains (Haworth *et al.*, 1928). In this way 3–5% of 2,3-dimethylglucose and 3–5% of 2,3,4,6-tetramethylglucose is found, the remainder (90–94%) being 2,3,6-trimethylglucose. As may be seen from Fig. B–13, the tetramethylated molecules are derived from the end members, the trimethylated ones from ordinary chain members, and the dimethylated ones from branching members. It has been found that the branching bond is an α-1,6-linkage (Myrbäck and Ahlborg, 1942).

If dichotomous branching is admitted, for m end members there are m–1 branching members, and twice as many chain segments (see Fig. B–14a). This is in accordance with observed equal amounts of tetramethyl- and dimethyl-glucose residues (K. H. Meyer, 1943 and Table B–IIIa). Assuming that all chain segments are of equal length, an amylopectin molecule having a molecular weight of say 1,000,000, *i.e.* with a degree of polymerisation equal to 6200 and with 3.0% of end members, would then have the following features: 186 end members, 185 branching members, and chain segments with 6200/371 = 17 glucose residues. This theoretical value agrees satisfactorily with experimental data.

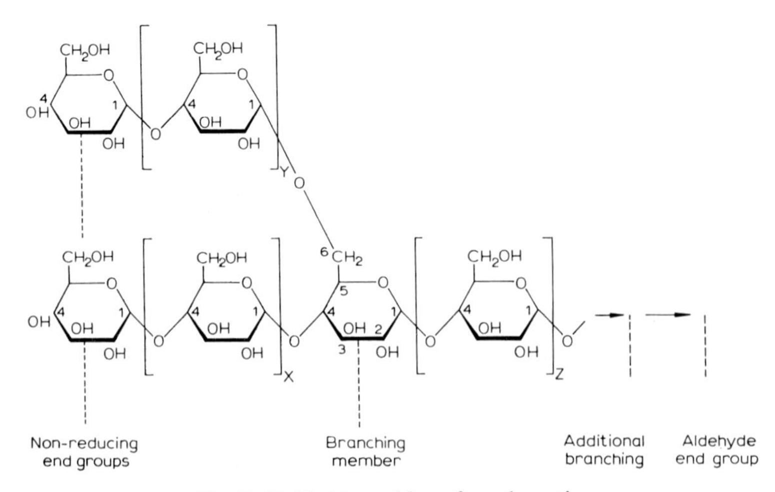

Fig. B–13. End branching of amylopectin.

TABLE B–IIIb

IDEALISED MOLECULES OF AMYLOPECTIN
(see Fig. B–14)

Number of stories m	End-groups $2^{(m-1)}$	Branching members $2^{(m-1)}-1$	Chain segments 2^m-1	DP $20 \times (2^m-1)$	MW $162 \times$ DP	Chain length $m \times 70$ Å
1	1	0	1	20	3240	70
2	2	1	3	60	9720	140
3	4	3	7	140	22,680	210
4	8	7	15	300	48,600	280
5	16	15	31	620	100×10^3	350
6	32	31	63	1260	0.2×10^6	420
7	64	63	127	2540	0.4×10^6	490
8	128	127	255	5100	0.8×10^6	560
9	256	255	511	10,200	1.6×10^6	630
10	512	511	1×10^3	20×10^3	3.2×10^6	700
11	1×10^3	1×10^3	2×10^3	40×10^3	6.4×10^6	770
12	2×10^3	2×10^3	4×10^3	80×10^3	12.8×10^6	840
13	4×10^3	4×10^3	8×10^3	160×10^3	25.6×10^6	910
14	8×10^3	8×10^3	16×10^3	320×10^3	51.2×10^6	980
15	16×10^3	16×10^3	32×10^3	640×10^3	100×10^6	1050

In cereal starch, amylopectins have been found with chain segments containing about 20 glucose residues (wheat 19, oat 20, maize 25–26) and with molecular weights (MW) of up to 80–100 $\times$ 10^6 corresponding to degrees of polymerisation (DP) of more than half a million (Aspinall and Greenwood, 1962). The features of such molecules are collected in Table B–IIIb, based on regular bifurcations of chain segments holding 20 glucose residues. The Table demonstrates that such an amylopectin molecule with MW = 100 millions would display 16,000 non-aldehydic end groups (with only one aldehyde group), an equal amount of bifurcations, and as many as 32,000 chain segments arranged in 15 stories (see Fig. B–14). Since a chain segment of 20 members is $20 \times (10.6/3) = 70$ Å long, the chain length from the aldehyde group to one of the non-aldehydic end groups can be calculated. This amounts to $15 \times 70 = 1050$ Å. As such a chain is not straight but angular, and is bent due to branching, the above calculated figure somewhat exceeds the actual length of the molecule. The approximate minimum width of the molecule can be found by calculating the number of end groups along a line running across the molecule. This is $\sqrt{16{,}000} = 130$. Since glucose residues have a mean width of about 6 Å, the idealised tree-like molecule 1000 Å long whould have a diameter of the order of 780 Å at its top.

Amylopectin is less easy to saccharify enzymatically than is amylose. The sugar-forming β-amylase can for instance transform amylose completely into maltose, rendering it soluble in cold water, whilst from amylopectin it can only split off a certain number of maltose units, leaving an insoluble remainder known as dextrin. The enzyme starts from the non-reducing ends of the branches, and proceeds close to the bifurcations. Another enzyme is necessary to attack the branching members, and in its absence the reaction comes to an end (Fig. B–11b).

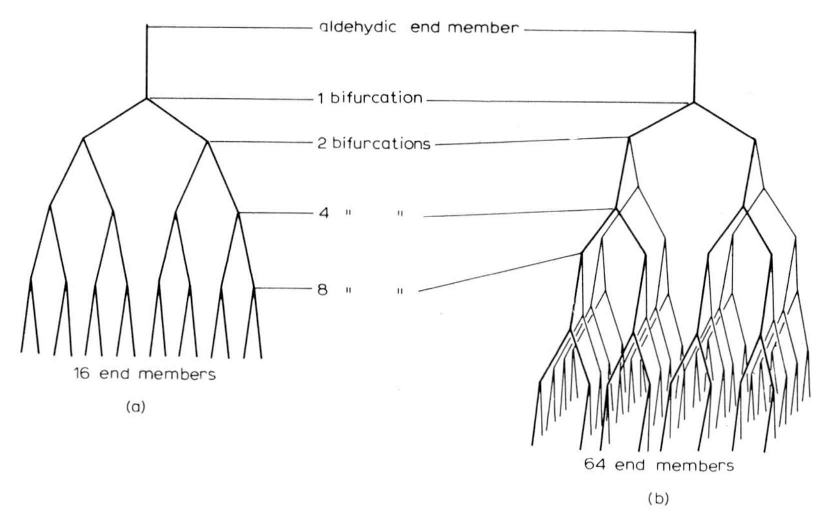

Fig. B–14. Dichotomous branching of amylopectin, (a) in a plane, (b) in space (Frey-Wyssling, 1948a).

The amount of amylopectin in starch is in general considerably larger than that of amylose. Table B–IV provides some information of this relation. As can be seen, the proportions of these two constituents vary for different types of starch. Since the ratio is governed by the action of genes, it may be altered by mutations. Thus high-amylose maize flour (52% amylose) and waxy maize (< 1% amylose) are known, in addition to the normal maize flour which contains 21% of amylose. These figures and those in Table B–IV are based on the standard extraction of water-soluble starch at 70° C. The percentages increase at higher temperatures, e.g. 61% of amylose is formed in barley at 90° C instead of 26%, and 81% in wheat at 98° C instead of 24%. This indicates that there is no sharp transition between the unbranched chains of amylose and the fully branched amylopectin with its short chains of 20–28 glucose units, and that all intermediate forms occur (Aspinall and Greenwood, 1962).

TABLE B–IV

THE AMYLOSE CONTENT OF STARCH
(Bates *et al.*, 1943)

Starch	% Amylose
Ketan *(Oryza sativa* f. *glutinosa)*	0
Waxy corn *(Zea mays* f. *saccharata)*	0
Tapioca *(Manihot utilissima)*	17
Rice *(Oryza sativa)*	17
Banana *(Musa sapientium)*	20.5
Corn *(Zea mays)*	21
Potato *(Solanum tuberosum)*	22
Wheat *(Triticum aestivum)*	24
Sago (*Metroxylon* spec.)	27
Lily bulb (*Lilium* spec.)	34

Starch granules have a spherite texture and do not therefore yield unambigous X-ray patterns with diffraction spots, but only diffraction rings. In 1933, Katz and Derksen have established that different kinds of starch do not produce the same ring pattern. For example, the cereal starch of wheat, rice, maize, barley, oats, produces what is known as an *A* spectrum, whereas potato starch has a *B* spectrum, and both, when formed into a paste, produce a third type called the *V* spectrum. Starches with a *B* spectrum (tuber starch) have been converted by heat to the *A* type; it has also been shown that the *V* spectrum reverts to the *B* spectrum in the so-called retrogression of paste. The quantity of the bound water plays a certain part in this process. Thus the following conversions may be observed in pasted up and retrogressed wheat starch: $A \rightarrow V \rightarrow B$.

The unit cell in the lattice of crystalline starch could not be deduced from the X-ray spectra mentioned above; this only became feasible when Kreger (1951) succeeded in shooting X-rays across the parallel textured parts of a single starch granule (Fig. B–15) from the orchid *Phajus grandifolius* (starch with a *B* diagram). This method yielded fibre patterns from which the dimensions of the unit cell could be calculated. The cell has a rhombic symmetry, with $a = 9.0$, $b = 10.6$, and $c = 15.6$ Å. As the ratio $a:c$ is equal to $1:\sqrt{3}$, the lattice is hexagonal. The period of 10.6 Å corresponds to the chain axis; it is a threefold helix comprising three glucose residues. The rhombic unit cell holds 3, and the enlarged hexagonal cell 18 amylose chains (Fig. B–16), with 54 glucose residues and 54 water molecules. From these data a density of 1.80 is calculated. Since starch granules are in fact considerably less dense (in water 1.62, in xylene 1.43), Kreger attributes this difference to the poor crystallinity of amylopectin.

Since most starch granules consist of only 25% of amylose (Table B–IV) and 75% of amylopectin, the branched starch molecule must also fit into the observed lattice. This is possible if symmetrical threefold branching is assumed in place of bifurcations (Fig. B–14). Fig. B–17a shows how this is possible (Frey-Wyssling,

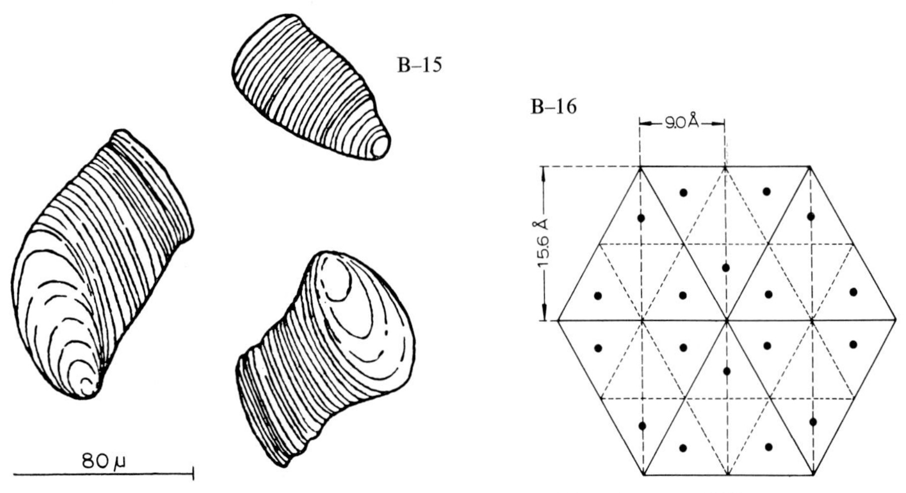

Fig. B–15. Starch granules of *Phajus grandifolius* (Kreger, 1951).
Fig. B–16. Possible unit cell of crystallised starch (Kreger, 1951).

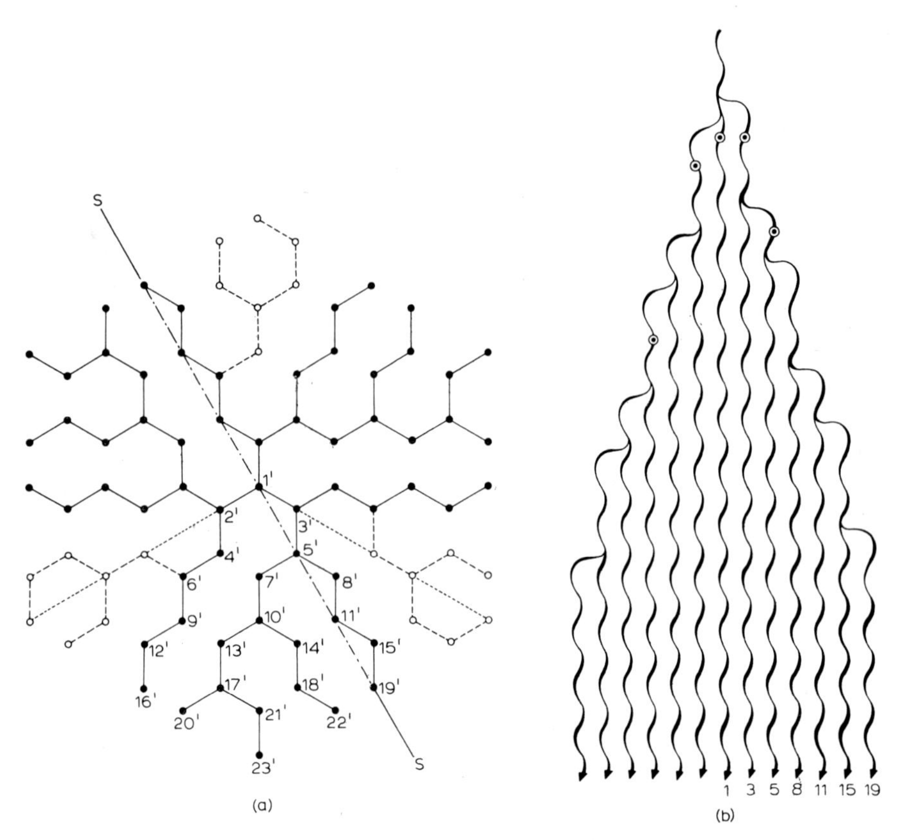

Fig. B–17. Possible lattice of amylopectin; (a) plan view, combination of bifurcation with threefold rotational symmetry; (b) longitudinal section.

1957). A longitudinal section across this pattern looks like Fig. B–17b which represents a more densely packed molecule than Fig. B–14 based on dichotomous branching. In any case, the amylopectin molecule is oblong and brushlike, in contrast to glycogen which has a related chemical structure (Fig. B–11).

(*e*) *Glycogen*

Certain heterotrophic plants (Fungi) accumulate glycogen in place of starch, and behave like animals in this respect. In yeast, the glycogen content may rise to 30% by dry weight.

Glycogen is a polyglucosan similar to amylopectin, but the ramification is more highly developed and the chain segments between the bifurcations are shorter, so that the molecule is more globular, (Fig. B–11c) and is soluble in water. It is not synthesised in plastids as are the starch granules, but in the cytoplasm where diffuse deposits of glycogen occur.

The molecules of isolated glycogen are visible under the electron microscope. As a matter of fact, the first macromolecules ever imaged were *p*-iodobenzoylated glycogen (Husemann and Ruska, 1940), with a molecular weight of 6×10^6 (the glycogen involved had MW = 1.5×10^6, and DP about 10,000), and the molecules were spherical, with a diameter of 150 to 300 Å. In recent times (Drochmans, 1962) glycogen isolated from rat liver was investigated by the negative contrast method. Large particles, called α-particles, with diameters of 600–2000 Å were found. They are of a composite nature, consisting of β-particles 300 Å in diameter. At pH < 4.5, the α-particles disaggregate; the same is true of the β-particles, which disintegrate at pH < 3.0 into rodlike γ-particles 30 Å in diameter and 200 Å in length. As the volume of a glucose residue amounts to 169 Å^3 (in cellulose), the following approximate DP's can be calculated: α-particles 3×10^6, β-particles 2×10^4, γ-particles 2×10^2. It seems that the macromolecules photographed by Husemann and Ruska in 1940 were the β-particles.

In 1963 Moor and Mühlethaler found similarly aggregated bodies in the cytoplasm of yeast. It therefore seems that the extremely high molecular weights found for this carbohydrate refer to associated particles, consisting of a certain number of smaller molecules.

This is probably also true of amylopectin, although no such morphological features have as yet been found by electron microscopy. It is however very likely that molecular weights as high as 100×10^6 (found by turbidity measurements) refer to aggregated particles (Aspinall and Greenwood, 1962).

(*f*) *Chitin*

Chitin is a skeleton substance characteristic of Arthropoda (Crustacea, Insecta), and also forms the framework substance in the cell walls of Fungi (R. Frey, 1950). The behaviour of vegetable and animal chitin is identical, as has been proved for the sporangiophores of *Phycomyces* chemically, optically, and by X-rays (Diehl and van Iterson, 1935; van Iterson *et al.*, 1936).

Although chitin contains nitrogen, it belongs to the class of carbohydrates. Similarly to cellulose, it forms unbranched high-polymer chains based on acetylglucosamine. In this compound the hydroxyl group on the second carbon in β-glucose (Fig. B–6) is replaced by an acetylated amino group. The residues of this glucosamine are united through β-glucosidic linkages (Fig. B–18a).

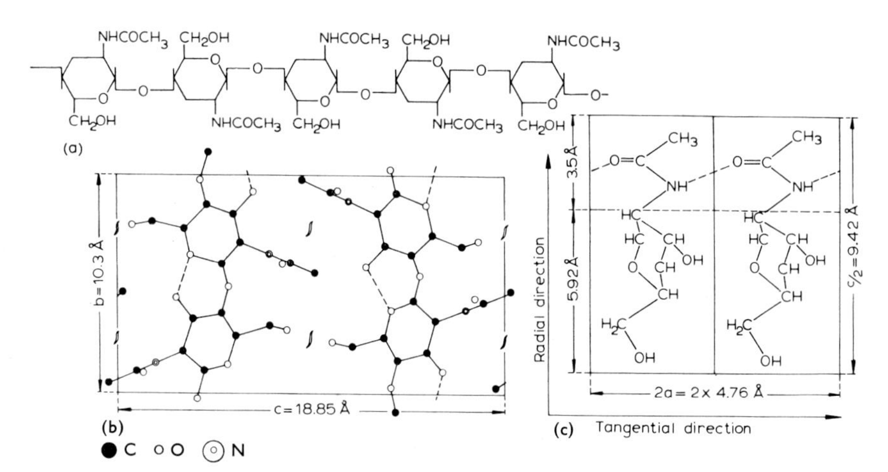

Fig. B–18. Chitin; (a) chain molecule, (b) unit cell of the crystal lattice (Carlström, 1957), (c) orientation in the cell wall of *Phycomyces*.

X-ray analysis of lobster tendons yields a fibre pattern of a chain lattice whose unit cell has the following dimensions: $a = 4.76$, $b = 10.28$, $c = 18.85$ Å (Carlström, 1957). Two antiparallel chains run across the unit cell, embodying a twofold helical screw axis. The fibre period is 10.28 Å as in cellulose, and also embraces two glucosamine residues (Fig. B–18b). The chains are zig-zagged by the glucosidic bond angle, and it is possible that, as in cellulose, the chains are less straight than is indicated in Fig. B–6. A hydrogen bond exists between the hydroxyl of the sixth carbon atom and the next aldehyde oxygen. The separation of two parallel chains is 4.76 Å, corresponding to the so-called backbone spacing between polypeptide chains of 4.6 Å (see p. 62). This distance is bridged by N—H···O=C bonds (Fig. B–18c). In the cell wall of the *Phycomyces* sporangiophore, the b-axis of this chain lattice runs longitudinally in a helical manner, with changing helical angle (helical growth). Heyn (1936) has shown that the c-axis is oriented radially.

2. PROTEINS

(*a*) *Polypeptides*

The basic substances of proteins, isolated by means of hydrolysis, are compounds known as α-amino acids, which possess the structure given in Fig. B–19a, in which R represents a group often containing a large number of carbon atoms. To be exact, the NH_2— and COOH— groups should be bound to the carbon as individual groups, as shown in Fig. B–19b.

Two amino acids can eliminate a molecule of water, to form a so-called dipeptide. This process can be repeated many times, giving a long polypeptide chain, the ends of which have been left open in Fig. B–19c. Similarly to the case of paraffins, the chain is kinked. X-ray analysis of crystalline fibre proteins showed that the distance between two equivalent groups is 3.5 Å. Only the >CO and >NH groups are similar along the whole length of the chain; R differs according to the kind of amino acids which cause the great variety found in the class of proteins. The zig-zag chain drawn in Fig. B–19c can be considered as a relatively inert frame, which cannot be responsible for the characteristic chemical lability of proteins. This unusual reactivity is due to the side chains R.

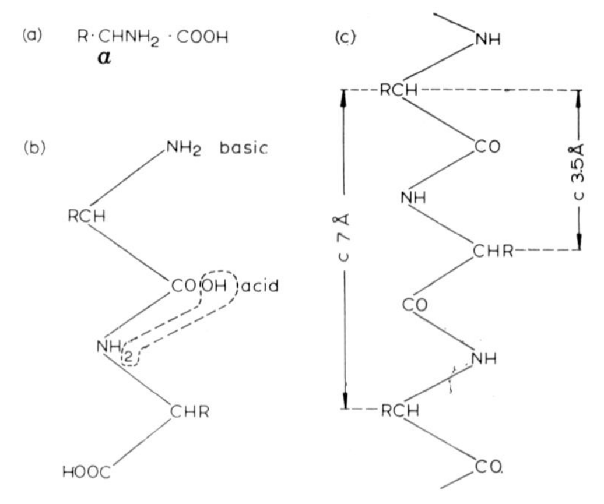

Fig. B–19. Polypeptides; (a) α-amino acid, (b) peptide formation, (c) polypeptide chain.

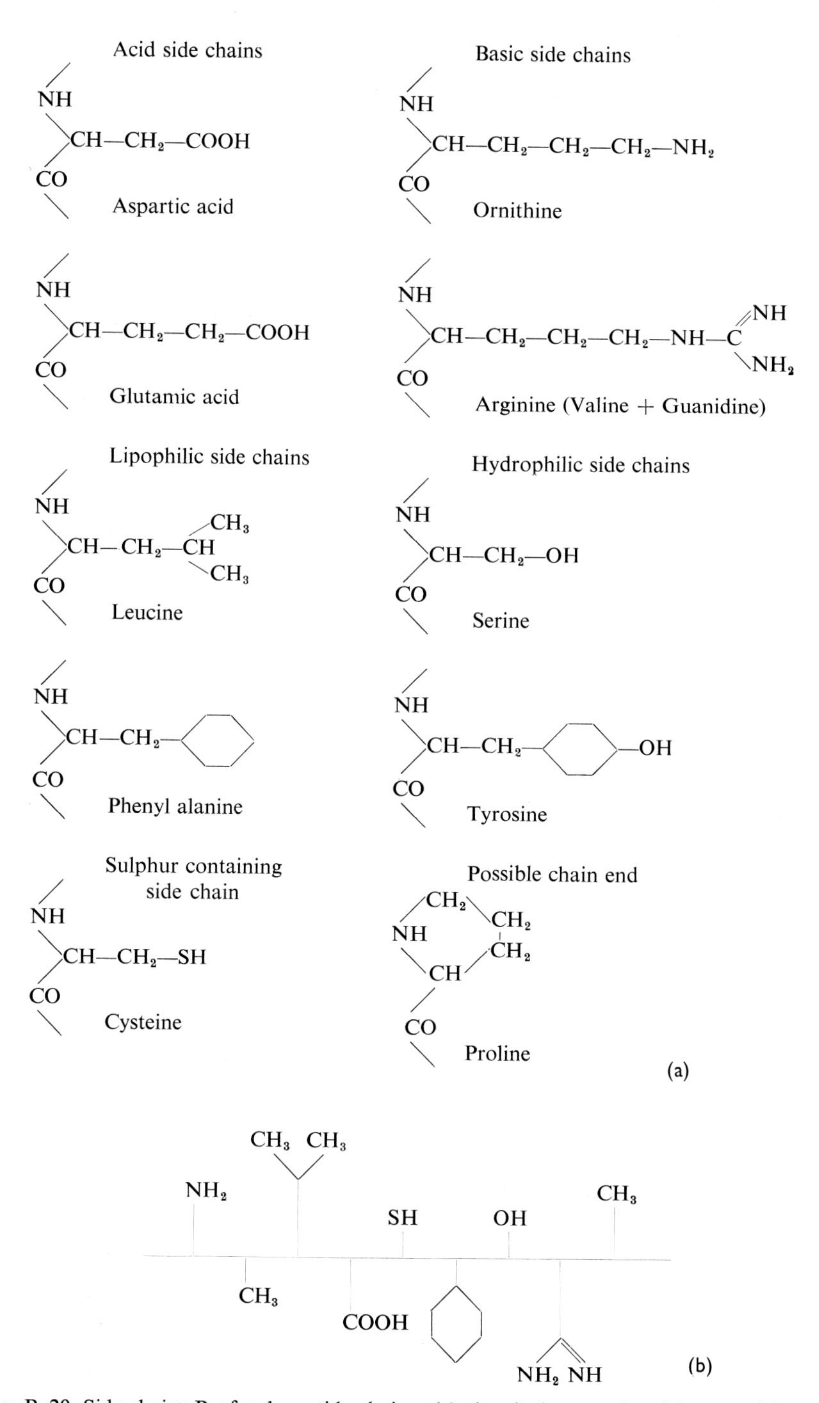

Fig. B–20. Side chains R of polypeptide chains; (a) chemical properties, (b) unequal length.

The amphoteric behaviour of proteins is often explained in chemical textbooks by the fact that amino acids possess both acidic and basic groups (Fig. B–19b). However, it follows from the structural picture of the polypeptide chain that these groups disappear in the condensation, thus losing their capacity for dissociation. If in spite of this the proteins clearly show acid or basic properties, this must be ascribed to the side chains, which carry free COOH— or NH_2— groups. This happens when some members of the polypeptide chains consist of dicarboxylic amino acids, or of diamino acids (Fig. B–20).

There are 24 amino acids, 22 of which are listed in Table B–V, with their symbols and the formulae of groups R according to the number of the C atoms in their aliphatic chains. Substituents other than H are added between brackets. Hydroxyglutamic acid (Glu—OH) is not included in the Table B–V, because it

TABLE B-V

AMINO ACIDS

Size	Name	Symbols	R
C_2	Glycine	Gly	—H
C_3	Alanine	Ala	$—CH_3$
$C_3(O)$	Serine	Ser	$—CH_2OH$
$C_3(S)$	Cysteine	Cys	—CHSH
$C_4(O)$	Threonine	Thr	$—CHOH \cdot CH_3$
$C_4(SC)$	Methionine	Met	$—(CH_2)_2 \cdot S \cdot CH_3$
$C_4(O_2)$	Aspartic acid	Asp	$—CH \cdot COOH$
$C_4(ON)$	Asparagine	Asn	$—CH \cdot CONH_2$
C_5	Valine	Val	$—CH{:}(CH_3)_2$
C_5	Norvaline	Nval	$—(CH_2)_2 \cdot CH_3$
C_5	Proline (= cycloNval)	Pro	$—(CH_2)_3—$
$C_5(O)$	Hydroxyproline	Opro	$—CH_2 \cdot CHOH \cdot CH_2—$
$C_5(O_2)$	Glutamic acid	Glu	$—(CH_2)_2 \cdot COOH$
$C_5(ON)$	Glutamine	Gln	$—(CH_2)_2 \cdot CONH_2$
$C_5(NCN_2)$	Arginine	Arg	$—(CH_2)_3NH \cdot C(=NH)NH_2$
C_6	Leucine	Leu	$—CH_2 \cdot CH{:}(CH_3)_2$
C_6	Isoleucine	Ileu	$—CH(CH_2 \cdot CH_3)(CH_3)$
$C_6(N)$	Lysine	Lys	$—(CH_2)_4 \cdot NH_2$
$C_3 \cdot C_6$	Phenylalanine	Phe	$—CH_2 \cdot C_6H_5$
$C_3 \cdot C_6(O)$	Tyrosine	Tyr	$—CH_2 \cdot C_6H_4OH$
$C_3 \cdot C_3N_2$	Histidine (= Imidazoalanine)	His	$—CH_2$–C(=CH–NH–CH=N–) (imidazole ring)
$C_3 \cdot C_8N$	Tryptophan (= Indolealanine)	Try	$—CH_2$–C=CH–NH– (indole ring with benzene)

has no symbol of its own. The same is true of cystine (Cys—Cys), a molecule consisting of two cysteine units which can be split by reduction:

$$COOH \cdot CHNH_2 \cdot CH_2S—SCH_2 \cdot CHNH_2 \cdot COOH \underset{-H_2}{\overset{+H_2}{\rightleftharpoons}} 2\,COOH \cdot CHNH_2.CH_2SH$$

In the oxidised form, this system can tie together neighbouring polypeptide chains (see Fig. B–21).

A classification of amino acids which is more judicious than that in Table B–V is obtained if the chemical character of the lateral groups in the side chains R is considered (Fig. B–20). The dicarboxylic amino acids (Asp, Glu) contain acidic groups, the diamino acids (Arg, Lys, His) basic groups, the ordinary amino acids (Ala, Val, Nval, Leu, Ileu, Phe) lipidic groups, and the oxidised specimens (Ser, Thr, Opro) hydrophilic side groups. In two-dimensional paper chromatography, these properties cause the amino acids to separate (according to their molecular weight) along four different lines corresponding to the above-mentioned four categories. The amino acid ornithine, listed in Fig. B–20a is a degradation product of arginine and does not exist in nature as a polypeptide residue.

According to their character, the various side groups linked to a polypeptide chain may react with water, anions, cations, or corresponding active groups of other polypeptide chains; this is indicated in Fig. B–21.

Although the amino acids differ considerably in size, a mean molecular weight of a peptide residue can be calculated and, taking the density of proteins into account, it is possible to derive the average space it needs. As a rule, proteins hold 16% of nitrogen. This means that the protein content in a preparation free from other nitrogen compounds may be calculated by multiplying the amount of nitrogen

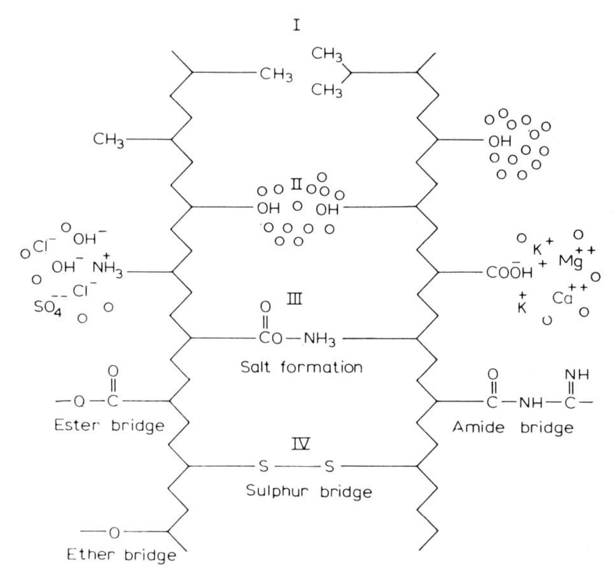

Fig. B–21. Schematic representation of possible junctions between neighbouring polypeptide chains. o = water molecules.

determined by Kjeldahl's method by $\frac{100}{16}$, that is by 6.25. If all amino acids would contain but one nitrogen atom, their average molecular weight should be $6.25 \times 14 = 87.5$. Since however the basic amino acids possess two or even four (Arg) nitrogens, the mean factor indicating the average N-content of a peptide residue is 1.37. Exceptions with smaller factors are insulin (1.33), lactoglobulin (1.27) (Jordan, 1947), and especially gelatin, elastin, and silk fibroin, with almost no basic amino acids. With the factor of 1.37, the mean molecular weight of an amino acid is $6.25 \times 1.37 \times 14 = 120$.

The space required by a peptide residue may be found by dividing the molecular weight by the density of the compound involved, and by the number of molecules in 1 gram-molecule (Avogadro's number). Since the density of most proteins is found to be 1.3, the average space for an amino acid is $120/(1.3 \times 6.02 \times 10^{23}) = 154 \times 10^{-24}$ cm^3 = 154 Å^3. This value is close to the volume found for a peptide residue by X-ray analysis of fibrous proteins (see p. 62), *i.e.* $3.5 \times 4.6 \times 10 = 161$ Å^3. Fig. B–22 shows the space requirement found in this way. Similar calculations based on helical polypeptides (Pauling and Corey, 1954) are listed in Table B–VI.

TABLE B–VI

AVERAGE SPACE REQUIREMENT OF AN AMINO ACID RESIDUE

Polypeptide	Average molecular weight	Density of protein	Calculated volume
Expanded chain	120	1.3	154 Å^3
α-Helix	105	1.33	131 Å^3
γ-Helix	110	1.30	141 Å^3

Similarly to the polysaccharides, the polypeptide chains are characterised by chemically defined repeating units. It is thus legitimate to determine the mean size of such a unit (Table B–VI), to obtain an idea of the space requirements of the chain as a whole if its degree of polymerisation is known. On the other hand, the polypeptides are unique in that 24 different monomers can be inserted into the chain in various proportions and sequences.

To summarise, the structure of polypeptide chains is governed by two seemingly contradictory features.

(1) The principle of repetition which is known in biology as segmentation or metamerism. Most high polymers are built according to this principle. In the majority of such compounds, however, the monomeric groups are identical whilst in the polypeptide chains the side groups R, which occur at regular spacings of 3.5 Å, have different constitutions.

(2) The principle of specificity. As a result of the great number of possible side chains R and their unlimited number of possible arrangements along the polypeptide chains, an infinite number of polypeptides is conceivable, distinguished only by slight chemical alterations which become apparent in the specific properties of the proteins.

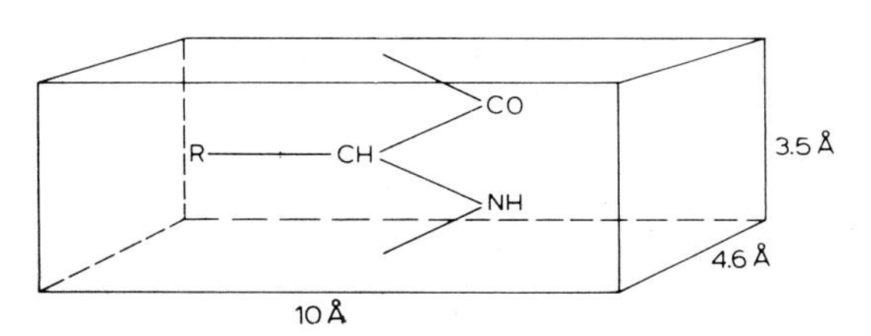

Fig. B–22. The space requirement of an amino acid residue.

(*b*) *The sequence of amino acids*

Since the specificity of proteins is due to the sequence of amino acids in the polypeptides, it is of considerable physiological interest to find the exact position of every residue in the peptide chain. For a long time this seemed an almost impossible task. Since however chromatography has facilitated the analysis of amino acids, methods have been developed to dismantle the polypeptides step by step.

The first requirement is to start with a pure polypeptide having a definite molecular weight. The number of peptide bonds and thus the degree of polymerisation can then be determined. The next step is to hydrolyse the chain completely to establish the amino acids present, e.g. by paper chromatography. The procedure will be explained on a relatively simple compound elucidated by Harris and Roos in 1956. This is the melanophore-stimulating hormone isolated from hog posterior pituitary extracts, which turned out to be an octadecapeptide. Its 18 amino acids are Arg, 2 Asp, 2 Glu, 2 Gly, His, 2 Lys, Met, Phe, 3 Pro, Ser, Try and Tyr.

Each polypeptide chain has an N-terminal and a C-terminal. These —NH_2 or —COOH groups are not involved in the peptide bonding —NH—CO— and are thus free to react chemically. Their amount in a preparation can therefore be analysed, and by such end-group determination the molecular weight can be calculated, as long as the peptide is not too highly polymerised. Furthermore, the N-terminal group can be combined with organic micromolecules in such a way that the new bond is stronger than the peptide bond which links the end-group to the chain. The terminal amino acid compound can therefore be split off and analysed individually. When this is done, the second amino acid of the chain becomes terminal, and can be treated in the same way. The peptidic chain can also be degraded from its acid end by the enzyme carboxypeptidase, which splits the peptidic bonds next to a free carboxyl group.

Successive degradations such as described above permit the amino acid sequence in oligopeptides to be established. For high polymers containing dozens of amino acids, however, this work turns out too laborious. The high polymer chain is therefore broken down into a mixture of oligomers by specific enzymes. There are many such catalysts, which split the peptidic bond of definite amino acids; only two of them can be mentioned here. One of them, trypsin, attacks the chain between the carbonyl of the basic residues lysine or arginine and the NH-group of the next amino acid. With this enzyme the melanophore-stimulating hormone is split into three oligopeptides which have been indexed T_{1a}, T_{4a}, T_3. On the other hand, chemotrypsin splits the peptidic bond preferentially on the carbonyl side of phenylalanine, tyrosine, or tryptophan, all three with an aromatic

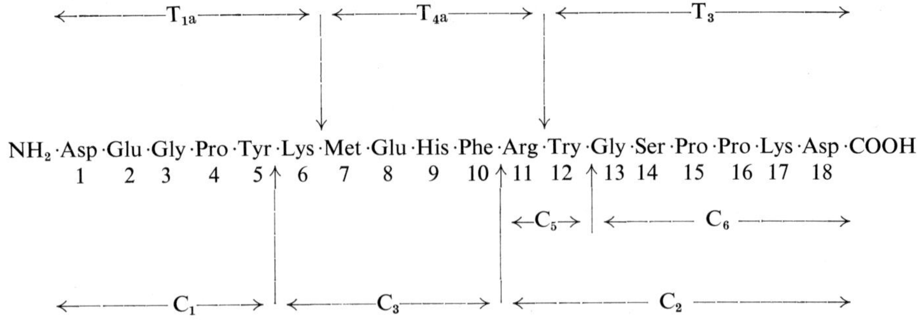

Fig. B–23. Amino acid sequence of a melanophore-stimulating peptide (Harris and Roos, 1956).

ring in their structure; it must be added that methionine, leucine, and histidine are also attacked in certain cases. In our example three oligopeptides C_1, C_2, and C_3 are obtained with chemotrypsin, and a prolonged digestion degrades C_2 further into C_5 and C_6 (Fig. B–23).

In Fig. B–23 the 18 amino acids are numbered starting from the N—terminal group. Both the N— and the C— end-groups are asparagine. Within the indicated oligopeptides, the sequence is established by splitting off the amino acids one by one. Corresponding sequences are then found in the T— and C—oligopeptides, indicating how the whole chain must be composed. Physiologically, certain partial sequences are of particular interest. It has thus been found that the sequence Met · Glu · His · Phe · Arg · Try · Gly (positions 7–13) also occurs in the corticotropins, suggesting that there is a relation between the structures and the biological activities of the posterior pituitary hormones.

The first protein analysed for its amino acid sequence was the hormone insulin (Sanger and Tuppy, 1951; Tuppy, 1959). Insulin may be called a protein, because it consists of two different polypeptide chains A and B, with 21 and 30 amino acids respectively – *i.e.* with 51 amino acids in all. The two chains can be separated by oxidation, because they are linked together by two S—S bridges (Fig. B–24). The A-chain harbours 4 and the B-chain only two cysteine residues. As a result a cysteine bridge exists in the A-chain between the cysteine in position 6 and that in position 11, so that a loop is formed. The sequence within this loop is different in the insulin molecule of different mammals (Harris *et al.*, 1956) as shown in Fig. B–25. This is the reason why the insulins of those mammals are specifically different. On the other hand, positions 8, 9 and 10 in the A-chain do not seem to be important for the hormonal activity of the insulin.

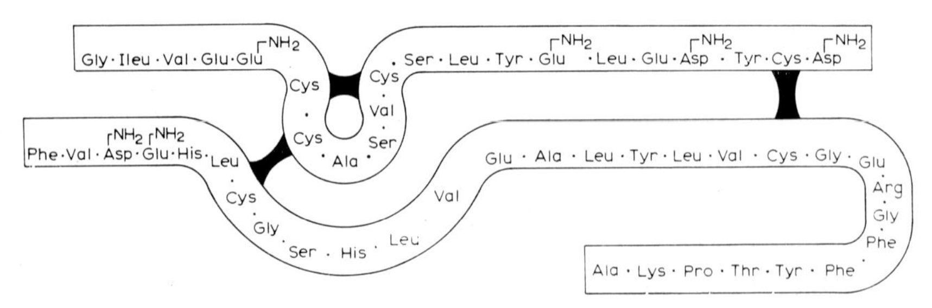

Fig. B–24. Amino acid sequence in insulin (Ryle *et al.*, 1955).

A more complicated compound, whose amino acid sequence has also been elucidated is the ribonuclease molecule (Smyth *et al.*, 1963). This is a single polypeptide chain containing 124 amino acids. Its six cysteine residues give rise to three intramolecular S—S bridges, so that the chain is quite irregularly curved (Fig. B–26). The enzyme subtilisin splits this chain only between positions 20 and 21 (arrow in Fig. B–26), and the resulting 20-member and 104-member polypeptides are no longer capable of attacking ribonucleic polyacids. The activity may however be restored by dissolving the two peptides together (Richards, 1958). This experiment proves that the entire sequence of 124 amino acids is necessary for the ribonuclease molecule to exert its enzymological activity.

Bovine	—Cys—	Cys—	Ala—	Ser—	Val—	Cys—
Pig	—Cys—	Cys—	Thr—	Ser—	Ileu—	Cys—
Sheep	—Cys—	Cys—	Ala—	Gly—	Val—	Cys—
Horse	—Cys—	Cys—	Thr—	Gly—	Ileu—	Cys—
Whale	—Cys—	Cys—	Thr—	Ser—	Ileu—	Cys—
	6	7	8	9	10	11

Fig. B–25. Amino acid sequence in the loop of the A–chain in insulin (Harris *et al.*, 1956).

Before the sequence of amino acids could be established, there was much discussion as to whether there was a periodicity by short repeating sequences along the polypeptidic chain (Bergmann and Niemann, 1937). No such regularity could however be discovered. The analysed physiologically active proteins always show an irregular sequence, as exemplified in Figs. B–23, B–24 and B–26. Intramolecular periodicities occur only in some fibrous proteins with a poor inventory of amino acids (see p. 61).

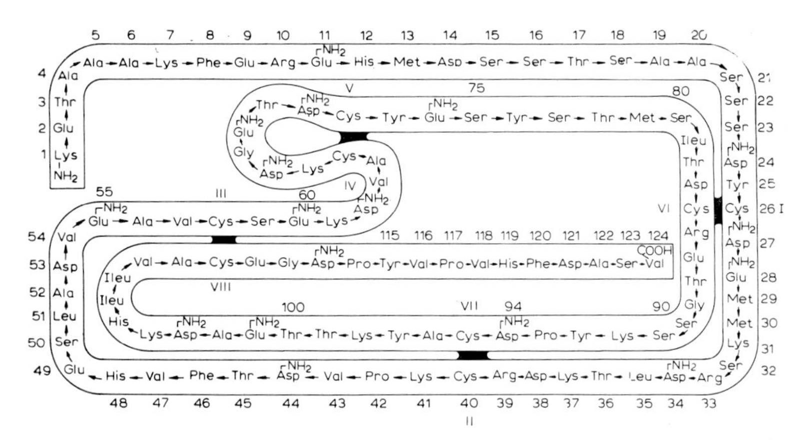

Fig. B–26. Amino acid sequence in ribonuclease (according to Smyth *et al.*, 1963).

(*c*) *The classification of proteins*

No clear cut dividing line may be drawn between proteins and polypeptides. The polypeptides are considered to be building units of proteins, which in addition can react with carbohydrates (gluco- and mucoproteins), lipids (lipoproteins), pigments (chromoproteins), or nucleic acids (nucleoproteins). By definition, a protein molecule ought to contain more than one polypeptide chain. The aggregation may consist of identical (silk fibroin, Fig. B–29) or different polypeptides. In insulin there are two such chains, united by sulphur bridges (Fig. B–24); this is the reason why this hormone is regarded as a protein, although its molecular weight of about 6000 is rather low. In most 'simple' proteins the molecular weights are of the order of 15,000–18,000. The enzyme nuclease, with its 124 amino acids and a molecular weight of 15,300 (Cohn and Edsall, 1943) falls within this group; although its molecule consists of a single polypeptide chain (Fig. B–26), it is nevertheless called a protein.

Since the very definition of a protein molecule is uncertain, their chemical classification cannot be easy and physical methods such as solubility measurements and particle size determinations in the ultracentrifuge are used. For plant proteins, the classification based on solubility (adopted by Osborne in 1924) is still valid. Osborne distinguished:

(1) albumins, soluble in water and dilute salt solutions,
(2) globulins, soluble in neutral salt solutions,
(3) prolamins (gliadins) soluble in 70–80% ethanol, but insoluble in water or pure alcohol,
(4) glutelins, soluble in dilute acids or dilute alkalis, but insoluble in the solvents already mentioned.

More chemical definitions are used for the histones and protamines found in nuclei. Both are soluble in water, but are distinguished from albumins by their high contents of basic amino acids (histones: diamino acids up to 30%, protamines: arginine up to 87%). Histones have a higher molecular weight and can be precipitated by solutions of salts or ammonia, whilst protamines have small molecules that cannot be coagulated by heat.

Physiological classifications distinguish between reserve proteins which, as indicated, are arranged according to their solubility, plasma proteins, which have not yet been crystallised, and scleroproteins such as keratin, collagen, elastin, silk fibroin, etc., which possess mechanical functions.

The morphological classification is particularly important for our considerations, because physiological and functional properties of proteins depend not only on their chemical composition, but in certain instances even more so on the shape of their molecules. In this respect two types of proteins have been distinguished: fibrous proteins with extended straight or helical chains, and globular proteins with coiled polypeptides in spherical molecules. The fibrous type is covered by the scleroproteins just mentioned, which crystallise in chain lattices forming fibrils; the globular proteins such as enzymes, reserve proteins, etc. on the other hand yield more or less isodiametric crystals, with individual macromolecular particles in their lattices. The molecular threads of the fibrous type are held transversely together along their total length by hydrogen bonds, whilst the spheres of the globular types touch each other only at certain points, according to their coordination number, leaving freely accessible interstices. These structural differences make the globular proteins more readily soluble and digestible than the fibrous proteins.

It has been found that certain globular proteins can be converted into the

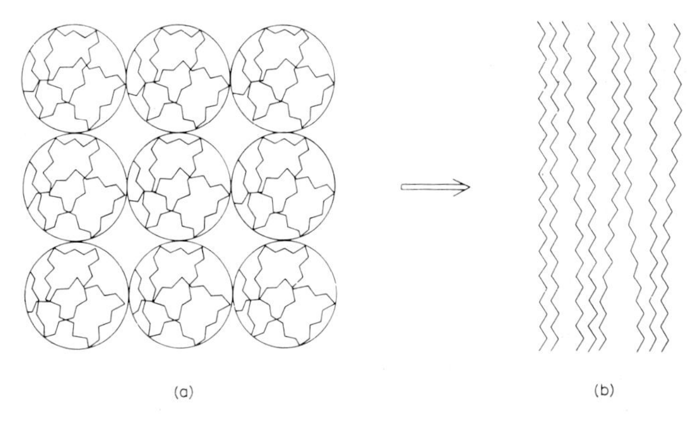

Fig. B–27. Denaturation of globular to fibrous proteins. (a) Folded polypeptide chains are transformed into (b) expanded chains.

fibrous form (Fig. B–27), by heating or even by simple spreading as a monomolecular film on water. The solubility and digestibility is then strikingly decreased, and the proteins also lose mostly their physiological activity (Waldschmidt-Leitz, 1958, p. 298), owing to disturbance of the specifically arranged polypeptide chains in sinuosities, loops (see Fig. B–24) and winding convolutions by the extending molecular chains. In the extended state many of the active side groups are linked together by hydrogen or other bonds so that they lose their reactivity; in contrast, in the globular molecule these groups were readily accessible and free to react. Because of this loss of digestibility and physiological reactivity the above transformation from the globular to the extended form has been termed denaturation. The process is not reversible, and the globular state must be therefore considered to possess higher energy (and thus be less stable) than the extended form.

In silk fibroin the polypeptide chains are fully extended and consequently their elasticity is low. On the other hand, keratin, (the protein of hair) is astonishingly extensible and under favourable conditions may be stretched almost by up to 100%. It was therefore believed that polypeptide chains could fold and unfold (Astbury, 1933). Since helical structures within a chain lattice offer less geometrical difficulties for the extension and contraction of fibres than unfolding and folding chains, proteins with helical polypeptide chains have been postulated, and their existence could be demonstrated (see p. 62). The morphological classification based on the shape of the polypeptide chains is thus:

1 – fibrous proteins { a – with linear chains,
b – with helical chains,

2 – globular proteins { c – with coiled chains.

Three grades are distinguished in the structure of the most complicated protein molecules. The basic chemical structure, *i.e.* the amino acid sequence throughout the whole macromolecule, with all the end and side groups of its polypeptidic chains, is called the primary structure of the protein. The morphological features of the chains, *i.e.* whether they are fully extended or helical, are considered to be the secondary structure. The coiling of the peptid chains and helix segments within the globular particles is called the tertiary structure. Finally, account is taken of the spatial relationships between the different polypeptide units: whether they are running parallel or antiparallel, and how they are twisted when extended, or how arranged in triplets, tetrads etc. when globular; this could be termed the quaternary structure. It represents the ultrastructure which may be disclosed in the electron microscope.

The secondary structures of some well known proteins will be described in the succeeding three paragraphs.

(*d*) *Proteins with linear chains*

In fully extended polypeptides the chain runs in a zig-zag line along a straight axis. The interatomic distances and the bond angles have been established from X-ray studies of oligopeptide crystals (Corey and Donohue, 1950; Pauling and Corey, 1954). As may be seen in Fig. B–28, the interatomic distances and the bond

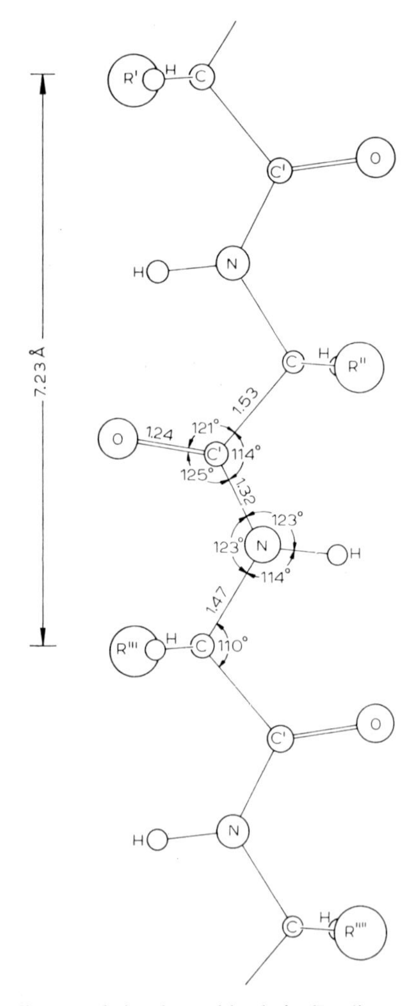

Fig. B–28. Fully extended polypeptide chain (Pauling and Corey, 1954).

angle of 110° around the $\rangle C \langle^{R}_{H}$ group correspond to standard values (see Table A–I), whilst around the central atoms of the carbonyl $\rangle C{=}O$ and the imino $\rangle NH$ groups the bond angles add up to 360°, so that the three substituents are coplanar. The chain period is 7.23 Å, which value equals the fibre period of silk.

The best known linear protein is silk fibroin, yielding beautiful X-ray fibre diagrams identified as chain lattice patterns by Meyer and Mark (1928) and by Kratky and Kuriyama (1931). There is an amorphous form of fibroin which obscures the sets of diffraction spots, but being more soluble than the crystalline fibroin this interfering component can be removed. For the crystalline protein the following unit cell is found, through which run two pairs of fibroin chains, in an antiparallel sense (Table B–VII).

TABLE B–VII

THE UNIT CELL OF SILK FIBROIN

	a	b	c	β
Kratky *et al.*, 1931	9.60	6.95	9.02	74°
Marsh *et al.*, 1955	9.40	6.97	9.20	90°

Kratky *et al.* (1931) assumed a monoclinic, and Marsh *et al.* (1955) an orthorhombic unit cell. The fibre axis b is 7 Å. Since this is twice as long as an amino acid residue (Fig. B–19c), the unit cell must contain 4 such residues. The chain consists of 44.7% Gly, 25.7% Ala, 11.9% Ser, and 5.4% Tyr, with minor amounts of other amino acids (Neurath and Bailey, 1953). The following approximate molar proportions have been deduced from these figures: Gly:Ala:Ser = 3:2:1 or Gly:(Ala + Ser) = 1:1. It thus appears that there is a repeating unit of [—Gly—Ala (or Ser)—] in the chain, and two such doublets, running in opposite directions, must fill the unit cell. Since the mean residue weight of the above amino acids is 67, the density of silk fibroin $\varrho = \frac{8 \times 67}{VN} = 1.47$ (V = volume of the unit cell, N = Avogadro's number). It is difficult to measure experimentally the density of a fibrillar body possessing ultrastructural capillaries (Hegetschweiler, 1949), but the density can be calculated from the refractive indices of silk and the molecular refractions of the amino acids involved. A value of $\varrho = 1.464$ has been found in this way (Frey-Wyssling, 1955a).

The antiparallel chains (Fig. B–29) are held together by hydrogen bonds $\rangle CO{\cdots}H{-}N\langle$ (see p. 23), so that pleated sheets of molecular chains are formed. These sheets, extending along the b-axis (fibre axis), were originally placed parallel to the c-axis of the unit cell. Marsh *et al.* (1955) suggested that they ran parallel to the a-axis (Fig. B–29c). Since the unit cell is almost tetragonal, this does not appreciably affect the chain spacings in the pleated sheets, which were 4.5 Å according to Meyer (1940) and which have now been found to be 4.7 Å. Separations of the chains have been called backbone spacing by Astbury (1933).

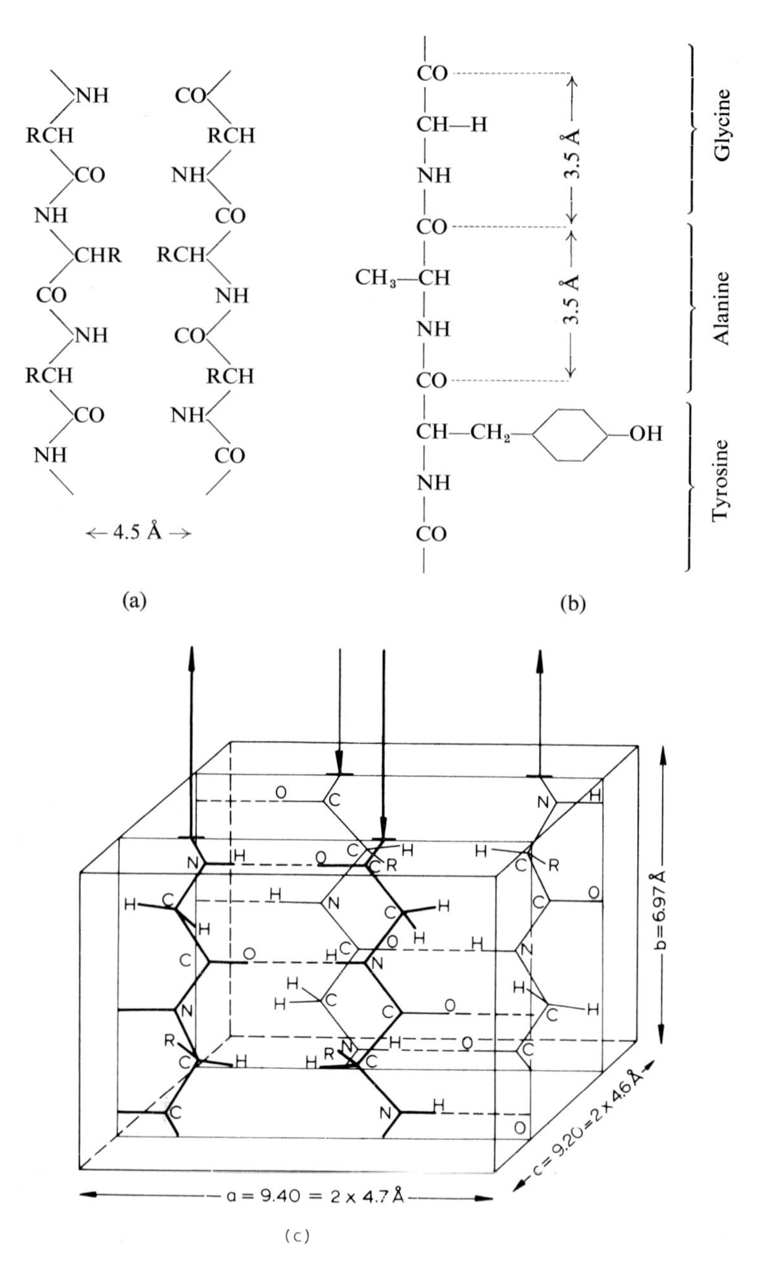

Fig. B–29. Silk fibroin; (a) profile of a chain, (b) front view of a chain, (c) unit cell (Marsh *et al.*, 1955).

The side groups of the polypeptide molecule are more or less perpendicular to the molecular sheets, and the sheet spacing is accordingly termed side chain spacing. This distance measures 5.0 or 4.6 Å (Table B–VIII), corresponding to the length of an alanyl or serinyl residue. The hydrogen bonds CO---H—N measure 2.8 Å (Fig. B–29c).

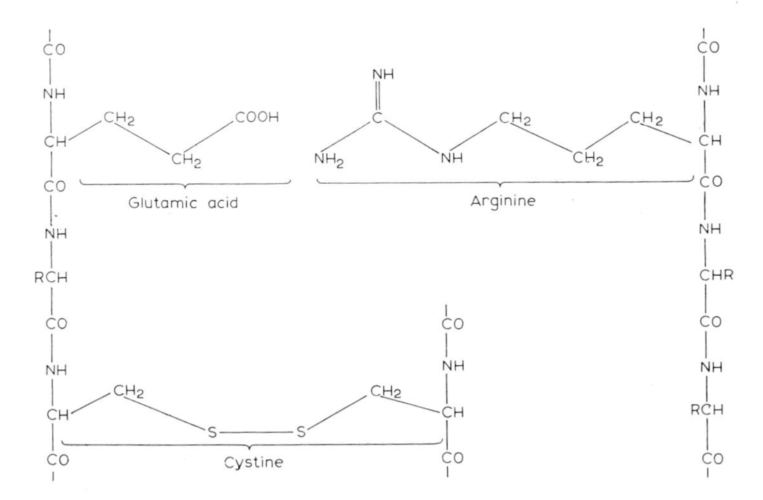

Fig. B–30. Side chains of keratin.

The tyrosine residues found in silk fibroin are too large for the established unit cell (Fig. B–29), and it is therefore believed that they belong to the amorphous part of silk fibroin or that the chain lattice loses its crystallinity at the sites at which tyrosine occurs in the chain sequence. Waldschmidt-Leitz (1958) has found the following sequence for silk fibroin:

$$(\text{Gly—Ala—Ser—Gly—Ala—Gly})_7\text{—Tyr},$$

n agreement with the [Gly—Ala (or Ser)] repeating set of the unit cell disclosed by X-ray analysis. There is a larger unit of 6 residues which, after seven repeats, is followed by a tyrosyl residue. This sequence would imply long-range spacings of 6×3.5 Å $= 21$ Å and of $(42 + 1) \times 3.5$ Å $= 150.3$ Å. Instead of these, a long range period of 70 Å has been found by Friedrich-Freksa *et al.* (1944), corresponding to about half of the proposed sequence. This observation may possibly indicate that neighbouring polypeptide chains are displaced lengthwise by one-half of their 150 Å period.

Apart from silk fibroin, wool keratin can also occur as a linear protein. In 1933 Astbury discovered that X-ray irradiation of hair yields diffraction patterns which differ according to whether the pattern is taken from normal samples or from hairs stretched to the maximum (about 100%). Astbury gave the name α-keratin to hair protein in normal conditions, and β-keratin to the protein in its extended state. X-ray analysis of β-keratin gives similar spacings for the fibre and backbone periods as in silk fibroin (Table B–VIII), whilst the side chain spacing appeared to be about twice as long. This is due to the fact that keratin contains glutamic acid and arginine residues (Fig. B–30), whose chains are considerably longer than those of the alanine and serine groups in silk fibroin. Since similar values of about 10 Å have been found in blood fibrin (Table B–VIII) and for the diameter *i.e.*, the spacing of α-helices (see p. 64), the average space requirement of an amino acid corresponds to the volume of Fig. B–22.

(*e*) *Proteins with helical chains*

Astbury found a fibre period of about 5 Å for α-keratin (Table B–VIII). In a hair stretched by 100%, this period ought to be extended to about 10 Å. Since three amino-acid residues cover a distance of 10.5 Å, it was originally believed that the linear β-keratin chain must contract by folding in such a way that a sequence of three amino acids would occupy a spacing only half of that in the extended chain. The reversible $\alpha \rightleftharpoons \beta$ transformation was found with muscle and epidermal proteins, as well as with blood fibrin, and these were designated the keratin–myosin–epidermin–fibrin class (k–m–e–f–proteins) by Astbury (1947, 1949).

This theory of unfolding and reversible folding soon met with difficulties, and a way for the interpretation of the established facts was looked for in the field of extensible helical chains. Taking the known atomic distances, bond angles, and free rotation in a polypeptide chain (Fig. B–28) into account, it is possible to arrive at a helix with a pitch of 5.33 Å and 3.6 amino-acid residues per turn (Pauling and Corey, 1954). This represents the secondary structure of α-keratin and has therefore been termed the α-helix; it has also been found in the proteins listed under that heading in Table B–VIII.

Such a helix is shown in Fig. B–31, and in Fig. B–32 we see the effect obtained when its cylindrical surface is opened into a plane, so that the helical form becomes a straight line rising with a pitch angle of 25°. After five turns, the 18th amino-acid residue is reached, standing vertically above the zero number of the chain. The spacing of each circumvolution above the other is about 4.8 Å. It is bridged by a hydrogen bond, with a length calculated at 2.79 Å. These spacings correspond to those in the chain lattice of silk fibroin (Fig. B–29c), but in the latter case inter-

TABLE B–VIII

FIBROUS PROTEINS

	Fibre period (half) Å	Back-bone spacing Å	Side chain spacing Å	Long-range spacing Å	
Linear chains					
Silk fibroin	3.5	4.5	5.0	70	Kratky and Kuriyama, 1931
Silk fibroin stretched	3.5	4.7	4.6		Marsh *et al.*, 1955
Blood fibrin stretched	3.35	4.7	10.1	240	Bailey *et al.*, 1943
Feather keratin	3.1	4.65	9.8		Astbury and Marwick, 1932
Wool β-kreatin	3.38	4.65	9.8		Astbury, 1933
α-Helix					
Wool α-keratin	5.03			66	Astbury, 1933
				200	McArthur, 1943
Porcupine quill	5.18			198	Astbury, 1949
Frog muscle	5.13				Astbury, 1949
Bacterial flagella	5.1				Astbury and Weibull, 1949
Collagens					
Vertebrates	2.8			640	Astbury and Atkin, 1933
Echinodermata	2.8			670	Marks *et al.*, 1949
Porifera (spongin)	2.9				Marks *et al.*, 1949

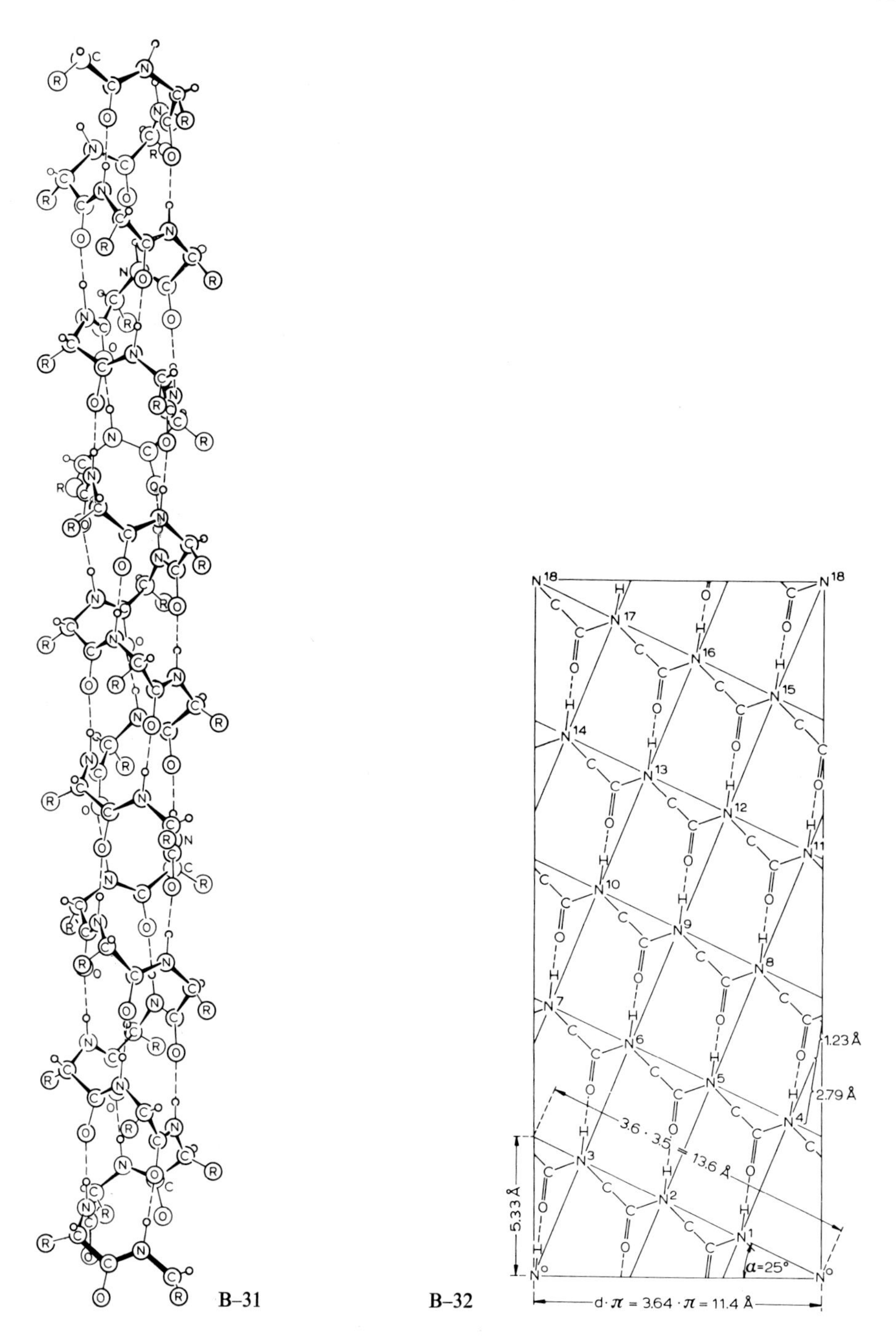

Fig. B–31. α-Helix of a polypeptide chain (Pauling and Corey, 1954).

Fig. B–32. Diagrammatic representation of an α-helix unrolled into a plane.

molecular hydrogen bridges occur between adjacent polypeptide chains, whereas intramolecular hydrogen bonds are present in the α-helix.

From a consideration of Fig. B–32 it follows that an elongation of over 100% must be theoretically possible in the $\alpha \rightarrow \beta$ transformation of the keratins, for when the chain is placed in the vertical position the pitch of 5.33 Å would be more than doubled (3.6 $\times$ 3.5 Å $=$ 12.6 Å). To stabilise the elongated β-keratin, the intramolecular hydrogen bonds of the α-helix would also have to be changed into intermolecular bridges between adjacent β-chains, and the side chains would have to develop bonds with one another (Fig. B–30). As this is not the case, a stretched hair returns elastically to its original length when released, as determined by its α-structure.

From the circumference of 11.4 Å of the cylinder in Fig. B–32, the diameter d of the α-helix is calculated as 3.64 Å. If, however, this figure is calculated from the average space requirement of an amino acid of 131 Å^3 (Table B–VI), with an axial length of 1.48 Å per residue, then a cross-section of 89 Å^2 and a diameter D of 10.6 Å result. The striking difference between these two figures is due to the fact that the chain with the diameter d is very narrowly wound, so that all side chains point outwards and thus occupy a much larger cross-section area with diameter D = 10.6 Å (Fig. B–33).

In phyllotaxis, the helical arrangement of the leaves on a stem is described by a fraction (e.g. 3/8), in which the numerator indicates how many times the helical

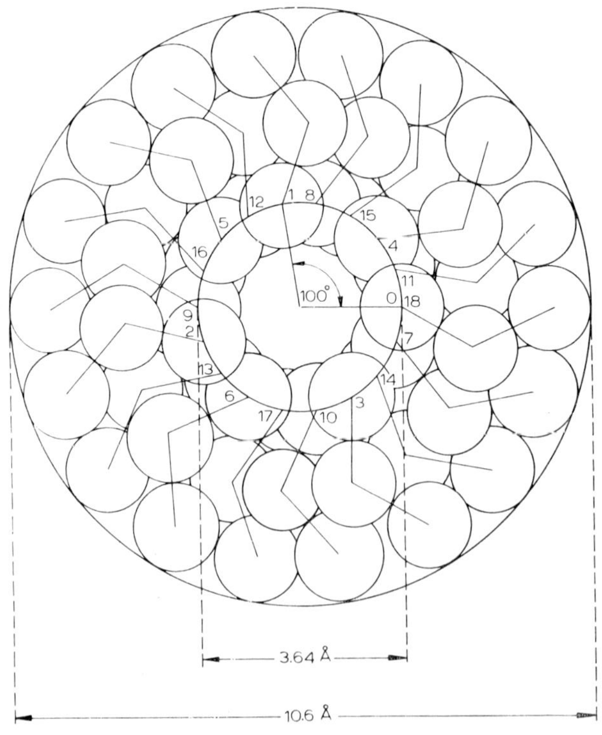

Fig. B–33. Projection of an α-helix onto its base. Outer diameter 10.6 Å, inner diameter 3.64 Å.

axis must be encircled until a leaf is reached that is vertically above the initial leaf, while the denominator indicates the number of leaves encountered during these circumvolutions. This fraction also indicates what angles (fraction of 360°) form vertical planes with one another through successive leaves; it is therefore also termed a divergence. If we apply this concept to the α-helix of Fig. B–32, then a 5/18 helix is involved, for in five circumvolutions 18 amino acids are encountered until a member is reached which lies vertically above the starting point. The divergence of successive members amounts to 100°.

In addition to the 5/18 helix, other chains have been found in polypeptide crystals, with divergences of 8/29 and 13/47 (Bamford *et al.*, 1954). Moreover, Pauling and Corey (1954, p. 206) state that a 3/11 helix is possible by slight changes in the bond angle. If these figures are arranged in sequence, it will be found that the sum of successive numerators is equal to the numerator of the next member, and the same applies to the denominators (Frey-Wyssling, 1954). The sequence can therefore also be extrapolated for smaller figures. We then have the α-helix sequence

$$\left(\frac{1}{3}\ \frac{1}{4}\ \frac{2}{7}\right)\ \frac{3}{11}\ \frac{5}{18}\ \frac{8}{29}\ \frac{13}{47}\ \cdots\cdots\cdots\cdots \quad \begin{array}{l}\text{lim } 0.276\\ \text{divergence } 99^\circ\ 20'\end{array}$$

This recalls the well-known phyllotaxis sequence (Church, 1904; Schoute, 1913):

$$\frac{1}{2}\ \frac{1}{3}\ \frac{2}{5}\ \frac{3}{8}\ \frac{5}{13}\ \frac{8}{21}\ \frac{13}{34}\ \cdots\cdots\cdots\cdots \quad \begin{array}{l}\text{lim } 0.382\\ \text{divergence } 137^\circ\ 30'\end{array}$$

In both sequences, the numerators represent Fibonacci numbers (Coxeter, 1961, p. 165). In the phyllotaxis sequence, this series also occurs in the denominators, but displaced two members to the left. In the α-helix sequence, however, the denominators constitute another series, the values of which are known as Lucas numbers. Both series lead to a limiting value which amounts to 0.382 for the phyllotaxis sequence and is thus in relationship ($0.382 = 1 - 1/\tau$) to the constant $\tau = 1.618$ of the golden section.

It is difficult to say what is the significance of these sequences, and one can but assume that they are connected with the sectoral space requirements of the individual helix members.

Vertical lines along a shoot are termed orthostichies in phyllotaxis: point 0 and point 18 in Fig. B–32 lie on an orthostichy. Inclined lines, such as those joining points 0, 3, 6, 9 ... or 2, 6, 10 ..., are termed parastichies. As it is frequently found that in helical structures no point is exactly vertical above another, parastichies play a more important role than orthostichies in the morphological description of helices. In normal phyllotaxis, parastichies with the point series 0, 2, 4, 6 ... and 0, 3, 6, 9 ... are characteristic; this is known as a 2 + 3 helical system. For the polypeptide helices, in contrast, a 3 + 4 system applies. Fig. B–32 shows that this system, designed for the 5/18 divergence, can easily change into the 3/11 or even the 2/7 divergence.

According to Schimper-Braun's hypothesis, the Fibonacci sequence arises (see Church, 1904) if, starting from the 1/2 position characteristic for the first two leaves of dicotyledons, more leaves are interpolated per circumvolution during growth. The same is also applicable to the Lucas sequence, except that then we must start from a 1/3 position. If one assumes that, in the case of polymerisation-growth of polypeptide chains, the first three members show a divergence of 120° and that for reasons of space more than three but less than four members occupy each turn, then in the polar growth of the chain positions of 5/18, 8/29 or 13/47 are possible.

In addition to the α-helix of polypeptide chains, a flatter helix can also be constructed. Such a γ-helix (Pauling and Corey, 1954) possesses 5.2 amino-acid residues per turn and has a pitch of 4.95 Å (0.95 Å per residue). It corresponds to a divergence of 5/26, and is thus unsuited to either of the sequences described above. As the γ-helix is not free from tension in all respects, and its existence has not therefore been unequivocally demonstrated, it will not be considered further.

(*f*) *Fibrous proteins*

The secondary structure of fibrous proteins as helical or extended polypeptide chains is below the resolving power of the electron microscope; only their quaternary structures are of such dimensions as to be shown by that method.

In silk fibroin, this quaternary structure consists of the joining of adjacent polypeptide chains to monomolecular sheets through hydrogen bonds, and these together form stacks of sheets which can apparently easily be split, parallel to the molecular chains, into elementary fibrils. Fig. B–34 shows how silk fibroin appears to be mechanically divided into flat microbands and these in turn into elementary fibrils (Hegetschweiler, 1949).

In the proteins with α-helices, which include Astbury's k–m–e–f class, the long-range spacings detected by the use of X-rays permit the postulation of tertiary structures. According to Pauling and Corey (1953), the α-keratin helical chains

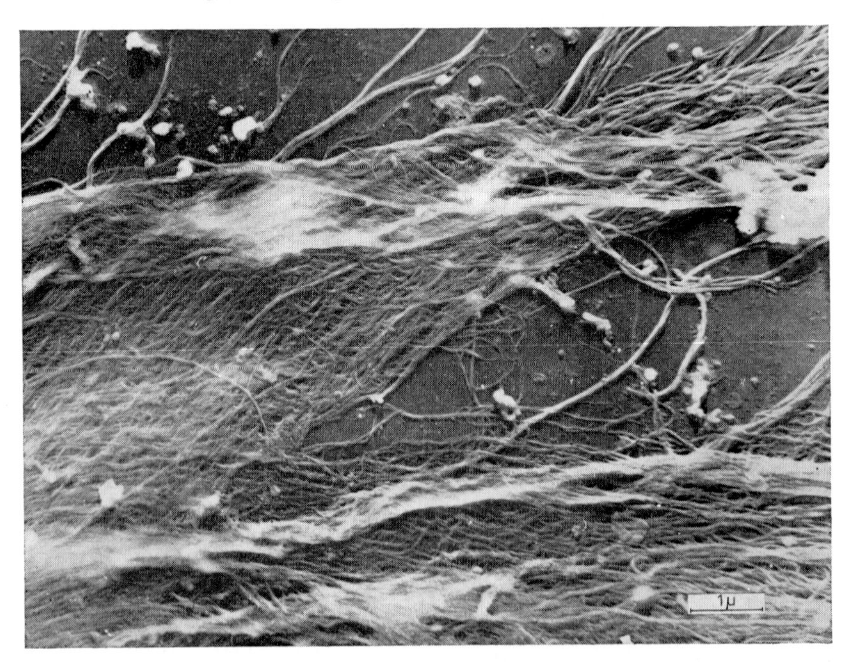

Fig. B–34. Microfibrils of silk fibroin (Hegetschweiler, 1949).

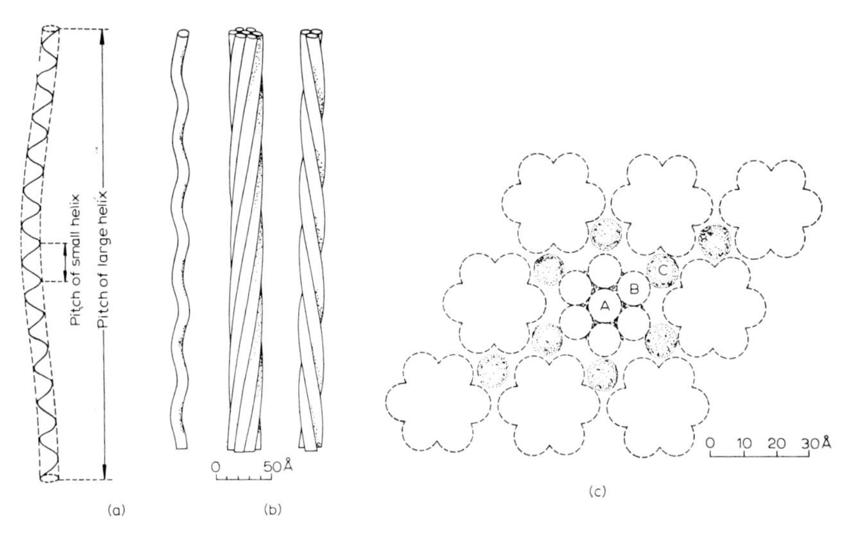

Fig. B–35. Helical structure of α-proteins; (a) α-helix with a twist to form a large helix (secondary structure), (b) aggregation of large helices (tertiary structure), (c) cross-section of a tertiary structure: cables AB_6 with a third interstitial polypeptide helix C (Pauling and Corey, 1953).

themselves twist, so that a large helix is formed (Fig. B–35a) with a pitch 12.5 times that of the 5.33 pitch of the small helix. As a result, a spacing of 66–68 Å is obtained, characteristic for α-keratins (see Table B–VIII); it would include 45 amino acids. The large helices can be taken as twisted into a quaternary structure of threes or sixes. We then have long-range spacings of 3×67 Å $\sim$ 200 Å (compare wool – McArthur, 1943 – and porcupine-quill keratin, Table B–VIII) or 6×67 Å $\sim$ 400 Å (Pauling and Corey, 1953). In the case of the six-stranded twist, there is room in the centre for a seventh large helical cord, which must however be wound in the opposite direction to that of the six outer chains (Fig. B–35b). The structural proof for such helical cables is very difficult to adduce. The intensities of X-ray interference are calculated for such a structural model theoretically, and are then compared with the radiographs actually obtained. The probable structure is thus inferred by trial and error rather than by strict deduction.

Pauling and Corey (1953) call the central strand of their postulated cable an A-chain and the outer strands B-chains, so that the seven-strand cable is denoted by AB_6. The construction of this model is probably influenced by the electron microscopic discovery of sublight-microscopic flagellate structures with two central strands, A_2B_9 (see Fig. K–1 and K–4). If several AB_6 cables are arranged parallel to each other, a hexagonal package results with interstices, which may harbour a third type of helical chains. There would then be three different α-polypeptide chains, A, B, and C (Fig. B–35c). The latter authors discuss the question whether, for example, the model AB_6C_2 from Fig. B–35c can correspond to the muscle protein, actomyosin, since the ratio of myosin to actin is about 7:2.

Blood fibrin is of interest because the development of long molecular chains from shorter units has been demonstrated in this fibrous protein. Fibrinogen is present in the blood and can polymerise to fibrin threads under the influence of thrombin, and it was demonstrated by physico-chemical methods (ultracentrifuge,

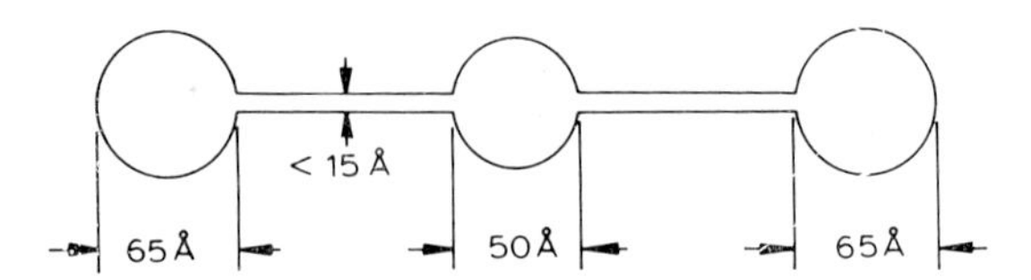

Fig. B–36. Model of a fibrinogen molecule (Hall and Slayter, 1959). Av.L. = average length.

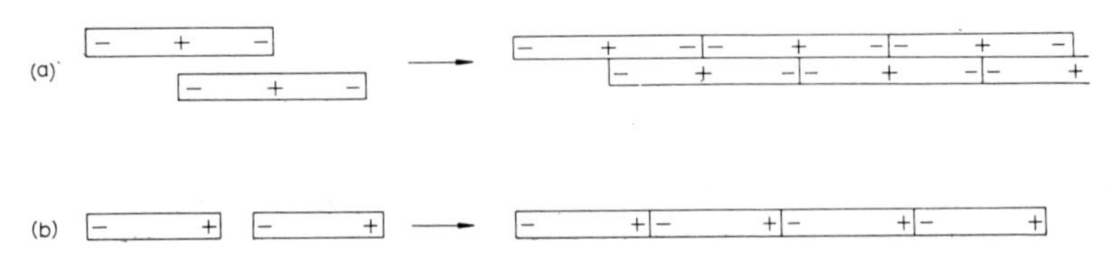

Fig. B–37. Polymerisation of elongated macromolecules to elementary fibrils; (a) by side to side attraction, (b) by end to end attraction.

nephelometry, flow birefringence, etc.) that the fibrinogen molecule consists of elongated particles with a molecular weight of 330,000, a length of about 500 Å, and an axial ratio of 5. The morphological structure of these particles could be elucidated under the electron microscope (Hall and Slayter, 1959). Contrary to expectations, fibrinogen molecules proved to be not elliptical, but triads of spheres joined together by a strand that could not be resolved under the electron microscope, the central one being somewhat smaller than the two at the ends (Fig. B–36). It must therefore be assumed that the fibrin α-helices are wound up in the spheres in some manner, whilst the joining strands consist of one or more straight helical threads. The length of the molecule is shorter than predicted, namely 360 Å.

Before these molecules can polymerise, they must be activated by thrombin, which takes place by means of a limited proteolysis (Scheraga, 1961), followed by association into a chain. With long particles, two ways of aggregation are possible: they either arrange themselves staggered sideways (side to side polymerisation) or end to end (Fig. B–37). The first of these arrangements is the more probable, as many more contact surfaces are thus available (Ferry *et al.*, 1954) and because staggering can lead to a helical arrangement of the monomers (Oosawa and Kasai, 1962). However, as the electron microscope shows, the triads arrange themselves in simple beaded chains; thus the arrangement of the fibrinogen polypeptide to the much longer chains of fibrin occurs by the end to end polymerisation.

Lateral fasciation of the molecule threads is the third step in the coagulation of blood, giving rise to microfibrils which appear striated under the electron microscope (Fig. B–38a). The spacing of the striations is 240 Å, which is divided by a finer band at a distance of 120 Å. The striations are more substantial than the intermediate regions, and they take up more metal when fixed with such electron stains as osmic or phosphotungstic acids. In 1959 Hall and Slayter observed that in the fasciation of monomolecular thread molecules the corresponding spheres of the beaded chains attract one another and become associated. These authors explain the ultrastructural striation of blood fibrin by the scheme reproduced in

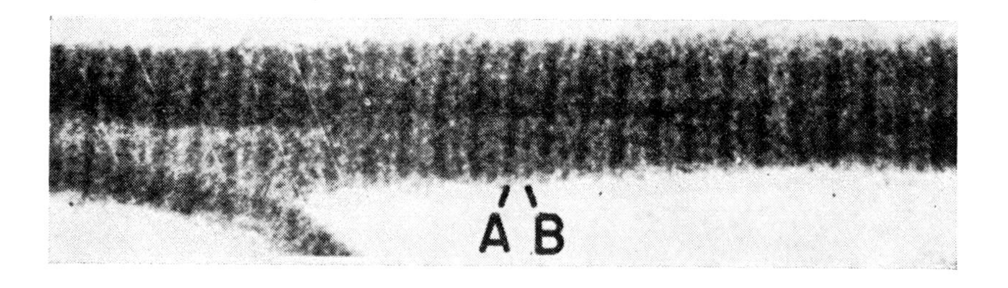

(a)

(b)

A B
240 Å
90 Å
50Å
50 Å

Model of the Structure of Fibrin

Fig. B–38. Polymerisation and fasciation of fibrinogen molecules to fibrils with a cross-striation of 240 Å periodicity, (a) as seen in the electron microscope (Hall and Slayter, 1959), (b) model of fine structure. Based on recent X-ray evidence (Stryer *et al.*, 1963), this model will be improved by taking into account substructures in the simplified nodules (Hall, privat communication, 1963). A = strong band, B = narrow band.

Fig. B–38b, but the explanation is rendered difficult by the fact that the fibrinogen molecules are 360 Å long whereas the periodicity of the fibrin microfibrils is 240 Å. A pronounced shortening of the fibrinogen molecule to $\frac{2}{3}$ of its original length must therefore take place during polymerisation, in which the α-helix contracts markedly.

Collagen will be considered as a last example of a fibrous protein. Proteins of the collagen types with a fibre period of 2.8 Å (Table B–VIII) appear to be a separate family, when compared with the k–m–e–f proteins, with fibre spacings of 5.3 Å in the α- and 3.5 Å in the β-forms (Marks *et al.*, 1949). This protein is present in tendons and decalcified bones. Glue and gelatin are relatively unchanged decomposition products of these insoluble structural substances, which become water-soluble after slight hydrolytic degradation.

The chemical composition of these materials is very different from those of other proteins, for they contain large quantities of proline and hydroxyproline, whilst tyrosine, tryptophan, cysteine, and methionine are absent. In 1957 Huggins reported the occurrence of 33.2% glycine, and a similar quantity of the joint compounds proline, hydroxyproline, aspartic acid, and glutamic acid, so that the rest of the amino acids account for the remaining $\sim$ 33%. Incorporation of the proline five-membered ring into the polypeptide chain is assumed to cause peptide bonds to appear in the tautomeric configuration —N=C(OH)— rather than as —NH—CO— bridges. The double bond should cause a shortening of the polypeptide

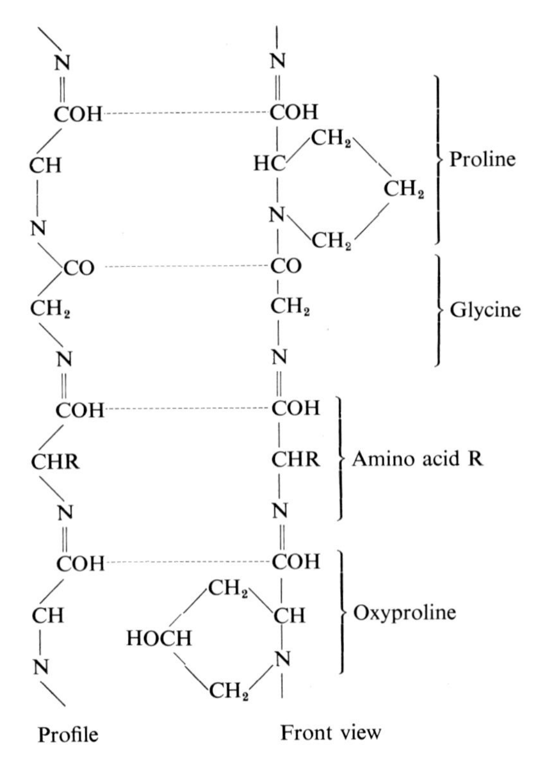

Fig. B–39. Diagram of a gelatin chain.

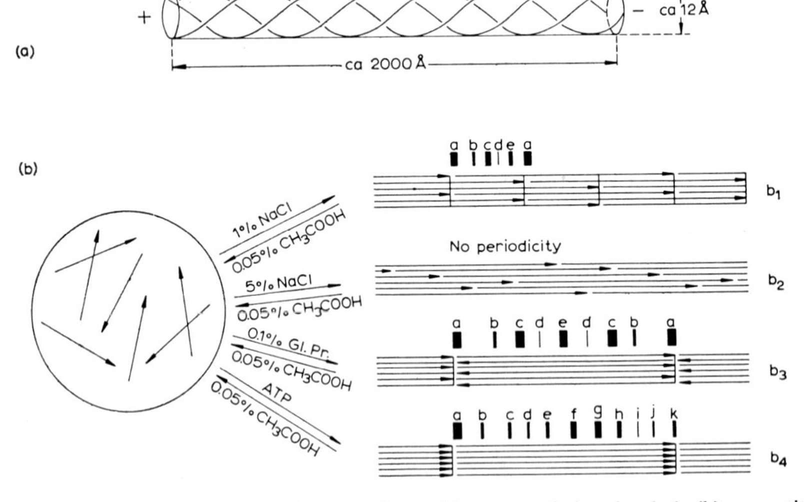

Fig. B–40. Collagen; (a) molecule of tropocollagen (three-stranded and polar), (b) reconstitution of collagen from a solution of tropocollagen. *a–e* striations visible in the electron mıscroscope; b_1 native collagen, parallel arrangement staggered by 1/3; b_2 fibrils without striations, random staggering; b_3 fibrous long spacing, antiparallel arrangement (Gl. Pr. = glucoprotein); b_4 segment long spacing, parallel arrangement without staggering (Schmitt *et al.*, 1953, 1955).

chains. There are also two stereoisomeric possibilities, since bonding can appear either in the *trans*- or *cis*-form; the more probable *trans*-form is shown in Fig. B–39.

X-ray analysis shows a hexagonal assembly of regular cylinders with diameters of about 12 Å. These should contain three polypeptide chains in steep helices (Rich and Crick, 1958). The cylindrical rods, about 2000 Å in length, form particles with a molecular weight of 360,000, which have been termed tropocollagen by Schmitt *et al.* (1953). It is obtained from native collagen by solution in 0.05% acetic acid.

Dissolved tropocollagen can be reconverted into fibrillar collagen by salting out. It is assumed that the rodlike molecules possess a dipole character (Fig. B–40a) and are constructed polarly; they are therefore depicted as long arrows. On salting-out they aggregate by end-to-end polymerisation into elementary or protofibrils, which associate together through lateral aggregation into microfibrils showing characteristic series of striations; these make collagen easy to identify under the electron microscope.

Native collagen reveals five cross-streaks (a, b, c, d, e in Fig. B–40b_1) which can be resolved into 2–3 fine lines; this set is repeated at spacings of 640 Å. Since this spacing is one third the length of the tropocollagen molecules, Schmitt *et al.* (1955) assumed that the rod molecules in the collagen fibrils are each staggered by about a third of their length. This concept is supported by the fact that the reconstitution of the fibrils shows quite different bandings, according to the manner in which the tropocollagen is precipitated.

By increasing the ionic strength during the salting-out process from 1 to 5% NaCl, fibrils are obtained which show no bands under the electron microscope. Dialytic precipitations, in the presence of glucoprotein, yield more interesting results. Fibrils are formed with spacings of 2000 Å, with a symmetrical rather than a polar arrangement of the striations. It was therefore assumed that in this case tropocollagen molecules were not staggered, so that the periodicity reflects the full rod-length, and that they were arranged in an antiparallel manner since there is no polarity (Fig. B–40b_3). Finally, by means of ATP it has been possible to obtain polar constructed segments 2000 Å long, which did not polymerise to fibrils; they do however show an exceedingly rich pattern of striations, from which it is assumed that they reflect the amino-acid sequence (Fig. B–40b_4). The cationic (Arg) or anionic (Glu, Asp) amino-acid residues are shown up by the type of staining employed, whether with phosphotungstic acid or with basic uranyl or chromic salts (Schmitt, 1960).

The question of the reconstruction of collagen has been discussed in some detail, because it demonstrates effectively the consequences of parallel or antiparallel arrangements of the rod molecules (see p. 40). There is a definite polarity in parallel alignment which is unknown in, for example, cellulose or silk fibres. We are therefore justified in assuming that there is an antiparallel arrangement of the polar thread molecules in the chain lattice of those fibres.

(*g*) *Globular proteins*

As has already been pointed out, the fibrillar proteins produce fibrous patterns in the X-ray camera and show micro- and elementary fibrils under the electron microscope, which contrast with the globular or sphero-proteins. These are more easily brought into solution than the fibrous proteins and produce sols with spherical colloidal particles. These particles can be made visible under the electron microscope by spraying such sols on filmed grids (De Robertis *et al.*, 1953). In contrast to fibrous proteins, they do not crystallise in strongly anisotropic chain lattices, but mostly produce isodiametric, only slightly (if at all) anisotropic crystals of the orthorhombic, rhombohedral, or cubic systems. Their particle shape and size may be determined by methods which we shall consider when discussing virus particles (see p. 120).

The existence of sphero-proteins was first indicated by the ultracentrifuge studies of Svedberg (1938), who found in their most varied representatives particle weights which he thought of as multiples of the molecular weight 17,600. Particles with this weight were therefore considered as the basic bodies of the globular proteins and designated as the 'Svedberg-unit'. Such multiple series in certain very highly polymerised proteins, as e.g. in the haemocyanin of the snail *Helix pomatia*, have been observed under the electron microscope (Polson and Wyckoff, 1947). It appeared that such giant molecules with a diameter of 400 Å can be divided into two or four sub-units. However, more recent work (see Van Bruggen *et al.*, 1962) shows that only division into two units is possible, since the cylindrical molecule has a fivefold axis and can, therefore, only divide in half perpendicularly and not parallel to this axis. The use of negative staining has also shown the presence of

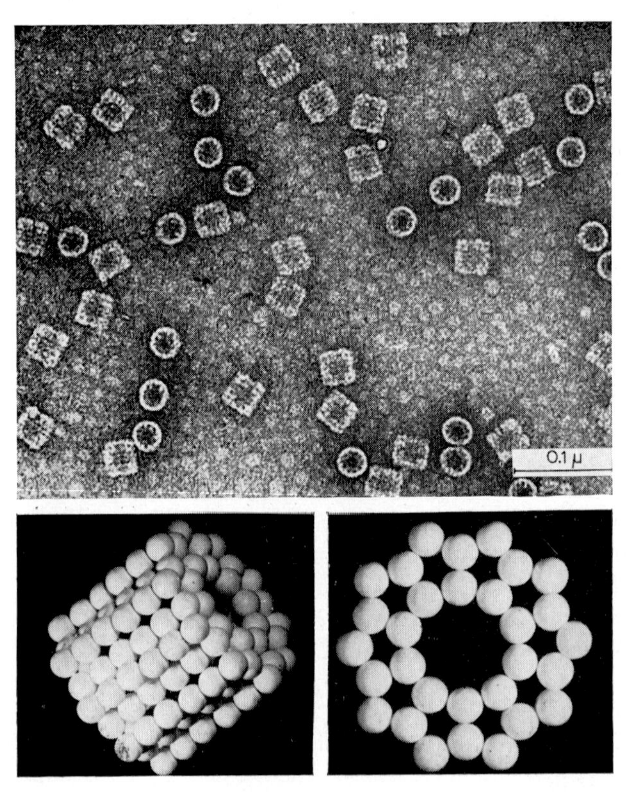

Fig. B–41. Haemocyanin molecule (van Bruggen, *et al.*, 1962).

Fig. B–42. A crystal of the rhombic type of tobacco necrosis virus (Labaw and Wyckoff, 1958).

much smaller globular sub-units which build up the whole molecule in six sheets or layers each of 30 globular particles; their diameter is 55 Å and their particle weight 37,000 – which is only a 180th of the molecular weight of 6.6 million found for haemocyanin (Fig. B–41).

This basic figure 37,000 is about twice the size of a Svedberg-unit. In a similar way it turned out that the molecular weights of most proteins differ widely from those of Svedberg's multiple series (see Table B–X), and this theory is therefore now obsolete.

Globular proteins crystallise as more or less isodiametric polyhedra. Such crystals have long been known to botanists from seeds (Nägeli, 1862). They are striking because of their ability to swell and to stain. Even then it was recognised that they had a porous structure, permitting the entry of micromolecules into the solid crystal, and were distinguished from 'true crystals' as 'crystalloids'.

Today we know, from X-ray analysis, that the lattice of the protein crystal is constructed similarly to the micromolecular lattice. The size of the protein molecules, however, causes interstices to occur between the spheres (see Fig. B–42), and these can be penetrated by water and other micromolecules. These mostly arrange themselves in the crystal lattice, but are so loosely bound that they can be removed by warming or dissolving (zeolitic water). On the basis of their lattice structure, crystalline proteins are today regarded as true crystals.

Data on the protein crystals so far studied are assembled in Table B–X. The lattice constants a, b, and c, which at the most are only a few Å in inorganic crystals, are of an imposing size. The volume V of the elementary cell is also correspondingly very large. The molecular weight of the protein in question can be calculated from the formula $M = \frac{V\varrho N}{n}$, where N is Avogadro's number $=$ 0.602×10^{24}, n the number of its molecules in the elementary cell and ϱ its density; if V is given in cm^3, $M = \frac{V\varrho}{1.66\,n}$. The molecular weights obtained agree satisfactorily with values obtained elsewhere (e.g. from ultracentrifuge measurements) or are multiples of it.

The diameter d of the globular protein particles can be calculated according to the formula $d = \sqrt[3]{\frac{6\,V}{\pi n}}$ from the elementary volume V; this is shown in the last column of Table B–X. It is recognised that all globular proteins have molecular dimensions which can be resolved under the electron miscroscope. As these macromolecules are completely dehydrated in vacuum, only the sizes of the anhydrous molecules are calculated in Table B–X. These are theoretical values, for in the course of drying for examination under the electron microscope the globular particles are broadened by flattening into ellipsoids of rotation.

It is possible to compute that the obsolete Svedberg unit would have a particle diameter of about 35 Å. Molecular weights increase in proportion to the third power of the particle diameter, as can be seen in Table B–IX.

TABLE B–IX

DIAMETER d OF GLOBULAR PROTEIN PARTICLES DEPENDING ON THEIR MOLECULAR WEIGHT

MW ($\varrho = 1.34$)	d in Å
15,700	33
125,000	66
1,000,000	132
8,000,000	264

Bernal's school has found that the crystal lattice produces diagrams much richer in interference if the crystals are studied in their mother liquor (Bernal and Crowfoot, 1934). The smallest spacings in moist crystals are often only 2 Å, from which it follows that in the spacious elementary cells (which may contain from $n = 2$ to $n = 54$ separate molecules) an astonishing order must prevail. The easily displaced molecules of the water of crystallisation assume quite specific positions, but as soon as the crystals are dehydrated this order disappears and only a few interferences are obtained which betray much coarser intramolecular interferences of, say, 20 Å. It will be seen from Table B–X that the crystal lattice shrivels on dehydration, so that the symmetry of the crystal can decrease (e.g. lactoglobulin passes from tetragonal to orthorhombic) and the angle of monoclinic crystals changes.

The globular proteins frequently crystallise in highly symmetrical rhombohedral or cubic space groups (Fig. B–42). Tetragonal or orthorhombic crystals which indicate protein molecules in the form of rotational or triaxial ellipsoids are less frequent, and the low-symmetry monoclinic system rarely occurs.

TABLE B–X

PROTEIN CRYSTALS
(after Cohn and Edsall, 1943, p. 328)

		Crystal system	Unit cell a Å	b Å	c Å	β degree	Volume Å^3	n	Density ϱ	Molecular weight $\frac{V\varrho}{1.66\,n}$	without residual water	Particle diameter $d = \sqrt[3]{\frac{6V}{\pi n}}$ Å
Ribonuclease	dry	orthorhombic	36.6	40.5	52.3	90	77,300	4	1.341	15,700	13,700	33
Insulin	wet	hexagonal[1]	144	83	34	90	404,000	6	1.28	52,400	—	—
	dry	hexagonal[1]	130	74.8	30.9	90	298,000	6	1.315	39,500	37,400	46
Lactoglobulin	wet	tetragonal	67.5	67.5	154	90	702,000	8	1.257	67,000	—	—
	dry	orthorhombic	60	63	110	90	416,000	8	1.27	40,000	—	46
Chymotrypsin	wet	monoclinic	49.6	67.8	66.5	102	219,000	2	1.277	84,500	—	—
	dry	monoclinic	45	62.5	57.5	112	151,000	2	1.31	60,000	54,000	52.5
Pepsin	wet	hexagonal[1]	116	67	461	90	3,580,000	54	1.32	53,000	—	—
Horse methaemoglobin	wet	monoclinic	110	63.8	54.2	112	352,000	2	1.242	132,000	—	—
	dry	monoclinic	102	51	47	130	188,000	2	1.27	72,000	66,700	56.5
Horse serum albumin	wet	hexagonal	96.7	—	145	120	1,170,000	6	1.27	150,000	—	—
	dry	hexagonal	74.5	—	130	120	610,000	6	1.34	82,800	—	58
Tobacco seed globulin	dry	cubic[2]	123	123	123	90	1,860,000	4	1.287	362,000	322,000	96
Excelsin (*Bertholetia excelsa*)	dry	hexagonal[1]	149	86	208	90	2,670,000	6	1.31(?)	350,000	305,800	95
Bushy stunt virus	wet	cubic[3]	394	394	394	90	61,000,000	2	1.286	24,000,000	—	—
	dry	cubic[3]	318	318	318	90	32,000,000	2	1.35	13,000,000	—	310

[1] The hexagonal system is based on a rhombic elementary cell $a:b:c$, where $a:b = \sqrt{3}:1$ (see Fig. A–7a); the space group of insulin and excelsin is rhombohedral.

[2] Unit cell face-centred (closest globular packing, see Fig. A–7b).

[3] Unit cell body-centred (see Fig. A–7c).

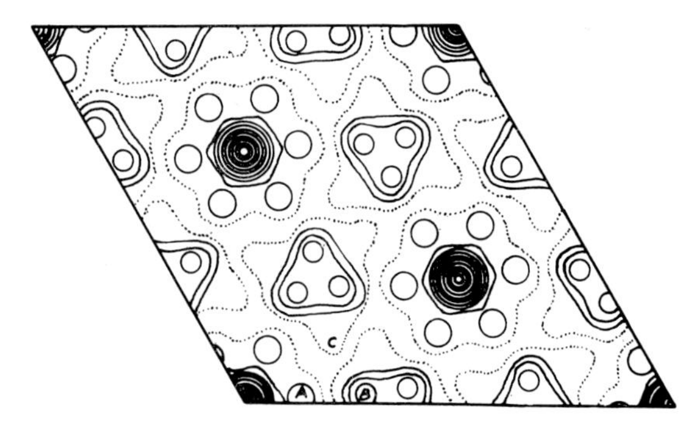

Fig. B–43. Patterson-projection of the insulin crystal lattice with three molecules (threefold symmetry). Edge of the unit cell 144 Å (Crowfoot, 1938).

The density distribution in the crystal lattice can be shown by the so-called Patterson–Fourier analysis. Fig. B–43 shows such a Patterson projection for insulin, according to Crowfoot (1938). We see that a threefold symmetry exists inside the unit cell. The domain depicted corresponds to half of the unit cell shown in Table B–X; therefore, it must contain three insulin molecules localised at the sites of the accumulations of mass, indicated by circles in Fig. B–43.

By a refinement of Patterson analysis, it has been possible to determine the density distribution in the various planes of the crystal lattice. In this way the spatial position of the individual atoms as mass centres can be located, and it becomes possible to explain their arrangement in the molecular lattices of protein crystals. Of course, studies of this kind require a lot of time, for unlikely results have to be excluded by trial and error from the general run of results obtained. In this context, the presence of a large, heavy atom whose position in the crystal lattice in relation to C, O, and N can be unequivocally determined, is a very great help.

This is the case with myoglobin, which, like haemoglobin, contains an iron-

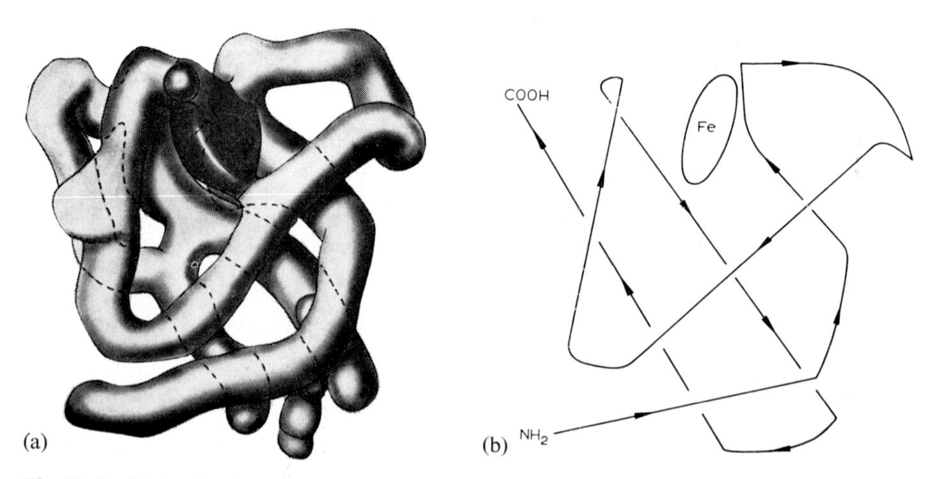

Fig. B–44. Molecule of myoglobin (Kendrew *et al.*, 1960); (a) tertiary structure, (b) course of polypeptide chain and location of the haem (Fe).

bearing haem. Both have the property of being able to bond oxygen reversibly to their haem. While the task of haemoglobin is to transport oxygen in the blood-stream, that of myoglobin is to store oxygen temporarily in the cells. This is especially important in diving animals, such as whales, seals, and penguins. In contrast to haemoglobin with a molecular weight of 67,000, that of myoglobin is only 18,000, with about 152 amino acid residues.

Owing to the favourable circumstances enumerated above, Kendrew *et al.* (1960) succeeded in explaining fully the tertiary structure of myoglobin. A single, irregularly coiled polypeptide chain is present (Fig. B–44a, b); the haem containing the iron atom lies asymmetrically sideways in the molecule and does not occupy, as expected, a central position.

As Fig. B–44b shows, the terminal (—NH_2) and (—COOH) positions of the polypeptide chain are known, but so far the amino acid sequences have not been absolutely elucidated. Whereas the primary and not the tertiary structures of such globular proteins as insulin (Fig. B–24) und nuclease (Fig. B–26) are known, it is the other way round with myoglobin, in that the much more difficult analysis of the chain convolutions has been successfully accomplished, however, without being able to indicate the exact sequence of all 152 amino acid residues. At present, therefore, none of the proteins has been fully elucidated from the point of view of its primary, secondary, and tertiary structures, but the solution of this problem is merely a matter of time.

Iron-containing proteins are of special importance in electron microscopy, for their heavy atom content provides a stronger contrast than that given by proteins, which contain only C, N, and O. In this respect, ferritin, which can be obtained from horse spleen, has become well known (Farrant, 1954). The crystalline preparation contains 23% Fe in the form of a 'ferric hydroxide phosphate', $(FeOOH)_8\ FeOPO_3H_2$, that makes up even 40% of the ferritin. The iron complex is easily separable. Apoferritin is then obtained, with a molecular weight of 465,000 and a particle diameter of 95 Å (see Table B–IX). Something like 2000 Fe atoms associate themselves with this protein, in the form of four symmetrical centres which are visible under the electron microscope; each centre measures 27 Å and the quadruple group has a diameter of 54 Å. If this highly characteristic molecule can be introduced into the cell (see p. 217), the physiological fate of the iron can be followed with the electron microscope; this is particularly important in the study of the iron metabolism in blood (Bessis, 1958). A similar iron protein named phytoferritin has been discovered by Hyde *et al.* (1962) in the radicles of germinating peas.

(*h*) *Protein films*

Not all parts of the molecules of globular proteins soluble in water or salt solutions show an affinity for water. The polypeptide chains carry hydrophilic and lipophilic (hydrophobic) side groups, the former striving for contact with water and the latter repelling it. Hence, proteins can be spread as molecular films on the surface of water (Adam, 1941, p. 87). One milligram of protein can cover a surface of 1 to $> 2.5\ m^2$, and assuming a density of 1.33 this means that films are some 7.5 to 3 Å thick, corresponding to the width of individual polypeptide chains. Globular proteins do not spread as spherical bodies on a hydrophilic surface, but unfold

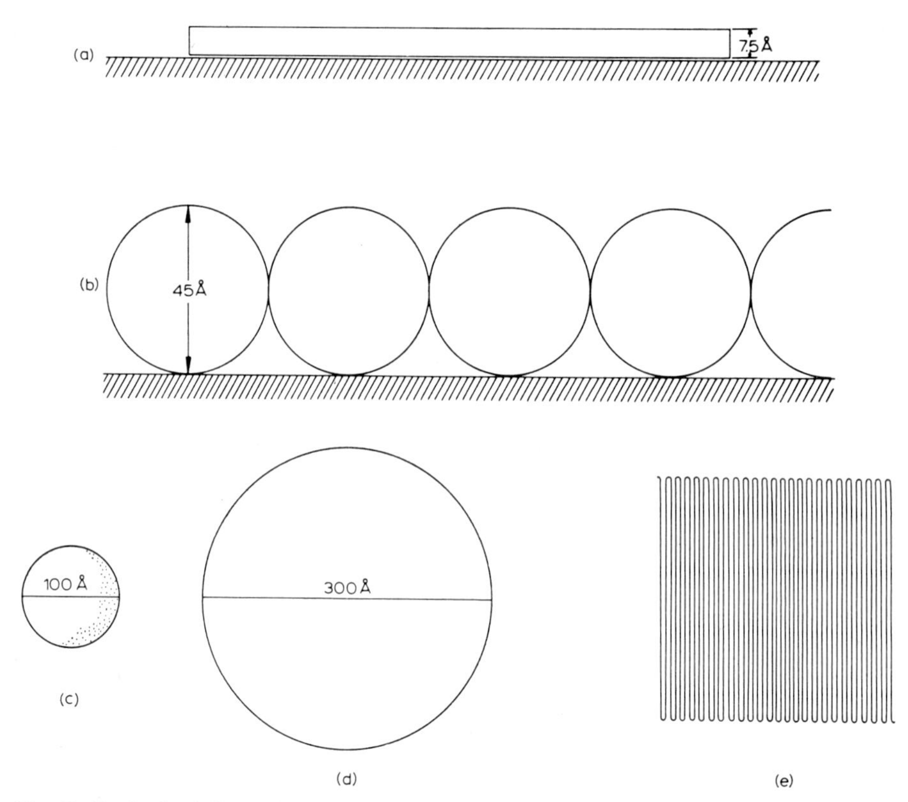

Fig. B–45. Surface films of proteins. Thickness of film, (a) with unfolded chains, (b) with globular molecules. Denaturation of (c) a globular molecule 100 Å in diameter to (d) a spread 7.5 Å film 300 Å in diameter with (e) a polypeptide chain of ca. 1 μ length.

their polypeptide chains. As long as the surface film of 1 mg protein is larger than 1 m^2, it is liquid, *i.e.* the flattened molecules remain movable and may change their position with regard to one another on the water surface. They also lose their physiological activity as enzymes or hormones, which seems to be intimately related to the secondary helical and tertiary globular structures that cannot be altered without completely changing the properties of the protein. However, as soon as the film is compressed to 1 m^2 it becomes solid and rigid, and in some cases the protein may even recover its physiological properties.

In a solid film, the molecules lose their individuality and, often aggregate to the state of fibrous proteins. This change causes the denaturation of globular proteins, and often the mere shaking of a protein solution suffices to form a foam of insoluble denatured protein.

Figs. B–45a and b show the sizes of monomolecular protein films in the actual unfolded and in a hypothetically stabilised globular state.

When globules (Fig. B–45c) with diameter d flatten to a film 7.5 Å thick, they cover a circular area whose diameter D is calculated in Table B–XI and shown in Fig. B–45d. In this area, the polypeptide chains of length L lie fully extended (Fig. B–45e).

TABLE B–XI

PROTEIN FILMS

Diameter d of the globular molecule	Diameter D of the spread film (7.5 Å thick)	Length L of the polypeptide chain		Molecular weight
in Å	in Å	in Å	in μ	
35	62	490	0.05	18,000
50	105	1,420	0.14	52,500
100	300	11,400	1.14	420,000
200	850	91,000	9.10	3,350,000

Protein films have the remarkable property of being able to lower considerably the surface tension (see p. 143) at interfaces. Table B–XII shows that oil drops from mackerel eggs develop a surface tension of 7 dynes/cm against sea-water, but in the protein-containing aqueous egg content the tension of such an oil drop is lowered to a tenth of that value (Danielli and Harvey, 1935). It has been shown experimentally (Danielli, 1938) that the lowering of interfacial tension between hydrophilic and lipophilic phases is a general feature of proteins (ovalbumin, pepsin, trypsin, haemoglobin). Since the surface tension of protoplasm against the solution in which the cells are cultivated is much lower than that of any lipid/water interface (see Table B–XII), it is concluded that the plasma membrane or plasmalemma cannot be a pure lipid layer, but that it must be covered by a protein film (see p. 146).

TABLE B–XII

SURFACE TENSION IN VARIOUS INTERFACES
(Danielli and Harvey, 1935)

	dynes/cm
Air/water	72
Oil/sea-water	7
Oil/egg content	0.8
Sea-urchin egg/sea-water	0.2

This hypothesis may help to elucidate the molecular structure of the plasma membrane, which is visible under the electron microscope. This membrane is considered to be of lipoproteinic nature, and it is therefore important to know something of the appearance of protein films without lipids. Mercer (1957) fixed such films of egg albumin with osmic acid, sectioned them, and found the protein layer to be 50 Å thick. This is five times thicker than an expanded film of about 10 Å, but on the other hand it is only about half the thickness of the plasmalemma (see p. 146). It is not known whether the layers resolved by Mercer were monolayers or multilayers, because of the distortion of the film caused during preparation to make it visible. An important point to be borne in mind is that protein films absorb osmium tetroxide on both surfaces and thus appear dark.

3. LIPOPROTEINS

In biological structures proteins are frequently associated with lipids into so-called lipoproteins, which play an important part in the construction of sublight-microscopic membranes and lamellar systems. Such structures usually break down when the lipids are extracted. Liquids which dissolve lipids, as for example trichloroacetic acid, acetone, and alcohol, thus act as precipitators of proteins and are unsuitable for the fixation of submicroscopic cell structures. Only preservatives such as osmic acid, which harden the lipids *in situ* are suitable fixatives for use in electron microscopy. Zeiger (1949, 1960) believes that lipids form a sort of 'façade', the removal of which results in the collapse of the protein structure.

Little is known about the stoichiometric relationships of the lipids to proteins or about the forces keeping the two components together; these may be ionogenic or perhaps only coordination and Van der Waals' forces, which bring about the association. There are but few cases in which lipoproteins can be perfectly isolated and correctly depicted, as an example of which we will consider the lipoproteins of the blood. Two such fractions are found in the blood serum and they migrate with the α- and β-globulins in electrophoresis; consequently they have been designated as α-lipoprotein (molecular weight 200,000, with 25% lipid, 50 Å × 300 Å) and β-lipoprotein (molecular weight 1,300,000, with 75% lipid, 185 Å in diameter). As these particles remain water-soluble despite their high lipid content, it is thought that they act as lipid carriers in the blood. The particle sizes and forms given in Fig. B–46 have been derived from the behaviour of solutions in the ultracentrifuge, as well as from measurements of viscosity and turbidity (Cohn, 1947). It is agreed that, in comparison with other macromolecules of the blood, the lipoprotein particles are especially large, and consequently they can be regarded as molecular associations.

Extracted lipids consist of micromolecules which should be considered briefly here, in order to estimate accurately their size in an association with the macromolecular protein partner.

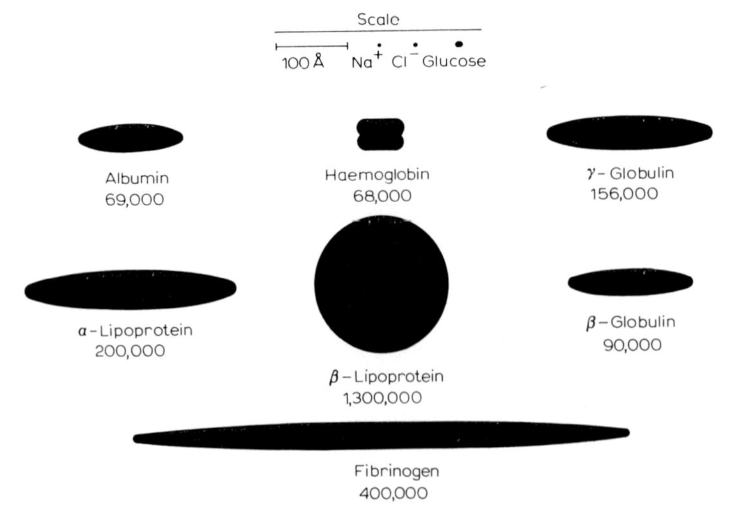

Fig. B–46. Relative dimensions and molecular weights of various proteins of blood (Cohn, 1947).

(*a*) *Lipids*

The biological concept of lipids embraces all hydrophobic substances. Lipids are thus characterised more by a negative property (insolubility in water) rather than by a positive one (solubility in organic solvents), and consequently include different classes of substances, such as terpenes, waxes, fats, sterols, etc.

True lipids are characterised by the fact that all their free end groups consist of typically lipophilic groups. This is particularly obvious in the fats, which are the esters of the trifunctional alcohol, glycerol, with fatty or oleic acids. The hydrophilic groups of the original compounds are screened as a result of esterification, as shown in Fig. B–47. In the same way, the hydrophilic groups are masked in waxes, which are formed by the esterification of higher alcohols with higher fatty acids. It is difficult to say why they are screened in the course of metabolism, but in any case these lipids contrast strongly with the hydrophilic compounds of living cytoplasm, and if they are formed in excess we observe the well known phenomenon of fatty degeneration of the protoplasm (lipophanerosis). A correct balance between hydrophilic and lipophilic compounds is essential in living matter.

In contrast to fats, most lipophilic compounds of the cytoplasm carry at least one hydrophilic group, which serves to bring about contact with other molecules. This applies particularly to the fatty acids, aliphatic alcohols, and sterols.

Such molecules consist of a lipophilic body on which the hydrophilic group (—OH, —COOH) confers a pronounced polarity. In contact with water, they gather along the phase boundary, as shown in Fig. B–48. There is only a relative accumulation of such molecules if their lipophilic parts do not impose complete insolubility (Figs. B–48a, b), and insoluble monomolecular surface films are formed if the lipophilic body is considerably larger than the hydrophilic one. The aliphatic chains may be bent, as in the case of dicarboxylic acids (Fig. B–48d), but they are straight when there is only one hydrophilic pole (Fig. B–48c).

The size and space requirements of such straight chains can be estimated from the distances found in paraffin crystals and from the thickness of condensed monomolecular films on water. The distance of the chains in the zig-zag plane is found to be 7.45 Å, and perpendicular to it, 4.97 Å (see Fig. A–8b). The cross-section of such a chain is therefore $\frac{7.45 \times 4.97}{2} = 18.5$ Å^2. The approximate length of the chain depends upon the number n of its methylene groups; since two (CH_2)

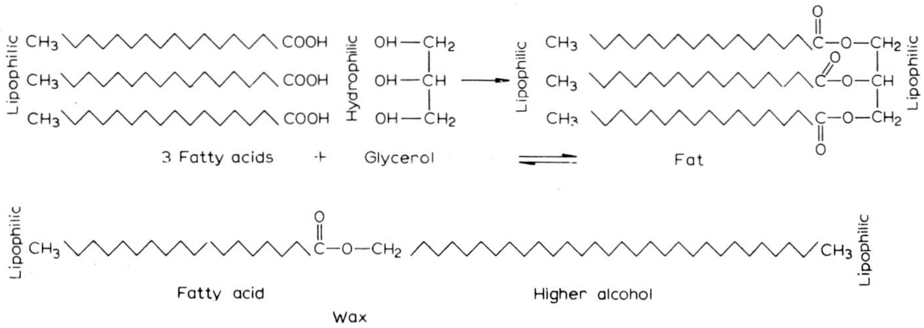

Fig. B–47. Molecular structure of lipid chains.

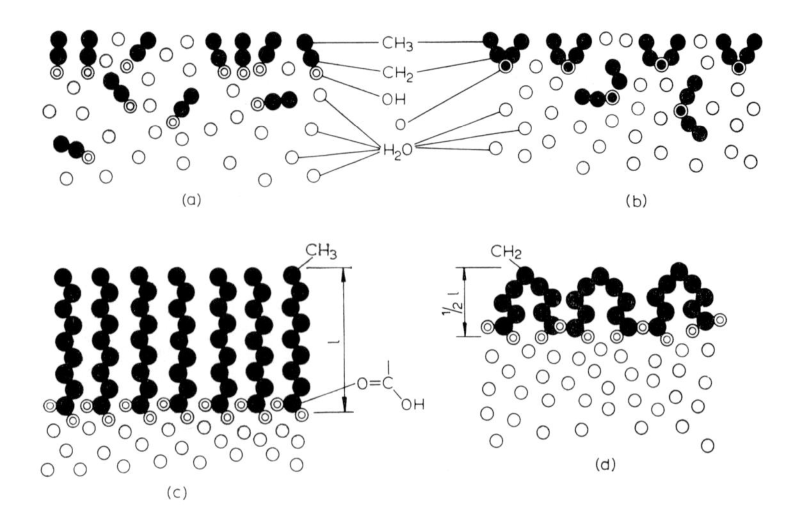

Fig. B–48. Molecular surface structure of aqueous solutions. Accumulation at the surface of (a) ethanol, (b) ethyl ether. Monomolecular films of (c) fatty acids, (d) di-basic acids. – ○ = water; hydrophilic groups white; lipophilic groups black; oxygen encircled; *l* length of a fatty acid chain.

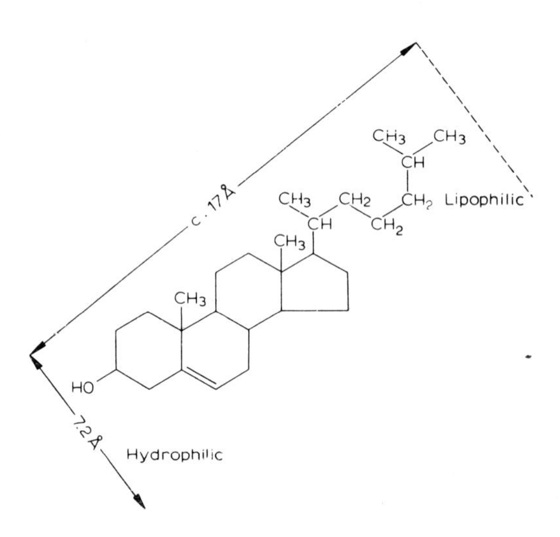

B–49. Molecular structure of cholesterol.

groups furnish an increment of 2.54 Å (see Fig. A–8b), the approximate length is $n/2 \times 2.5$ Å, and an aliphatic chain with 16 members is thus about 20 Å long.

Sterols (alcohols related to the terpenes and often associated with fatty acids) have a more complicated structure. Cholesterol may serve as an example (Fig. B–49). According to X-ray analysis (Bernal, 1932), the length of this molecule is 17 Å and its width 7.2 Å. These dimensions permit the incorporation of such molecules in monomolecular films of fats and fatty acids.

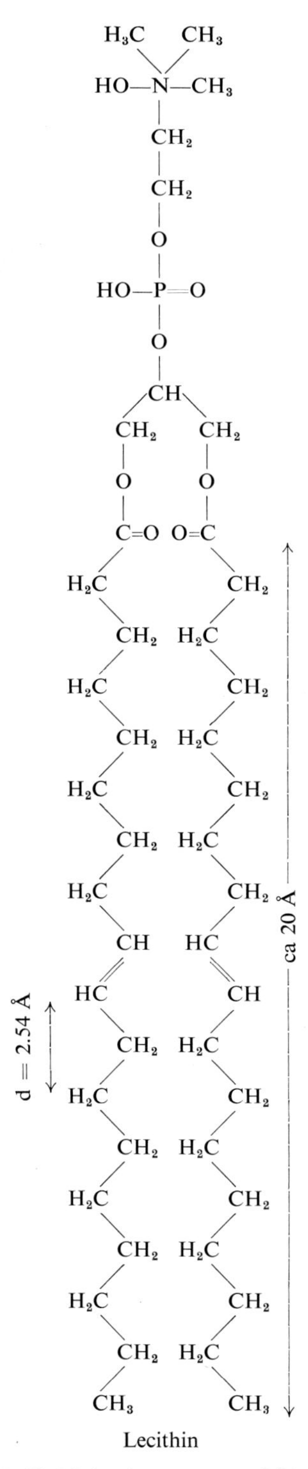

Fig. B–50. Molecular structure of β-lecithin.

(*b*) *Phospholipids*

Phospholipids (or phosphatides) are soluble in ether and other liposolvents, but in addition to their lipid character they possess a marked tendency towards hydrophilic character, shown by their sorption of water and the occurrence of myelin forms. Phospholipids thus represent molecules intermediate between hydrophobic and hydrophilic substances, and for this reason are among the most important mediators between the representatives of these two extreme groups in the protoplasm.

Let us consider lecithin as an example of this type of compound. Similarly to the fats, lecithin consists partly of glycerol and fatty acids, but in this case only two of the three hydroxyls are occupied by fatty or oleic acids, and the remaining OH is esterified by phosphoric acid and the latter in its turn by the amino alcohol, choline (Fig. B–50).

Choline ($HOCH_2—CH_2—N(CH_3)_3OH$) is a base the hydroxyl group of which is attached to a methylated ammonium group. It might be thought that the three methyl groups endow the end group $—N(CH_3)_3OH$ with lipophilic character in spite of the hydrophilic OH group, but this is not the case, for alkyl groups ($—CH_3$, $—C_2H_5$) bound to the ammonium nitrogen assume polar properties and therefore (like methyl bound to an oxonium oxygen, which makes pectic acid and methylcellulose soluble in water, see p. 34) exhibit hydrophilic behaviour. For this reason, the ammonium end group tends to escape from the neighbourhood of the lipophilic end groups of the fatty acids. The lecithin molecule may, in consequence, be said to resemble a tuning fork (Fig. B–50), in contrast to fats which can be represented schematically as a three-pronged fork without a handle. The prongs of the fork represent the lipophilic pole and the handle the opposite hydrophilic pole of the lecithin molecule.

The 'handle' of lecithin may be esterified with the central OH-groups of the glycerol, as in Fig. B–50 (β-lecithin), or with one of the terminal OH-groups (α-lecithin). Owing to its derivation from dihydroxyacetone phosphoric acid, α-lecithin is the form synthesised by living cells (Fig. B–53). Partial transformation into β-lecithin may occur on extraction.

Phospholipids may react with polypeptide chains by combining with either the lipophilic or the hydrophilic end groups of the side chains, as indicated in Fig. B–51. This attachment is not of a chemical nature, for the phospholipids can be extracted from the cytoplasm with ether. Nevertheless, phospholipid molecules occupy definite positions, according to the character of the side groups on the polypeptide molecules. Lipids without hydrophilic groups, such as fats, can combine only with the lipophilic side groups, and for this reason their possible combinations with protein chains are limited. As shown in Fig. B–51, they can only enter into relation with hydrophilic side chains by the interposition of phospholipids or other intermediates.

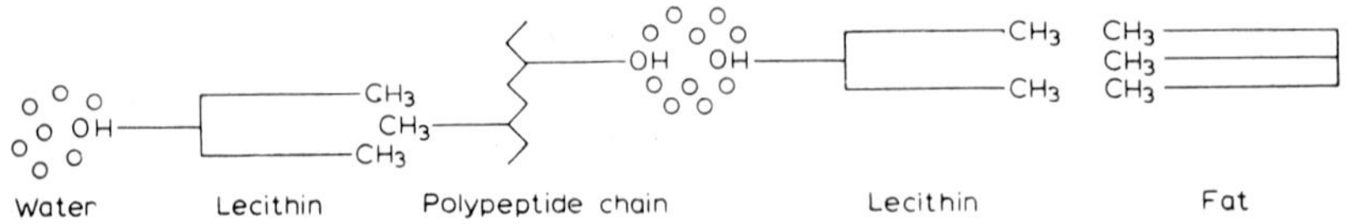

Water Lecithin Polypeptide chain Lecithin Fat

Fig. B–51. Relation between polypeptide side chains and lecithin. ○ = water molecules.

TABLE B–XIII

PHOSPHOLIPIDS AND GLYCOLIPIDS

	First alcoholic component	Second alcoholic component	P:N
1. Lecithins	Glycerol diester	Choline	1:1
2. Cephalins	Glycerol diester	Ethanolamine, serine	1:1
3. Phosphoinositol	Glycerol diester	Inositol	no N
4. Phosphoglycerol	Glycerol diester	Glycerol	no N
5. Plasmalogene	Glycerol ester-enolic ether	Ethanolamine, choline	1:1
6. Phytosphingomyelins	*N*-acylsphingosine	Choline	1:2
7. Cerebrosides	*N*-acylsphingosine	Galactose, glucose	no P

Various other phospholipids are assembled in Table B–XIII (after Karlson, 1961). Fatty acid chains of different length and nature can occur in all these compounds, giving rise to a series of slightly different compounds within each type.

In contrast to lecithins, the cephalins are based not on choline but on its demethylated precursor ethanolamine [$OH \cdot CH_2 \cdot CH_2NH_2$], or serine from which ethanolamine is derived by decarboxylation. The inositolphosphatides contain esterified cyclic hexafunctional alcohol $C_6H_6(OH)_6$ and glycerol, but no amino alcohol. In the plasmalogenes, one of the two fatty acids esterified with glycerol is replaced by a suitable enol alcohol ($R \cdot CH{=}CHOH$), and under certain conditions this enol is released as aldehyde $R \cdot CH_2 \cdot CH{=}O$, which is responsible for the histochemical aldehyde reaction in the plasma (plasmal reaction). The sphingomyelins contain the C_{18} compound sphingosine [$CH_3 \cdot (CH_2)_{12} \cdot CH{=}CH \cdot CHOH \cdot CHNH_2 \cdot CH_2OH$] the last three groups of which behave as an amino dialcohol rather than as glycerol; the amino group binds an acyl residue and the end alcohol group is esterified with phosphoric acid. In phytosphingomyelin, the C_{18} compound occurs as saturated dihydrosphingosine [$CH_3 \cdot (CH_2)_{14} \cdot CHOH \cdot CHNH_2 \cdot CH_2OH$]. Cerebrosides are glucosidic lipids without phosphorus, and for this reason are termed glycolipids.

(*c*) *Myelin forms*

Polar lipids such as fatty acids, oleic acid, cholesterol, or phospholipids give rise to the protrusion of the so-called myelin forms when brought into contact with water, or freed from the cell structure by gentle hydrolysis with very dilute ammonia. The name is derived from the lipidic substance, myelin, in the sheath of medullated nerves (Gr. *myelos* = marrow).

The addition of water to such nerve fibres causes threads to issue from the nerve sheath; these bend and curl and finally grow into irregular entanglements. The active substances causing these structures are the phospholipids present in the myelin sheath, for exactly the same phenomena are observed when water is added to isolated lecithin. Very beautiful myelin forms were obtained by Gicklhorn (1932), in the sap of the epidermal cells of onion bulbs, by the addition of ammonia or sodium hydroxide. The variety of shapes presented by these peculiar structures is well illustrated in Fig. B–52.

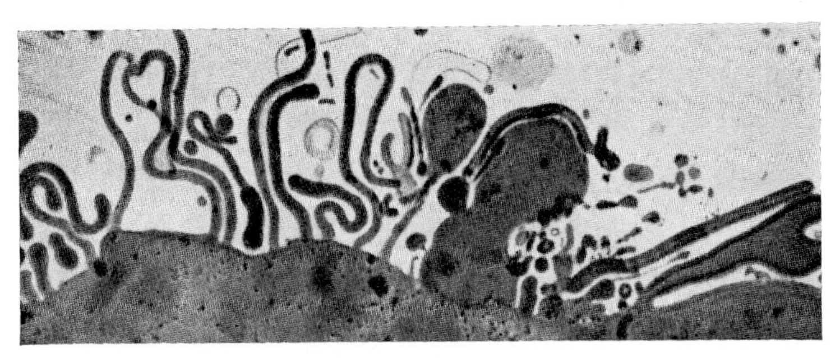

Fig. B–52. Myelin forms in the epidermal cells of *Allium* (from Gicklhorn, 1932).

Although myelin forms are usually designated as liquid crystals, it should be pointed out that there is a fundamental difference between such structures and the crystalline liquid state. In the latter case we are dealing with a special state of aggregation of a single substance, whereas at least two components take part in the development of myelin forms. One of these components is water. It is further essential that the molecules, which must have a chain-like shape, are heteropolar, as mentioned above. Myelin forms can occur if these conditions are fulfilled, provided that the molecules are sufficiently mobile.

The apparent growth is a swelling due to the absorption of water; the hydrophilic groups are surrounded by water, whilst the hydrophobic groups avoid this zone. The resultant orientation is represented in Fig. B–53; the lecithin underlying this scheme is α-lecithin, in which the phosphoric acid is attached to one of the terminal OH-groups of the glycerol molecule. Obviously, water penetrating into lecithin causes the molecules to arrange themselves in layers similar to surface films, except that here it is not a matter of mono- or oligomolecular layers but of huge, microscopically visible structures consisting of bimolecular lamellae. Assuming a length of about 44 Å to a pair of lecithin molecules, the wall of a myelin tube 5μ thick consists of some 1000 double layers (Fig. B–54). As long as not all the hydrophilic groups are saturated with water, more water can be absorbed, causing the structures to grow further. In the course of time, myelin forms will penetrate right across the field of view under the coverslip of the microscopic preparation.

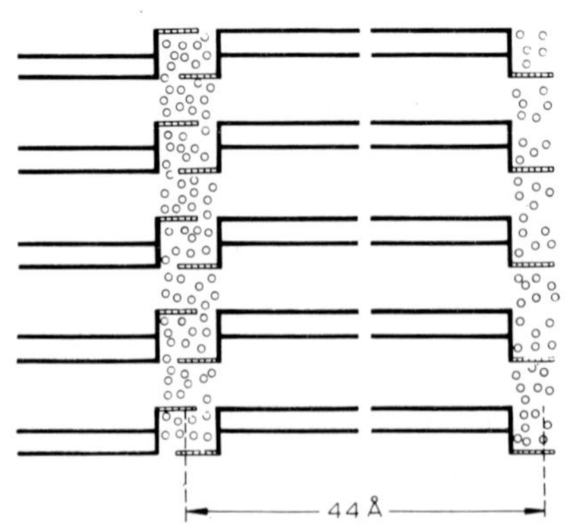

Fig. B–53. Orientation of α-lecithin in myelin forms.

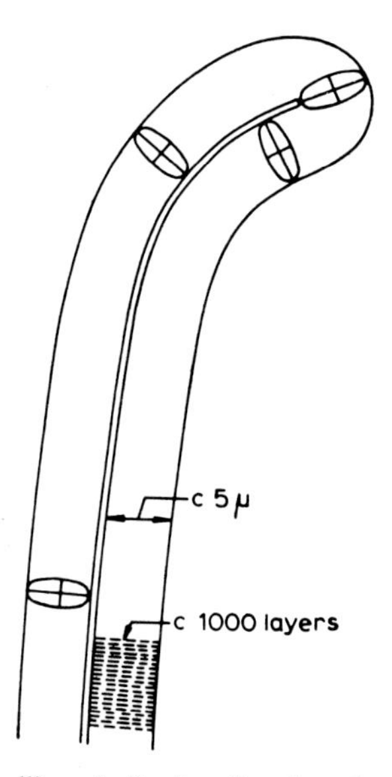

Fig. B–54. Optics of a myelin tube, ellipses indicating direction of major and minor refractive indices.

The orientation of lecithin molecules in myelin tubes can be proved by optical means, as indicated in Fig. B–54, for the molecules appear to be optically positive in a flowing solution. Since the myelin forms are optically positive with respect to their radius, it follows that the lecithin chains are perpendicular to the tube axis. The birefringence of myelin forms is composed of this positive intrinsic anisotropy

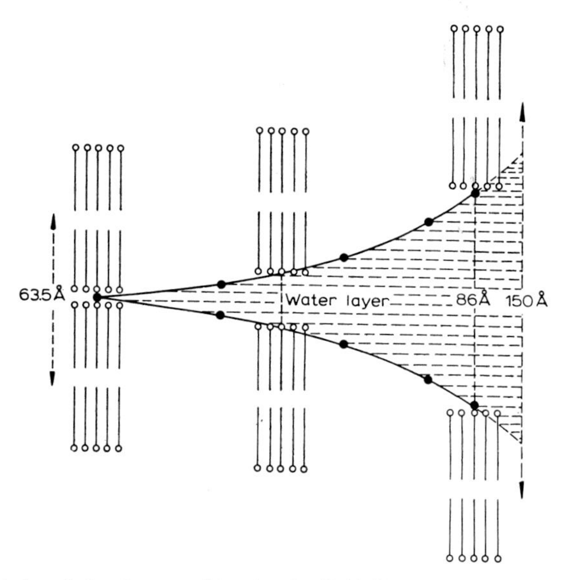

Fig. B–55. Water intercalation between bimolecular lipid films. Size of the adsorbed water layer with increasing water content. The black points correspond to 0%, 25%, 50%, 67% and 75% of water content (Schmitt and Palmer, 1940).

of the orientated molecules and of a negative form birefringence due to their lamellar structure. At first the positive birefringence prevails, but with increasing absorption of water the lamellar structure of the myelin forms becomes more and more pronounced. Finally, the intrinsic positive double refraction of the molecules is overcompensated by the negative double refraction due to the lamellar texture, and the sign of the myelin birefringence undergoes a reversal.

The absorption of water can be followed by X-ray analysis. The dry myelin substances obtained by evaporating the benzene solution give X-ray interferences which correspond to twice the chain length (lecithin and cephalin 44 Å, sterol 34 Å, sphingomyelins 63–67 Å; Schmitt and Palmer, 1940). If water is added to these lipids, the X-ray periods are enlarged and thus allow the thickness of the water lamellae to be evaluated. It will be seen from Fig. B–55 that the original period of 63.5 Å of mixed nerve lipids becomes 150 Å at a water content of 75%, indicating a water layer 86 Å thick between the bimolecular lipid layers.

The myelin forms are a good example of the manner in which complicated microscopic structures can result from a simple arrangement of ultrastructural entities. On the other hand, they also show that no coordinated growth is possible as a result of such a process, for myelin forms 'grow' at random, aimlessly, and without purpose in the substrate, and the final outcome is a state of chaos rather than an illustration of organised life.

Ultrathin sections of myelin figures can be made if they are fixed in osmic acid, and spacing of the bimolecular layers is then visible. The spacing of lecithin is 40 Å (Stoeckenius, 1960). A contraction of 4 Å appears to have occurred in the vacuum of the electron microscope, as compared with the dimension given in the X-ray chamber. Stoeckenius (1962) supposes that the reduction of OsO_4 is due to the double bonds of the oleic acid chains, whereupon the metallic osmium would be

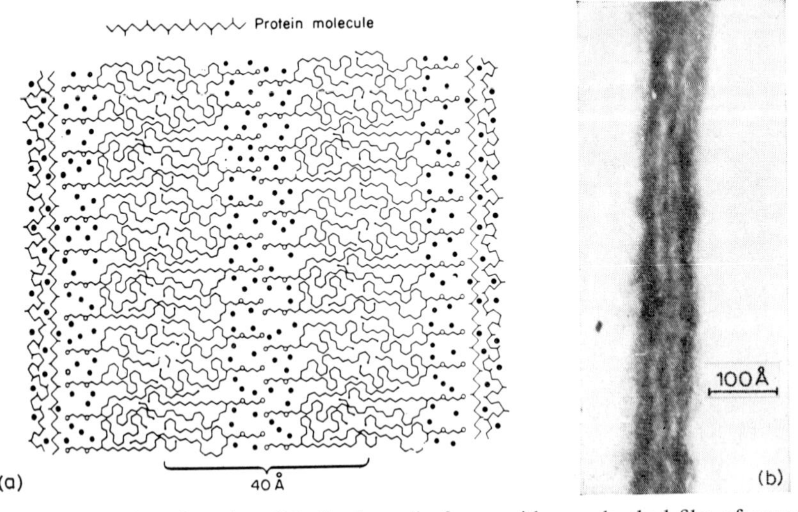

Fig. B–56. (a) Location of osmium ● in fixed myelin forms with an adsorbed film of expanded polypeptide chains (Stoeckenius, 1962). (b) Electron micrograph of a preparation of phospholipid in a globin solution, after fixation with OsO_4. Shown is a structure that corresponds to the schematic drawing in (a) (Stoeckenius, 1962).

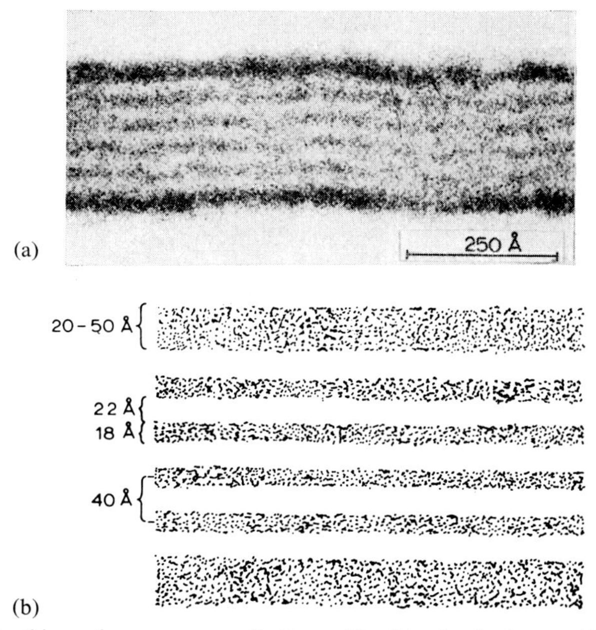

Fig. B–57. Ultrathin section across a myelin form of five bimolecular layers with globin molecules adsorbed on either side (Stoeckenius, 1960); (a) electron micrograph, (b) breadth of period and width of lines.

stored in the hydrophilic layers between the bimolecular layers of the lamellar packet (Fig. B–56).

Oligomolecular layers can be depicted (Fig. B–57) and protein (globulin) can be adsorbed to their surface from solutions; this does not alter the periodicity of the system, which remains at 40 Å. The lipid/protein contact surface shows wider stripes of reduced metal. Whilst the blackish bands in the lipid are about 18 Å wide (see Fig. B–57a, b), at the phase boundary with the protein they measure 25–50 Å. These observations show that more metal is accumulated in the mixed zone of protein and lipid than between lipid films. Stoeckenius (1960, 1962) proposed that the protein is adsorbed as an expanded polypeptide chain; we wonder, however, whether it is not more likely that the protein is present as a helix, or in globular form, in which the hydrophilic poles of the lecithin would be rooted in the loosely built globular molecules (see pp. 149 and 152).

(*d*) *Medullary nerve sheath*

A special kind of lipoprotein occurs in the myelin sheath of medullary nerves. On the basis of the dry weight, it contains about 50% lipids (Leuthardt, 1961, p. 622) embedded in a protein framework. The system shows lamellar form birefringence, from which a sublight-microscopic arrangement of alternating lipid and protein lamellae was predicted (Schmidt, 1936).

The extracted lipids consist of about one half phospholipids (lecithin, cephalin, sphingomyelin), one third sterols (cholesterol) and one sixth phosphorus-free glycolipids, *i.e.* the cerebrosides kerasin and phrenosin (see Table B–XIII). All these myelin lipids form plastic mesophases, which consist of bimolecular layers. The

TABLE B–XIV

THICKNESS OF BIMOLECULAR LAYERS OF LIPIDS IN NEURAL MYELIN
(after Bear *et al.*, 1941)

Substance	Spacing in Å	
	Determined by X-ray	Calculated from atomic distances
Lecithin	43.4	52
Cephalin	43.8	52
Sphingomyelin	66.2	65
Kerasin	66.1	64
Phrenosin	50.0	64
Cholesterol	34	34

length of their molecules can therefore be measured by X-rays. Data of this kind are presented in Table B–XIV, after Bear *et al.* (1941). They are compared with the molecular lengths calculated from atomic spacings. Since the bimolecular layers are as a rule thinner than the length of two molecules, it is concluded that the molecules in the double layers partly overlap or are not fully extended.

Less information is available on the protein component of these lipoproteins. This is termed neurokeratin, as it is similar to keratin, but the relationships of the basic amino acids in these two proteins are somewhat different. After extraction of the lipids, the refractive index of neurokeratin can be determined by the imbibition method from the minimum in the curve of its lamellar form birefringence (Schmidt, 1937a); it is $n_D = 1.58$. A similarly high refractive index is characteristic of myosin and the protein of extracted chloroplasts (see p. 253); it is due to the aromatic amino acids phenylalanine and tyrosine.

X-ray spacings of 171 Å (Finean, 1953) are obtained with fresh nerve sheaths, and of 162 Å when fixed in OsO_4 (Finean *et al.*, 1953). These spacings are reduced under high vacuum and, according to Robertson (1960) measure 150 Å in Os stained thin sections. The ontogeny of the myelin layer shows that the neural sheath arises from spirally pseudopodial growth of the Schwann cell around the neural axon (Fig. B–58). The nucleus of the cell is situated behind the point of the pseudopodium, and the cytoplasm is withdrawn from the inner convolutions. In this way, a spiral is formed, consisting entirely of plasma membrane termed plasmalemma (see p. 140).

The plasmalemma is about 75 Å thick and shows three strata under the electron microscope, the middle one of which appears light and the two outside ones dark. Moreover, when fixed with osmic acid, the black zone on the plasma side is wider than the black stripe facing the culture medium, which in some cases even appears broken and discontinuous.

Robertson (1960) describes this plasmalemma as a unit membrane. This author was also able to show that the above-mentioned spacing of 150 Å includes two such unit membranes; indeed, a halving of the spacing by a dotted line can be observed in neural myelin after fixation with OsO_4. The more pronounced blacken-

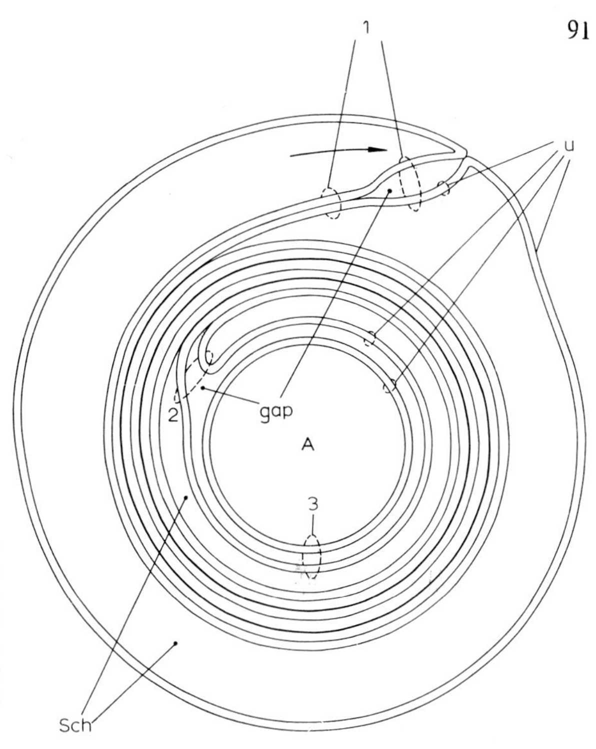

Fig. B–58. Ontogeny of the nerve myelin sheath by spiral growth of the Schwann cell (*Sch*). → direction of pseudopodial growth; *A* = neural axon; *u* = unit membrane; 1, 2, 3 double membranes of different types (Robertson, 1960).

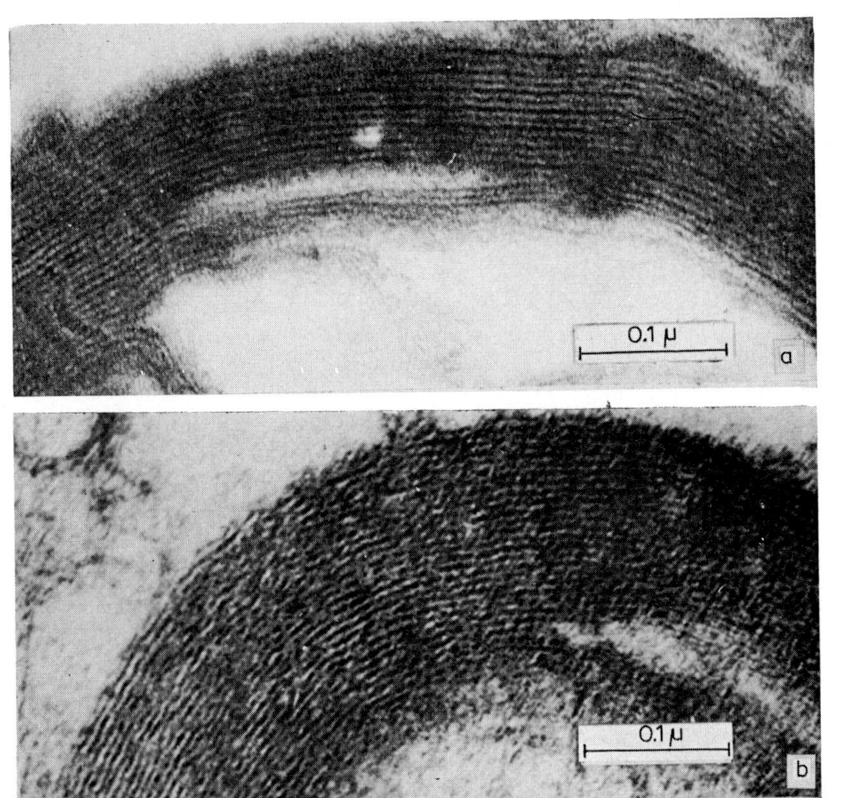

Fig. B–59. Lamination of myelin (human brain); (a) fixed with osmic acid, (b) with permanganate (courtesy of A. Vogel).

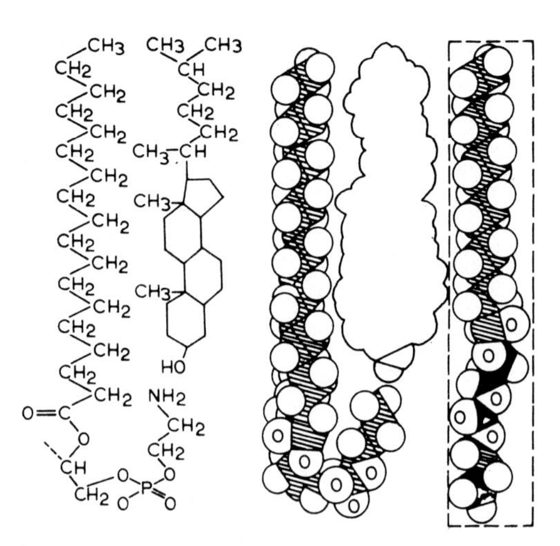

Fig. B–60. Cholesterol–phospholipid complex (Finean, 1953).

ing appears where the two inner sides of the plasmalemma touch, and the interspacing shows the place where the outer sides of the successive membranes lie one upon another. This important difference, which assumes the existence of polarity in the plasmalemma, disappears when $KMnO_4$ is used as the fixative, for then equally darkened 75 Å spacings result (Fig. B–59).

The following views exist as to the spacing and orientation of lipids and proteins in the unit membrane. The general opinion is that submicroscopic protein layers enclose between them a bimolecular leaflet of lipids. This sandwich model cannot, therefore, be significantly thicker than 75 Å, but, since according to Table B–XIV the double layers of some neurolipids require more than 60 Å, little space remains for the protein layers. In view of the fact that cholesterol molecules (17 Å) are shorter than lecithin molecules (25 Å), Finean (1953) accepted lipid complexes, consisting of bent phosphatides and cholesterol molecules (Fig. B–60); hence, a bimolecular layer of 43 Å results, as found in mixed neurolipids.

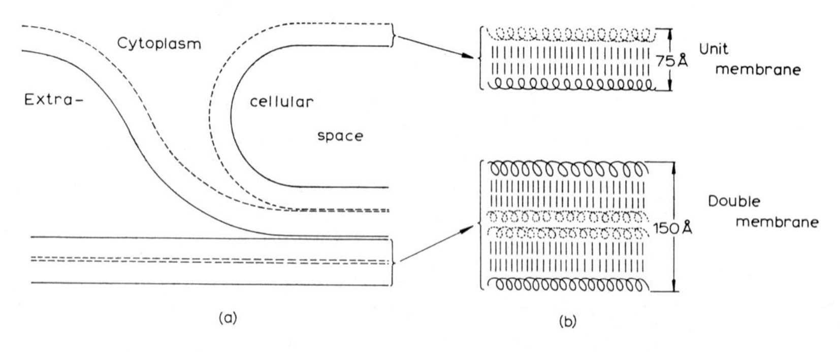

Fig. B–61. Unit membrane and double membranes in nerve myelin (Robertson, 1960); (a) as seen in electron micrograph, (b) molecular structure: protein = α-helix, lipid = rods.

A space 75–43 = 32 Å wide thus remains for the two protein films, *i.e.* the individual molecular protein layers would be only 16 Å thick. Since globular protein molecules have diameters of at least 35 Å (see Tables B–X and B–XI), expanded polypeptide chains are assumed for neurokeratin. The α-helix having a diameter of 10.6 Å (see Fig. B–33) could also be accomodated in the available space. As, in the first instance, neurokeratin has the function of a structural protein, it is probably correct to consider it as a type of fibrous protein. Robertson (1960) has therefore given the scheme in Fig. B–61b for his unit membrane.

The polarity of the unit membrane is not expressed in this molecular scheme, but it may be assumed that the proteins on both sides of the double layer are different from each other and behave differently towards OsO_4.

(*e*) *Unit membrane*

Other ultrastructural protein/lipid layer systems, such as those of retinal rods (Schmidt, 1937b) and chloroplasts, behave similarly to the neural myelin sheath. Layer packets are present in both cases and are visible under the electron microscope as main periods subdivided into two interperiods. The plasmalemma of some objects is also 75 Å thick, and with suitable fixation shows three different strata, more metal being reduced on the plasma side than on the surface; Robertson (1960) therefore came to the conclusion that the ultrastructure of his unit membrane constituted a general principle, applicable to all cytomembranes. Such a generalisation can hardly be accepted, since morphologically different lamellar systems exhibit markedly different physiological properties, and can scarcely be brought together under the same heading. The plasmalemma can be more than 100 Å thick (see p. 146) and the functions of the lipoproteins in chloroplasts can in no way be compared with the more passive tasks of the myelin sheath as an isolating coating for the neural axon.

Despite these limitations, the term 'unit membrane' is useful, for 'double membranes' frequently occur as envelopes of cell organelles, from which it is desirable to distinguish simple membranes. Unit membranes usually show the three dark–light–dark strata and have a thickness ranging from 75 to more than 100 Å. Fig. B–62 shows how unit and double membranes appear under the electron microscope, although deviations from these types are possible.

A controversy has arisen over the interpretation of the accumulation of metals in the outer strata and their absence in the unstained central stratum of the fixed unit membrane. Sjöstrand (1953) has always expressed the view that proteins are blackened by OsO_4 whereas lipids remain unstained, but we find in histochemistry, to the contrary, that as a rule fat droplets reduce osmic acid and turn black whilst proteins may remain white (see, for example, Sitte, 1958). As the reduction of an oxide is by no means specific but may be brought about by a great variety of reducing agents, both proteins (with —SH cysteine groups) and lipids (with unsaturated fatty acids) which are darkened by osmic acid (Bahr, 1954) are found in *in vitro* experiments. Stoeckenius (1960) thus assumed that the 40 Å spacing in myelin from lecithin indicated the zone of the double bonds in the oleic chain (see Fig. B–56).

Although the unit membrane model (Fig. B–61b) corresponds to Sjöstrand's views, it is by no means clear why the double lipid layer is completely indifferent to

OsO_4 and $KMnO_4$, for it is difficult to understand why lecithin extracted from nerves reduces osmic acid and yet loses that property in the myelin of the medullary nerves.

It is thus possible that reduction of the fixative is primarily a surface reaction, and that the reduced metal then penetrates more or less deeply into both sides of the membrane. To obtain the photographs in Fig. B–59, the ability to reduce OsO_4 had to be smaller on the outer side than on the inner side. According to Bahr (1955), the lipid fixed by OsO_4 on the phase boundary of oil droplets prevents further penetration of this reagent.

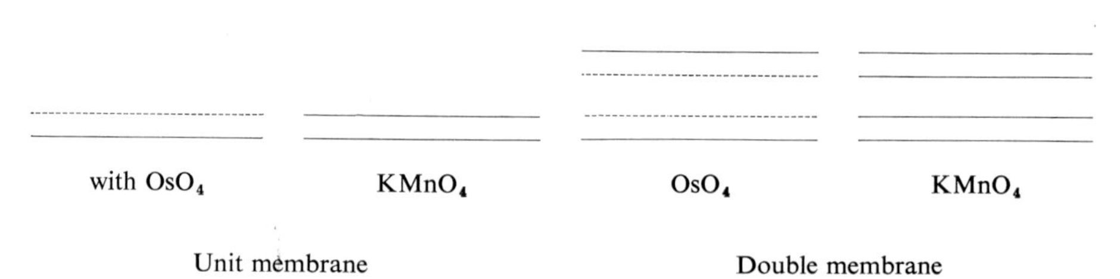

Fig. B–62. Fixation of unit and double membranes.

4. CHROMOPROTEINS

Pigments as well as lipids may be associated with specific proteins, which are then referred to as chromoproteins or holochromes. The pigments are not defined on the basis of their chemical properties, as for example are the lipids by their fat-solubility, but by their ability to absorb certain areas of the visible spectrum. As this property can appear in the most diverse types of substances, the chromoproteins form a less unified group than the lipoproteins. As there are also pigmented lipids, such as the carotenes, in combination with proteins, it follows that the name chromoprotein is a term of convenience rather than a scientific definition.

Nevertheless, the majority of organic pigments possess a common chemical property in that they are unsaturated compounds. In their molecular structure, double bonds usually alternate with single bonds: —C=C—C=C—. Sequences of this type are termed conjugated double bonds, and it has been found that certain frequencies of visible light are in resonance with such systems, which therefore absorb the corresponding wavelengths. The benzene, naphthene, and anthracene rings are conjugated systems of this type, and consequently many pigments occur among the aromatic compounds. Attention must here be drawn to two other groups, namely the carotenoids and the porphyrins.

(*a*) *Carotenoid pigments*

The carotenoids derive their name from carotene, the orange-red pigment of carrot (*Daucus carota*). They are compounds of the formula $C_{40}H_{56}$, and belong to the class of terpenes. They are derived from eight isoprene residues, C_5H_7, *i.e.* our terpene residues, C_{10}, and are therefore designated as tetraterpenes. As shown by the formula, they are highly unsaturated and highly lipophilic. They occur

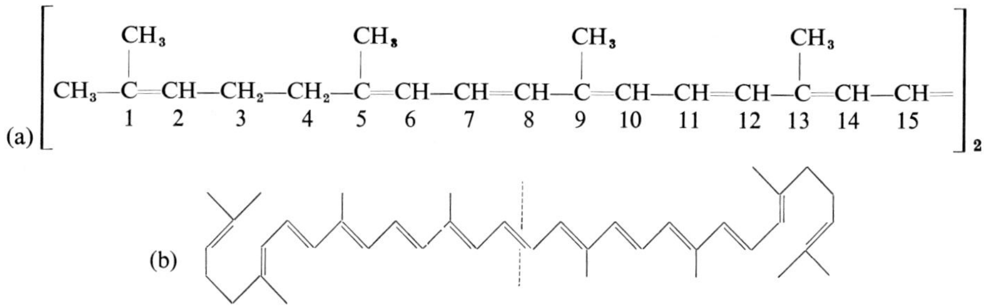

$$\left[\begin{array}{l} \overset{CH_3}{\underset{1}{\overset{|}{C}}}{=}\underset{2}{CH}-\underset{3}{CH_2}-\underset{4}{CH_2}-\overset{CH_3}{\underset{5}{\overset{|}{C}}}{=}\underset{6}{CH}-\underset{7}{CH}{=}\underset{8}{CH}-\overset{CH_3}{\underset{9}{\overset{|}{C}}}{=}\underset{10}{CH}-\underset{11}{CH}{=}\underset{12}{CH}-\overset{CH_3}{\underset{13}{\overset{|}{C}}}{=}\underset{14}{CH}-\underset{15}{CH}{=} \end{array}\right]_2$$

(a) CH_3- (chain above)

(b)

Fig. B–63. Lycopene; (a) molecular structure of half of the chain, (b) short hand formula of $C_{40}H_{56}$.

frequently stored in fat-droplets in cells, and their fat-solubility has led to the use of a synonym, lipochromes.

The carotenoids are partly hydrocarbon compounds (lycopene and carotene), partly they may also contain oxygen (xanthophylls). Their chemical structure is best demonstrated with lycopene, the pigment of tomato. This is a hydrocarbon chain with 11 conjugated double bonds, which yield an orange-red colour. As the chain is bilaterally symmetrical, only one half will be depicted (Fig. B–63a). The construction from four C_5-groups can be recognised, although this is not expressed in the usual numbering of the C atoms. The carotenoids are symbolized diagrammatically by a shorthand formula as shown in Fig. B–63b. The terminals are bent round, since they are formed into rings in the β-carotene of leaves (Fig. B–64). In the ring of α-carotene in carrot, the double bond between carbons 5 and 6 is shifted to the bond between atoms 4 and 5.

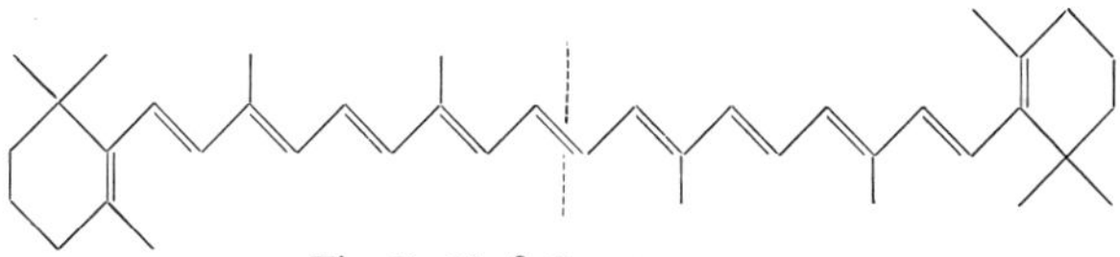

Fig. B–64. β-Carotene.

This has interesting physiological consequences. For these carotenes are precursors of vitamin A, $C_{20}H_{29}OH$, which arises in the animal body through halving of the chain and addition of water. As only the configuration of the terminal ring as it occurs in β-carotene has the specificity of the vitamin, for the cure of vitamin A-avitaminoses, α-carotene with two different terminal rings has but half the effect of that given by enrichment with β-carotene (Karrer 1935).

Hydrocarbon chains containing conjugated double bonds are termed polyenes and are somewhat shorter than the corresponding paraffin chains. Stuart (1934) gives 1.35 Å for the C=C interval in polyenes, whereas the C—C bond in paraffins measures 1.54 Å (see Table A–I); it is also considerably less than the 1.42 Å in aromatic rings. Eichhorn and McGillavry (1959) found 1.48 Å for the conjugated C—C bond. The length of polyene chains depends also on whether the double bond configurations are *cis* or *trans*. As Fig. B–65 shows, the length increment for each C_2 group, assuming a valency angle between 110° and 120° and a C—C distance of 1.48 Å, amounts to about 2.4 Å for the *trans*-position and about 2.0 Å for the

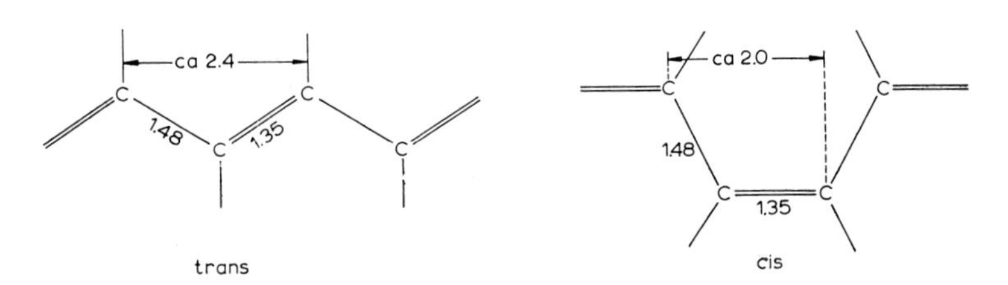

Fig. B–65. Length increment of a group of C_2 in polyenes.

cis-position. In comparison with paraffin chains, where this increment amounts to 2.5 Å (see Fig. A–8b), only the *cis*-configuration causes a significant shortening of the chain. As the carotenoids are usually all-*trans*-chains, and bearing in mind that 11 conjugated double bonds of about 2.4 Å and two terminal half rings with a diameter of about 3 Å are present, their length is usually of the order of 30 Å.

Besides β-carotene, the leaves contain carotenoids with oxygen groups termed xanthophylls. As is shown in Table B–XV, the number of O atoms may vary from one to six. They can occur as the hydroxyl group OH, the keto group C=O or in epoxides as the O bridge [–C—C–] (with O bonded to both C atoms) between two C atoms in place of a double bond. In the hydroxy and epoxy compounds, the hydrogen content $C_{40}H_{56}$ remains unchanged. Only in ketocarotenoids do dehydro compounds appear, as for example in rhodoxanthine ($C_{40}H_{52}O_2$) found in hibernating leaves of conifers.

TABLE B–XV

CAROTENES AND XANTOPHYLLS

α-Carotene	$C_{40}H_{56}$	carrot root
β-Carotene	$C_{40}H_{56}$	leaves
Lycopene	$C_{40}H_{56}$	tomato and capsicum fruit
Cryptoxanthine	$C_{40}H_{56}O$	leaves
Lutein	$C_{40}H_{56}O_2$	leaves
Zeaxanthine	$C_{40}H_{56}O_2$	maize grain
Flavoxanthine	$C_{40}H_{56}O_3$	leaves
Chrysanthemaxanthine	$C_{40}H_{56}O_3$	chrysanthemum flowers
Violaxanthine	$C_{40}H_{56}O_4$	pansy flowers
Taraxaxanthine	$C_{40}H_{56}O_4$	dandelion flowers
Fucoxanthine	$C_{40}H_{56}O_6$	brown algae

Composition of the xanthophyll complex in lucerne leaves is given in Table B–XVI.

In assimilating leaves, the molecular ratio of the xanthophylls to carotenes is approximately 2:1. These carotenoids have a physiological function as secondary pigments and are associated with proteins. The nature of their mutual bonding cannot be ascertained as yet. With carotene $C_{40}H_{56}$, which contains no chemically active groups, bonding is very loose, so that this pigment can easily be extracted quantitatively with fat-solvents. Polar solvents must be used for the extraction of

TABLE B–XVI

THE MAJOR XANTOPHYLLS OF LUCERNE LEAVES
(Goodwin, 1960)

Pigment	Structure	Relative amounts
Cryptoxanthine	3-hydroxy-β-carotene	4
Lutein	3,3-dihydroxy-α-carotene	40
Zeaxanthine	3,3-dihydroxy-β-carotene	2
Violaxanthine	5,6,5′,6′-diepoxyzeaxanthine	34
Neoxanthine	$C_{40}H_{56}O$ (exact structure unknown)	19

xanthophylls, and even then certain residues often remain in the leaf. It is therefore suspected that hydroxyl groups form ether or ester bonds with the functional proteins, and this seems probable, since the xanthophylls that become functionless in the chromoplasts of flower petals and in yellowing foliage are also esterified, and in fact appear as fatty acid esters during autumnal coloration.

The conditions in the chromoprotein known as visual purple, which facilitates light perception in the eye, are somewhat better known. It is termed rhodopsin, and when irradiated changes into the protein opsin and the pigment retinene ($C_{20}H_{28}O$), which corresponds to the aldehyde of vitamin A ($C_{20}H_{29}OH$). To regenerate visual purple, the double bond in the all-*trans*-retinene at the eleventh carbon atom must be converted to the *cis*-configuration; only this 11-*cis*-retinene can combine with opsin to give rhodopsin. On irradiation, the 11-*cis*-bond changes back to the *trans*-configuration (see Karlson, 1961, p. 204). These facts indicate that in the carotenoid chromoproteins double bonds are also responsible for the association between chromogen and protein carrier.

(*b*) *Porphyrin pigments*

The porphyrin pigments are chromoproteins containing the porphyrin ring as a prosthetic group. This ring consists of four pyrrole molecules arranged symmetrically in a tetracycle with 20 C and 4 N atoms (Fig. B–66). The porphyrin ring is synthesised in the cell by the condensation of four molecules of glycine with four molecules of succinyl-coenzyme A, with the elimination of 4 CO_2 molecules (Gibson *et al.*, 1961). It contains 11 conjugated double bonds, which give this molecule a striking red colour (Greek, *porphyra* = purple). The sites of the double bonds are not fixed as in the carotenoids but, as in the benzene ring, are oscillating around the molecule by resonance (see Fig. B–67a).

The presence of a metal atom in the centre of the porphyrin ring is most important from the point of view of electron transfer. Metal porphyrins thus constitute the prosthetic group in many important enzymes. A number of such porphyrin chromoproteins are presented in Table B–XVII. The colour of the compound is determined by the nature and the state of oxidation of the central metallic atom, so that changes in valency can be followed spectroscopically. Divalent iron occurs in haemoglobin (Fig. B–66); it assumes the coordination number 6. Such a porphyrin ring containing ferrous iron is known as haem (Fig. B–66). Five of the coordination sites are occupied by nitrogen atoms, *i.e.* with nitrogens of the four

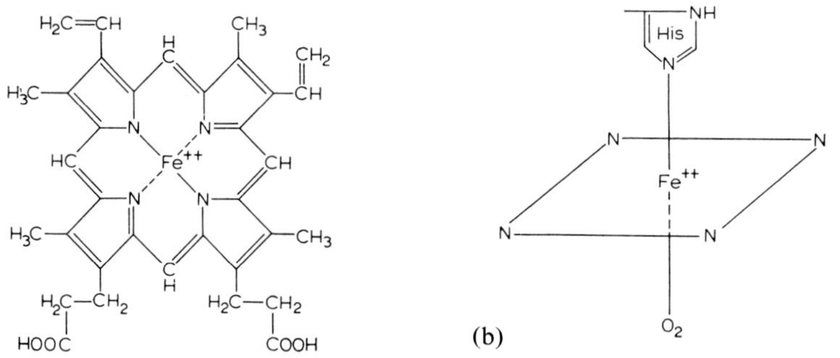

Fig. B–66. Haem; (a) porphyrin ring, (b) Fe^{2+} with an affinity to the carrier protein and O_2.

pyrrole ring and with one of a histidine residue of the carrier protein. The oxygen is stored at the sixth coordination site. There is no change in the valency of the iron during the absorption and release of molecular oxygen by the haemoglobin (Karlson, 1961, p. 145), because the oxygen in the haemoglobin of venous blood is replaced by H_2O.

As may be seen from Table B–XVII, in the enzymes peroxidase and catalase the iron is trivalent, and the corresponding porphyrin ring is known as haemin. Here also there is apparently no change of valency during enzyme activity, whereas such a change does occur in the cytochromes, which act as electron transmitters.

Chlorophyll is a porphyrin as well; however, the central porphyrin atom is magnesium, Mg, and the number of conjugated double bonds is only 10 instead of 11 (dihydroporphyrin). Although in this case the magnesium does not change its valency during photosynthesis, chlorophyll can in fact release and absorb electrons, as depicted in Fig. B–67a. This is because owing to resonance of the conjugated double bonds, the nitrogen atoms of adjacent pyrrole rings can temporarily be-

TABLE B–XVII

PORPHYRIN PROTEINS

Chromoprotein	Function	Prosthetic group	
		Name	Metal in porphyrin
Haemoglobin	O_2 transfer in vertebrates	haem	Fe^{2+}
Haemocyanin	O_2 transfer in molluscs		Cu
Myoglobin	O_2 accumulation in muscle	haem	Fe^{2+}
Potato oxidase	polyphenol oxidase		Cu
Plant peroxidase	oxidation with H_2O_2	haemin	Fe^{3+}
Catalase	$H_2O_2 \rightarrow H_2O + 1/2\ O_2$	haemin	Fe^{3+}
Cytochrome a	cytochromoxidase	cytohaemin	$Fe^{3+} \rightleftharpoons Fe^{2+}$
Cytochrome b	electron transfer	haemin	$Fe^{3+} \rightleftharpoons Fe^{2+}$
Cytochrome c	electron transfer	haemin	$Fe^{3+} \rightleftharpoons Fe^{2+}$
Chlorolipoprotein	energy capture and energy transfer	chlorophyll a chlorophyll b	Mg Mg

come cationic or anionic, and can then release or receive electrons from other sources, such as cytochromes (Lundegårdh, 1962) giving rise to electron flow. An isocyclic (*i.e.* consisting of C atoms alone) 5-membered ring carrying a carboxyl group esterified with methyl is attached to the pyrrole ring III (Fig. B–67b); its

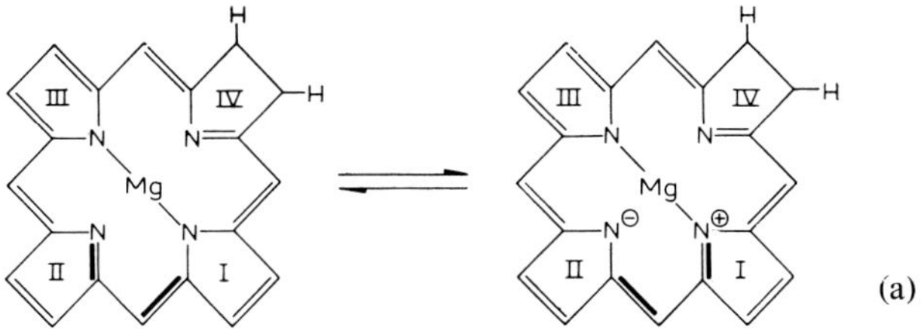

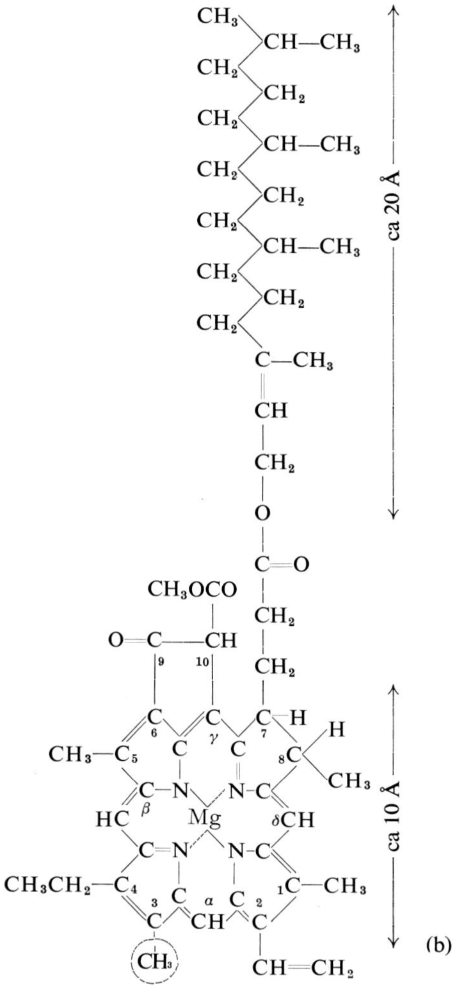

Fig. B–67. (a) Dihydroporphyrin of chlorophyll, as source and sink for electrons, (b) chlorophyll *a*.

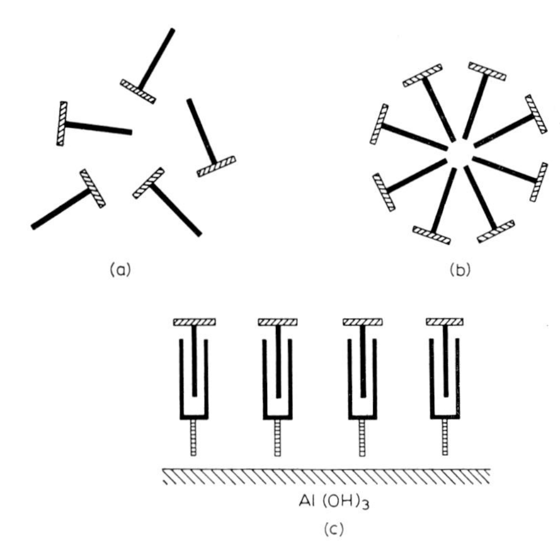

Fig. B–68. Chlorophyll; (a) molecular dispersion, (b) colloidal particle, (c) adsorbed on aluminum hydroxyde by lecithin.

significance, and that of the hydrogenated double bond between carbons 7 and 8 in ring IV (in contradistinction to protochlorophyll) is not clear (Stoll, 1936; Aronoff, 1960). On the other hand, the role of phytol ($C_{20}H_{39}OH$) esterified with the propionic acid residue at carbon 7 is obvious. Except for a single double bond, this is a saturated diterpene chain with pronounced lipophilic character, and imparts a polar character to the chlorophyll molecule, with the porphyrin ring being hydrophilic and the phytyl chain highly hydrophobic. As a result, chlorophyll dissolves only in polar solvents (alcohol, acetone) and forms colloidal solutions in water (Fig. B–68b). On suitable adsorbents, moreover, chlorophyll can be spread as a monomolecular film.

The molecule of chlorophyll is shaped like a tadpole with a large head and a long tail (Fig. B–67b). The porphyrin ring has a diameter of about 10 Å and the phytyl chain is about 20 Å long. It was formerly assumed that the lipophilic tail was perpendicular to the plane of the ring, but it has later been shown to form an acute angle with it (Fig. H–29b).

Chlorophyll *b* differs from chlorophyll *a* solely by the fact that the third C atom has an aldehyde group instead of a methyl one, and this difference is of no account morphologically.

Protochlorophyll, the faintly coloured precursor of chlorophyll, which is produced in the dark, has a double bond between the carbon atoms 7 and 8 in ring IV. This is reduced under the influence of light, when protochlorophyll is converted to chlorophyll *a*.

The secondary pigments of the blue and red algae contain the dyes phycocyanobilin and phycoerythrobilin. The bilins have a carbon skeleton, arising from the opening of a porphyrin ring, as in the decomposition of the blood pigment (Fig. B–69). The open chain can, however, also be formed along a secondary path in porphyrin synthesis. The length of the bilin chain is 20 Å (Fig. B–69).

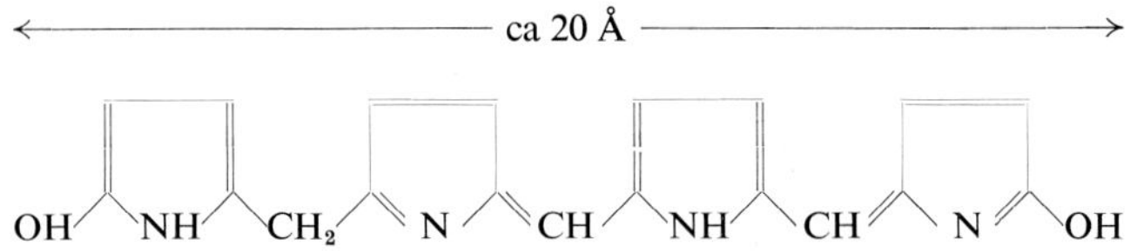

ca 20 Å

OH NH CH_2 N CH NH CH N OH

Fig. B–69. Molecule of phycocyanobilin.

More is known of the association of porphyrin molecules with polypeptide chains in chromoproteins than of the bonding in lipoproteins. In myoglobin (molecular weight 18,000), the position of porphyrin in the coils of the polypeptide chain is known exactly (see Fig. B–44; Kendrew *et al.*, 1960). The structure of haemoglobin (molecular weight 67,000) has also been elucidated (Perutz *et al.*, 1960). Its globin is composed of four polypeptide chains identical in length; although their amino acid sequence is not yet established in full, it is known that two pairs of them are identical, and that each in coordination with a histidine residue carries a haem (see Fig. B–66). The four units are combined into a spheroid measuring 64 × 55 × 50 Å, and these macromolecules are visible under the electron microscope.

Unfortunately the green chromoprotein of the photosynthetic apparatus in plants is much less well known. It falls so readily into its components that it is difficult to detect. Stoll *et al.* (1941) isolated a chloroplastin with a molecular weight of 5,000,000 in the ultra centrifuge, and Takashima (1952) obtained a crystalline green lipoprotein with a molecular weight of 19,000. The protochlorophyll-holochrome of Smith and Kupke (1956) must be mentioned as another chromoprotein.

TABLE B-XVIII

PORPHYRIN PROTEINS

Chromoprotein	Reference	Molecular weight	Particle size Å	Units	Porphyrins	Porphyrin per unit
Myoglobin	Kendrew *et al.*, 1960	18,000	35	1	1	1:1
Haemoglobin	Perutz *et al.*, 1960	67,000	55	4	4	1:1
Green lipoprotein	Takashima, 1952	19,000	35	1	2	1:2
Photoreceptor	Wolken, 1956a	21,000	17 × 121	1	1	1:1
Chlorolipoprotein*	Frey-Wyssling, 1957	68,000 + lipids	65	4	16	1:4
Chlorophyll-holochrome	Smith, 1960	1,000,000				
Chloroplastin	Stoll *et al.*, 1941	5,000,000	220	280	420	1:1.5**
Protochlorophyll-holochrome	Smith and Kupke, 1956	400,000	100	22	2	1:0.1

* Called quantasome by Calvin (1962); size 100 × 200 Å (Park and Pon, 1961).

** Wolken (1956a) gives a ratio of 1:1 for chloroplastin.

Globular particles with a diameter of about 65 Å were found in shaded chloroplast lamellae (Frey-Wyssling, 1957) which today are called quantasomes. To compare these different sizes with one another, they must be brought into relationship with myo- and haemoglobin (Table B–XVIII).

Both chromoglobins possess polypeptide chains, giving with the haem a molecular weight of 17,000–18,000, which measurement was formerly termed a Svedberg unit (see p. 72). Although Svedberg's hypothesis has been shown to be incorrect, a similar unit can be used as a reference size for comparative purposes, from which it will be seen that Takashima's green lipochrome corresponds to about one unit while Stoll's chloroplastin corresponds to 280. On the basis of the chlorophyll content of these proteins, they contain 1 to 1.5 molecules of chlorophyll per unit. From complete analyses of the chloroplasts, however, we obtain 3 to 4 chlorophyll molecules for the protein reference size of 17,000. Now, since the chlorolipoprotein particles (diameter 65 Å) are similar in size to the haemoglobin molecules (diameter 55 Å), they also must contain about four units and therefore 16 chlorophyll molecules (see Table B–XVIII). Thus, the chlorolipoprotein, according to its function as a light energy receptor, would be richer in pigment than the oxygen-transmitting porphyrin globins.

5. NUCLEIC ACIDS

The nucleic acids are high polymer chains built up from monomeric units termed nucleotides. Such units consist of a phosphorylated pentose molecule bearing an aromatic nitrogen base bound *N*-glucosidically:

$$\left[\begin{array}{c} | \\ \text{Phosphoric acid} \\ | \\ \text{Pentose} - N\text{-Base} \\ | \end{array}\right]_n$$

The length and width of a group of this type are 7.5 and 9 Å, respectively. Two different C_5 sugars and at least five different nitrogen bases occur in the nucleic acids.

(*a*) *Nucleic pentoses*

The pentoses have the steric configuration of β-ribose, in which the hydroxyl group on the second C atom may be replaced by hydrogen (Fig. B–70).

Esterification with phosphoric acid takes place at the hydroxyl groups on carbons 3 and 5, whilst the *N*-glucoside bonding of the *N*-bases takes place at carbon 1. Ribose therefore retains in the nucleic acids a free hydroxyl at carbon 2, whereas deoxyribose is devoid of any free alcohol group.

Either ribonucleotides alone or deoxyribonucleotides alone participate in polymerisation. There are, therefore, two classes: ribonucleic and deoxyribonucleic acids, differing considerably from each other. In biochemical and cytological usage, they are referred to as RNA and DNA, respectively.

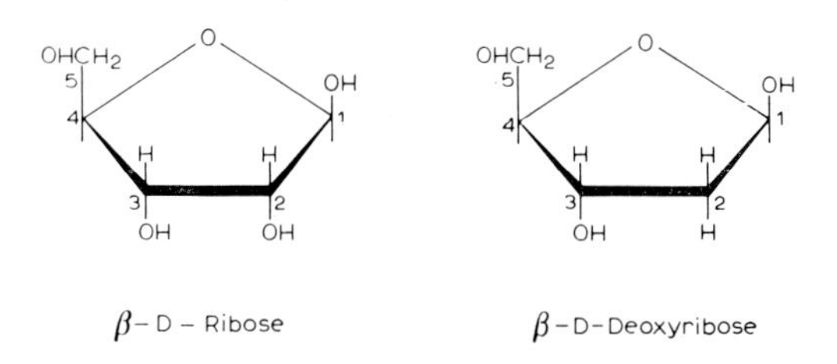

Fig. B–70. β-D-ribose and β-D-deoxyribose.

(*b*) *Nucleosides*

DNA and RNA differ not only in their pentoses, but also in the nature of their *N*-bases. These exhibit pyrimidine or purine structures (Fig. B–72a, b). There are five such bases distributed amongst the nucleic acids, and their names (like those of the amino acids in polypeptide chains) are abbreviated (see Table E–VI). They are presented diagrammatically in Fig. B–71. A few other bases also occur infrequently, and these have not been considered here.

Three of the five classical *N*-bases occur in both types of nucleic acids: adenine A, guanine G, and cytosine C; thymine T occurs only in DNA, and is represented in RNA by its demethylated form as uracil U. Since these bases can be identified by paper chromatography of hydrolysed nucleic acids, DNA or RNA can be determined by the presence or absence of T. Thymine also plays a major role as a tracer group that can be introduced as tritiated thymidine into the genetically active DNA.

In Fig. B–71, the *N*-bases which bear amino groups are shown in the top row, while those in the bottom row are only provided with hydroxyls in their lateral groups (except for T, which has an additional methyl group). This method of depicting them was chosen because the amino groups can be transposed to the corresponding oxy-compounds by oxidation with nitrous acid. By this method, chemical mutations can be achieved, for the bases hypoxanthine or xanthine

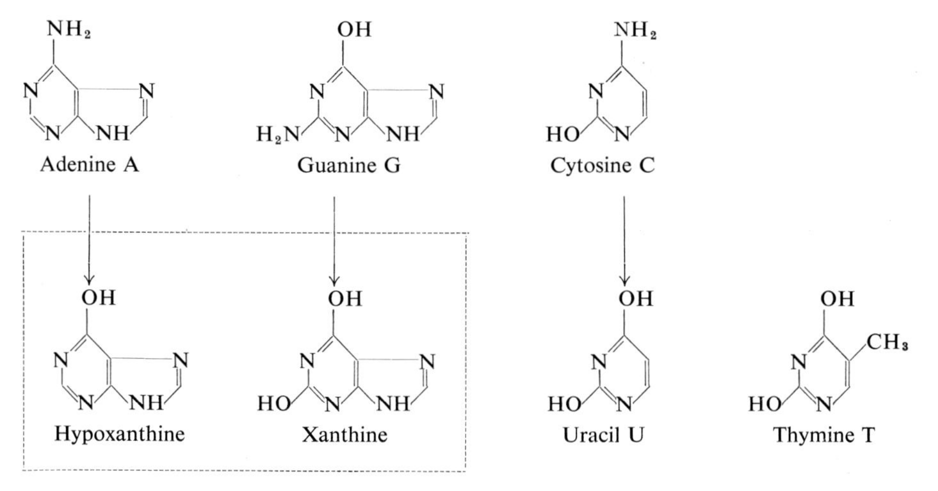

Fig. B–71. *N*-bases of nucleic acids: A, G, C, U and T.

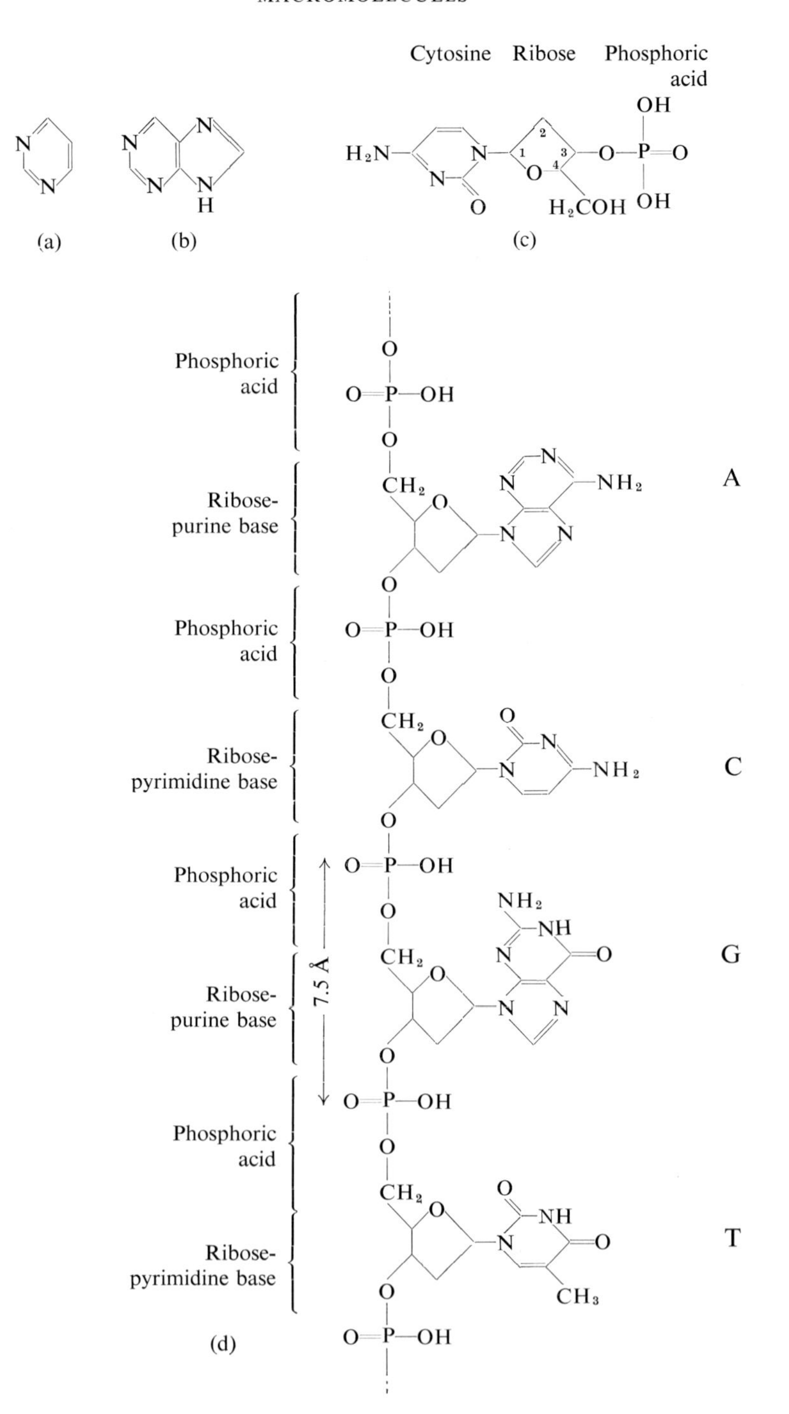

Fig. B–72. Primary structure of nucleic acids; (a) pyrimidine base, (b) purine base, (c) cytidylic acid = nucleotide cytosine-ribose-phosphoric acid, (d) deoxyribonucleic acid = polynucleotide. For C, G, and T the tautomeric formulae of Fig. B–71 are used.

TABLE B–XIX

TERMINOLOGY OF NUCLEOTIDES

Base	Nucleoside	Nucleotide	Base	Nucleoside	Nucleotide
adenine	adenosine	adenylic acid	cytosine	cytidine	cytidylic acid
guanine	guanosine	guanylic acid	uracil	uridine	uridylic acid
[hypoxanthine	inosine	inosylic acid]	thymine	thymidine	thymidylic acid

(which do not occur in native nucleic acids) as well as uracil U act differently in the gene-bearing DNA from the originally aminated bases (see p. 178).

Combinations of the *N*-bases described with pentoses are termed nucleosides; these have names ending in '-idine' when they are derived from pyrimidines and '-osine' when derived from purines (Table B–XIX).

(*c*) *Nucleotides*

Nucleotides are formed when pentoses of a nucleoside are esterified on carbon 1 with phosphoric acid. They were originally called nucleosylic acids, but as nucleosides can combine with polyphosphoric acids, we today speak of nucleoside phosphates, as, for example, adenosine monophosphate (AMP), diphosphate (ADP), or triphosphate (ATP). The letter P is used as a symbol for phosphoric acid.

In addition to their part in the formation of the structural high-polymeric nucleic acids, the nucleotides also play a major role physiologically as mono- or dimolecular units. As such, they can interact with the labile supermolecular structures, which can only be maintained by a continual in-feeding of energy, or be anchored in the apoenzymes carrying prosthetic groups of the coenzymes. They must receive brief mention here.

Adenosine triphosphate (ATP) has proved to be one of the most important compounds in physiological chemistry, acting as an energy transmitter in CO_2 assimilation (photosynthesis), respiration, and many other metabolic cycles. When energy is released, ATP turns into the diphosphate (ADP), which can be reconverted to ATP by certain sources of energy, such as light or respiration. The system

$$\text{ADP} + \text{P} + \text{energy} \rightleftharpoons \text{ATP}$$

is the carrier and storehouse of chemical energy in physiological processes. It acts at the same time as a transmitter of phosphoric acid in phosphorylation. A corresponding system based on uridine (UMP, UDP, UTP) is known, which plays a part in the isomerisation of sugar and in the synthesis of high-polymer carbohydrates (see Kessler *et al.*, 1960). The monophosphates of these two systems, are, as nucleotides, at the same time building bricks of the nucleic acids, so that they assume most important functions in different life processes.

Adenosine nucleotide also appears in the coenzyme of hydrogen-transmitting enzymes, e.g. as a partner of the prosthetic group of codehydrase I, which customarily is called diphosphopyridine-nucleotide (DPN), although more correctly we are dealing with a dinucleotide:

nicotinic acid amide – ribose – P – P – ribose – adenine.

The hydrogen is stored in the pyridine ring of the nicotinic acid amide and hydrogen transmission may therefore be expressed as follows: $DPN^+ + H_2 \rightleftharpoons DPNH + H^+$. As the coenzyme consists of two nucleotides, it would be more correct to call it nicotinic acid amide adenine-dinucleotide (NAD^+); efforts are being made to introduce this change of name.

In the coenzyme of flavoproteins the nucleotide-like prosthetic group riboflavin is also coupled with AMP. As pointed out it would therefore appear that adenosine nucleotide plays a common part in the fixing of prosthetic groups onto the apo-enzyme of dehydrase coenzymes.

(*d*) *Ribonucleic acid* (RNA) *and deoxyribonucleic acid* (DNA)

The two kinds of nucleic acids are distinguished by their different stability. DNA has a higher degree of polymerisation and is much more stable than the shorter and more easily transformed RNA. Molecular weights of up to 10,000,000 are given for DNA (Karlson, 1961, p. 97), corresponding to a degree of polymerisation of about 30,000 (the average molecular weight of nucleotides amounts to 330), whereas the molecular weight of RNA varies between 20,000 and 2,000,000 (degree of polymerisation 60–6,000).

In 1960 Todd proposed an interesting explanation for this difference between the two nucleic acids. As phosphoric esters, they are readily hydrolysed by OH^- ions. Todd showed that the diesters representing the nucleic acids (Fig. B–73), in contrast to the triesters, are protected against this breakdown when their free hydroxyls on the phosphoric acid are dissociated. The resulting negative charge prevents the equally negatively charged alkaline OH^- ion from approaching the chain and starting saponification. This applies fully to the DNA chain, but in RNA the free alcohol group lies so close to the free hydroxyl of the phosphoric acid that its anionic properties are weakened by an H-bridge. This is shown exaggeratedly in Fig. B–73 by omitting dissociation. As in this case saponifying hydroxyl ions have unimpeded access for their action, RNA is more easily depolymerised, under similar conditions, than is DNA.

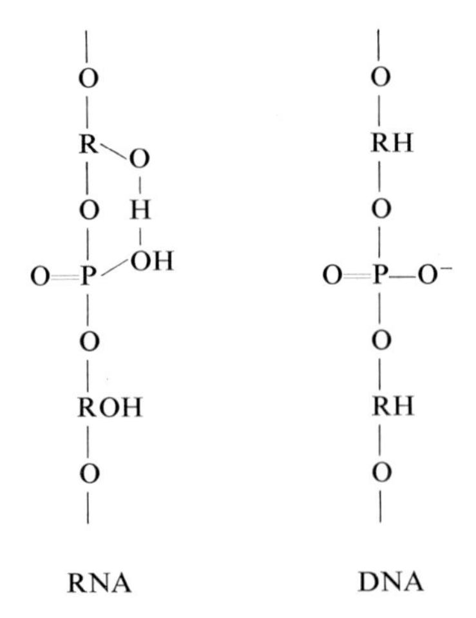

Fig. B–73. Explanation of DNA stability (R = pentose).

The different stability of the two types of nucleic acids corresponds to their biological functions in the cell. The more stable DNA plays the role of gene carrier and the transmitter of heredity characteristics, whilst the more labile and more reactive RNA participates actively in metabolism by catalysing the synthesis of specific proteins. It is surprising that the apparently insignificant difference between ribose and deoxyribose, *i.e.* the presence or absence of the OH group, brings about such far-reaching functional differences.

The specificity of ribonuclease must be attributed to the OH group of the ribonucleotide already mentioned. This enzyme causes a transphosphorylation of the phosphorus bridge arising from carbon 3 in the ribose of pyrimidine ribonucleotides onto the free OH group at carbon 2 (Karlson, 1961, p. 111). As a result, the polynucleotide chain is split into as many pieces as it contains pyrimidine nucleotides. As deoxyribose shows no such transfer sites, DNA cannot be attacked by ribonuclease.

Whether the absence of the hydroxyl in question in DNA is also responsible for the specificity of the Feulgen test (Feulgen and Rossenbeck, 1924) still remains obscure. The fuchsin-sulphurous acid used in this test is the Schiff aldehyde reagent. Free aldehyde groups —CH=O can convert the leuco-pigment bleached with sulphurous acid back into the bright red fuchsin. Only free aldehyde groups undergo this reaction, and radicals closed by glucoside bonds are inactive. To bring the nucleotides to reaction, therefore, they must be partly hydrolysed with hydrochloric acid. By this means the bases are split off, whereupon certain pentose rings open and can react as aldehydes. As only DNA shows the Feulgen reaction, such opening of the ring occurs only with deoxyribose, although it is not very clear why this does not happen with ribose. Certain fluorochrome dyes, e.g. auramine O (Bosshard, 1964) are also specific stains for DNA.

Unfortunately, there is no specific dye for RNA. It has been observed that from a methyl green-pyronine mixture the green pigment tends to be taken up by cell organelles containing RNA and the violet pyronine preferentially by those containing DNA, but this differentiation, which possibly represents a colloidal reaction dependent on particle size or tertiary structure, is not strictly specific and is therefore unreliable. The same is true for the fluorescence of acridin orange, which is said to be yellow with DNA and orange-red with RNA. Spectroscopic measurements by Loeser and West (1962) did not reveal, however, any different emission peaks on the registered intensity curves.

All nucleic acids, which as ionising polyacids have a very low isoelectric point in the cell at about pH 2 (Pischinger, 1937), readily take up basic stains. The basophilic cellular components, therefore, are usually rich in nucleic acids (nuclei, chromosomes, ergastoplasm).

The elegant method of quantitative ultraviolet microspectroscopy also covers both nucleic acids (Casperson, 1950; Ruch, 1961) and allows no differentiation between them. The characteristic absorption maximum at 2600 Å is in fact determined by the aromatic *N*-bases and not by the pentoses. A similar situation occurs with proteins, which have a non-specific absorption maximum of 2800 Å, due to the cyclic amino acids tyrosine, tryptophan, and histidine, present in the majority of polypeptides.

For these reasons, RNA can only be determined indirectly in cells, by measuring the total nucleic acids by one of the methods mentioned (colorimetry, microspectrophotometry), decomposing the RNA by the specific ribonuclease (see p. 56), dissolving it out and determining how much nucleic acid remains in the form of DNA (Brachet, 1953).

Whilst it was formerly believed that DNA was localised in the nucleus and RNA in the cytoplasm, it can be demonstrated by these methods that RNA also occurs in the nucleus (see p. 182) and DNA possibly in small quantities outside the nucleus. The former view partly arose from the fact that the first DNA preparations were isolated from the nuclei of the thymus gland (thymonucleic acid) and those of RNA from yeast protoplasts (yeast nucleic acid).

(e) *Secondary structure and function of nucleic acids*

So far, the secondary structure of DNA alone has been successfully explained. The very high molecular primary chain length makes it possible to spin threads of isolated DNA, which appear optically negative (in relation to the direction of the fibres) in the polarising microscope (Schmidt, 1932) and give excellent X-ray diagrams (Astbury and Bell, 1938). Precise analysis of the latter leads to the conclusion that nucleic acid chains are not extended but helical, and as a similar result is obtained from the heads of sperm of cuttlefish *Sepia* (Wilkins and Randall, 1953), it may be accepted that DNA in histochemical association shows the same secondary structure as it does in the isolated threads.

Various authors (Watson and Crick, 1953; Franklin and Gosling, 1953; Wilkins *et al.*, 1953a) concluded that two helices were present. Fig. B–74 shows the

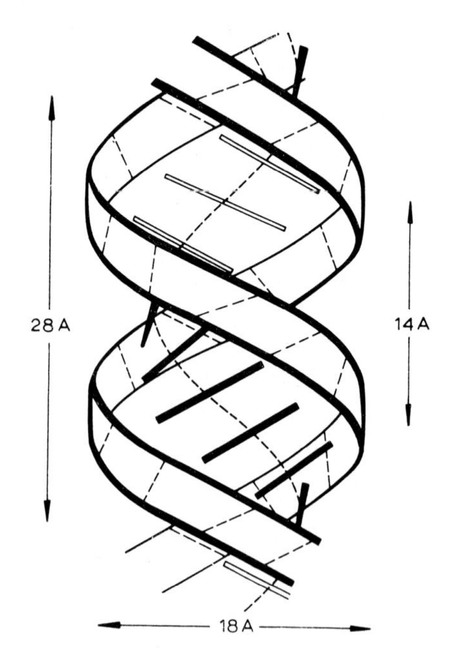

Fig. B–74. Secondary structure of DNA: double helix stabilised by bridges of bases (Wilkins *et al.*, 1953a).

model constructed by Wilkins *et al.* These authors found a monoclinic elementary cell with $a = 22$, $b = 40$, $c = 28$ Å, and with $\beta = 97°$. Two single polynucleotide chains twined around each other in a double helix would fit into the observed dimensions, each with a pitch of 28 Å and containing eleven nucleotide residues. These would require a height of 2.5 Å in relation to the helical axis. Arndt and Riley (1953) also found eleven residues per turn, with a pitch of between 26.4 and 37.4 Å, so that the divergence of the nucleic acid helical chain amounts to 1/11.

Fig. B–74 shows the pitch of the helical bands (28 Å), their distance from one another parallel to the helical axis ($1/2 \times 28$ Å $= 14$ Å), as well as the diameter of the double helix (18 Å). Both bands are connected together by rung-like cross-bars, which do not lie horizontally but are inclined at about 65° to the helical axis. Their centres form a further helical line, with a diameter of 10 Å. Watson and Crick (1953) interpret these cross-bars as the two basic side chains of nucleotides, held together by hydrogen bonds (Watson–Crick model). In Fig. B–75 the exact measurements of these hydrogen bridges are given according to Pauling (1960, p. 502).

It will be seen that the cross-bars are 11 Å in length. Of this, the shorter pyrimidine residue requires about 4 Å and the longer purine residue about 7 Å or, excluding the 3 Å long H-bridge, 2.5 and 5.5 Å, respectively. As the cross-bars are all equal in length, in combining the bases, a purine can be paired with a pyrimidine only. It will also be seen that the most favourable spacing for a hydrogen bridge —N—H····O= (Fig. B–75) is realised in the combinations:

thymine – adenine (T – A) and cytosine – guanine (C – G)

It is therefore assumed that the two nucleic acid helices are joined together by these two types of bridge.

Since, as has been shown, the hereditary substance (hereditary factors, genes) consists of DNA, the latter must possess the ability of transmitting characteristics, of identical replication, and of mutation. For the transmission of characteristics, instructions are required by the cell nuclei derived from the fertilised egg nucleus, and it is proved that this genetic information is passed on by DNA from cell to cell. The DNA code is analogous to the Morse code in telegraphy, but whilst in the latter example only two signs (dot and dash) are available for encoding the infor-

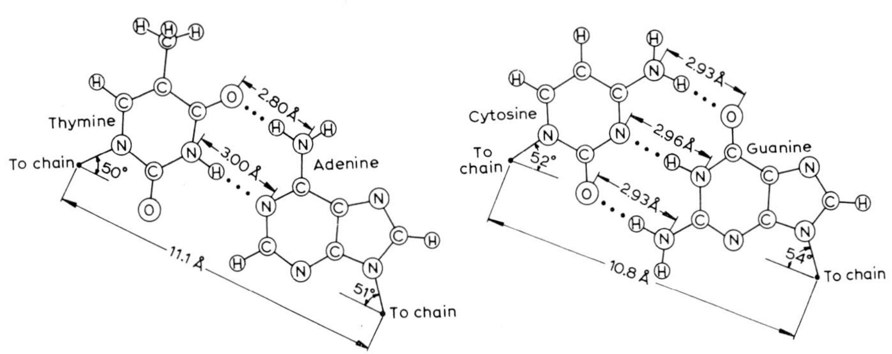

Fig. B–75. Hydrogen-bonded pair of bases on neighbouring DNA chains (Watson and Crick, 1953. from Pauling, 1960).

mation, there are four symbols (T, A, C, and G) in the genetic code of DNA. If, for sake of example, the combination T, A, C, G, A of chain I in Fig. B–76 were to transmit the characteristic for red flowers, then perhaps a slight variation (say T, A, C, G, and T) could transmit the characteristic for white blooms. The number of possible combinations is immense and far exceeds the number of possible genes, for the DNA chains are highly polymeric and the arrangement, repetition, and grouping of the four signs along the chain can be varied at will.

The phenomenon of mutation can also be plausibly explained by this theory, without difficulty. As is known, mutations are induced by energy-rich rays (cosmic rays, β-radiation, X-rays, ultraviolet) or by chemical reactions. One can easily imagine how any of the signs of the DNA code could be changed through such quantum reactions or by oxidation (see Fig. B–71), affecting the genetic pattern (mutation). This is in fact found in nature, and explains why mutations are usually unfavourable (loss mutation), for a given combination group of bases needed for normal development is altered and drops out.

The keystone of the theory whether heredity is really located in the double helices of DNA is the realisation of identical replication. The code established in the chromosomes of the fertilised egg cell must be identically duplicated at each chromosome cleavage during cell division and distributed to the two daughter nuclei. Mitosis serves as the apparatus of division. It is known that in this process the double strands of which the chromosomes are composed separate from one another and are then reconstituted into two new double strands. This is now also accepted for the double helices of DNA. From Fig. B–76 it can be seen that both

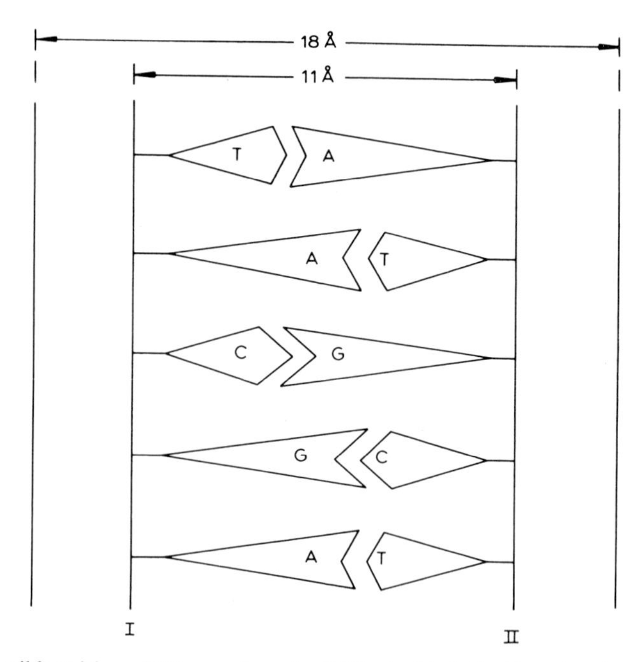

Fig. B–76. Possible pairing of N-bases along the DNA double helix. T = thymine, A = adenine, C = cytosine, G = guanidine.

chains I and II can serve as a template for each other. Each simple chain, therefore can be reconverted into an identical double chain. This represents the mechanism of the all-important maintenance of gene stability.

It must not be concealed, however, that this excellent and illuminating theory of heredity is faced with a number of difficulties. For the time being, there is still no proof that the double helices of DNA do in fact divide in the same manner as the helical chromosome threads (see p. 205). Molecular biologists, who have never themselves prepared chromosomes, easily succumb to the temptation to treat chromosome and nucleic acid threads the same way; yet the former have a size of the order of 2000 Å and the latter are but 18 Å thick! The question of the so-called 'despiralisation' of the threads remains an unsolved problem (see Fig. F–16). Chromosomal threads are paired and coiled in a way that during mitosis they are able to separate from each other and be extended in meiosis to straight parallel threads. This is not so with the Watson–Crick model; the two chains of DNA are so intertwined that they cannot be laterally separated, and for 'despiralisation' the double helices would have to twine around the central axis, and to do this, the macromolecule being many thousands of angstroms long, would require a special untwining mechanism.

The two DNA chains can however be relatively easily separated by the action of heat. Between 80° and 100° C, the double helix of DNA is broken down by cleavage of the hydrogen bonds, and the threadlike form of the double helix, which is resolvable under the electron microscope, separates into two spherical particles of coiled single chains. The process was discovered by ultraviolet spectroscopy, since the absorption at 2600 Å undergoes a sudden 50% increase as a result of the formation of chains (Marmur and Doty, 1959). It is interesting to note that the individual DNA chains can be coupled with RNA chains if the latter have the complementary base sequence. This reaction is referred to as the hybridization of the two types of nucleic acids. The process is used in genetic studies to localise the formation of specific RNA chains in the cell (Yanofsky and Spiegelmann, 1962).

We only have indications from aqueous solutions to guide us on the tertiary structure of DNA. It is known that a paracrystalline order prevails in such systems.

The individual double helices can rotate freely against each other, yet in spite of this they associate together in micelle groups. The X-ray studies of Riley and Oster (1951) revealed that each group of seven double helical chains forms micelle strings with one another, in a hexagonal arrangement. Wilkins *et al.* (1953b) also found a hexagonal arrangement of helices 20 Å wide, which retained free rotation. The micelles composed of seven helices arrange themselves in higher regular groups, whose lattice constants *a* depend on the degree of hydration and the pre-treatment of the system. The micelles within these groups are also freely rotatory (Fig. B–77).

It has been possible to show the thread form of the nucleic acid particles under the electron microscope (see Fig. C–13). Kahler and Lloyd (1953) found small rods and fibrils with a diameter of 15 ± 5 Å, which would be in agreement with the figure of 18 Å given in Fig. B–74. According to Hall and Litt (1958), DNA from salmon sperm, with a molecular weight of 1 million, consists of 20 Å threads and has a high degree of stiffness. Decomposition of the secondary structure leads to dissociation of the double helices and the individual chains shrivel and form amorphous patches.

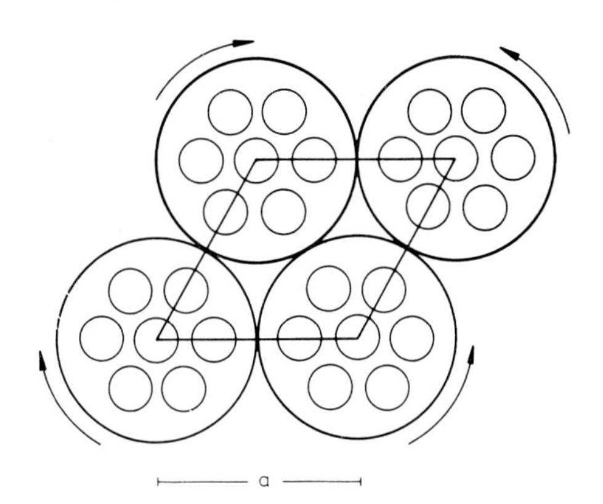

Fig. B–77. Tertiary structure of DNA in highly concentrated solution. Association of double helices (represented by circles) to seven fold strands which are arranged in hexagonal close packing. a = identity period (Riley and Oster, 1951).

Our ideas concerning the secondary structure of RNA are considerably less clear than those of the ultrastructure of DNA, because there are two classes of ribonucleic acid: 'soluble', which exists free in the cell plasma (molecular weight, 20,000 to 40,000), and 'bound', which is associated with the proteins of the ribosomes (see p. 176) (molecular weight 1/2 to 2 million) when the cell components are fractionated in the ultracentrifuge. It is not easy to prepare RNA completely free from protein contamination and for this reason, and because of its greater lability as compared with DNA, (see Fig. B–73), it is not possible to produce preparations sufficiently crystalline for thorough X-ray analysis. It is assumed that RNA tends, in contrast to DNA, more towards the formation of cyclic tertiary structures and thus of globular particles.

We are somewhat better informed about the function of RNA, since ribosomes have been recognised as the sites for the biosynthesis of proteins. If an amino acid mixture is added as a substrate to ribosomes *in vitro*, it is built up into polypeptide chains whose specificity of the amino acid sequence being determined by the code of the base sequence of the RNA. Since this code is fixed in the genes of the nucleus, an indirect connection must exist between the DNA of the nucleus and the RNA of the ribosomes. It is believed that the sequence of bases in RNA is built up in contact with the specific sequence of DNA. The RNA thus formed then has to migrate as messenger RNA from the nucleus to the site of its operation in the plasma (see p. 177).

There is clear evidence that the RNA intervenes actively in the sequence determination of the polypeptide chains, since specific polynucleotides have been successfully synthesised by this process. If polyuridylic acid is used in a cell-free system of *Escherichia coli*, pure polyphenylalanine-peptide is synthetised (Nirenberg and Matthaei, 1961; Wittmann, 1961). A poly–(G + U + A) nucleotide is necessary for the incorporation of glutamic acid. For the exact sequence of the three nucleotides there are six possibilities: GUA, GAU, UGA, UAG, AGU, and AUG.

Such experimental results are likely to show a way to the decoding of the genetic code. It is known that mutative changes in the amino acid sequence can lead to hereditary afflictions; thus for example the substitution of a Glu residue by a Val residue in haemoglobin results in the development of sickle cells instead of normal erythrocytes, because the haemoglobin crystallises out due, apparently, to the loss of the negatively-charged Glu (Ingram, 1957). On the other hand, it might seem that in view of the abundance of specific proteins and the infinite possibilities of gene mutations, it would be impossible to decipher the code; but the following reflections show that inspite of these difficulties the problem can be attacked successfully with the help of mutation studies.

As shown by chromatography, the proteins consist of 24 different amino acids. As there are only four RNA bases, the different amino acid types must be determined by at least three base residues, since if they were combined in pairs the bases would offer only $4^2 = 16$ possibilities. To gain an insight into which of the $4^3 = 64$ triplets induces the incorporation of a specific amino acid, the RNA of a virus is treated with nitrite and an investigation is then made of the mutations that occur in the virus protein. According to Fig. B–71, the RNA bases are changed as follows by nitrite oxidation: adenine A and guanine G are converted to hypoxanthine and xanthine, respectively, which are not present in native nucleotides. On the other hand, cytosine C is changed to uracil U nucleotide and a conversion of adenine A into guanidine G by oxidation is also possible (Ochoa, 1964a). As Glu at certain positions in the TMV protein appears to be replaced by Gly in such investigations, the incorporation of Gly is probably due to a poly–(G + U + G) nucleotide. Instead of the six possible sequences of bases mentioned above for the incorporation of Glu, for Gly only three possibilities GUG, GGU, and UGG remain. In this way the number of possibilities can be reduced, until finally the exact sequence of bases for the incorporation of any specific amino acid can be determined (see Table E–VI).

To sum up, the following comments may be made concerning the role of nucleic acids in heredity: DNA preserves the genetic information through its sequence of bases (code); mutations arise through local changes in the bases (gene mutations); identical replication of the information takes place by a template process, in which the single DNA chain, by a complementary chain, is reshaped into a new double helix (Watson–Crick model).

Transmission of information is carried out by RNA which takes over the code from DNA and transfers it to the sites of synthesis (ribosomes) for specific proteins whose characteristic amino acid sequence is induced by the sequence of the bases along the RNA chains.

However, RNA can also function as a code carrier in viruses which contain only this nucleic acid (see p. 115). Its structure consists of a double helix (Ochoa, 1964b) of the Watson-Crick type and it can act as a template also. This fact proves that for a molecule to function as a code carrier, its general structure is more important than its specific chemistry, since thymine T is seemingly not absolutely necessary for the mechanism of heredity.

(*f*) *Nucleoproteins*

Proteins associated with DNA in the nucleus are basic in nature (protamine, histone). Their isoelectric point lies in the alkaline region, e.g. at pH 8.5 (Pischinger, 1937). As the isoelectric state of nucleic acids is in the acidic region (e.g. at pH 2), nucleoproteins have always been regarded as saltlike compounds formed between nucleic acids and polypeptide chains. Astbury and Bell (1938) believed that they had demonstrated in the DNA chain a nucleotide repeat similar to the polypeptide chain spacing of 3.5 Å, while Zubay and Doty (1959) found on the electron micrograph of a deoxynucleoprotein histone threads (molecular weight 18.5 million) as α-helices, and DNA (molecular weight 8 million) as extended double helices, indicating that the histone α-helix fits into the grooves of the double helix.

The difficulty experienced in separating RNA from the ribosome proteins is also an indication of a direct bonding of molecule to molecule in these two compounds.

This view of the association of individual oppositely charged molecular chains is not substantiated by electron microscopic studies of viruses. Viewed chemically, the virus particles consist of nucleoproteins but under the electron microscope, as will be shown in Section C, we find that the nucleic acid is surrounded by a sheath consisting of a considerable number of globular protein sub-units. In view of these discoveries it is impossible to regard viruses and phages as nucleoproteins in the usual sense. Although stoichiometric proportions are present, a compound of two oppositely charged macromolecules is not involved, but rather a separable association of a nucleic acid chain with numerous macroglobular protein molecules tied together in this way.

C. Virus Particles

Viruses are ultrastructural pathogens, whose size varies considerably (Stanley, 1936): diameters of 2500 Å are given for the smallpox virus, 900 Å for that of fowl pest, 280 Å for the polio virus, and 100 Å for the virus of foot-and-mouth disease (Table III, where dimensions are given in nm). The smallest viruses are thus no larger than macromolecules of known constitution, such as glycogen or globular protein molecules, with molecular weights of more than 500,000.

Chemically, viruses are composed of nucleic acids and proteins. Tobacco mosaic virus, for example, contains 5.1% of ribonucleic acid. When the nucleic acid is split off, the virus loses its pathogenicity and its reproductive ability, from which it is deduced that nucleic acids are essential for virus reproduction. The nucleic acid content, which in general is larger than that of the mitochondria (some %, Table G–I), can be as much as that of the ribosomes (up to 50%, p. 175).

Many viruses such as tobacco mosaic (TMV), tomato bushy stunt (BSV), etc., contain only RNA (see Table C–I).

Morphologically, we have rod-shaped and globular virus forms. To the former type belong tobacco mosaic virus, potato virus X, sugar-cane chlorosis virus (Leyon, 1951), broken tulip virus (Bakker, 1953), etc. The elongated forms of viruses of this type were already recognised before the advent of the electron microscope by such indirect methods as flow birefringence, small-angle scattering of X-rays, nephelometry, and so on. Today, on the basis of X-ray studies and negative staining techniques under the electron microscope, we know that their structure is helical.

To the globular viruses belong the pathogens of the human diseases smallpox, influenza, yellow fever, poliomyelitis and the viruses of foot-and-mouth disease, tobacco necrosis, tomato bushy stunt, carnation mosaic, etc.

A special group, which will be considered separately because of its particular morphology, consists of the bacteriophages, *i.e.* viruses that attack and decompose bacteria.

1. HELICAL VIRUSES

(*a*) *Tobacco mosaic virus*

As Klug and Caspar (1960) stated the tobacco mosaic virus is in many respects unique. It was the first virus to be discovered (Iwanowski, 1892); it was the first to be obtained in pure state (Stanley, 1936), and in which sub-units were found (Bernal and Fankuchen, 1941), and also from which it was shown that nucleic acid is the only vector of infection (Gierer and Schramm, 1956; Fraenkel–Conrat, 1956). Moreover, the molecular morphology of this virus has been so well explained that TMV is used here as a model for the rod-shaped viruses with a helical structure.

Formerly, the ultrastructural TMV rods (Fig. C–1) were assumed to be undivided macromolecules with a molecular weight of 39×10^6, 3000 Å long, and,

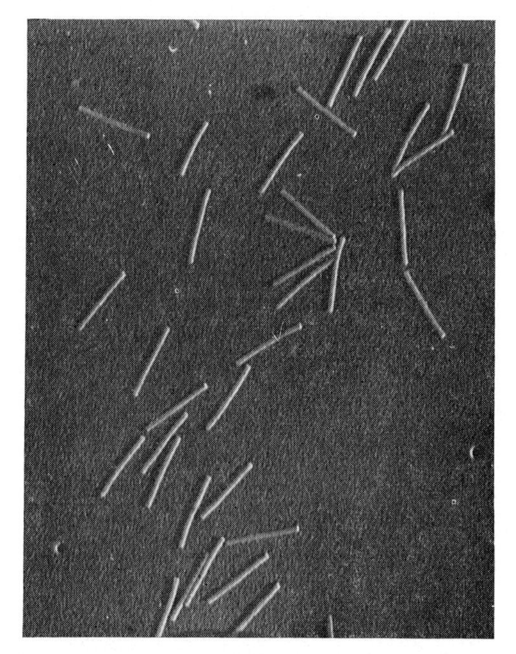

Fig. C–1. Tobacco mosaic virus (TMV) (Steere, 1963).

in close packing, 150 Å wide (Bernal and Fankuchen, 1941; Wyckoff, 1949; Steere, 1963). Refined X-ray analysis has however shown the presence of helically arranged sub-units (Watson, 1954; Franklin, 1955). The helical structure has a divergence of 3/49 (divergence angle $\pm 22°$), *i.e.* the helical line twines round the rod axis three times before a sub-unit is again vertically over the point of departure; from layer lines in the X-ray diagram, such a spacing comprises $3n + 1$ sub-units. In TMV n was found to be 16, *i.e.* the number of sub-units of each period is 49. The helical period amounts to 69 Å and the height of a single turn with n sub-units is 23 Å (Franklin and Holmes, 1956).

The sub-units are globular protein molecules with a molecular weight of 17,400. They are elliptical in shape, with a long axis of 70 Å and a short axis of 20–25 Å. Owing to the form of the sub-units, the helical surface appears humpy. The protuberances rise about 15 Å over the average diameter of the columns, producing a maximum particle radius of 90 Å (Fig. C–2). With close packing in concentrated gels and pseudocrystals, one must imagine the surfaces of the neighbouring parallel rods to be dovetailed into one another, so that the axes of the rods are an average of 150 Å apart.

The humpy surface of TMV rods can be made visible under the electron microscope by the so-called negative staining technique (Brenner and Horne, 1959). For this purpose the virus particles are embedded in phosphotungstic acid, whereupon they appear as light objects against a dark background, like bacteria treated with Indian ink under the light microscope. The finest details of surface relief show in clear silhouette, and by this procedure it has been established that the cross-section of the virus rod has 16 subunits and a central hole (Fig. C–2a).

It had previously been observed that the rods were traversed by a central

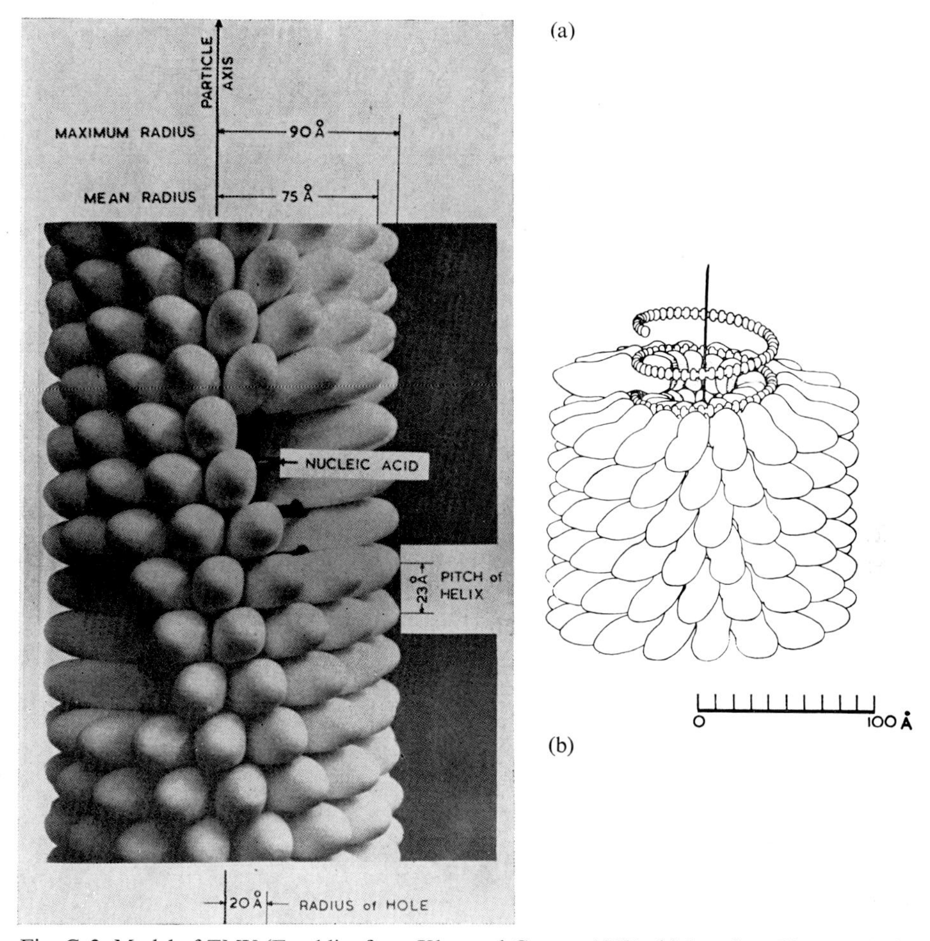

Fig. C–2. Model of TMV (Franklin, from Klug and Caspar, 1960); (a) based on X-ray analysis and negative staining, (b) shows RNA helix.

thread, and it was therefore thought that an RNA core was involved (Schramm *et al.*, 1955). This inhomogeneity has now, however, been found to be the bore of a tube, thanks to the application of X-ray mass-determination methods which made it possible to detect the mass distribution along the radius of the cylinder (Fig. C–4). This curve revealed no mass up to a distance of 20 Å from the helical axis, so that an empty cylinder with a lumen diameter of 40 Å must be present. This internal canal, filled with liquid, permits reacting micromolecules to reach the inner side of the sub-units.

The high mass maximum of the curve in Fig. C–4 with a distance of 40 Å from the helical axis corresponds to the position of RNA with its heavy P atoms. This is shown by the fact that virus rods freed from RNA exhibit instead of a wave crest an equally wide trough at that place. The nucleic acid is therefore not in the centre of the rods, but is inserted on the cylinder surface, at a distance of 40 Å from the centre. A simple RNA chain is involved, following the 3/49 helix of the protein sub-units, and containing three nucleotides per sub-unit (Fig. C–2a, b). This presupposes that the P—P distance is not 7.5 Å as in an extended nucleic acid chain (Fig. B–72d), but only 5 Å, which means that the RNA helix must be slightly folded. It is also of interest that the P—P distance on the DNA helices of the Watson–Crick model is likewise shortened to 5 Å.

The primary structure of the sub-units has been elucidated to a large extent. The 158 amino acid residues that make up such a particle are known (Anderer and Handschuh, 1962). Only 16 of the possible 24 amino acids occur. Some are numerous in the polypeptide chain (e.g. 18 Asp, 16 Glu, 16 Ser, 16 Thr, 15 Ala, 11 Arg), whilst others occur infrequently (such as 4 Try, 2 Lys); histidine is completely absent. The presence of a cysteine residue is of particular importance; it reacts with methyl mercuric nitrate, whereby a heavy metal can be introduced into the particle; this was of great value when the structure was being examined radiometrically. On the other hand, the —SH group does not react with larger organic molecules, as was found by the nitroprusside test. It is therefore assumed that the cysteine group is located inside the sub-units and that access to it is only possible through the small interstitial canals running radially between them (Fig. C–3c). Small molecules, such as I_2 or $Hg(NO_3)(OCH_3)$ can circulate in these spaces, but larger ones are stopped at the entrance, being as it were sieved out. In Fig. C–3b, the position of the sulphhydryl group is indicated by S.

Lead can also be introduced, which is assumed to react with amino acids. X-ray investigations have shown that it accumulates at distances of 40 and 84 Å from the axis of the helix, but as only one Pb is adsorbed per sub-unit, it is uncertain how the lead ions are arranged. In Fig. C–3b, they are distributed alternately, at the two radial spacings found. They consolidate the helical structure, so that a virus charged with lead is more stable towards decomposition reactions than is the natural virus.

The amino acid sequence is known. Decomposition of trypsin gives 12 oligopeptides the sequence of which is established. The secondary maxima in Fig. C–4 indicate that the polypeptide chain is convoluted apparently corresponding to the α-helix.

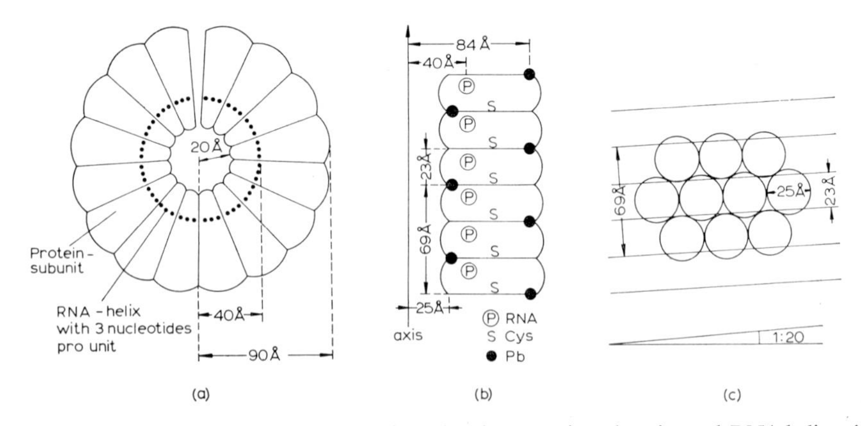

Fig. C–3. Tobacco mosaic virus; (a) top view, showing protein sub-units and RNA helix with 3 nucleotides per unit (see also Fig. C–2b); (b) radial section, S position of Cys, ● Pb atoms introduced, fixed by —COOH groups; (c) front view, pitch of helix 23 Å, pitch angle 1:20.

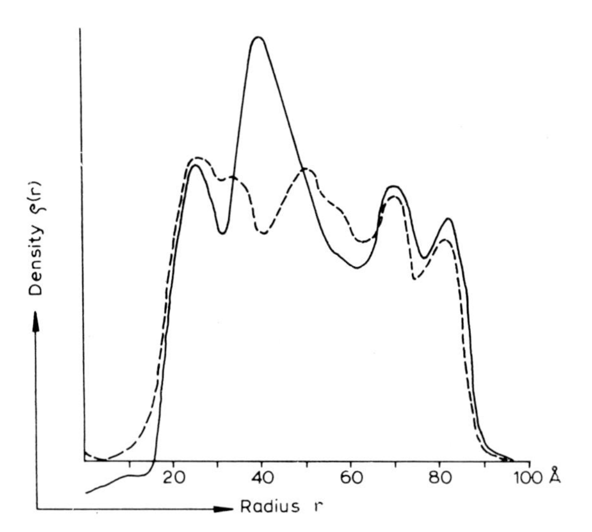

Fig. C–4. Radial density distribution of TMV rods; ——— nucleoprotein, - - - - protein without RNA (Klug and Caspar, 1960).

The RNA chain runs within a groove of the sub-units (Fig. C–2b). The molar ratios of the nucleotides are A:G:C:U = 0.29:0.26:0.18:0.27. The whole chain contains 6400 nucleotides, but their sequence it not known. It is demonstrably responsible for the behaviour of the virus. Oxidation of cytosine residues to uracil produces mutations, and can destroy the infective capacity of the chain. The three nucleotides which fall as a share to each protein sub-unit must form different triplets from unit to unit. The sub-units can therefore hardly be identical with one another in every particular. The fact that a triplet must always contain either two purine or pyrimidine nucleotides requires the presence of at least two types of sub-units with different large grooves – one with a larger space for nucleotide triplets of the purine–purine–pyrimidine type and a lesser one for the pyrimidine–pyrimidine–purine type.

The following is a summary of our knowledge of the structure of the 3000 Å long TMV particles: 2130 protein sub-units, which may be regarded as identical for the time being, are placed on a flat-wound helix. The RNA, which shows 6400 nucleotides in a specific order, runs in a groove of this beaded helix. Thus the structure consists of numerous globular protein macromolecules, paraded on a single polynucleotide chain. The capacity for infection, assimilation of foreign proteins, and identical replication of the virus belongs to the RNA chain, which evidently stores the code for these activities in its nucleotide arrangement and is therefore able to transmit the instructions for the specific behaviour of the virus. In view of this, one wonders what may be the function of the protein sub-units. It might be thought that they serve exclusively as a sheath for the stabilisation of the labile RNA chain, but such oversimplification is certainly incorrect. The specificity of the virus protein, which can be determined exactly serologically by methods of immunochemistry, must play a role in virus transmission as a vehicle of the infectious RNA.

(*b*) *Other rod-shaped viruses*

It has been established that other rod-shaped viruses have the same structure as TMV. There is a species of TMV infecting beans, that cannot be distinguished morphologically from the classical strains, except that its amino acid composition is slightly different. The same applies to cucumber virus 4 (Franklin, 1956), which is distinguished only by the absence of the cysteine residue from the sub-units. Potato virus X, with a rod size of 100 × 5000 Å, likewise shows a structure similar to that of TMV with helically arranged sub-units, and a helical structure is also indicated for the very long (100 × 12,500 Å) beet yellow virus (Burghardt and Brandes, 1957) and the virus of mumps (Horne and Waterson, 1959).

2. GLOBULAR VIRUSES

(*a*) *Small viruses*

At first sight, the globular virus particles look like spheres, and their particle size is therefore given as a spherical diameter. If this falls within the range of 200–300 Å they are termed small viruses (Table C–I), in contrast to large viruses with diameters of 1000 Å and above (Table C–III). The size specifications vary by 10–20%, according to whether the particles are determined in a dissolved or 'dry' state. Methods for the measurement of particle sizes in solution comprise studies on the scattering of light (measurement of turbidity), X-ray low-angle diffraction, diffusion rates, and viscosity behaviour. As in these ways the size of the mobile particles is determined together with their shells of hydration, we speak of the 'hydrodynamic' diameter (Lauffer, 1950). This value differs considerably from the 'dry' diameter observed with the electron microscope, as in the high vacuum the particles lose their hydration water and in the process of drying the globular particles become flattened; this must be allowed for when the measurements are made (Hall, 1960).

TABLE C–I

VIRUS CRYSTALS
(figures from Klug and Caspar, 1960)

Globular virus	Structure body-centered	Identity period Å	Particle distance Å	RNA %	Molecular weight × 10^6	Number of sub-units
Tomato bushy stunt (BSV)	cubic	386	334	17	9	2 × 60
Turnip yellow mosaic (TYMV)	cubic*	703	304	40	5	2 × 60
Polio virus (POLV)	orthorhombic	353:378:320	± 304	25–30	6.7	60
Southern bean mosaic (SBMV)	orthorhombic**	295:508:474	295	—	—	60?

* With 16 particles per unit cell, made up of two interpenetrating diamond-type lattices.

** Since $b/a = \sqrt{3}$ and $c/a = 1.61 \approx \sqrt{8/3}$, the lattice is approximately hexagonal close-packed.

The particle spacing in virus crystals, which are examined in their mother liquor, corresponds approximately to the hydrodynamic diameter, but the particle size in crystals examined in the dry state, with the lattice broken down, falls even below the measurement given as the 'dry' diameter. For tomato bushy stunt virus, for example, the hydrodynamic diameter (determined by low-angle scattering) is 309 Å and 334 Å in the hydrated crystal, as compared with 300 Å for the 'dry' diameter and 267 Å for the dried crystals.

The globular virus particles only appear to be spherical in form. Actually, they are isodiametric polyhedra, and it could further be shown that they consist of sub-units like the helically-built virus rods. Both X-ray low-angle diffraction and electron microscopy show that these sub-units are arranged around a central core. Fig. C–5 shows the density measured from the centre of the particles along the radius, by the radiometric method.

It will be seen that the particle must be hollow within. Maximum density is found 50 Å from the centre, which is due to the RNA since particles without nucleic acid show a low density at this point. Thus, as with TMV, there is a central hollow, the surface of which is formed of nucleic acid, which is covered by sub-units towards the outside. The undulating rise of the dotted curve with its small maxima in Fig. C–5 shows that the sub-units apparently consist of helically-built polypeptide chains (α-helix?).

Crick and Watson (1956) have proposed a reasonable theory accounting for the arrangement of the sub-units on the surface of the particle. If we assume that all sub-units are identical and asymmetric, and that they are arranged regularly on the particle surface, then their arrangement would require cubic point symmetry. The assumption of asymmetric sub-units means that the mirror planes would be

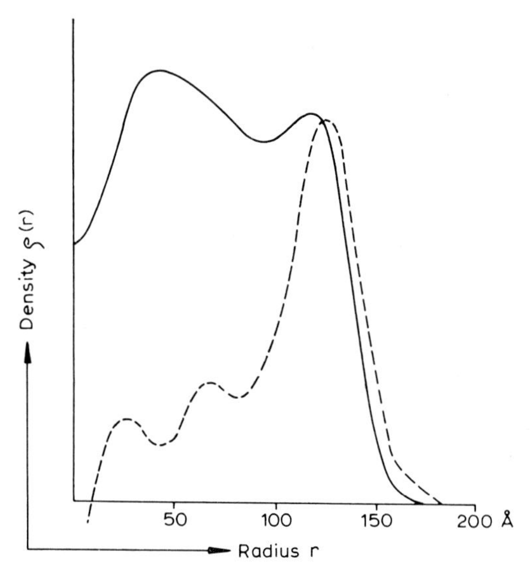

Fig. C–5. Radial density distribution of wild cucumber mosaic virus (WCMV); ——— nucleoprotein, - - - - protein without RNA (Klug and Caspar, 1960).

TABLE C–II

THE THREE POSSIBLE CUBIC POINT GROUPS FOR A GLOBULAR VIRUS
(Crick and Watson, 1956)

Symmetry symbol	Number and type of rotation axes present	Number of asymmetric sub-units	Resulting Platonic polyhedra
23	3 2-fold 4 3-fold	12	tetrahedron
432	3 4-fold 4 3-fold 6 2-fold	24	octahedron cube
532	6 5-fold 10 3-fold 15 2-fold	60	icosahedron dodecahedron

omitted as symmetry elements, and that the symmetry would be governed only by the axes of rotation. In the cubic system, we therefore have the possibilities listed in Table C–II.

Any point on the surface of a Platonic solid, *i.e.* a polyhedron whose facets have edges of equal length, must repeat itself the number of times given by the characteristic of the rotation axes (e.g. twofold, threefold, etc.), and the point system arising from such operations is termed a point group.

As is shown by the symmetry symbol of the point groups (left-hand columns in Table C–2), two-, three-, four- and fivefold rotation axes are possible in this case, in contrast to the two-, three-, four- and sixfold axes which occur in crystallographic space systems (crystal lattices). This difference is due to the fact that the symmetry of the space groups (see page 10) *i.e.* of the point arrangements in the space lattice is so constructed that an *uninterrupted* occupation of space results from translation in the three main directions. Since regular pentagons cannot be attached to one another uninterrruptedly, fivefold rotation axes are impossible in space groups, yet they can occur in the point groups which show the point distribution on the surface of a body.

In the cubic system, two- and threefold rotation axes alone result in tetrahedral symmetry, but if four- or fivefold axes are present, the symmetry becomes respectively octahedral or icosahedral. The three polyhedral types are represented in Fig. C–6, in elevation and in plan.

The projection of the 60° angle of the equilateral triangle at the vertex is 120° for the tetrahedron, 90° for the octahedron, and 72° for the icosahedron (see the lower part of Fig. C–6). No further Platonic polyhedra with facets in the shape of equilateral triangles, can thus occur, since with the angle of 60° required by a sixfold rotation axis the six triangles would lie in the same plane, so that no spatial entity would arise; axes of higher multiplicity are impossible, for the horizontal projection of the angle of the triangle can only be $\geqslant 60°$.

The number of sub-units which can occur on the surface of a polyhedron may be arrived at by taking account of the fact that each point on the surface of a

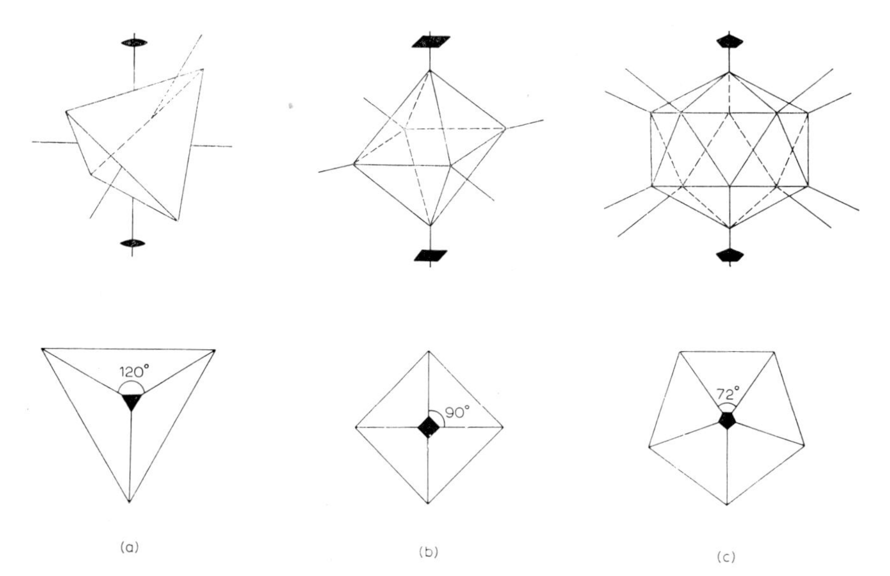

Fig. C–6. Platonic polyhedra with regular triangles as facets; (a) tetrahedron, symmetry 23, (b) octahedron, symmetry 432, (c) icosahedron, symmetry 532.

triangle is trebled by its threefold rotation axis, so that there are three times as many places as there are triangular facets: 12 in the case of a tetrahedron, 24 in the case of an octahedron, and 60 in the case of an icosahedron (Fig. C–7a).

It is interesting to note that X-ray analysis of virus crystals has shown that only particles with a point symmetry 532 and a corresponding number of 60 or 120 sub-units are found in the small viruses (Klug and Caspar, 1960). There is a special reason for this preference to icosahedral symmetry. Since in cubic point symmetry the largest possible number of surface sites is 60, it is assumed that an effort is made to surround the nucleic acid with the largest possible number of sub-units: this is only possible in the case of icosahedral symmetry. Of the simpler point symmetries, which have not so far been found to occur in viruses, octahedral symmetry 432 is probably found in ferritin, which has a molecule similar to that of the small viruses, with a protein shell surrounding the iron core.

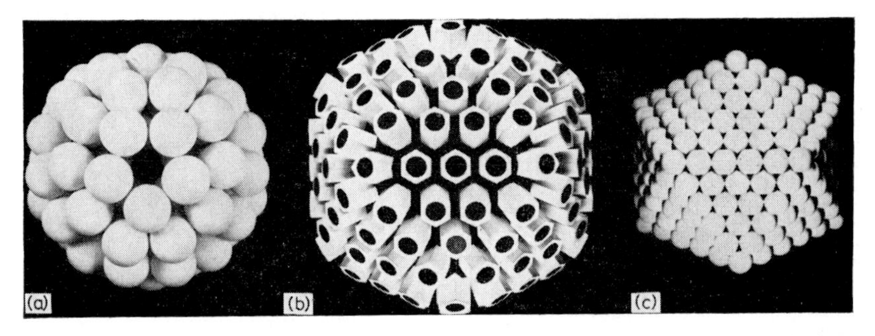

Fig. C–7. 532 symmetry of virus particles; (a) icosahedron with 60 sub-units, (b) model of Herpes virus, (c) icosahedron with 252 sub-units (Wildy and Horne, 1960).

Icosahedral virus particles form crystal lattices; however, their shape precludes complete filling of space, and the endproduct is a gap lattice. This may be likened to the regular packing of spheres, with large amounts of water of crystallisation more or less strongly bound in the spaces between them. It is assumed that the icosahedra in the crystal lattice are almost in contact, so that the particle spacing may be said to give, within about 10%, a measure of the particle diameter (see Table C–I). The forces bonding the crystallising particles are weak, and comprise hydrogen bonds, hydrophobic Van der Waals' forces, and salt linkages; long-range forces (see p. 26) also occur with particles whose size is of the order of 1000 Å.

Different lattice types (see Table C–I), breaking down when the water of crystallisation is removed, appear under X-ray analysis, whereas only the closest spherical packing (see Fig. A–7a, b) has been observed with the electron microscope (Labaw and Wyckoff, 1955). The binding forces on which the quaternary structure (arrangement of the sub-units) of virus particles is based, are therefore considerably stronger than the crystallisation forces responsible for the arrangement of the virus particles in the crystal lattice.

(*b*) *Large viruses*

The sub-units of globular virus particles may be rendered visible under the electron microscope, by means of phosphotungstic negative staining, and, although in the case of the small viruses it is scarcely possible to count them, their point group and the number 60 resulting from it have been determined by X-ray analysis. In the case of large viruses, both the number and the arrangement of the sub-units can be observed directly. It has even been possible to establish that the 'sub-units' themselves consist of still smaller elements, which according to their molecular weight of about 20,000 correspond roughly to the sub-units of the small viruses. As the proteinic coat of virus particles is termed the capsid, it has been suggested by Wildy and Horne (1960) that the knobs observed with the aid of the electron miscroscope should be known as capsomeres.

Point symmetry (532) has also been established in the structures of large viruses studied so far, but the numbers of capsomeres do not follow the expected series of 60 succession (120, 180, 240, etc.); instead the numbers as given in Table C–III, are observed. This results from the fact that point number of 60 is achieved

TABLE C–III

CAPSID STRUCTURE OF VIRUSES
(Wildy and Horne, 1960; Horne and Wildy, 1961)

Virus	Number of capsomeres	n	Particle size Å
Bacteriophage φX-174	12	2	250
Polyoma virus	42	3	
Herpes simplex	162	5	1000
Adenovirus	252	6	800
Tipula iridescent virus	812	10	1300

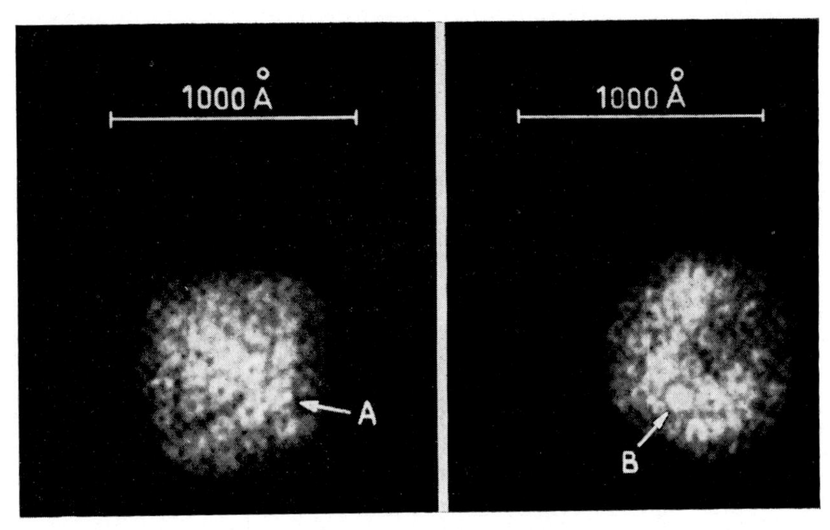

Fig. C–8. Symmetry of Herpes virus; (A) capsomere with 5-fold symmetry, (B) capsomere with 6-fold symmetry (Wildy and Horne, 1960).

only when the points lie in the icosahedral planes; when they occur on the edges, their number is reduced since they are then shared by two adjacent triangles. An imaginary point on a vertex would only be multiplied by a factor of twelve by the rotation operations, since its appearing would be restricted to the twelve vertices of an icosahedron. If another point was situated in the middle of a connecting edge, then 42 point situations would occur (Howatson, 1962), *i.e.* 12 points on the vertices and 30 points on the edges. The number of capsomeres may be calculated from $10(n-1)^2+2$, where n is the number of points on the icosahedral edges. The quadratic term of this formula indicates that further capsomeres must occur on the planes when $n > 3$. Examples of the different values of n are given in Table C–III.

The occurrence of capsomeres on the edges and vertices of the icosahedron suggests that these are not asymmetric, as was at first postulated, but rather that they possess rotational symmetry about the corresponding axes, such as fivefold rotation at the vertex, twofold on the edges, and threefold in the centres of triangular facets.

The actual occurrence of such conditions has been established by Wildy and Horne (1960) for the Herpes virus. Their illustrations (Fig. C–8) show that the capsomeres consist of short tubes constructed from rods lying parallel to one another, and that according to their position on the surface of the icosahedron they are formed from five (having fivefold rotation symmetry) or six (having twofold rotation symmetry) such sub-units. These conditions are illustrated schematically in the model shown in Fig. C–7b.

The Herpes virus has 162 capsomeres (Fig. C–7b: $n = 5$). If n is increased, the spherical form of the particles modulates more and more into definite polyhedra (Fig. C–7c), but the icosahedral form cannot immediately be recognised under the electron microscope because such particles exhibit a hexagonal contour. There are in fact three different cubic polyhedra that appear hexagonal when resting on any given facet: the octahedron, the rhombododecahedron, and the icosahedron.

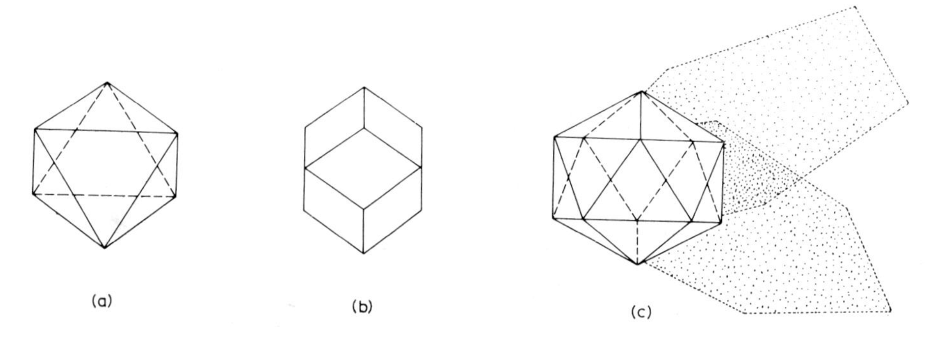

Fig. C–9. Cubic polyhedra with a hexagonal contour; (a) octahedron, (b) dodecahedron, (c) icosahedron, shadowed in different directions.

The identity of the polyhedral form can, however, be determined by strong contrast shadowing (Kaesberg, 1956) or, preferably, by suitable double shadowing (Fig. C–9c). Smith and Williams (1958) were successful in unequivocally determining the icosahedral form of the Tipula virus (from the larvae of *T. paludosa*) in this way. The facets of the particle appear to be almost smooth, for in this case the number of capsomeres per icosahedral edge is 10.

While, with the research methods available, we have built up a considerable volume of information concerning the morphology of the protein components of globular viruses, very little indeed is as yet known of the arrangement of the nucleic acid chain. The fact that the globular viruses are hollow did not fit with the original view that they consisted of an RNA core surrounded by a shell of protein. Apparently the nucleic acid chain with its code runs, as in TMV, at a certain distance from the particle centre, only its arrangement is not in this case helical, but stands in some relationship to the (532) symmetry of the capsomere arrangement. Viruses such as the bacteriophage φX-174, which contain DNA instead of RNA, are of particular interest. In this bacteriophage, the DNA occurs as a single-stranded helix and not as a double one. Although this behaviour imitates the state of RNA in viruses, we have to face a serious ambiguity. Thus if DNA represents the stabilised hereditary substance and RNA the versatile vehicle of protein synthesis, then it is not clear why the DNA viruses should be able to synthesise protein from lysed host cells just as RNA viruses. It therefore appears that the physiological significance of the two nucleic acids cannot be so strictly interpreted as would seem to be the case on p. 113.

(*c*) *Phages*

Bacteriophages are the viruses which attack bacteria. From their appearance as spherical particles under the electron microscope they were at first regarded as nucleoprotein molecules, but it was soon discovered that many of them possessed tailed forms. These appendages were stumpy and rigid in the viruses T2, T4 and T6 of the coli bacteria (*Escherichia coli*), or long, thin, and pliable in T1 and T5 (Wyckoff, 1948), and were found to occur in many other cases, such as in *Phagus lacticola* of *Mycobacterium battaglini* (Penso, 1955).

On the occasion of infection, these phages attach themselves with their tails to the surface of the bacterium and empty their contents into it. The protein of the host bacterium is then modified into virus protein, the quantity of nucleic acid is increased, and the cell then becomes filled with new, tailed, phage particles. Finally, the cell membrane dissolves (undergoes lysis) and the infectious virus particles are released into the surrounding nutrient solution (Herčik, 1955). This cycle is shown diagrammatically in Fig. C–10.

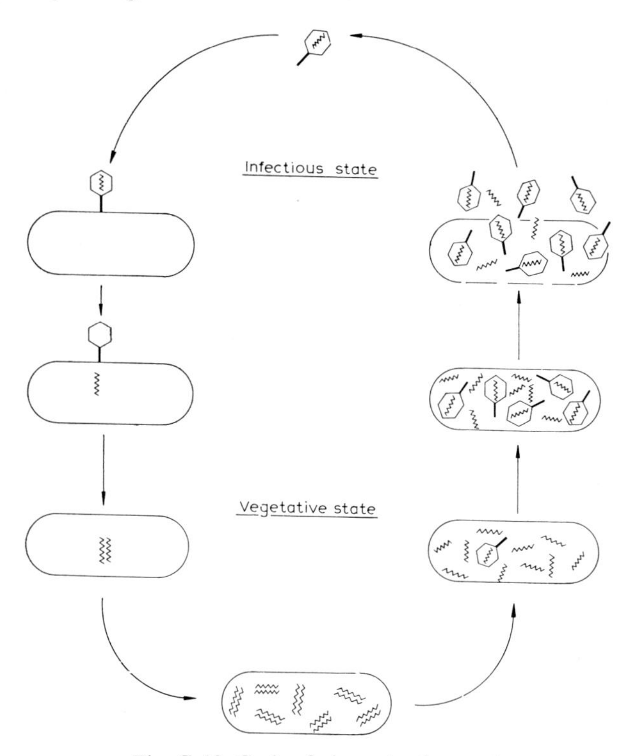

Fig. C–10. Cycle of phage development.

It is now known that only the nucleic acid is injected into the bacterium, whilst the protein layer of the phage head remains outside the bacterial cell as a ghost (Hershey and Chase, 1952; Kellenberger, 1957). The virus protein therefore plays no part in virus reproduction, although it is essential for infection, and reacts specifically with the cell membrane in an antigenic manner. The bacterial membrane can be broken down into globular protein macromolecules visible under the electron microscope. These macromolecules unite in vitro with the tail ends of the phages by adsorption, and the phage then loses its infectiousness (Weidel and Kellenberger, 1955). Thus the specificity of the virus infection is clearly explained by the observation that phages are adsorbed by specific membrane proteins.

Figure C–11 shows two particles of the coli-phage T5, one of which has made contact with a receptor particle: the DNA in the head has emerged by a trigger mechanism from the narrow gap between the tip of the tail and the receptor particle and appears as a single thread $34 \pm 1.7\,\mu$ long (!), equivalent to a molecular weight of 66×10^6 (Frank *et al.*, 1963).

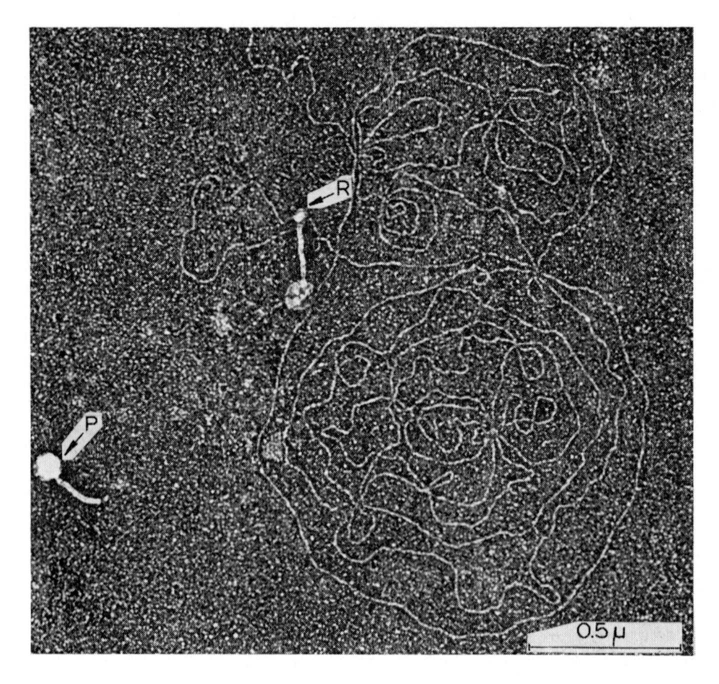

Fig. C–11. T5–phage tail attached to receptor particle *R*, whereupon DNA ejection is triggered. *P* = phagus particle without *R*. (Courtesy of W. Weidel.)

Development of the technique of negative contrast with phosphotungstic acid (PTA) in the preparation of material for examination under the electron microscope has led to the discovery of many additional details of bacteriophage structure. The particles are so extensively articulated that it is no longer possible to consider them as macromolecules, but rather as groups of different types of molecules in their own right (sub-units). Phage weight, as determined in the ultracentrifuge, is thus referred to as particle weight instead of molecular weight.

We shall here discuss, as an example, the ultrastructure of the coli-phage T2, which has a particle weight of about 200 million (Brenner *et al.*, 1959). The head and tail may be separated, and are composed of different proteins for they show immunologically different antigenic reactions in suspension. The head is hollow, has a wall thickness of 35 Å, and contains the nucleic acid (Fig. C–13). The tail has a peculiarly complex structure. The hollow cylindrical core, 800 Å long and 70 Å in diameter, has a central canal 25 Å wide (Fig. C–12b); it is surrounded by a contractile helical sheath, with the helical thread enveloping the core in 25 turns with a pitch of 40 Å. The helix is a beaded chain of some 200 sub-units, about eight occurring in each turn. In cross-section, the sheath has the appearance of a cog wheel. It is 165 Å in diameter and the internal diameter of the tube formed by it is equal to the core diameter of 70 Å. The sheath contracts under the action of such chemicals as hydrogen peroxide, and then envelops the core as a short sleeve. The core is terminated by a hexagonal plate with six tail fibrils (Fig. C–12a).

The proteins of the head, sheath, and tail fibrils have been isolated and characterised chemically. The wall of the head consists of many sub-units with a

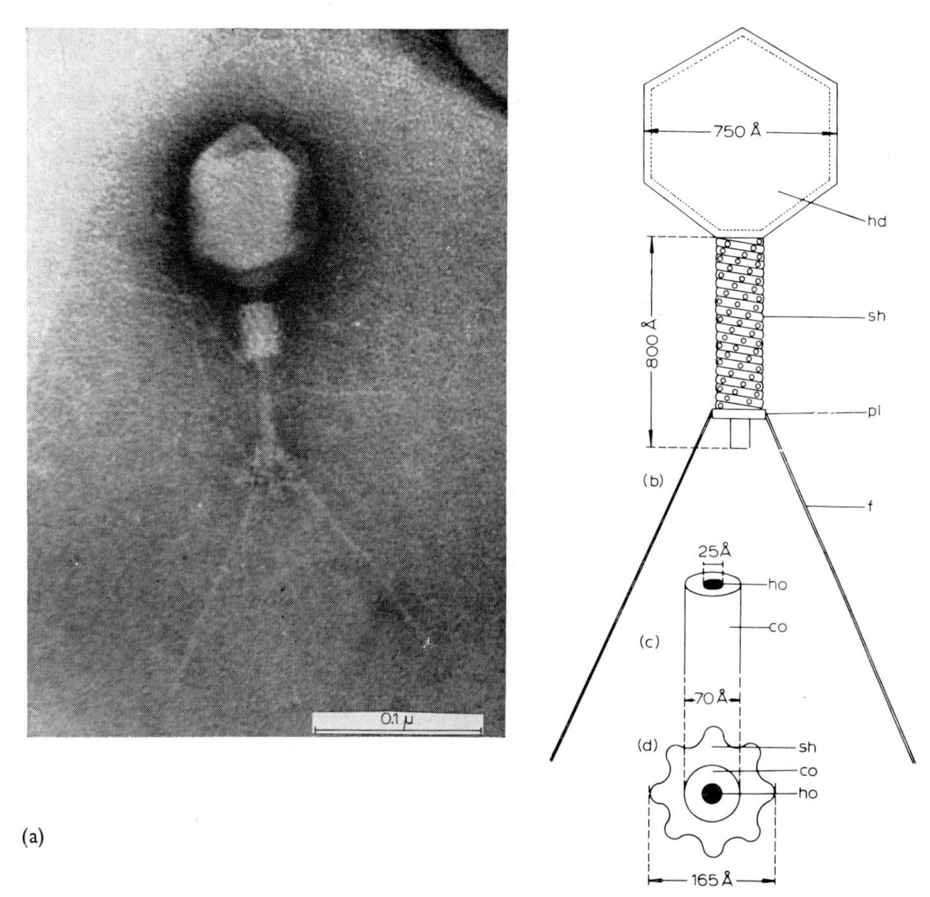

Fig. C–12. Coli-phage T2. (a) Electron micrograph (Brenner and Horne, 1959); (b) diagram: *hd* = head (750 Å) with wall (35 Å), *sh* = sheath of tail (helix with 25 turns of 40 Å spacing), *pl* = hexagonal plate, *f* = tail fibres (1300 × 20 Å); (c) core of tail (70 × 800 Å) with central hole (25 Å); (d) section across the tail, with *sh* = sheath, 200 sub-units (8 per turn), *co* = core and *ho* = hole.

molecular weight of 80,000. These should have a diameter of about 60 Å, so that the empty envelopes found, with a regular wall thickness of 35 Å, are apparently shrunken. The sub-units of the sheath have a molecular weight of 20,000 and therefore have a diameter equal to the pitch of the sheath helix. It has not as yet been found possible to isolate the protein of the core. The tail fibrils are composed of a fibrillar protein with a molecular weight of more than 100,000, which corresponds to the mass of whole fibrils measuring 20 × 1300 Å. The three proteins give different oligopeptide spectra when broken down by trypsin, and are thus chemicallydifferent.

Morphology of the nucleic acids of phages is less clearly understood, but they probably consist of single-thread molecules and not of double strands. It is of considerable importance that only DNA or only RNA with multifunctional capacities is involved; thus in the phages DNA not only transfers the information but also apparently participates in protein synthesis, a function which is usually attributed to RNA. No stoichiometric relationship, as in TMV (three nucleotides/

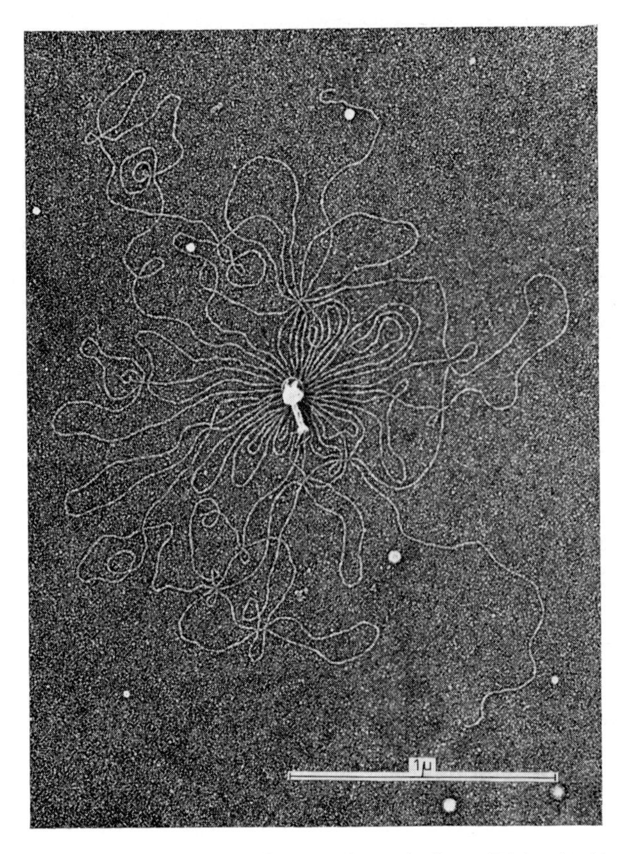

Fig. C–13. Thread of DNA discharged from T2-bacteriophage (Kleinschmidt *et al.*, 1962).

protein sub-unit), has been determined. The DNA chain must in some manner be contained within the head of the phage, for if the latter's wall is broken, a mass of threads is ejected (Weidel, 1957; Kleinschmidt *et al.*, 1962, Fig. C–13). The mechanism by which these threads are discharged through the narrow canal of the core (25 Å) is unknown.

The nucleic acid content of bacteriophages is so high that it can be disclosed under the polarisation microscope. Whilst the flow birefringence of TMV is based solely on form birefringence (Lauffer, 1938), strains of T2 show a high negative intrinsic birefringence caused by the nucleic acid chains coiled parallel to the axial direction of the head (Bendet *et al.*, 1960).

As a result of the outstanding discoveries on the life cycle of the bacteriophages, nucleic acid alone (and not the nucleoproteins) has been shown to be the gene carrier. The protein shell of the viruses is thus probably simply a protective wall guarding the sensitive nucleic acid chains against external influences; on the other hand, the high specificity of the virus proteins would not be needed for such a passive role. Importance should therefore be placed on the finding that this very specificity plays a part in the infectiousness of the bacteriophages, although such a function is only apparent where the cells are attacked by specific adsorption of the infective particles on their surfaces, as in the case of the phages and of animal and human viruses.

Plant viruses penetrate passively into damaged cells, as for instance broken plant hairs, or via the proboscis of sucking insects, and the role of the protein coating in this case is not understood. Another point which requires further elucidation is the fact that bacteriophages synthesise very different proteins, only one of which (that of the final plate of the tail core) is a likely candidate for specific adsorption on the surface of the coli bacteria. We must therefore look forward with keen anticipation to the results of further research on the viruses, which are no living organisms in so far as they must 'borrow' the energy for their metabolism and their replication from living cells.

PART II

Cytological Morphology

D. Organelles of the Plant Cell

The successful elucidation of the structure, nature, and function of the chromosomes with light microscopy has resulted formerly in karyology becoming undoubtedly the most important branch of cytology. On the other hand, the light microscope has failed us with regard to other components of the cell: for the cytoplasm appeared to be homogeneous and 'optically void' (Guilliermond *et al.*, 1933, p. 386), and it was not possible to distinguish the developing plastids from mitochondria so that their ontogeny was obscure. The question of whether the so-called Golgi apparatus was a reality or an artefact remained unresolved until very recent times (Baker, 1957).

The situation has been changed dramatically by the use of the electron microscope. Its application, in conjunction with such methods as heavy metal fixation (osmic acid, potassium permanganate) and the preparation of ultrathin sections down to a thickness of 200 Å (0.02 μ), has made it possible to resolve new structures in the cell. A whole new world has been discovered in the cytoplasm shown previously as homogeneous under the light microscope (Fig. D–1).

Unfortunately, we still await elucidation of the structure of the cell nucleus, for it appears unexpectedly homogeneous even in ultrathin sections, and the threads or fibril structures of the chromosomes cannot be observed under the electron microscope. Work on the development of the theory of nucleus structure has therefore been involuntarily neglected, while the cytoplasm, previously unapproachable by direct methods of investigation, now freely yields its secrets. What the immersion objective has done for the rapid advance of knowledge of the cell nucleus since 1878, is now being achieved by the electron microscope in the field of cytoplasmic research. Cytostructural science is therefore not developing uniformly in all directions, but proceeds irregularly in different sectors depending on new instruments and methods of research evolved.

The great advance in cytology is best appreciated by comparing the cell components resolvable under a light microscope with those shown under the electron microscope. Fig. D–2a shows a meristematic plant cell with its conventional features. In addition to the nucleus and the cell wall, mitochondria and proplastids can be disclosed by the use of suitable staining techniques; these two types of cell inclusions were indistinguishable under the light microscope, although Guilliermond (1922) had already recognised that during cell differentiation some of the particles developed into chloroplasts (as shown in Fig. D–2a′) while others remained small and became typical mitochondria. Simultaneously with the development of these cell constituents, usually an elongation of the meristematic cells occurs, with the formation of a giant cell vacuole which presses the groundplasm (whith its inclusions) into a thin layer of plasma against the cell wall.

Specific functions can be attributed to such components of the cell, and these are then known as cell organelles. Classical plant cytology distinguished the cytoplasm, cell nucleus, plastids, mitochondria, and the cell wall, but the knowledge of

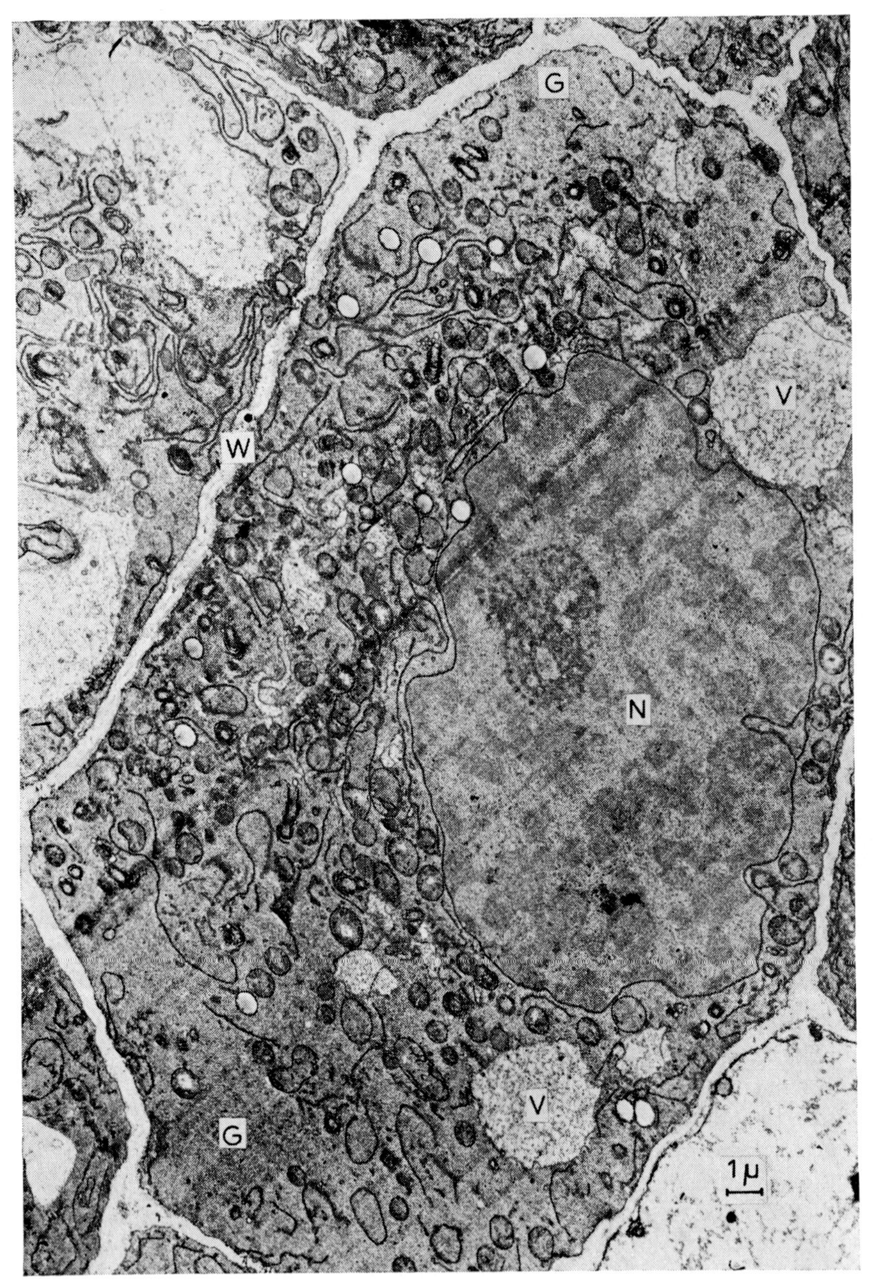

Fig. D–1. Meristematic cells from the growing tip of the onion root (see Fig. D–2b), fixed with potassium permanganate. N = nucleus; V = developing vacuole; G = groundplasm with inclusions; W = cell wall.

their significance was rather uncomplete. For example, so little was known about the plant cell mitochondria until about the middle of the present century that there

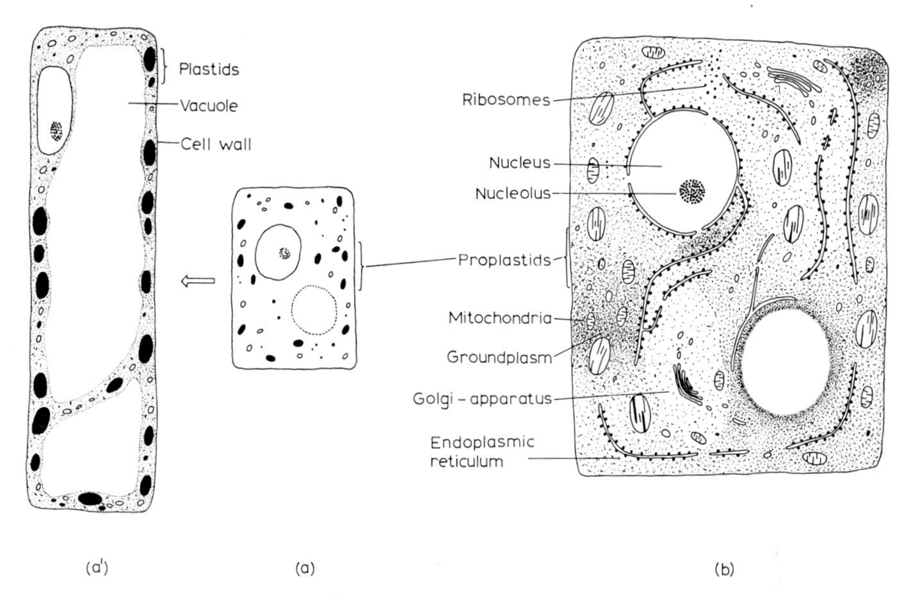

Fig. D–2. Cytological inventory of the plant cell. (a) Meristematic cell; (a′) permanent cell arising from elongation and differentiation (proplastids and chloroplasts shown in black); (b) the same cell as in (a) as seen under the electron microscope (see Fig. D–1).

was no special chapter devoted to them in Küster's standard work, "Die Pflanzenzelle", published in 1951.

Electron microscopy has now revealed organelles in the cytoplasm, in addition to the mitochondria and plastids, which had previously been unknown or overlooked (Fig. D–2b). The most important discovery has been that of the so-called endoplasmic reticulum (designated by the generally accepted abbreviation ER) (Porter, 1948; Palade, 1956), consisting of a system of ultrastructural canals, cisternae and concavities; another was the general resolution of the Golgi apparatus into stacks of ultrastructural double lamellae which are as a rule invisible under the light microscope. A further surprise has been the general detection of ribosomes, *i.e.* ribonucleic acid protein bodies about 150 Å in diameter, which have been found in all cells so far examined.

From the above observations it is clear that the use and definition of the term 'cytoplasm' has now become somewhat problematic. If all organelles enumerated in the previous paragraphs were withdrawn from the cell, we would be left with a finely-grained basic mass, which is best referred to as the groundplasm or matrix. The use of the word 'cytoplasm' for this matrix would appear to be unjustified, for the term would become increasingly restricted, and finally evanescent, as in the future the nature of yet finer grains within it will be established and defined. It is therefore preferable to retain the term 'cytoplasm' in the classical sense, but excluding from it the mitochondria (chondriome) and plastid (plastidome) systems. Such a classification appears to be justified, because plastids and mitochondria are self-contained, individually organised particles with a double membrane, whereas the other organelles are structures of variable forms merged in the groundplasm.

E. Cytoplasm

1. GROUNDPLASM (MATRIX)

The term groundplasm will here denote the homogeneous plasma remaining after all organelles and particles visible with the electron microscope have been excluded. Although such a definition is not based on existing but on lacking properties, so that its meaning is negative rather than positive, it is useful in practice, because a ground mass of this type is demonstrable in all cells. With perfect fixation, the groundplasm appears to have a very fine granular structure (Fig. E–1), the diameters of the grains varying from over 100 Å down to the limits of resolution of the electron microscope. The grains are uniform for a given object, but vary in size according to the method of fixation used, so that exact measurements are not of particular value. Even with identical fixation, they also vary in different protoplasts. It is probable that globular protein macromolecules are involved (see Table B–IX, p. 74), their diameters varying according to the method of precipitation and dehydration used in their preparation. In this connection, it would be interesting to know whether the Fraction-I protein, with globular particles 100–200 Å in diameter, isolated by Wildman and Cohen (1955) from tobacco leaves, corresponds to the groundplasm.

Classical cytology distinguishes between peripheral homogeneous ectoplasm,

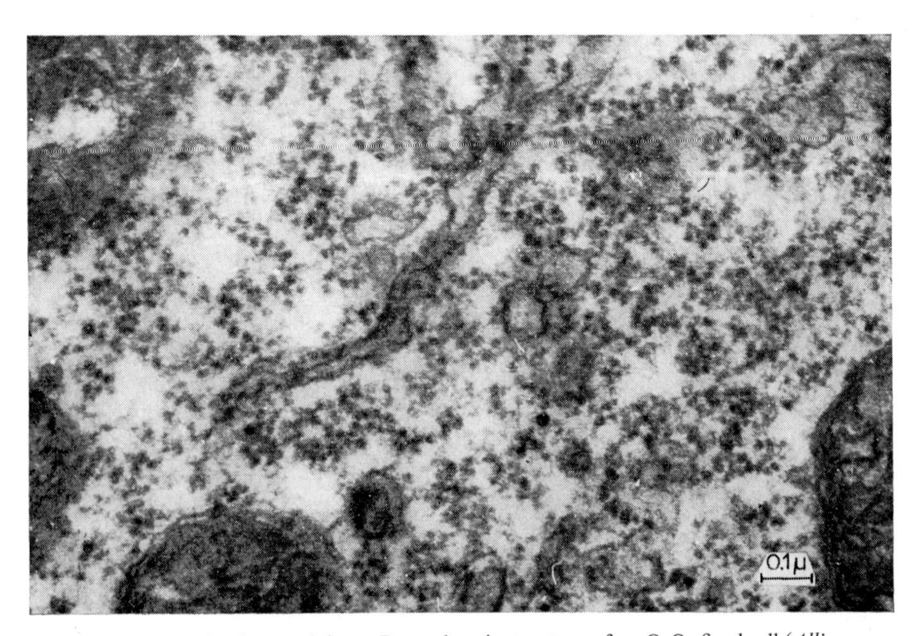

Fig. E–1. An example of groundplasm. Protoplasmic structure of an OsO_4-fixed cell (*Allium cepa*, root). The dark particles represent ribosomes.

or hyaloplasm (a term used by Pfeffer in 1890), and endoplasm or granular plasma within the cell (Pringsheim, 1854; Hofmeister, 1867). The groundplasm now under discussion must appear as 'hyaloplasm' under a light microscope. It is known to be capable of reversible sol $\rightleftharpoons$ gel transformation and that, in the liquid state, it is a 'gel solution' disobeying Newton's law by exhibiting elasticity and Poiseuille's flow equation by the dependence of its viscosity on pressure (see Frey-Wyssling, 1953, pp. 163 ff.). It is particularly interesting that the cytoplasmic gel does not solidify under high hydrostatic pressures (e.g. 300 atmospheres), but becomes reversibly liquid (see p. 28). As a result of this liquefaction, which may be observed in a high-pressure chamber with a light microscope, the flow of plasma ceases temporarily, until atmospheric pressure – and therefore the condition for resetting to a gel – is re-established (Marsland, 1942).

Although all such physico-chemical measurements have been carried out on cytoplasm containing all its organelles, both visible and invisible under the light microscope, and not on pure groundplasm, the special properties of flow elasticity and volume increase on setting to a gel must be attributed to the granular structure of the cytoplasmic matrix. The grains would appear to have the ability to form a gel structure or a liquid sol by the formation or destruction of junctions. Filamentous elements which facilitate gelation at low concentrations do not seem to be primarily present in the plasma sol, and if they do appear in electron micrographs, they must have arisen secondarily by the formation of beaded chains (see p. 29) during gelling. The helical filaments (cytonemata) postulated by Strugger (1956/57) could likewise not be identified by other authors in the groundplasm.

Although the various theories of classical cytology concerning ultrastructure of the cytoplasm are today invalid, the plasma system being far too heterogeneous for conclusive findings to be possible at the time, it is still of interest to see which structures have played a part in the experiments performed during that period. Berthold's emulsion theory (1886), of droplets embedded in a semi-liquid phase, was probably based on the observation of mitochondria, proplastids, and other particles, with dimensions at the limit of resolution of the light microscope, the nature of all of which was at that time unknown. Bütschli's much vaunted foam theory (1892) must today be dismissed as an artefact due to fixation.

Flemming's fibrillar theory (1882), according to which filamentous elements occur in the cytoplasm, was based on irreproachable observations made *in vivo*. The fact that heavy particles, wandering through the plasma under the influence of the gravitational field, are constantly obstructed by invisible threads and consequently show deviations from Stokes' law of velocity in liquids, is a major consideration. According to Scarth (1927), these particles behave as a handful of lead shot falling through a 'brush heap'. The falling particles repeatedly encounter (invisible) strands, lose speed, and change their direction. It is hence clear that the cytoplasm cannot be homogeneous, but rather must be full of invisible structures possessing a higher density. It is also not uniformly viscous, and the results derived from the fall method (Heilbronn, 1914) represent merely some kind of an average. It is likely that the strands of the endoplasmic reticulum (see p. 161) are in fact responsible for these irregularities, so that observations leading to the theory of fibrillar cytoplasmic ultrastructure are due to the embedded organelles rather than to the groundplasm itself.

At first sight, the grain theory due to Altmann (1894), according to which the cytoplasm has a granular structure of individual globular bodies, seems to fit best with Fig. E–1, especially since Heidenhain (1907) declared these elements to be too small to be visible under the light microscope. It should however be realised that these authors attributed physiological autonomy, autoreproduction, and other vital properties to their hypothetical granular bodies in the sense of a pangene theory, so that the coincidence of the postulate of a granular cytoplasm with the globular macromolecules detected with the electron microscope is merely fortuitous.

The functions of the groundplasm must be manifold – it has enzymatic faculties, it works under the control of the genes, and it must incorporate the enigmatic trends of morphogenesis. Experimentation on these lines is however, as yet impossible, since "pure" groundplasm separated from the embedded organelles is still unsuited for a biochemical approach because of its complexity.

2. PLASMALEMMA

All protoplasts are separated from their environment by a membrane possessing special characteristics, called the cell membrane (Nägeli and Cramer, 1855) or *plasmalemma*. The term plasmalemma was coined by Plowe in 1931, and is preferred in botanical cytology to the older term cell membrane because of the danger of confusion with the cell wall. In coated plant cells, the plasmalemma provides the contact between the inner plasma and the cell wall, while in naked cells it forms the barrier between the protoplasm and the culture medium. It was established long ago (Pfeffer, 1890) that the delimiting layer could not be differentiated under a light microscope from the inner plasma, and that it could scarcely be identified with the entire layer of the hyaline ectoplasm, which in the plasmodium of Myxomycetes can reach a thickness of 8 μ. It seemed more probable that an extremely thin film is involved, which is not perceptible with the light microscope, so that information regarding the nature of the plasmalemma was obtainable only by indirect chemical and physico-chemical techniques before the advent of electron microscopy.

(*a*) *Physico-chemical behaviour*

Overton's discovery in 1899 that, under otherwise identical conditions, lipophilic molecules penetrate cells more rapidly than do hydrophilic ones, led to the lipid theory of permeability, according to which the plasmalemma must consist of lipid substances and the uptake of the material is governed by the partition coefficient of the substance in question between the lipophilic phase and the hydrophilic culture medium (Collander, 1937). The assumption of the existence of a lipophilic boundary membrane was reinforced by physico-chemical evidence that polar molecules having a lipophilic pole accumulate in the surface of an aqueous phase (see Fig. B–48).

According to those views it was thought that some protoplasts are so deficient in lipids that there is only enough of the latter to cover the cells with two to four molecular layers (Danielli, 1936; Törnävä, 1939); all cellular lipids such as fatty acids, cholesterol, and phospholipid acids would then be accumulated at the cell surface.

Let us now consider the membrane of the erythrocyte, as an example of the way in which the molecular structure of a membrane used to be envisaged (Winkler and Bungenberg de Jong, 1940/41). By exact measurements of the migration velocities of red blood corpuscles in various electrolytes under the influence of an electric field, the above authors found that the cells behaved quantitatively in the same manner as phospholipid droplets. It was hence concluded that the surface of the erythrocyte is covered with a phospholipid film (layer I in Fig. E–2) which is stabilised by cholesterol. Since the isoelectric point of the stroma of haemolised cells (pH 5.2) is between those of the phospholipids (2.7) and of the peripheral protein stromatin (5.8), it was assumed that the phospholipids formed a complex with the stromatin (layer IV), with their positive choline groups linking to the anionic end groups of the protein (layer III). Haemolysis experiments have further shown that calcium ions consolidate and stabilise the erythrocyte membrane. The calcium ions in layer II, with their strong positive charge, are located between the negative phosphoric acid groups of the lecithin, thus giving more powerful ionic cohesion. So the membrane is regarded as a complex system of phospholipid-calcium ions and stromatin, whose definite arrangement and orientation are due to the regularly distributed charge pattern. The tricomplex system is completed by the assumed presence of a linkage of the haemoglobin (layer VI) with anionic end groups in layer V to cationic groups of the stromatin.

The scheme of Fig. E–2 is further complicated by layer *A*, which represents an incomplete film of polar lipids (fatty acids and possibly cholesterol) with their lipophilic ends turned towards the monomolecular phospholipid layer I and their hydrophilic ends facing outwards. The presence of this layer had to be assumed for

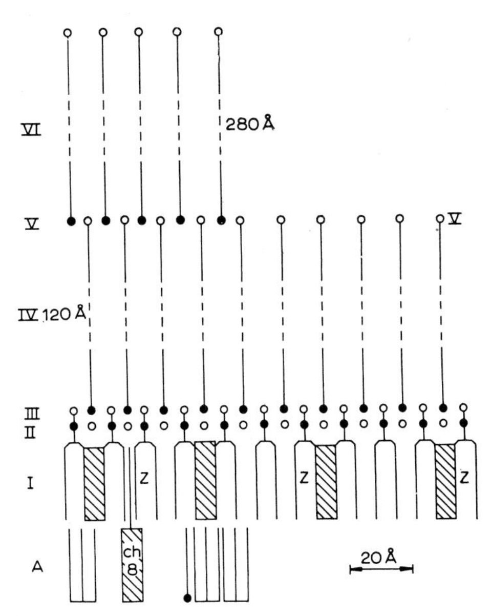

Fig. E–2. Molecular structure of the membrane of the red blood cell (Winkler and Bungenberg de Jong, 1940/41). ● = anionic groups; ○ = cationic groups or cations (Ca); Z = phospholipidic acid; ●——— = fatty acid; shaded = cholesterol; ch8 = cholesterol with a lipid tail.

without it the erythrocytes would agglutinate in aqueous solutions and pass into the lipid phase when shaken with paraffin oil, which is not the case.

Many of the properties of erythrocytes may be explained by the scheme proposed by Winkler and Bungenberg de Jong in 1941. The lipid filter theory of permeability is included, for there is a lipid film with molecular pores at the places where the cholesterol covering is fragmentary; the characteristic permeability of erythrocytes to anions is explained, as the Ca^{++} layer III blocks the cations, and the same layer, with its water of hydration, is responsible for the effects of hydrating and dehydrating ions upon the properties of the erythrocytes. In spite of such logical explanations, however, this scheme also presents a number of difficulties. Analysis of red cells does not reveal the presence of calcium (Ponder, 1961), and, according to Fricke (1926), the electrical properties of the erythrocyte wall are such that the presence of a non-conductive layer 33 Å thick must be assumed, corresponding to the lipidic part of the phospholipid layer I. On the basis of the lipid content of erythrocytes, Gorter and Grendel (1925) assumed the presence of a bimolecular lipid film, although the amount of lipids is somewhat too small for covering the whole cell surface with such a film.

From the stromatin and lipid contents of pig erythrocytes, Winkler and Bungenberg de Jong calculated that the oriented lipid molecules covered the surface of the cell in the manner indicated in Fig. E–2 with a layer of stromatin below them 120 Å thick.

Since it is unlikely that the stromatin molecules are in fact expanded rods as shown, and as the haemoglobin molecules of layer IV are known to be globular, only layers *A*, I, and II of Fig. E–2 remain to be considered. They consist of fats, fatty acids, cholesterol, and phospholipids, and form an incomplete double layer with, as stated above, a slightly hydrophilic outer surface.

The idea that the protoplast is coated by a lipidic film, which was held generally about 1930, was challenged by Danielli and Harvey (1935), who based their argument on surface tension phenomena. As the classical concept of a purely lipidic plasmalemma has been shaken by quantitative studies of the cellular surface tension, some comments concerning the theory and the methods involved will not be out of place.

The surface tension of a liquid owes its existence to the fact that whereas molecules within the bulk of the phase are surrounded on all sides by like molecules, at the phase boundary such similar molecules lie only on one side. Taking a liquid/gas interface as an example, we find that attraction forces due to the relatively few gas molecules present may be ignored for the time being, so that the molecules at the surface are subjected to a field of cohesive forces very different from that acting on molecules lying deeper within the liquid. As may be seen from Fig. E–3, the cohesive forces at the surface do not mutually cancel out, but give a resultant directed towards the liquid. The surface yields to this attraction as far as possible, and comes slightly closer to the deeper-lying molecules, resulting in an increase in density which is shown diagrammatically in Fig. E–3d. A surface 'skin' is thus formed, which merges on the inner face with the homogenous liquid.

This surface skin possesses a certain strength, because its molecules cannot move as freely as those in an ideal liquid. The degree of strength may be determined by stretching a film of the liquid (suspended in a frame) by means of a movable bar,

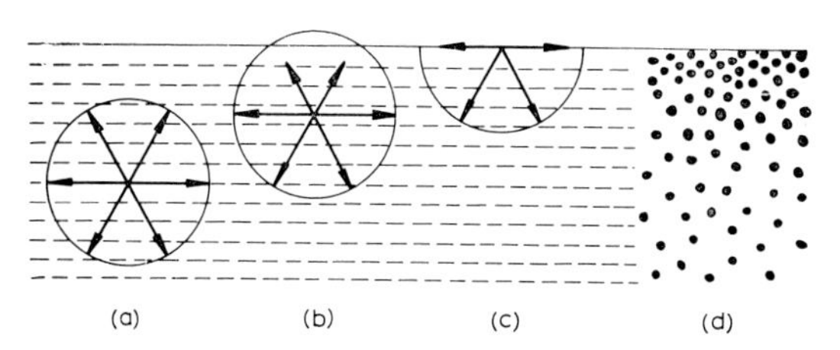

Fig. E–3. Inhomogeneity of the liquid/gas boundary. Cohesive forces are (a) symmetrical, (b) asymmetrical, and (c) directed inwards; (d) schematic representation of the inhomogeneous arrangement of molecules (greatly exaggerated, since the compressibility of liquids is very slight).

and measuring the weight required to break the film. This weight is independent of the film thickness, but is a linear function of the bar length l (e.g. a film twice as wide will sustain twice the weight) so that the strength of the surface is referred to a unit length. The minimum force required to break a lamellar surface 1 cm wide is called the surface tension γ of the liquid, and, as both the front and back surfaces of the film must be broken, this force, p, is given by $2\gamma l$ (Fig. E–4).

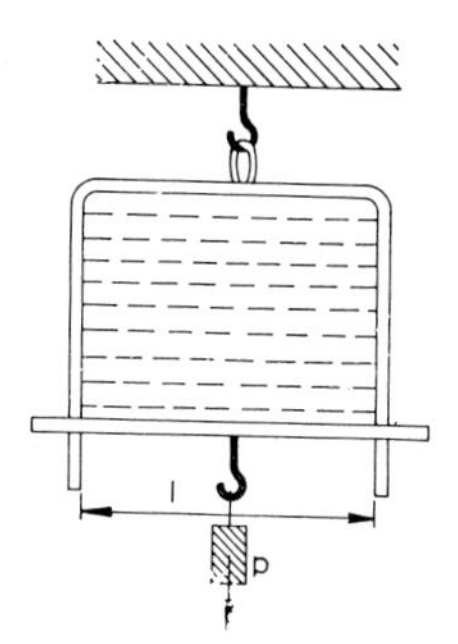

Fig. E–4. Measurement of the surface tension of a film.

This primitive method of measuring surface tension has been mentioned here, in preference to the more accurate methods based on capillary attraction (Höber, 1922, p. 154), because the definition of surface tension is founded on it, and because it demonstrates in a simple way why the dimension of surface tension is force/length. Surface tension is not, therefore, a tension in the usual sense of the term (force/length^2). The difference between these two quotients is shown in Fig. E–5; in order to rend a plane the cohesive forces have to be overcome along a line, whereas in the case of a rod they have to be overcome at a plane. In other words, Figs. E–5a and b are graphic representations of the definitions of surface tension (force/cm) and cohesive tension or pressure (force/cm^2), respectively.

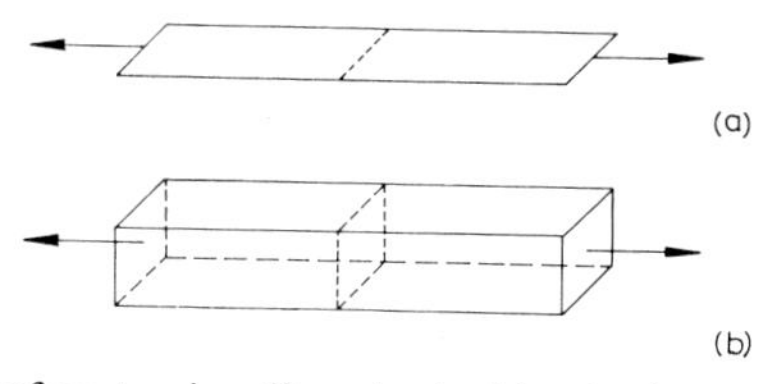

Fig. E–5. (a) Surface tension (force/cm); (b) cohesive tension (force/cm^2).

The product of surface tension and area has the dimension of $cm^2 \times force/cm = force \times cm$, *i.e.* the dimension of energy. The concept of surface energy rather than surface tension is therefore often used. If much work is required to increase the surface, as for instance in water or other liquids with many OH-groups in contact with air, the surface energy is large (see Table E–I).

TABLE E–I

SURFACE TENSION γ AGAINST AIR AT 15° C
(Höber, 1922, p. 167)

0.25-molar solutions	γ dynes/cm	Relative γ (γ water = 1)
Water	71.6	1.000
Cane sugar	72.1	1.007
Urea	71.6	1.000
Glycerol	71.5	0.999
Acetic acid	66.8	0.932
Ethyl alcohol	66.0	0.922
Ethyl ether (saturated)	53.1	0.742
Ethyl acetate	41.5	0.578
Iso-valeric acid	34.9	0.487
Iso-amyl alcohol	29.9	0.417

Surface tension is responsible for the formation of liquid droplets. When a drop falls from the orifice of a vertical capillary tube, its surface is rent, and the surface tension may then be calculated from its weight (stalagmometry). Unfortunately, the method cannot be used with viscous protoplasmic droplets. Naked cells with a more or less liquid cytoplasm assume a spherical shape when suspended in a culture medium, and it is believed that this effect is due simply to surface tension. In such a system the interface is formed between the cytoplasm and the surrounding liquid, and the surface tension is then considerably lowered.

Harvey (1936) devised methods by which the surface tension could be accurately measured under such conditions. When a semi-liquid cell rests on the bottom of a culture dish it may be regarded as a sessile drop flattened by gravity, and its surface tension γ is given by $\gamma = g(\varrho - \varrho')r^2F$, where g is the acceleration due to gravity, $(\varrho - \varrho')$ the difference in density between the drop and the medium, r the radius of the greatest flattening and F a function of the distance a in Fig. E–6a,

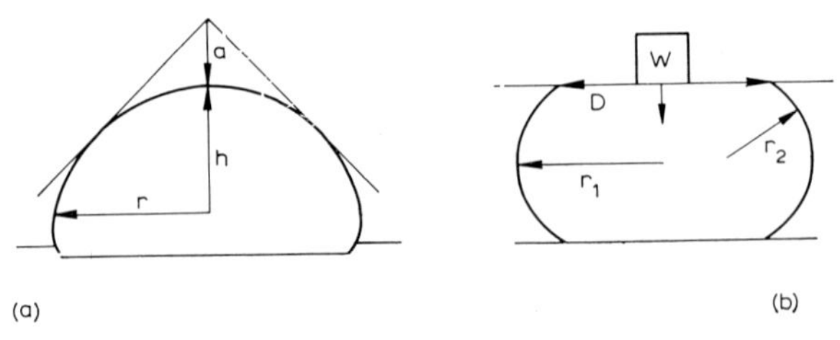

Fig. E–6. Measurement of surface tension; (a) sessile drop, (b) flattened drop (E. N. Harvey, 1936, 1937).

representing the flattening of the drop. Using this method, surface tensions of 0.5 and 0.1 dynes/cm were found for the eggs of a mollusc (*Busycon canaliculatum*) and a salamander (*Triturus viridescens*), respectively (see Table E–11).

TABLE E–II

(*cf.* Table B–XII)

SURFACE TENSION γ OF PROTOPLASM IN CULTURE SOLUTIONS

(after E. N. Harvey, 1937)

Naked protoplasts	γ dynes/cm	Medium
Leucocytes (*Lepus caniculus*)	2.0	Ringer's solution + serum
Leucocytes (*Rana pipiens*)	1.3	Ringer's solution + serum
Amoeba (*Amoeba dubia*)	1–3	Ringer's solution, diluted
Slime mould (*Physarum polycephalum*)	0.45	Ringer's solution, 250 × diluted
Sea-urchin egg (*Arbacia punctulata*)	0.2	sea-water
Salamander egg (*Triturus viridescens*)	0.1	pond-water + gum arabic

Mackerel eggs contain a large oil droplet that can be flattened against the rigid cell membrane when the egg is revolved at high speed in the centrifuge microscope. An oil/cytoplasm interfacial tension of 0.8 dynes/cm has been calculated from its shape (Table B–XII), and the fact that this tension remains unchanged when the centrifugal force is increased to 450 gravities shows that the surface is not elastic. This oil gives a tension of 7 dynes/cm in contact with sea-water, and the high value is explained by the rule that the interfacial tension between two immiscible liquids is equal to the difference between the individual surface tension of the two liquids against air. As the surface tension of water is 72 dynes/cm and that of oils is only about half as much, it is evident that the cell surface cannot be composed of pure lipids, for this would result in a higher interfacial tension between the surface of the cell and its culture medium. A surface with only 0.1 dynes/cm tension against the medium cannot consist of lipids alone (Table E–II); rather it must contain proteins having a certain affinity towards water.

If the cell under examination cannot be flattened by normal or increased gravity, then the flattening can be achieved by compressing it with a small piece of gold-leaf loaded with microweights (Harvey, 1936). The weight W divided by the area D of the flattened cell in contact with the beam gives the pressure, P from which the surface tension γ is derived by the formula

$$\gamma = \frac{P\, r_1 r_2}{(r_1 + r_2)}$$

where r_1 and r_2 are the two radii of the flattened cell as indicated in Fig. E–6b.

The unfertilised egg of a sea-urchin, *Arbacia punctulata*, gives a surface tension reading of 0.135 dynes/cm under a load of 2 μg, and lower values with lesser weights; extrapolation of the tension/compression curve gives 0.08 dynes/cm for the non-compressed egg. Because the surface tension is not constant but varies with the applied pressure, the surface displays elasticity, and this again is evidence for the presence of proteins in the cytoplasm at the surface, for a lipidic layer would be inelastic. Sols also have no elastic properties, and it is therefore evident that the proteins of the surface layer are in a gel-like state.

Hydrophilic proteins completely soluble in water are unable to form surface films, whereas polar proteins accumulating at the surface of an aqueous phase lower the surface tension. Depending on its nature, therefore, a proteinic interface can give rise to surface tension ranging from zero to several dynes/cm. For this reason, Danielli and Harvey (1935) concluded that the lipid film of the plasmalemma must be covered on the outside by a film of protein molecules (Fig. E–7). A complex film of this type would however be unstable, as in water the low surface tension on the protein side is opposed by a much higher tension from the lipid surface, and the films would buckle and disintegrate into droplets.

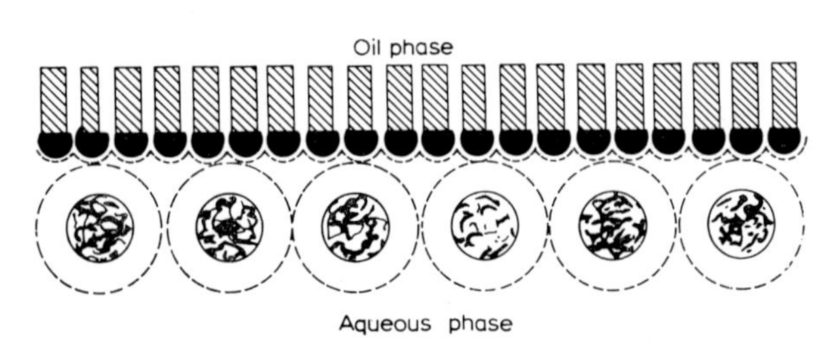

Fig. E–7. Scheme of the molecular conditions at an interface of oil/aqueous egg contents. Hydrated protein molecules below, lipid molecules above (Danielli and Harvey, 1935).

It was hence concluded that a stable film must be constructed symmetrically, and a bimolecular lipid membrane was assumed for the structure of the plasmalemma, covered on both surfaces by proteinic films (see Davson and Danielli, 1943, p. 64). Such a scheme was also adopted by Robertson (1960) for the molecular structure of his unit membrane (see p. 93). Very few biophysicists are aware that the scheme of the double lipid film covered with protein on both sides has been deduced indirectly from surface tension measurements on the plasmalemma rather than from direct observations, and it is indeed regrettable that comparative measurements against water have been made not on lipoprotein or phospholipid droplets but only on droplets of oil or bromobenzene (Danielli, 1936, 1938). As the polarity of fat molecules and halogenated benzenes is very slight as compared with that of most plasma constituents, it is possible to differ about the far-reaching generalisations of this postulate, from which a uniform construction for all unit membranes has been deduced.

(*b*) *Ultrastructure*

As the plasmalemma is invisible under the light microscope, it was for a lon time a matter for conjecture how deeply it extended into the cell. We can now stat that in a wide variety of objects the plasmalemma appears as a unit membran ca. 100 Å thick. It seems to be smooth in cross-section (Fig. E–8), but exhibits a granular structure when studied in surface view with the freeze-etching techniqu (Moor *et al.*, 1961). This granular structure is normally invisible, because th section thickness of 300–500 Å is several times greater than the diameter of indi vidual granules. When fixed with osmic acid, the external contour is more weakl

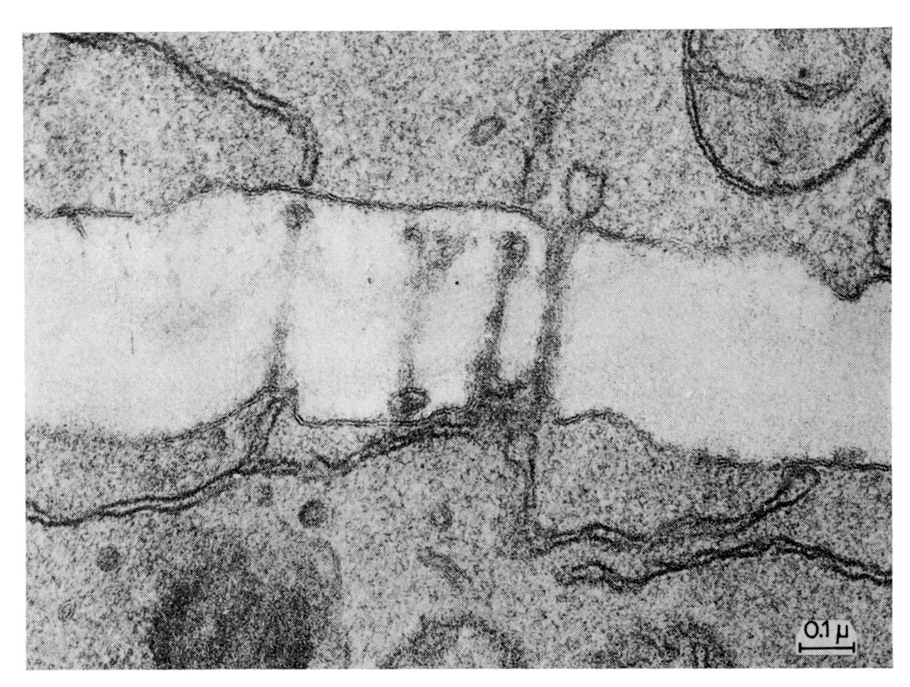

Fig. E–8. Cell wall exhibiting plasmodesmata and the plasmalemma (*Allium cepa*, root).

contrasted than the internal one (Finean, 1961, p. 65), indicating that the plasmalemma has a polar structure. According to Latta (1962), the external face of the plasmalemma of the glomerular epithelium of the kidney stains more strongly with the anionic electron stain phosphotungstic acid, whereas the cationic lead acetate does so on the internal face.

If Robertson's (1960) view, according to which all unit membranes have the same molecular structure, is accepted, then the plasmalemma would have to possess the structure shown in Fig. E–10b. Such a scheme corresponds to that proposed by Davson and Danielli (Fig. E–10a), although its perfect symmetry is opposed to the polarity of the plasmalemma, which is apparent in physiological processes (such as resorption and secretion) as well as by fixation with osmic acid.

In view of the fine granulation shown by the plasmalemma, we can envisage a structure such as that depicted in Fig. E–11a, which will be termed a block structure. This consists of a bimolecular layer of globular lipoprotein macromolecules 40–50 Å in diameter. The lipid molecules must be imagined as anchored in the loosely-constructed protein macromolecules with their helically wound polypeptide chains, and in the interstices between the globular macromolecules. The necessary polarity of the plasma membrane is given by the intramolecular polar construction of the globular macromolecules. Such a structure has been suggested not only by permeability studies, but also from observations made under the electron microscope with the freeze-etching technique. Unit membranes of the plasmalemma and the nuclear membrane, broken in the frozen state and then etched, show a dip, when shadowed, in the position of the central layer. This dip is best visualised as a trough between the structural units of a broken bimolecular layer of globular macromolecules, whose density is sharply reduced towards the middle (Fig. E–11b). In yeast, the plasmalemma shows folds of uniform length (Fig. E–9) which increase the surface by about 50% (Moor and Mühlethaler, 1963).

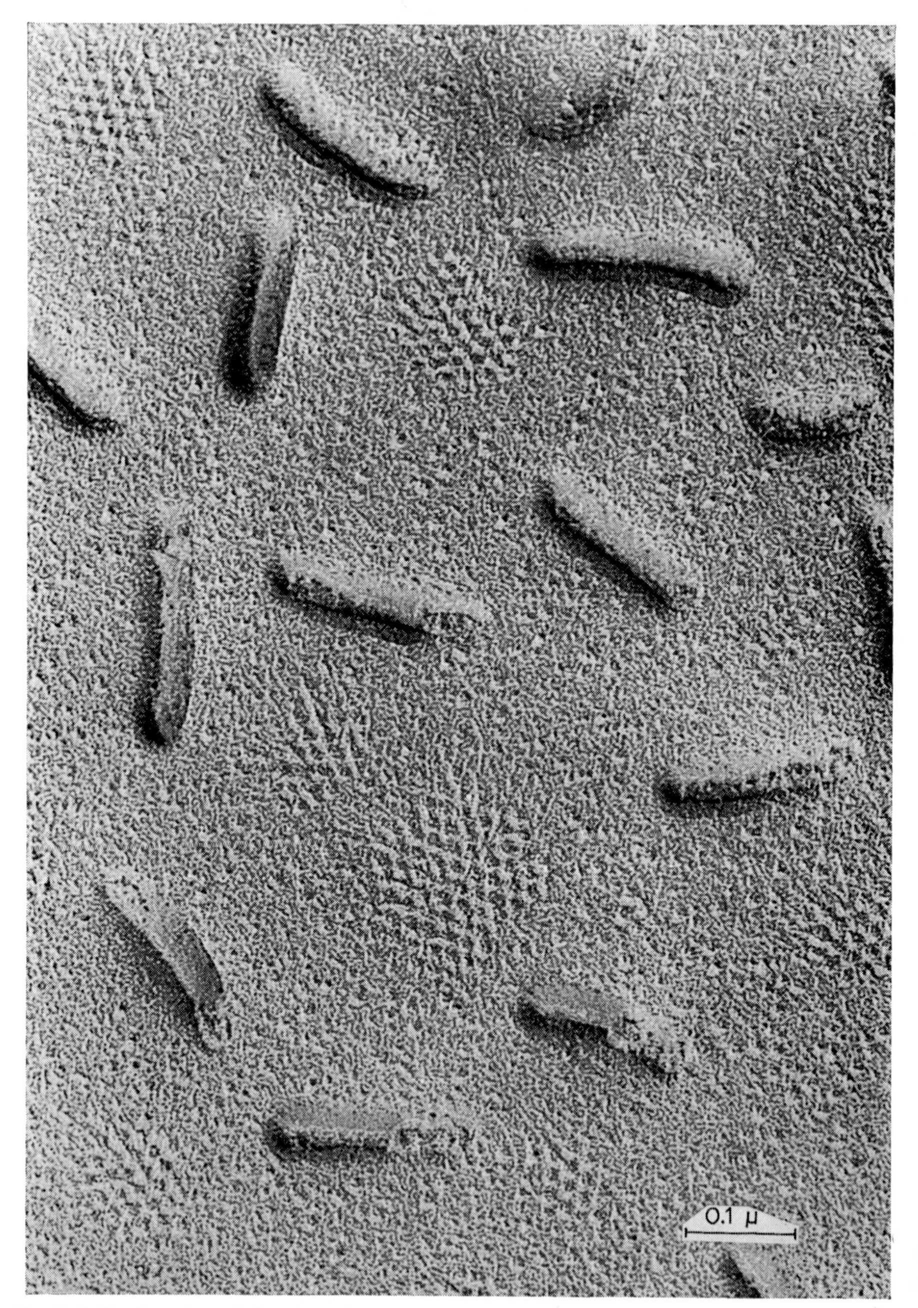

Fig. E–9. Surface view of the plasmalemma in yeast cells, with folds and close-packed clusters of giant particles 180 Å in diameter (photo H. Moor).

In plasmolysis with calcium salts, plasma threads are formed from the plasmodesmata which connect the shrinking protoplasts to the cell wall. These threads are surrounded by plasmalemma (Sitte, 1963b). This process is associated with a very large increase in the surface of the plasmalemma, which presupposes a rapid reconstitution of the plasma membrane and leads us to consider the plasmalemma structure as being far from static, but, rather, capable of fast dynamic change.

(*c*) *Functions*

The plasmalemma performs a large variety of functions: it controls the semi-permeability, resorption, excretion and secretion, leading to the formation of slime and a whole series of cell-wall substances, and it is also capable of breaking down substrates enzymatically.

Quite clearly, no such activities could be achieved by static structures similar to those shown in Figs. E–10 or E–11; only the passive function of semi-permeability would be possible. Energy-requiring processes, such as resorption and secretion, require a dynamic model, and above all the presence of ATP, which presumably participates temporarily in the ultrastructure, and for the enzymatic activity of the plasmalemma (such as digestion and synthesis) the presence of a mosaic structure of different desmoenzymes must be assumed. Since most of the enzymes known so far are globular proteins, the block structure of Fig. E–11 appears to be more probable for this activity than the scheme of Fig. E–10.

The electron micrograph E–9 shows that a mosaic structure does indeed occur in the plasmalemma of yeast. The fine granular pattern is interrupted locally by other structures in two fashions: firstly, the folds of the membrane give the appearance of having a sulcate surface, and secondly, areas of large particles arranged like crystal lattices (lattice spacing, 180 Å) are disclosed. Moor and Mühlethaler (1963)

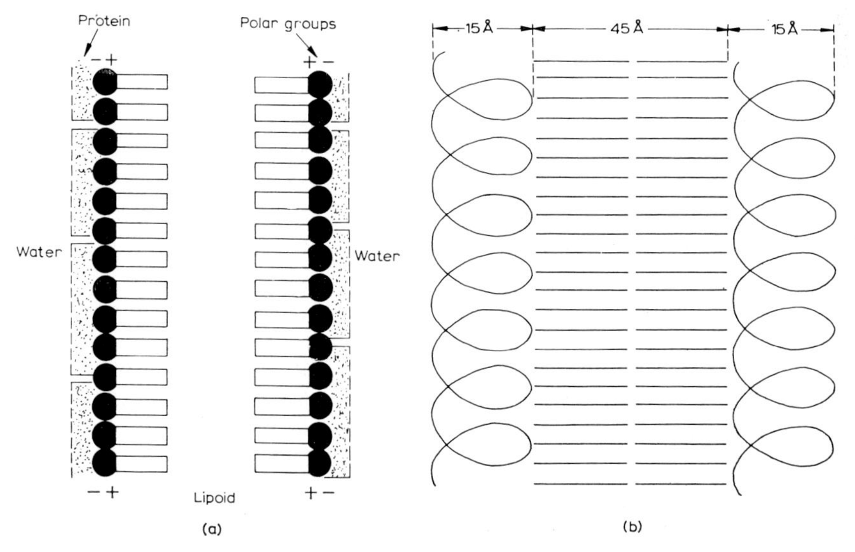

Fig. E–10. The sandwich model of the plasmalemma; (a) after Davson and Danielli (1943), (b) as a unit membrane.

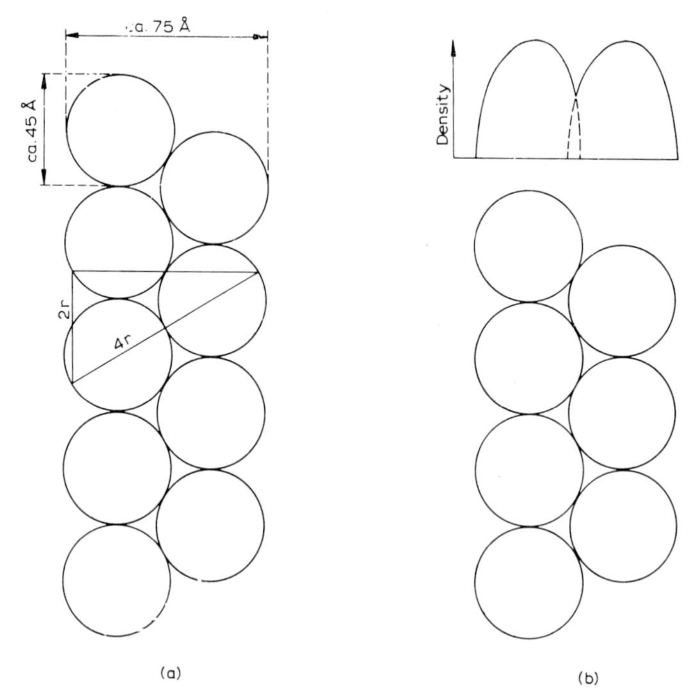

Fig. E–11. (a) The plasmalemma as a bimolecular film. (b) The relative density of such a film.

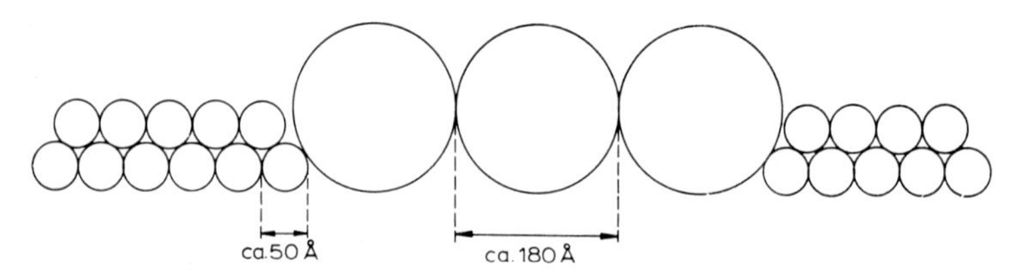

Fig. E–12. Size of particles in granular plasmalemma, as compared with the locally close-packed giant molecules (see also Fig. E–9).

were able to show that the regularly arranged groups of particles are centres of fibril formation (see p. 306). Fig. E–12 shows how these enormous enzyme particles might be built into the plasmalemma.

(*d*) *Permeability*

The semi-permeability of the plasmalemma is imperfect, for not only water and other small molecules such as urea or glycerol but also narcotics with molecular weights greater than that of sucrose can penetrate the cell. It must be made clear that permeability has nothing to do with the active resorption of nutritional molecules in which chemical processes (phosphorylation) and energy expenditure (ATP) play a role, for in permeation a passive exchange of molecules is involved, promoted by the existing differences in concentration. The process is thus one of diffusion slowed down by films of specific structure.

The basic diffusion equation is:

$$\frac{dm}{dt} = D \cdot q \cdot \frac{\Delta c}{\Delta s},$$

i.e. the diffused quantity *dm* of a dissolved substance in time *dt* is proportional to the concentration gradient $\Delta c/\Delta s$ of the substance, *q* being the diffusion cross-section and *D* the diffusion coefficient. The latter has the dimension of $cm^2 \cdot sec^{-1}$ and is usually measured in cm^2 per day. For example, for urea dissolved in water at room temperature, $D = 0.34$ cm^2/day.

An analogous equation can be written for permeation, with the difference that the distance Δs, for which the concentration difference is Δc, is unknown. The unknown distance enters into the constant, and we obtain:

$$\frac{dm}{dt} = P \cdot q \cdot \Delta c,$$

where *P* is the permeability coefficient.

P may be determined by allowing cells to shrink osmotically (water loss) and then observing how the volume of the cell regains its original size (entry of substance) by penetration of permeable molecules. In the case of cells with a measurable surface *q* (such as spherical and cylindrical cells) *P* may be determined in this way, and the water permeability can be calculated analogously by quantitative assessment of the water loss during shrinkage (Bochsler, 1948). We must bear in mind in studies of this type involving plant cells with large central vacuoles, that the molecules have not only to pass through the plasmalemma but also through the membrane of the vacuole (tonoplast) (see p. 157), so that both act as resistances to diffusion. The *P* values in cases involving plasmolysis refer therefore to the combined effect of two boundary layers, whereas no marked resistance to diffusion is shown by the groundplasm, and indeed non-permeable stains freely diffuse in the internal plasma when injected by microtechniques beneath the plasmalemma.

A few such *P* values, expressed in cm/h, calculated from measurements on algal cells, *Tolypellopsis* (Wartiovaara, 1944) and *Zygnema* (Bochsler, 1948), and of stem cortex cells of *Maianthemum* (Höfler, 1934) are listed in Table E–III. They

TABLE E–III

PERMEABILITY COEFFICIENTS *P* OF SMALL MOLECULES: WATER, METHANOL, ETHYLENE GLYCOL, UREA AND GLYCEROL

	Tolypellopsis cm/h	*Zygnema* cm/h	*Maianthemum* cm/h
H_2O	1.04	0.014–0.040	0.005–0.014
CH_3OH	0.85	—	—
$OHCH_2 \cdot CH_2OH$	0.010	0.0075	—
$CO(NH_2)_2$	0.00084	0.00016	0.00015
$CH_2OH \cdot CHOH \cdot CH_2OH$	0.000081	0.00021	0.000055

show that the water molecule can pass through the boundary membrane 200 to 10,000 times more quickly than that of glycerol, and when this difference becomes significantly larger still, as is the case with sugars, the ideal condition of semi-permeability is reached. The figures also enable us to appreciate the magnitude of the diffusion resistance of the boundary layers, and, knowing the thickness of the plasmalemma as measured under the electron microscope, the concentration gradient becomes a reality. If a material which does not occur in the cell is used, then Δc is equal to the external concentration. If D' denotes the diffusion coefficients in the plasma membrane, we obtain

$$D' = P \cdot \Delta s.$$

From the data in Table E–III, the permeability coefficient P of urea in cells of *Tolypellopsis* is 8.4×10^{-4} cm/h or about 2×10^{-2} cm/day. As Δs is of the order of 10^{-6} cm, varying with the thickness of the plasmalemma, we have 2×10^{-2} cm/day $\times\ 10^{-6}$ cm $= 2 \times 10^{-8}$ cm²/day for D', which may be compared with 0.34 or 3×10^{-1} cm²/day for the free diffusion of urea. It is thus evident that the plasmalemma very greatly (e.g. a million times) reduces the movement as compared with free diffusion. The boundary layer of the vacuole, the tonoplast, is a barrier of similar or even higher resistence to diffusion. As a consequence plasma membranes constitute major obstacles to the translocation of molecules, and it is thus understandable that resorption must occur metabolically with the assistence of chemical energy (ATP).

Various theories have attempted to explain the differences in P, (see Table E–III), of which we would mention Overton's (1899) lipid theory, Ruhland's (1912) ultrafilter theory and the mosaic theory of Nathanson (1904). According to Collander (1937), both the principles of lipid solubility and pore filtration apply, and so

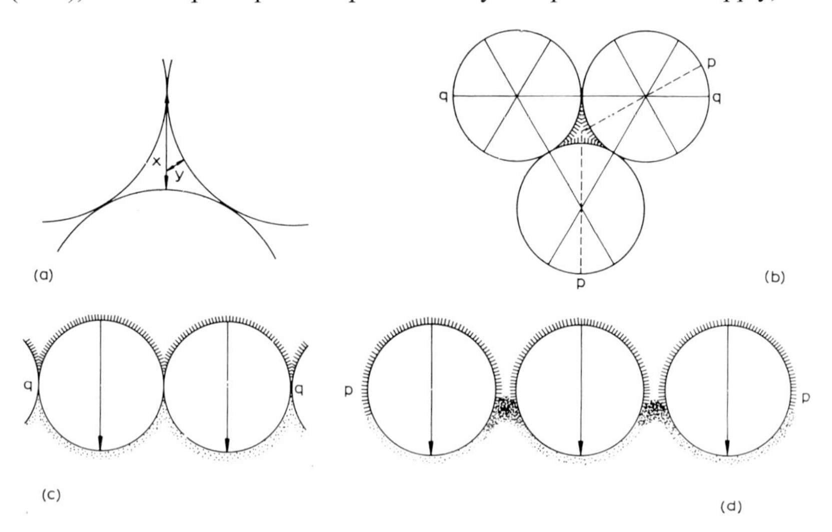

Fig. E–13. Pores in a film of globular protein molecules arranged in hexagonal array. (a) x, y dimensions of the interstitial pores; (b) surface view of the film: pores coated with lipophilic groups (see text); (c) section q–q; (d) section p–p across the film. Hatched: lipophilic groups; dotted: water of hydration; the arrows in (c) and (d) symbolise polarity of the molecules.

must the mosaic theory in a sense, if account is taken of the illustration given in Fig. E–9, for it presumed that different areas of the membrane exhibit different permeabilities. According to the concepts of the lipid filter theory, lipid solubility is involved for large molecules, while smaller ones (such as methanol, urea, and glycerol) permeate via pores.

If the plasmalemma is regarded as a film of globular macromolecules arranged in the spherical close packing, wedge-shaped pores will be found between the spheres (Fig. E–13b), of width depending on the size of the molecules. The radius y of the circle inscribed within the triangular pore is calculated for various particle sizes in Table E–IV, from which it will be seen that, for example, close-packed spheres with a diameter of 100 Å arranged in a plane enclose pores with a diameter of 15 Å; these would allow correspondingly smaller molecules to filter through.

TABLE E–IV

INTERSTITIAL SPACES (x, y) BETWEEN GLOBULAR MACROMOLECULES OF RADIUS r (*cf.* Fig. E–13a)

r	x	y
50 Å	36.6 Å	7.8 Å
100 Å	73.2 Å	15.4 Å
200 Å	146.4 Å	30.8 Å

According to the lipid filter theory, these pores are lined with lipid molecules. Ruhland and Heilmann (1951) were able to induce regular changes in the ultra-filter permeability in the sulphur bacterium *Beggiatoa mirabilis*, related to the Cyanophyceae, by treatment with alcohols of chain lengths increasing from CH_3OH (2.3 Å) to $CH_3(CH_2)_7CH_2OH$ (11.2 Å). From this, they concluded that the pores become equipped with a lipid brush, as is indicated in Figs. E–13b and d. The hydrophilic poles would be anchored to the spherical particles and the lipophilic chains would increasingly block the pore opening as their length was increased. Finally, only the smallest molecule, H_2O, can still filter through, while the remaining molecules would pass in a measure according to their lipid solubility. This experimentally based theory corresponds to the concept of a block structure for the plasmalemma (Fig. E–11) rather than to the lipid film ultrastructure of Fig. E–10.

(*e*) *Pinocytosis*

Whilst ultrastructural morphology can still make no contribution to the problem of how substances are actively conducted through the plasmalemma, with an expenditure of energy against the prevailing concentration gradient (nutrient intake, digestion), electron microscopy has revealed a method for this type of material transport. Ultrastructural vesicles invaginated into the surface of metabolically active cells are often observed. These are lined with plasmalemma (Fig. E–14). The vesicles migrate to the inside of the cell, and either remain there or, what is of greater interest, become dissolved (Wohlfarth–Bottermann, 1960; Weiling, 1961, 1962). It is often difficult to decide whether in fact the reverse process is not taking

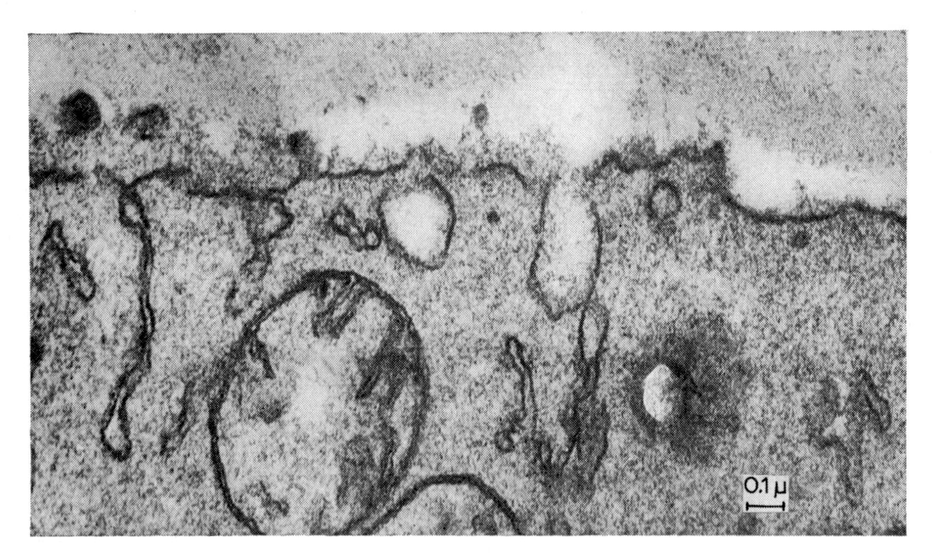

Fig. E–14. Pinocytosis in the cells of the root tip (*Ricinus communis*).

place, *i.e.* whether the vesicles do not form in the groundplasm, migrate to the surface, and there empty their contents like a pulsating vacuole.

A large scale intake of liquid by surface vesicles was discovered in amoebas (Lewis, 1931) and was termed pinocytosis (Greek: *pínein* = to drink) by analogy to phagocytosis (Greek: *phageĩn* = to eat). By marking the cell surface with adsorbed 'fluorescein-labelled protein' (Holter, 1959) it was shown that in pinocytosis the plasmalemma invaginates into the plasma. With regard to birefringence (Mitchison, 1950) and dichroism of the adsorbed dyestuffs, the lining of the vesicles shows the same properties as the plasma membrane (Rustad, 1959). A contractile plasma gel is assumed for the migration of such vesicles into the interior of the cell (Brandt, 1958).

Bennett (1956) observed that not only vesicles visible under the light microscope but also large quantities of ultramicroscopic ones can be taken up. Guttes and Guttes (1960) showed that, in the plasma of Myxomycetes, the vesicle penetrates into the more deeply located granular plasma and remains connected for some time to the exterior medium by a very fine channel which crosses the clear ectoplasm, until this tunnel is interrupted. Solid components are also frequently taken up by vacuoles of this kind. Since, according to the original definition given by Lewis (1931), only liquid is absorbed in pinocytosis, Policard (1958) has designated the absorption of the solid particles, which can be shown in the electron microscope, as rhopheocytosis (Greek: *rhopheĩn* = to swallow). This process can be followed with the aid of electron microscopy, if undigestible polystyrene particles (2200 Å in diameter; Sanders and Ashworth, 1961), or ferritin molecules (95 Å in diameter, *cf.* p. 77; Steinert and Novikoff, 1960) are supplied to the cell surface.

The functions of pinocytosis may be manifold. According to its definition, pinocytosis is the active intake of water, a concept which interferes with the osmotic theory of cellular water supply. However, in many cases the absorption of water by growing cells is astonishingly large so that the osmotic concentration of the cell would tend towards zero. Thus a starved amoeba can take up 30–40% of its own volume in a space of 2 hours, in the form of 1% γ-globulin solution (Holter, 1959). Even larger quantities of fluid are absorbed during the so-called extension growth of plant cells (Frey-Wyssling, 1945), whose length and volume can increase by a factor of 60–150 within a few hours (cells of filaments) or a few days (epidermal cells of coleoptiles). In 1961, Weiling demonstrated that this water intake is partly (or fully?) due to pinocytosis. It can also be shown that in certain instances the osmotic turgor pressure of expanding cells is insufficient to account for the necessary stretching of cell walls and tissues (e.g. in geotropic bending reactions; Frey–Wyssling, 1952), so that supplementary plasmatic pressure must be developed in addition to the osmotic turgor pressure. Whether pinocytosis is involved in this phenomenon must be decided experimentally.

In cells suspended in a culture solution, pinocytosis may be responsible for the intake of nutrients. Such a hypothesis may interfere with the various theories of active permeability, but it cannot replace them. It is remarkable that sugars, although important substrates for respiration, do not induce pinocytosis. Holter (1959) distinguished inducers such as salts (K^+, Na^+) and charged macromolecules (γ-globulin, gelatin, ribonuclease, Na-glutamate) and non-inducers: carbohydrates and nucleic acids.

It seems that a surface adsorption of charged particles is necessary to stimulate the plasmalemma to produce membrane flow and invagination. Solutions of neutral (sugars) or poorly dissociated (nucleic acids) molecules are not resorbed unless the culture medium contains the necessary inducer ions. Ultramicroscopic particles visible in the electron microscope, such as ferritin, can also act as inducers.

Pinocytosis could also explain the movement of substances through the cells.

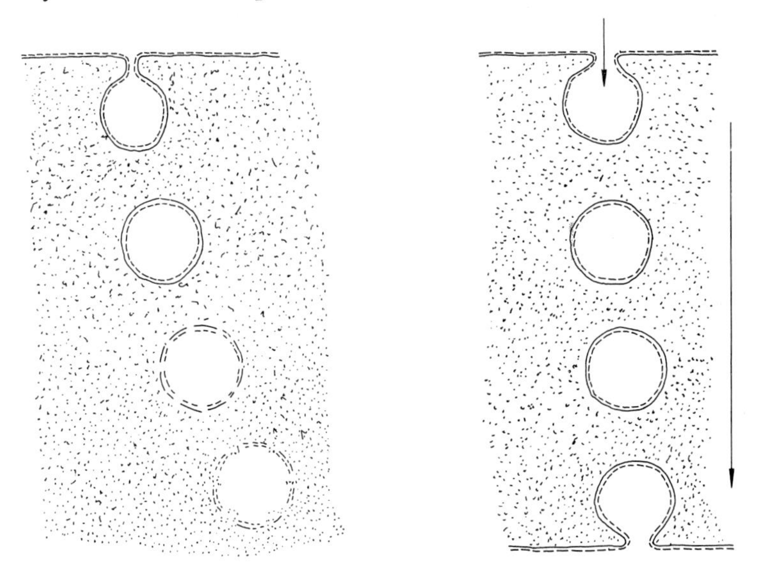

Fig. E–15 (left) and 16 (right). Origin and migration of pinocytotic vesicles. E–15. Digestion of its membrane. E–16. Translocation across a plasmatic layer.

Thus, in the endothelial cells lining the blood capillaries, it is possible to observe pinocytotic vesicles migrating from the capillary lumen through the cells and giving up the liquid resorbed from the blood serum to the surrounding tissue (Vogel, 1962).

In this process, that part of the plasmalemma, which by invagination had become the membrane lining the pinocytotic vesicle, is subsequently re-incorporated into the plasmalemma (Fig. E–15b). The outcome is different when the vesicle is dissolved in the groundplasm (Fig. E–15a). As long as this does not occur, the same difficulties exist regarding the intake of substances from the surrounding medium into the interior of the plasma, as when the substances are outside the cell, since in both cases the barrier of the plasmalemma must be surmounted. This obstacle disappears, when the vesicle membrane is dissolved, in which case the pinocytosed substances pass directly into the plasma. Individuality of the membrane is however then removed, since it merges with the groundplasm or is digested by it. We thus come to the question of whether the plasmalemma is an autonomous system or a form of the groundplasm.

(*f*) *Origin*

There are two possible interpretations regarding the nature of the plasmalemma: it is either an independent organelle which autonomously enlarges its surface during cellular growth, or simply the surface boundary of the groundplasm, which forms wherever the plasma borders on another phase.

In the second interpretation, the cytoplasm behaves passively and the surface is 'organized' according to the laws of accumulation of surface-active substances in the phase boundary. This type of plasmalemma formation is supported by the fact that a new plasma membrane is formed at once in the case of plasma surfaces formed freshly by micrurgy, plasmolysis, or by amoeboid motion, characterised by local ruptures of the ectoplasmic gel where portions of the endoplasmic sol flow out. In plasmolysis experiments, this membrane hardens after a certain time and is then termed a haptogen membrane (Küster, 1951, p. 99). According to this interpretation, the requisite quantity of surface-active lipoproteins must be present in the groundplasm to cover immediately the newly formed boundary areas. The observation that the plasmalemma at the posterior pole of moving amoebas (Wohlfarth–Botterman, 1960) or that pinocytotic vesicles can be dissolved in the groundplasm (see Fig. E–15a), may fit into such an interpretation.

The interesting ultrastructure and the manifold functions of the plasmalemma, however, tend to speak against its formation by purely physicochemical equilibrium processes. Since its semipermeability disappears when the cell dies, it must be assumed that the ultrastructure of the plasmalemma is only maintained with an expenditure of energy (ATP). The fine structure of the plasma membrane is therefore the result of a vital function which, however, should proceed at a similar rate as the establishment of a surface equilibrium, since otherwise the immediate covering of fresh plasma surfaces with new plasmalemma cannot be explained. Experiments on the extensibility of the plasmalemma also suggest a vital organisation. According to Törnävä (1939), cell surfaces stretched by endosmosis suddenly begin to leak after a certain degree of extension is achieved. This has been interpreted as exhaustion of the lipid reserves in the groundplasm, but in our view the phenomenon is rather due to rupture of the plasmalemma, for the electron microscope reveals that the plasmalemma is a firmly organised boundary membrane, and not a liquid boundary film.

The question has thus arisen as to the nature of the relationship of an autonomous plasmalemma to other cell organelles. In particular, a connection with the endoplasmic reticulum (see p. 161) has been postulated. According to McAlear and Edwards (1959), for example, there is a direct connection in conidiophores of the Deuteromycete *Stilbum* between the plasmalemma and the nuclear envelope (p. 182). The published electron micrographs are however too unclear to decide the matter and the plasmalemma is assumed to be a double membrane, which hardly agrees with the more recent findings. The question of whether the plasmalemma can form ER tubules is of great theoretical importance, since in this event the medium surrounding the cell would not only come into contact with the cell surface but could also penetrate directly through the ER canalicules into the interior of the plasma, and into the perinuclear space of the nuclear envelope. No clear pictures of such invaginations of the plasmalemma have ever been obtained. Most authors prefer to express themselves cautiously on this subject. No continuity is claimed between the plasmalemma and the ER membrane, but only a 'close relationship' (Shatkin and Tatum, 1959) or an inter-relationship 'by content at least' (Whaley *et al.*, 1959). In fact, ER tubules are often extended directly under the plasmalemma, and it may be supposed that the requisite nutrients are supplied for the maintenance of the labile structure of the plasma membrane in this way.

Whereas the plasmalemma does not appear to be directly connected with the ER membrane, there is no doubt about its clear relation to the Golgi membranes (see p. 279, Fig. J-4).

3. TONOPLAST AND VACUOLES

Fully developed plant cells usually possess a large central vacuole (Fig. D–2a′) enclosed by a semi-permeable membrane. This has been termed the tonoplast since it is held in tension by the osmotic properties of the cell sap. The term was first used by De Vries (1885) to designate an autonomous vacuole-producing cell particle. Although Pfeffer (1886) has shown that there is no such organelle, and that the vacuoles are formed by a striking hydration of certain areas in the cytoplasm, the term has been adopted by the cytologists in the specific sense of the vacuolar membrane. The tonoplast is frequently a rather firm envelope, which is mechanically much stronger than the plasmalemma, but in some cases its physical nature may also be similar to that of the external plasma membrane. In rare cases, as in sieve and latex tubes, the tonoplast may even be entirely lacking; and there is then no boundary detectable under the light microscope between the peripheral plasma and the central cell fluid, so that it is difficult to decide whether this is a true vacuolar sap or extremely diluted liquid groundplasm.

By plasmolysis (Fig. E–17a), which is accompanied by a reduction in the vacuole size, the tonoplast hardens as a result of a reduction in its surface area, and can then be isolated from the cell. This is accomplished by a micrurgical section through the forecourt of the plasmolysed cell, after which the tonoplast stretched by the

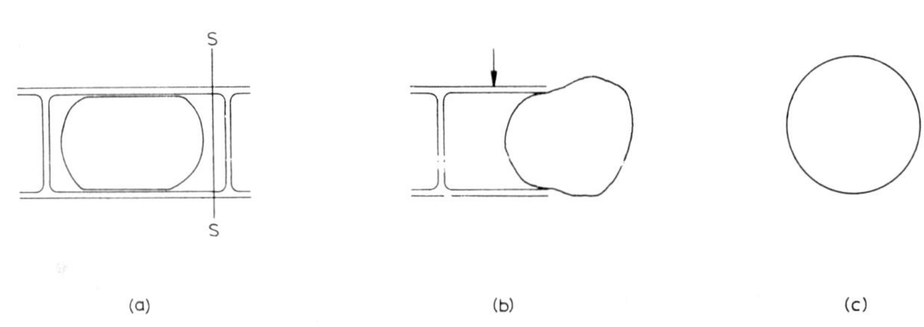

Fig. E–17. Isolation of the tonoplast from a plasmolysed cell. (a) Micrurgical section *s–s* through the forecourt; (b) squeezing out of the tonoplast by gentle pressure; (c) isolated tonoplast.

turgor pressure of its vacuolar content can be sequeezed out (Fig. E–17b). When freed from the cell wall, the tonoplast immediately assumes an ideal spherical shape and the cytoplasm adhering to it disintegrates, whereafter it can be observed under the light microscope as a passive "ghost" for many days in suitable suspension media. During this, the tonoplast retains its semi-permeability, and the sphere can be made to swell osmotically or to shrink by transfer to more diluted or to more concentrated solutions respectively.

(*a*) *Ultrastructure*

The electron microscope reveals the vacuolar envelope to be a unit membrane having the same appearance as the plasmalemma (Fig. E–8). When fixed with osmium, the thicker contour of its three strata is turned towards the plasma and the finer contour towards the vacuole. Although such a film can scarcely possess the mechanical properties of the vacuolar envelope described above, the term tonoplast has been adopted in electron microscopy for this unit membrane (Buvat, 1962).

There is no doubt that in a firm tonoplast additional layers of the protoplast participate in its construction, although no electron micrographs have as yet been obtained of a membrane reinforced in this manner. Nevertheless, it is known from indirect evidence that the vacuolar envelope must be very rich in polar lipids. By gentle hydrolysis with ammonium or sodium hydroxide, these can be induced to migrate; as has been shown by Gicklhorn (1932) on the epidermal cells of onion bulb scales (*Allium cepa*), this migration is accompanied by the formation of myelin figures, visible under the light microscope (Fig. B–52).

The retention of semi-permeability in isolated tonoplasts even after the adjacent cytoplasm has died is evidently due to this high lipid content. In this respect, there is a significant difference from the plasmalemma, which maintains its semi-permeability only as long as the protoplast is alive.

The permeability barrier of the tonoplast is generally considerably more pronounced than that of the plasmalemma. Thus, certain strongly swelling salts, such as potassium rhodanide, KCNS, which penetrate through the plasmalemma and swell the plasma, are unable to permeate through the tonoplast, so that the vacuole loses water and shrinks plasmolytically. Simultaneously, the plasma swells to cap-like calottes. The resulting light microscope picture is termed cap plasmolysis (Fig. E–18). On the basis of experiments of this kind, three different grades of permeability are distinguished in the theory of botanical permeability (Fig. E–18):

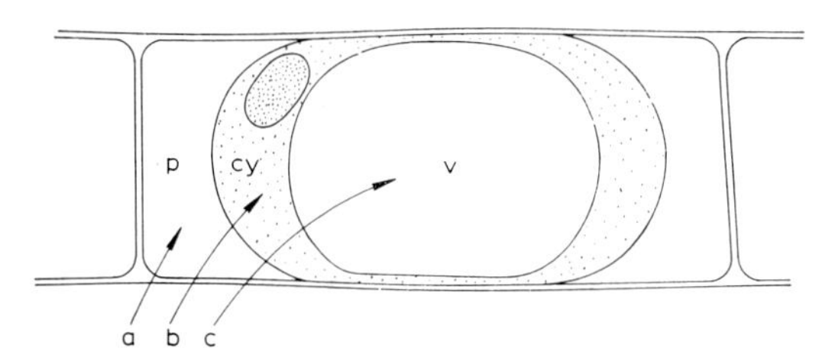

Fig. E–18. Cell with cap-plasmolysis, demonstrating the various types of permeability (after Höfler, 1932). (a) Holopermeability (*p* = plasmolysis forecourt); (b) intrability (*cy* = cytoplasm); (c) permeability (*v* = vacuole).

(a) holopermeability – no hindrance to diffusion by the cell wall, (b) intrability – selective hindrance to diffusion by the plasmalemma, and (c) permeability – hindrance to diffusion by the tonoplast membrane.

(*b*) *Functions*

The reason why expanding plant cells produce their enormous vacuoles is thought to be related to a lack of sufficient nitrogenous nutrients. As a matter of fact the plant cannot produce sufficient protein to fill all its expanded cells with cytoplasm; instead, the available space is filled by the watery cell sap of the central vacuole.

By virtue of the molar concentration of the cell sap and the semi-permeability of both the tonoplast and the plasmalemma, the vacuole coated by a layer of cytoplasm functions as an osmometer and gives the cell the necessary rigidity or turgescence.

The dissolved micromolecules which account for the osmotic concentration of the cell sap belong to two physiologically different categories. On the one hand they are the so-called secondary metabolites such as phenols, flavonols, anthocyanins, alkaloids, etc. Since the plant's ability to excrete those substances externally is limited, they are deposited inside the cell in the vacuoles; if the substances are not sufficiently water-soluble (e.g. anthocyanidins), they are transformed into glucosides (anthocyanins) for this purpose. The tonoplast can therefore be considered to be engaged in excretion processes. How its unit membrane accomplishes these energy-consuming processes cannot be analysed in detail; it is much the same problem as the active elimination of dissimilates by the plasmalemma.

Apart from these metabolites, valuable assimilates such as sugars and proteins are secreted into the vacuoles, where they are stored as reserves, being reactivated and re-introduced into the metabolism when required. Whilst sugars occur only as solutes, the proteins can be stored in the solid state. This is the case when aleurone grains are formed in the endosperm of seeds.

Originally, the aleurone grains are liquid vacuoles, which lose water by active dehydration. In this process the various vacuolar components precipitate according to their solubility. In the aleurone vacuole of the *Ricinus* seed, the almost insoluble magnesium-potassium salt of inositol phosphoric acid (phytin) is precipitated first as a body known as 'globoid'. The macromolecules of globulin proteins then begin to arrange themselves into the crystal lattice of a 'crystalloid' (see p. 73), and finally the last remnants of liquid, containing an easily soluble albumin, solidify into a homogeneous substance surrounding both globoid and crystalloid. On mobilisation of the reserve substances, dissolution proceeds in the reverse order: the albumin is dissolved first, followed by the globulin crystalloid, and finally by the mineral globoid.

Particularly large crystals of the globulin excelsin (Table B–X) are found in the aleurone grains of the seeds of *Bertholetia excelsa* (Brazil nut). The relatively small aleurone grains in the caryopsis of cereals consist only of a large globoid (Innamorati, 1963).

Here again the mode of translocation of these molecules across the unit membrane of the tonoplast into or out of the vacuole is open to discussion. If the vacuolar membrane has real secretory faculties, it must have a more dynamic ultrastructure than the isolated tonoplast with its amazing stability.

(*c*) *Origin*

The youngest meristematic cells are free from vacuoles. The formation of the vacuole at the beginning of cell differentiation can therefore be followed in the electron microscope.

Large areas are observed in the groundplasm, which lose mass (Mühlethaler, 1958/60). This could be interpreted in the sense of the hydration theory postulated by Pfeffer (see Figs. D–1 and E–19). Closer inspection reveals, however, that these vacuoles are seen to be surrounded by a unit membrane and therefore delimited from the cytoplasm. In actual fact, the vacuoles first appear as flat bladders, the contents of which swell up with time, as if by 'inflation'. Such elon-

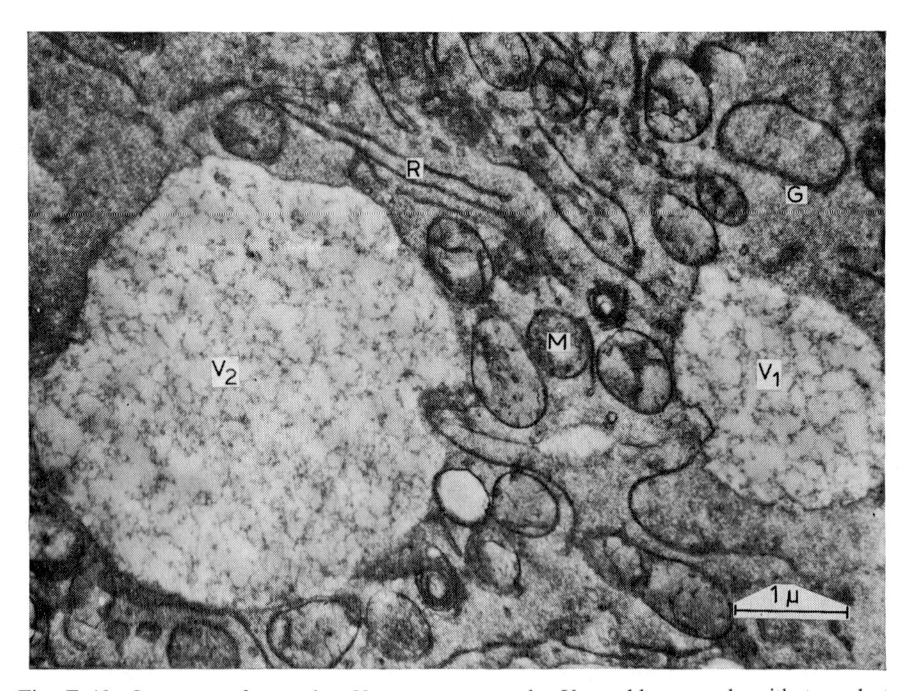

Fig. E–19. Ontogeny of vacuoles. V_1 = young vacuole, V_2 = older vacuole with tonoplast, G = groundplasm, R = strand of endoplasmic reticulum, M = mitochondrion.

gated narrow vacuoles have also been seen in the light microscope (Guilliermond *et al.*, 1933).

On the other hand, extremely minute bladders can be detected, which are interpreted as the forerunners of vacuoles (Poux, 1962a, b). It is difficult to differentiate these small vacuoles from pinocytotic vesicles, especially since according to Poux they are sometimes formed by invaginations of the plasmalemma. Furthermore, an open contact with some ER canals is postulated, and thus a connection between the ER and the vacuole system discussed (Buvat and Mousseau, 1960). According to this view (Buvat, 1962), the vacuoles would be inflated caverns of the reticulum.

Derivation of the vacuoles from the Golgi system has been demonstrated by Marinos (1963). In this case the tonoplast would be a Golgi membrane and therefore homologous to the plasmalemma.

4. ENDOPLASMIC RETICULUM

Even before it was technically possible to prepare ultrathin sections for the electron microscope, Porter *et al.* (1945) discovered a fine reticulum on drying up preparations fixed with OsO_4 in the endoplasm of thinly spread cells of chicken fibroblasts and in other cells. The above authors have consequently called it the endoplasmic reticulum (Porter, 1948). Later, it was found from sections that this is a system of intercommunicating sublight-microscopic canals, vesicles, and cisterns (Palade and Porter, 1954). In sections, however, the system does not show an actual net, this being only simulated in the projection of thicker layers by the strands which cross over one another in different planes. The term endoplasmic reticulum has therefore been criticised repeatedly (Sjöstrand, 1956), but since no better suggestion has been made, it has been generally accepted and the term endoplasmic reticulum with the usual abbreviation ER should be retained, even if they do not fit the morphological facts in every respect, especially since the most important feature of the system, *i.e.* its continuity, is properly implied in the concept 'reticulum'.

(*a*) *Ultrastructure*

ER strands mostly appear as double membranes with variable spacing. The strand can widen out locally into vesicles, and frequently chains of vesicles connected to one another by canalicules, or larger ventricles and caverns are formed in the groundplasm. These observations show that this is a hollow system, which interlaces the whole cell. Its continuity is not recognisable in ultrathin sections, as the system is sliced across so that transversely or obliquely cut strands appear as isolated circular or oval vesicles. It can be seen, however, from a series of sections that all these hollow spaces are interconnected. The capillary system is coated by a unit membrane. The characteristic 'double membranes' which occur frequently, must be regarded as being in fact pairs of unit membranes separated by a very narrow lumen. The freeze-etching technique, which allows a surface view of the cell organelles, shows that the ER system often consists of flat bladders openly connected with one another (Fig. E–20).

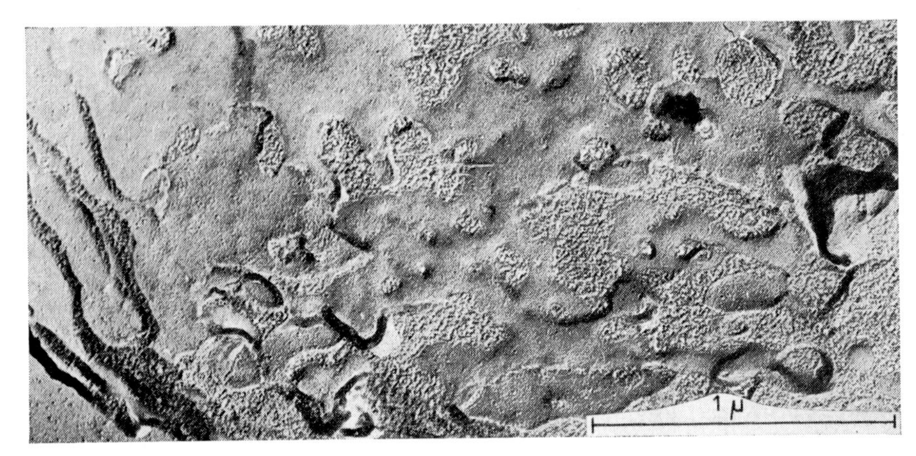

Fig. E–20. Surface view of the ER after freeze-etching (onion root tip; photo Branton).

The number and location of the ER strands vary. In meristematic cells, branches of the system pass directly under the plasmalemma, parallel to the cell surface. The relationships to the nucleus are of particular importance. It has been established that the nuclear envelope is part of the reticulum, since there is an open connection between the lumen of that double membrane and the hollow ER system (see p. 183).

The membranes of the ER often carry on their plasma side surfaces osmiophilic granules with a diameter of about 150 Å, which have been identified as ribosomes. In contrast to the membranes bearing ribosomes, those which are free of granules are referred to as smooth.

The hollow spaces of the ER system appear to be mass deficient in the electron microscope. They must be filled with a kind of serum. We would like to propose that this should be called enchylema, which term the classical cytologists (for example Hanstein, 1880) used to denote a kind of plasma sap.

Discovery of the ER has led to the explanation of light microscopic basophilic areas which can be shown by staining and which have been termed 'ergastoplasm' or 'accessory nucleus' by the classical cytologists.

The ergastoplasm stores basic dyestuffs (haematoxylin, toluidine blue). Such basophilic zones in the cytoplasm are particularly characteristic of the endocrine gland cells (e.g. the pancreas), while in plant cells they very rarely occur in this characteristic manner. In the electron microscope, the ergastoplasm shows itself as an accumulation of ribosomes, situated on the parallel lamellae of ER stacks or accumulating freely in the groundplasm. The basophilic behaviour is due to the RNA content of the ribosomes (see p. 174).

Accessory nuclei are diffuse zones of the cytoplasm, which like the ergastoplasm react to nuclear dyestuffs. In the plant world, accessory nuclei have been detected after anoxia in the shoot apex of *Elodea* (Wrischer, 1960), in the root tips of hyacinths (Gavaudan *et al.*, 1960) and in young sieve tubes (Falk, 1962b). They consist of concentric rings of ER lamellae, which are caused by the shortage of oxygen. For the present, it is unknown why the increased production of ER lamellae during anaerobiosis (see p. 216) leads to the formation of such rings.

(*b*) *Function*

Nothing definite is as yet known about the tasks which the ER performs as a cell organelle; there is, however, a number of hypotheses based on its morphological features (Porter, 1961).

The form and distribution of the cavity system indicate that ER plays a part in the migration and distribution of matter within the cell. Where an intensive consumption of matter takes place, as in the synthesis of the cell wall, ER capillaries run approximately parallel to the consumption sites. Individual ER strands are also known to pass through the plasmodesmata (see p. 147), and whole bunches of them through the pores of the sieve plates; similarly, the sieve tube plasma is densely packed with ER canals, all oriented parallel to the direction of movement of the assimilation stream. All this speaks in favour of the participation of ER in the displacement of assimilates. Nevertheless, we must not imagine that a mechanical pump principle is involved as the *fors motrix*, as is the case in the blood capillaries, since the membrane of the ER system does not display any contractile element. ER strands 100–200 Å in diameter and 10 Å sucrose molecules may be compared to the blood capillary/erythrocytes system, but if we consider protein molecules with a diameter of 35 Å, the frinctional forces become so large that migration through a pressure or concentration gradient is hardly conceivable. We should therefore rather consider chemical forces, and imagine that the molecules glide along the surface of the ER membranes in the manner of an ion ladder, by means of attractive forces which move quickly along the wall (see p. 217).

ER is involved not only in a possible translocation, but above all in the synthesis of assimilates. Since isolated ribosomes can synthesise proteins from amino acids *in vitro*, it is assumed that this also takes place with the ribosomes observed on the membrane of the ER. Whether the hollow ER system provides the necessary amino acids, or whether it translocates the protein molecules formed, cannot be decided. In both cases, the molecules in question would have to pass through the ER membrane. ER is also concerned with carbohydrate and enzyme synthesis. Glycogen is produced in its caverns in the liver, and zymogen granules in the pancreas. Plastid-like bodies coated by ER membranes are found in the glandular hairs of the insectivorous leaves of *Pinguicula* (Vogel, 1960).

Branching of the ER increases astonishingly in anaerobiosis. This is particularly noticeable in yeast, at the transition from aerobic respiration to alcoholic fermentation (Linnane *et al.*, 1962). In the nectar hairs of *Abutylon*, it is sufficient to bring the blossoms into a nitrogen atmosphere, to obtain an equally remarkable increase in the development of the ER, which then completely fills the plasma (Mercer and Rathgeber, 1962). It is as if the elimination of the function of mitochondria resulted in a transfer of the metabolic tasks to the ER.

(*c*) *Origin*

Endoplasmic reticulum is a most variable cell organelle. Its formation is moderate in very young meristematic cells, but it can develop strongly in growing cells and in mitosis. Here it can be observed how branches and ramifications grow out of the extant reticulum, and how vesicles are separated off, which can grow again into strands. One gains the impression that all parts of the system arise from the strands which were taken over from the mother cell at cell division.

The ER elements are sparse in mature cells without intensive metabolism, as if a part of the system, which had developed strongly during the growth phase, had melted into the groundplasm. If this is possible (and there can be little doubt about it, since the ER strands are not permanent organelles) we must ask ourselves whether, conversely, it is not also possible for reticulum membranes to be formed *de novo* from the groundplasm. No conclusive evidence regarding this matter has however been reported up to now.

Since the unit membranes of the ER look like the plasmalemma, the question was examined whether initials of the reticulum can originate by invagination of the plasmalemma (Palade, 1956). As has already been mentioned, some authors have in fact postulated a direct connection between the ER and the plasmalemma (e.g. McAlear and Edwards, 1959). On our pictures however, we were never able to establish a fusion of the ER membrane with the plasmalemma. Thus although the ER strands frequently passed very close under the cell surface, parallel to the plasmalemma, no actual contact was ever observed. We therefore come to the conclusion that only in pinocytosis are parts of the membrane of the cell surface taken up into the cytoplasm, whilst the endoplasmic reticulum is a system independent of the plasmalemma.

On the other hand, there is clearly a direct connection between the ER and the nuclear envelope. The caverns of these two systems are obviously interconnected. We can see how ER strands change into the nuclear envelope and how this forms a variety of evaginations. Both ER and the nuclear envelope belong therefore to the same cell organelle, and are treated here in two separate sections only on historical grounds. It could be said that the nuclear envelope constitutes a part of the ER. We may then ask whether the nuclear envelope is to be ontogenetically derived from the ER, or, conversely, whether the cytoplasmic ER is descended from the nucleus. Bearing in mind the position of the nucleus as the central organ of the cell and the processes occurring in mitosis (see p. 188), the concept of the nuclear envelope being the ontogenic source of the endoplasmic reticulum seems to be much more probable than its formation from the cytoplasm.

Formation of the ER in the endosperm of fertilised ovules following free nuclear division is of a special kind. Buttrose (1963a) found a marked inflation of the perinuclear space in the nuclei of the wheat endosperm cells. The enormously enlarged space is provided with vesicles of nuclear plasma by evaginations of the inner nuclear envelope. By the growth of these vesicles, whose contents now function as cytoplasmic groundplasm, the perinuclear space is broken up into septa and caverns having the character of ER.

5. GOLGI APPARATUS

On the basis of light microscopic observations, Golgi (1882–1885) described the existence of networks in nerve cells, which could be stained with silver nitrate or osmic acid, and whose occurrence in other cells led to the designation 'dictyosomes' (from the Greek *dictyes* = net). The questions of the structure and nature of this so-called Golgi apparatus led to a bitter free-for-all controversy among cytologists. Instead of the network, some workers found roundish bodies, surrounded by rings or croissants, which could be stained with neutral red (Guilliermond *et al.*, 1933, p. 334). Others explained the Golgi complex as a normal vacuole system (Küster, 1951, p. 475). Most authors agree that this cell organelle is very variable and is not always easy to detect. It was found fairly generally in animal but only seldom in plant cells, and until very recently it was even declared by certain cytologists as a fixation artefact (Baker, 1957).

This controversy has now been resolved by the electron microscope. Cytoplasmic packets of submicroscopic double lamellae, identified with the Golgi apparatus, have been found in practically all objects so far studied: in unicellular organisms (Schneider and Wohlfarth–Bottermann, 1959) animal cells (Afzelius, 1956a; Hirsch, 1958), and in lower (Heitz, 1958) and higher plants (Perner, 1958b). In fungi, which were thought to be exempt of Golgi bodies, they have so far been found in the Ascomycetes *Saccharomyces* (Moor and Mühlethaler, 1963) and *Neobulgaria* (Moore and McAlear, 1963b).

The Golgi network in the nerves is however different, as it consists of deposits of osmiophilic granules (Thomas, 1960) rather than lamellar stacks. Though perhaps the Golgi network of the nerve cells may be unrelated to the cell organelles now under discussion, recognised as being generally distributed, the term Golgi apparatus or Golgi complex is still retained.

(*a*) *Ultrastructure*

The Golgi apparatus (Fig. E–21) consists typically of a series of double membranes concentrically bent. They enclose sublight-microscopic spaces as in the endoplasmic reticulum which however are individualised. Along the periphery of the vesicles inflated extensions may be observed. These small round vesicles ligate, and may develop into large vacuoles. No network is visible anywhere, so that it would be best to drop the term dictyosome. It is believed that the controversy as to whether the Golgi apparatus is a network or an accumulation of small vacuoles arose from the fact that either a coagulum of osmiophilic lamellae (network) or

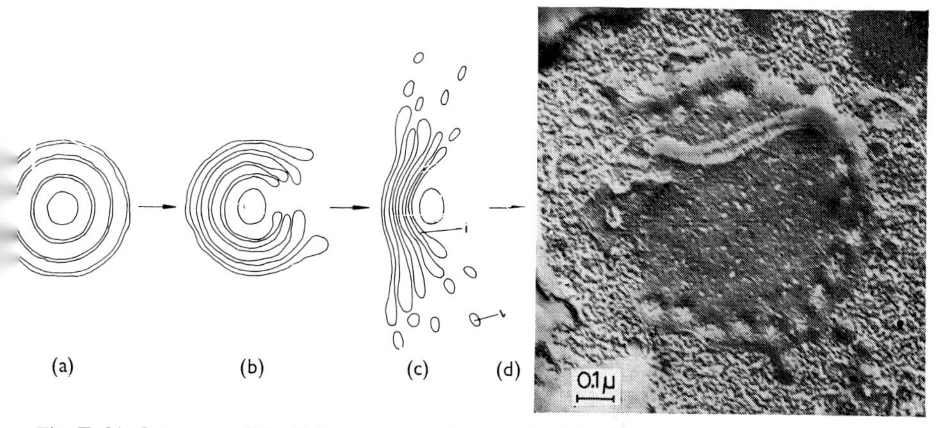

Fig. E–21. Ontogeny of the Golgi apparatus. l = stack of double membrane, v = Golgi vesicles. (a) Concentric double membranes; (b) opening of the bilamellar rings; (c) formation of vesicles budding from the rim of the saucer-like double membrane; (d) surface view of a 'saucer' with vesicles on the rim (photo Branton).

resolvable microvacuoles appeared in the light microscope (Pollister and Pollister, 1957).

(*b*) *Function*

The function of the Golgi complex was originally stated to be the synthesis of secretions, for example of zymogen granules in the pancreas. Localisation of the zymogen synthesis in the Golgi apparatus is however disputed, and has been transferred to the cisterns of the reticulum (Siekevitz and Palade, 1958). It was also established, in the salivary glands of aphids, that the prosecretion granules do not appear in the Golgi apparatus but again in the endoplasmic reticulum (Wohlfarth–Bottermann and Moericke, 1959).

The formation of vesicles as stages preceding lipid droplets is also discussed. As will be shown in the next section, on spherosomes, however, in oleagenous seeds ER is the site of fat synthesis. On the other hand, carbohydrates of the mucilage and hemicellulose types do appear to be produced in the Golgi vesicles (see p. 301).

In the glandular hairs of insectivorous plants (*Drosera, Drosophyllum*) and in the outer rootcap cells (Mollenhauer *et al.*, 1961, 1962), a quantitative relation was established between the secretion of mucilage and the number of Golgi structures (Schnepf, 1961). Drawert and Mix (1962a) believe that they have found a similar relation in the conjugate alga *Micrasterias*. The Golgi vesicle mucilage, which is strongly contrasted in the electron microscope after fixation with $KMnO_4$, is emptied into the space between the plasmalemma and the cell wall (Mollenhauer *et al.*, 1961) whence it diffuses through the wall. The elimination occurs the other way round to pinocytosis and recalls the activity of contractile vacuoles (Schnepf, 1963; Sievers, 1963). The formation and size of the Golgi vesicles in the root cap can be markedly increased by aeration (Falk, 1962c).

Chemistry of the mucilage suggests that the synthesis of further polymeric carbohydrates may take place in the Golgi apparatus. Nevertheless, this cannot apply to the cellulose, as its elementary fibrils are polymerised *in situ* by the plasmalemma on the surface of the cell wall. However, the Golgi apparatus plays a decisive part in the formation of the cell wall matrix which consists of hemicelluloses, since the vesicles from which the cell plate is formed during cell division (see p. 277) are the Golgi vesicles (Whaley and Mollenhauer, 1963; Frey-Wyssling *et al.*, 1964).

In placing special synthetic processes in the Golgi apparatus, we should always bear in mind that these organelles occur not only in secretion cells, but quite generally in all meristematic and (generally to a lesser extent) mature cells. It may thus appear that the function of the Golgi apparatus lies rather in the control of a general physiological process which is equally important and necessary for all cells, as is the case of its cooperation in the formation of vacuoles and the cell wall.

(*c*) *Origin*

In early stages of a growing meristematic cell, the number of Golgi bodies may increase and then it can decline strikingly during the differentiation to a mature cell. It would therefore appear simple to follow the origin and fate of this organelle. This, however, is not the case, as it is not easy to determine whether disappearing Golgi bodies merge into the groundplasm or whether they turn into ER vesicles

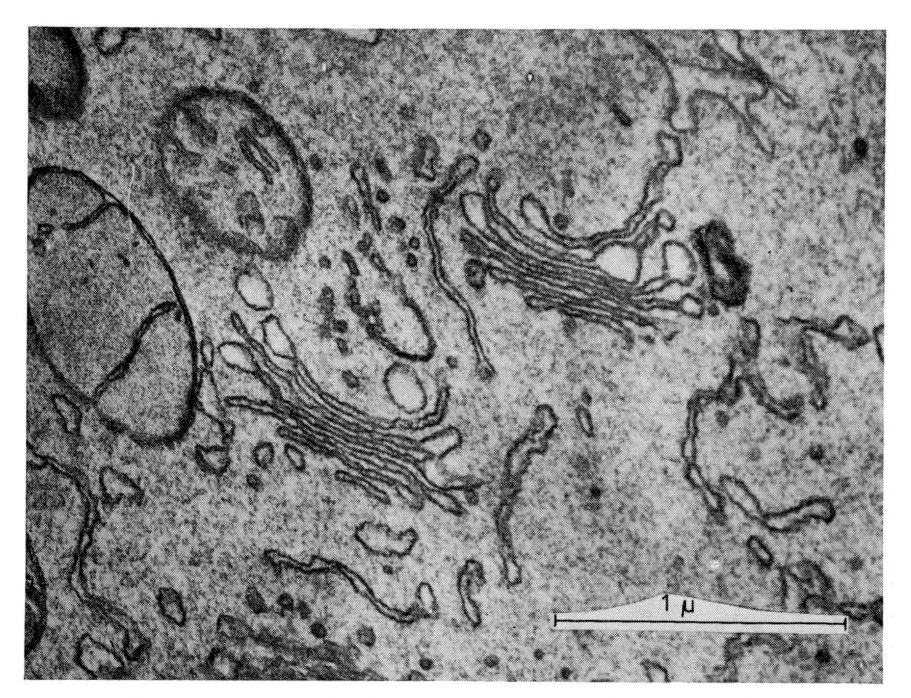

Fig. E–22. Two Golgi bundles in the root meristem of *Ricinus communis.*

and strands. In the first case, one should also have to attribute to the groundplasm the property of forming Golgi bodies *de novo*, whilst in the second case these would constitute a special differentiation of the reticulum. According to Moore and McAlear (1963b), the Golgi apparatus is formed from the nuclear envelope in the exciple of the apothecium of the Discomycete *Neobulgaria*. Its origin would therefore be the same as that of the endoplasmic reticulum. Only the future will show how far this discovery can be substantiated and generalised.

Thus, although the origin of the Golgi complex still remains problematic, a certain development and differentiation of these organelles can nevertheless be established in meristematic cells by electron microscopy. As mentioned before, Golgi bodies do not appear as lamellar packets from the very beginning, but as concentric lamellar systems. These cup-shaped vesicles then open and flatten into typical saucer-shaped Golgi lamellae (Fig. E–22).

The organelle is therefore composed of saucer-shaped layers, so that the detached vesicles do not in fact arise from club-shaped strand ends, but bud from a circular saucer rim. This particular spatial form of the Golgi apparatus is shown by the examination of serial sections or surface view after freeze-etching (Fig. E–21d).

6. SPHEROSOMES

In 1880, Hanstein described small, highly refractive bodies of 'denser substance' in the cytoplasm of plant cells, which he called microsomes. They are easily recognisable under dark field illumination as brightly shining, and in phase contrast as black, granules with a diameter of 0.5–1.0 μ. For decades these microsomes have served the plant cytologists as convenient objects for the study of Brownian movement and of the plasma streaming in the cell. The term microsome had however to be relinquished, after Claude (1943), unaware of its use as a botanical–cytological concept, had applied it to osmiophilic particles about 0.1 μ across, observed in liver homogenates. In its new application, the term 'microsome' was accepted so rapidly in biochemical publications that the original name due to Hanstein had to be changed. The choice (Perner, 1953) fell on 'spherosomes', as Dangeard (1919) had called the microsomes in the sense used by Hanstein.

The spherosomes stain like fat droplets with the usual fat dyes (Sudan black, Sudan III, Rhodamine B, Nile blue sulphate, Indophenol blue, Scarlet R, etc.). In addition, they show the Nadi reaction (synthesis of Indophenol blue from α-naphthol and dimethyl-*p*-phenylene diamine through O_2-transfer). From this, Perner (1958b) concluded that the spherosomes were not ergastic particles like fat droplets, but enzyme-active organelles, and that they contained cytochrome oxidase. It was however possible to show that the reaction in question was catalysed by the mitochondria and that the spherosomes merely acted as selective storage sites for the stained reaction product Indophenol blue (Drawert, 1953). The reduction product, formazan, formed by the H_2-transfer enzyme from tetrazolium chloride (TTC reaction) is also stored in the lipid rich (40%) spherosomes (Ziegler, 1953). After their lipids are extracted, the spherosomes leave behind a protein ghost which can be stained with pyronine, which is indicative of a proteinic stroma. The aromatic amino acid content (tyrosine, etc.) should perhaps be held responsible for the not only relatively but also absolutely high refractivity of the spherosomes.

(*a*) *Ultrastructure*

As droplets of oil and other lipids are difficult to distinguish from the spherosomes, an attempt was made to find characteristic differences in the electron microscope (Drawert and Mix, 1962b; Peveling, 1962). In comparative tests, the spherosomes in the epidermal cells of onion bulb scales of *Allium cepa* and in the tissue of the coleorrhiza of *Zea mays*, were compared with the oil droplets in the storage tissue of the oil seeds of *Sinapis alba* and *Brassica napus* (Frey-Wyssling *et al.*, 1963; Grieshaber, 1964).

It was found that both kinds of particles varied in size between 0.2 and 1.3 μ. They lose their spherical form as result of the preparative treatment, and appear irregularly shrunken (Fig. E–23a-c). Both are surrounded by a unit membrane. As will be shown, this membrane is derived from ER, in the cavern system of which the contents of the particles are synthesised (Fig. E–23d,e).

One difference was however found. Spherosomes fixed with osmic acid or permanganate show a fine granulation, whilst the oil droplets appear optically void. This granulation of the spherosomes indicates a proteinic stroma, with a definite affinity for the electron stains used. The oil droplets, on the other hand, seem to lack the ability to reduce OsO_4 or $KMnO_4$, so that they remain unstained with the usually short duration of the reaction, like starch granules. The oil droplets are formed in very large quantities and accumulate in the peripheral cytoplasm (Fig. E–26d), which is then termed fat plasma (elaeoplasm).

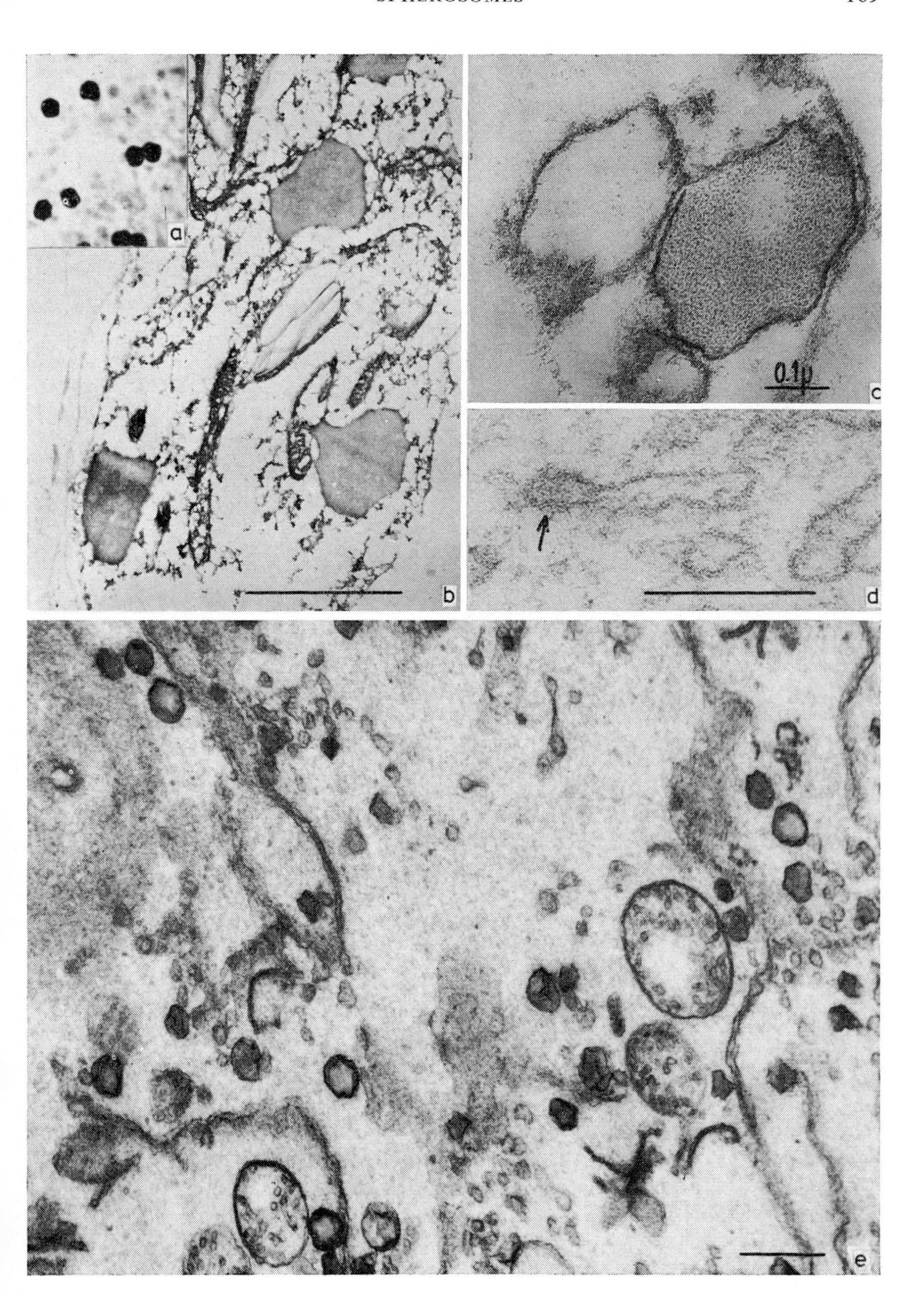

Fig. E–23. Spherosomes; (a, b) in the adaxial epiderm of onion scales (OsO_4), (a) light microscope, (b) electron miscroscope; (c, d, e) in the coleorrhiza of germinating maize ($KMnO_4$), (c) granulated stroma, (d) granulated ER bud, (e) clusters of juvenile spherosomes.

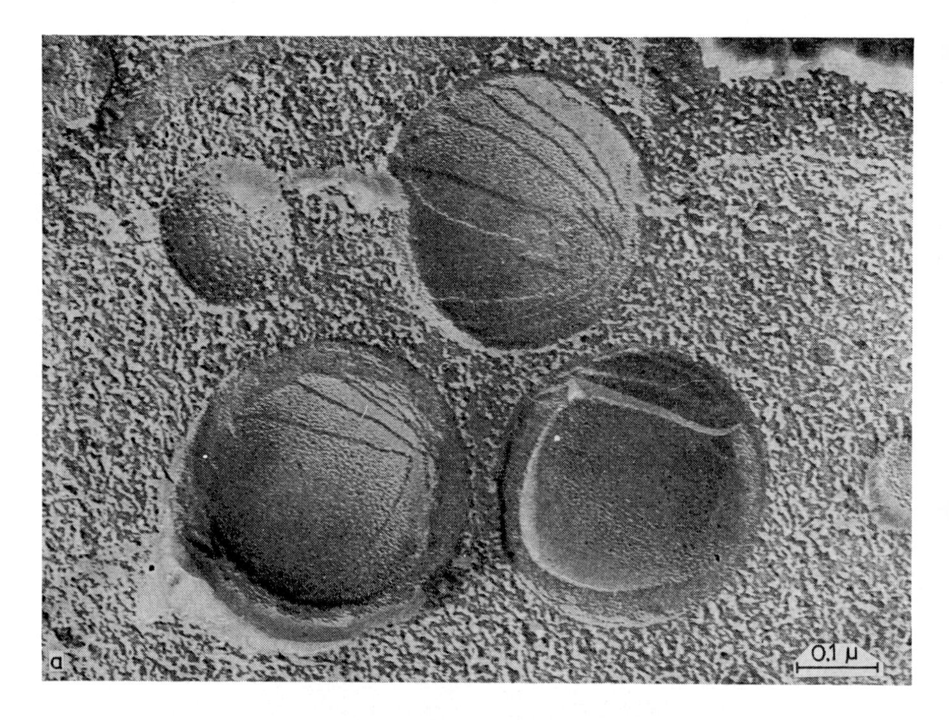

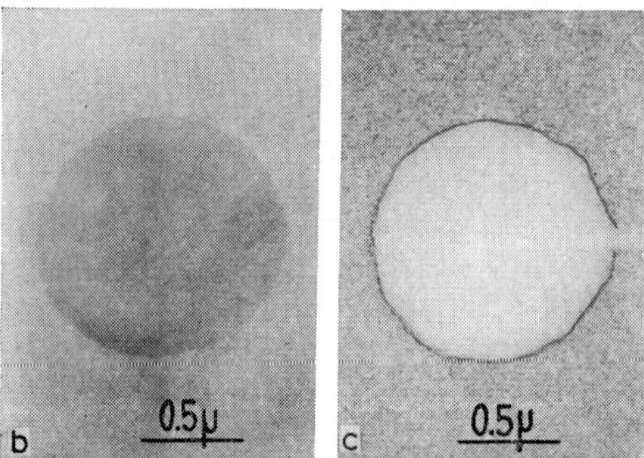

Fig. E–24. Lipid bodies. (a) Lipid particles in yeast. Bimolecular lamination. Freeze-etching (Moor and Mühlethaler, 1963). (b) Droplet of olive oil in agar, (c) in gelatin ($KMnO_4 + UO_2ac$).

A molecular layer structure of bimolecular lamellae has been detected in the fat bodies of yeast (Fig. E–24a). When olive oil is artificially emulsified in agar or gelatin, it forms droplets similar in size to those in the cell. In agar they remain bare (Fig. E–24b), while in gelatin they are surrounded by an osmiophilic layer which is visible in the electron microscope (Fig. E–24c). The ultrastructure of this boundary may correspond to an adsorption layer of globular protein molecules, but no unit membrane structure can be recognised.

(*b*) *Origin*

The development of the spherosomes and the fat droplets is similar. It begins by the accumulation of osmiophilic material in the end lobe of a reticulum strand.

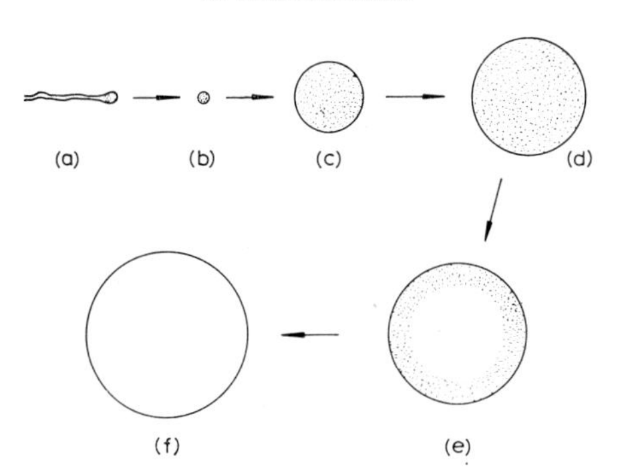

Fig. E–25. Development of spherosomes and of oil droplets; (a) terminal strand of ER, (b) de tached vesicle, (c) juvenile spherosome, (d) spherosome, (e) transition stage, (f) fat body (oil droplet)

Then by constriction a small body with a single membrane is cut off, which quickly grows to a diameter of 1000–1500 Å (Fig. E–25). In this stage, its granulated stroma is easily recognisable. Such particles have been called prospherosomes (Jarosch, 1961), although this designation is not absolutely necessary since in the course of further development these early stages undergo no structural changes until they attain the size of fully developed spherosomes (0.5–1 μ).

The oil droplets pass through the same development until they reach this stage, which means that they are formed from spherosomes. Then they begin to become dappled (Fig. E–26b) and their central part, apparently through fat storage, appears to become light (Fig. E–25e), whereupon the granulation progressively disappears like an opening iris diaphragm, so that finally a homogeneous droplet appears, which is only just enclosed by its unit membrane. This prevents coalescence of the droplets which can even touch each other in their tight packing in the elaeoplasm (Fig. E–26d) without losing their individuality.

If the spherosomes synthesise fat, they must contain the necessary enzymes. Of these, acid phosphatase recently has been demonstrated in the spherosomes of onion bulb scales by Walek–Czernecka (1962). Of all the constituents in the epidermal cells of *Allium* only the spherosomes proved to liberate phosphate from glycerol phosphate. The histochemical demonstration of phosphatase by converting the precipitated lead phosphate into black lead sulphide takes place inside the spherosomes. Neither mitochondria nor plastids show any such reaction. The objection put forward against the Nadi reaction of the spherosomes being a proof of their enzymatic activity does not apply to the demonstration of phosphatase. The attribution of mitochondrial enzymatic activity to the spherosomes seemed unlikely, but the presence of enzymes involved in fat metabolism will hardly be objected to. Thus at least the final step in fat synthesis, *i.e.* the transesterification of glycerol phosphate by exchanging phosphoric acid with fatty acids, can be attributed to the spherosomes.

If such enzymatic activity is admitted, the spherosomes may be compared with proplastids and leucoplasts which synthesise starch. There too, as in the case of the formation of fat bodies, the starch grain may grow and replace the synthesising stroma completely, so that ultimately, only the plastid membrane is left. On the basis of these considerations, the spherosome would not only be a simple precursor of the oil droplets, but a special organelle endowed with the function of fat production.

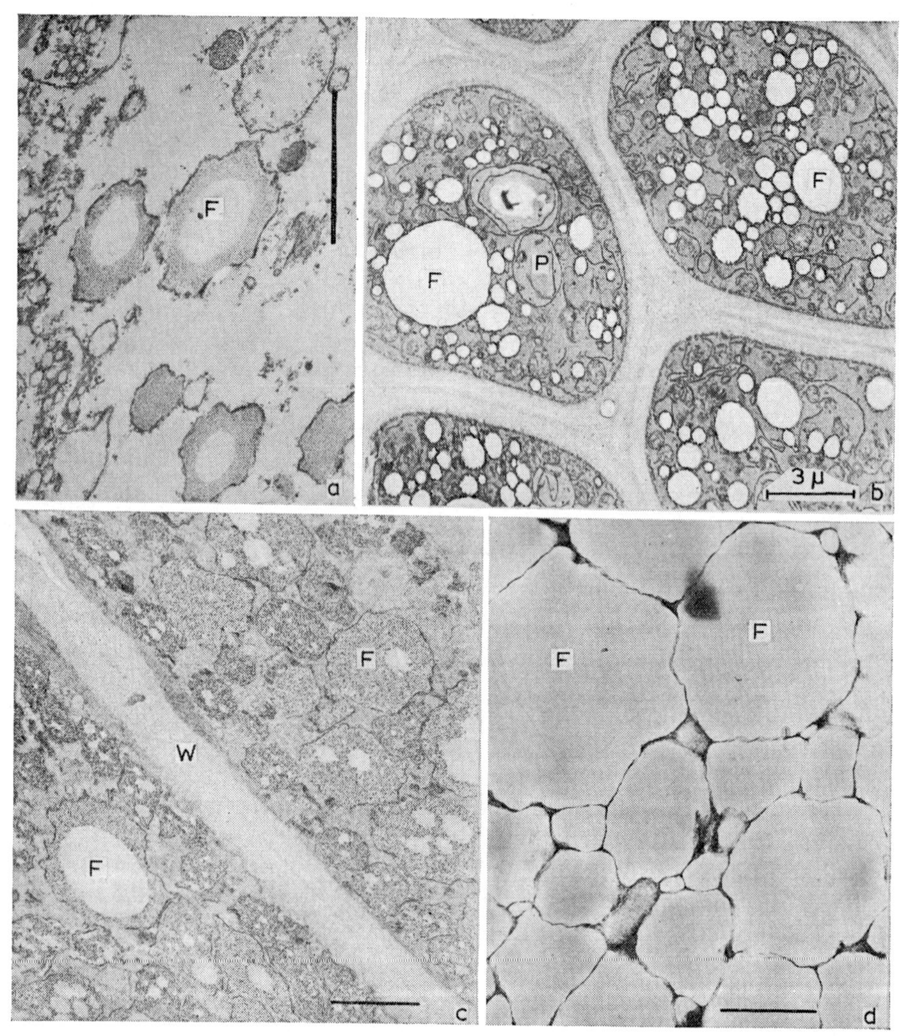

Fig. E–26. Fat bodies (oil droplets), $KMnO_4$ and UO_2ac. (a) Transition of spherosomes to fat bodies in the coleorrhiza of corn, (b) fat bodies in the scutellum of corn, (c) transition of spherosomes to fat bodies and (d) fully differentiated fat bodies in rape seed, coated by unit membranes, close packing. F = fat body, P = proplastid, W = cell wall; UO_2ac = uranyl acetate.

7. LYSOSOMES

Hepatic and renal cells contain a special group of particles which are rich in enzymes. These particles, about 0.4 μ in size can be concentrated from cell homogenates in the centrifuge. They have a lipoproteinic membrane, a densely granulated stroma (so that originally they were called 'dense bodies', Novikoff *et al.*, 1956) and a large central vacuole when fully differentiated. In this stage, they appear like small bags.

The particles contain acid phosphatase, acid ribonuclease, acid deoxyribonuclease, cathepsin, and β-glucuronase. When the particles are injured so that their membranes are broken, lysis of phosphate esters, nucleic acids, proteins, mucopolysaccharides, and sulphate esters can be performed *in vitro*. These 'enzyme bags' have consequently been called lysosomes by De Duve (1959).

Lysosomes have been found to be of general occurrence in animal cells (De Duve and Berthet, 1954; Novikoff, 1961; De Duve, 1963). Their function is manifold. In cooperation with phagosomes, they form digestive vacuoles. Autolysis or necrosis takes place if they discharge their enzymes into the cytoplasm, so that they have been termed popularly 'suicide bags'.

The very general distribution of lysosomes in the cells of various animal tissues suggests that they should also occur in plants. According to Matile (1964), the proteases, which the fungus *Neurospora* excretes in the culture medium for the extracellular digestion of proteins, are bound intracellularly to a particle fraction which can be concentrated by centrifuging, and whose activity can be studied *in vitro*. The particles in question are probably lysosomes.

As organelles storing enzymes (acid phosphatase, Avers and King, 1960; Walek–Czernecka, 1962), the spherosomes must be related to the lysosomes. Ahearn and Biesele (1965) actually believe them to be equivalent. This would be correct, if the lysosomes are defined merely as bags of enzymes. If, however, we consider their lytic capacity, it must first be ascertained whether the spherosomes are used only for the breakdown of cell substances (katabolism) or whether, as is probable (see p. 171) in the case of fat synthesis, they also participate in synthesising anabolism. As the phosphatases are effective not only in splitting off but also in the transfer of phosphate groups, the presence of acid phosphatase alone can scarcely be used for a definition of lysosomes. It is quite possible that spherosomes and lysosomes are phylogenetically the same organelle for storing enzymes away from their substrates in the cytoplasm. It would not however be surprising if these particles, according to the general principles of plant and animal metabolism, were also able to organise themselves for anabolic synthesis in plant tissues.

There is an increasing number of observations in metabolically active cells concerning 'unidentified' or 'dense' bodies which may be related to lysosomes. Buttrose (1963c) described them in developing aleuron cells of the wheat kernel and Schnepf (1964) in the glandular cells of the septal nectaries along the ovary of *Gasteria*.

8. RIBOSOMES

As was already mentioned in the enumeration of the ultrastructural inventory of the cell, osmiophilic granules which have been identified as ribonucleoprotein particles are always found in the cytoplasm. Their discovery goes back to the exact investigation of the basophilic (*i.e.* highly stainable with basic dyestuffs) granular plasma in animal gland cells. In 1943, by the separation of cell homogenates in the centrifuge, Claude was able to show that this basophilic plasma contains sublight-microscopic particles rich in nucleic acids. Claude termed these particles microsomes, thus bringing about the terminological confusion referred to on p. 168, as the name microsomes had been coined in 1880 by Hanstein for the microscopic particles in

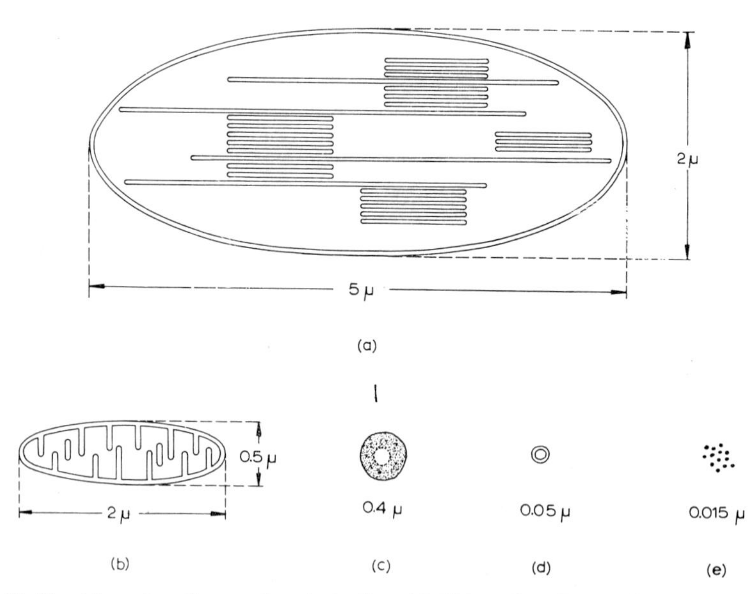

Fig. E–27. Size hierarchy of protoplasmic bodies. (a) Chloroplast ($2 \times 5\ \mu$), (b) mitochondrion ($0.5 \times 2\ \mu$), (c) spherosome (0.4–0.8 μ), (d) initial of proplastid (0.05–0.5 μ), (e) ribosomes (0.015 μ).

plant cells, which are today designated as spherosomes. To save the microsomes in Hanstein's sense of the word, Höfler (1957) proposed to change the name of Claude's microsomes into meiosomes. Subsequently however, it was found that Claude's so-called microsome fractions were not uniform with particles of the order of 0.1 μ, but still contain fragments and vesicles of the ER (Palade, 1955, 1958). By separating out impurities of this type, a uniform particle fraction could be obtained with particles having an average diameter of 150 Å; because of their high RNA content, these particles are today called ribosomes, thus disposing of the terminological controversy over the name microsomes. Figure E–27 shows the order of magnitude of the ribosomes as compared with other cell organelles.

(*a*) *Ultrastructure*

The ribosomes are osmiophilic and are therefore shown as black granules after fixation with osmic acid. On the other hand, when fixed with permanganate, they appear to be unable to store up manganese, so that the ribosome-containing cytoplasm does not show to be structurally granular, but rather homogeneous. As has already been said, in differentiated cells the ribosomes are found lined up on the plasma side surface of the ER membrane. In osmium preparations, therefore, the entire endoplasmic reticulum is frequently fringed with dark granules. This has led to the ribosomes being described as an integral part of the reticulum. However, in young meristematic plant cells, the ribosomes lie free in the cytoplasm and the boundary layers of the still sparingly developed reticulum show no evidence of being vestured with ribosomes. In the extent, however, as the reticulum grows through the entire cytoplasm, it comes up against the ribosomes, which then arrange themselves on the boundary layers, without growing firmly with them, since they can easily be detached from the membrane of the reticulum.

Falk (1961) has described a spiral arrangement of ribosomes in the rootcap of *Allium cepa*, just as they are reported to occur in the pancreas (Haguenau, 1958). This arrangement in a plane should not be confused with Strugger's picture (1956, 1957) obtained from the same subject, in which ribosomes appearing disposed in linear parallel rows were interpreted as sections through spatial helices with a diameter of 300–600 Å. According to this interpretation, helical threads termed cytonemata have been supposed to occur in the cytoplasm. This view is certainly incorrect, since with the great focal depth of the electron microscope we should be able to see the entire coils of the wrongly postulated micro-helices, and not only their optical sections.

In 1962, Warner *et al.* observed clusters composed on average of 5 ribosomes in preparations of lysed blood reticulocytes *in vitro* which they named polyribosomes, or briefly, polysomes. As the case may be, much more numerous particles can be united into a polysome (Fig. E–31). Since such ribosome groups, which appear to be of general occurrence, for example, in *Escherichia coli* (Huxley and Zubay, 1960) and in yeast (Koehler, 1962), actively participate in protein synthesis, polysomes occasionally have also been called ergosomes (Staehelin *et al.*, 1963).

Ribosome size does not appear uniform in the electron microscope. Porter (1961) quotes 150 Å for the ribosomes of liver and pancreas, and Szarkowski *et al.* (1960), in conformity with the data of many other authors, found an average of 125 Å with a variation of 100–150 Å for plant ribosomes. Thus until the cause of this discrepancy in the particle size is cleared up, the ribosome problem can scarcely be considered as being finally settled.

The internal structure can be studied in isolated ribosomes. Such isolation is achieved by fractional centrifuging of cell homogenates, as can be seen from Table E–V.

TABLE E–V

FRACTIONATION OF CELL HOMOGENATES FREE FROM CHLOROPLASTS
(Bonner, 1959)

Acceleration	Time	Sediment
300 G	a few min	cell fragments
1000–2000 G	a few min	nuclei
12,000 G	a few min	mitochondria
100,000 G	30–60 min	ribosomes

According to the magnesium content in the dispersion medium, isolated coli ribosomes fall into two sub-units, or aggregate to double particles; addition of a magnesium salt promotes the aggregation, and removal of Mg^{2+} causes the splitting (Bonner, 1960). The two sub-units differ in size and therefore exhibit different sedimentation constants, *S*, in the ultracentrifuge (Fig. E–28). After negative contrast staining, the larger unit appears round in the electron microscope and the smaller is flattened into the shape of a cap (Huxley and Zubay, 1960).

Since ribosomes are about 50% protein and 50% RNA, it is of interest whether a similar division of these two substances occurs here as in the case of the viruses (see p. 127). However, this is not the case. Both sub-units (50 *S* and 30 *S*, Fig. E–28) show the same protein/nucleic acid ratio and no concentration of RNA in the centre of the particle can be established (Watson, 1963); it appears to be regularly distributed. It can be separated from the ribosome protein and sedimented in the ultracentrifuge, the sedimentation constants obtained being about half as large as for the sub-units (Fig. E–28). The RNA of the sub-unit 30 *S* has a degree of polymerisation (DP) lying between 1500 and 2000, and that of the sub-unit 50 *S* between 3000 and 4000. Although the ribosomes have been recognised as the organelles of protein synthesis, it has not yet been possible to detect any special participation for this highly polymeric RNA in those processes. It is thus accepted as inert (Wittmann, 1963). Its activity is presumably blocked by bonding to proteins, so that ribosomal RNA plays more of a morphological role than a metabolic one, as a structural element, in analogy to the structural proteins.

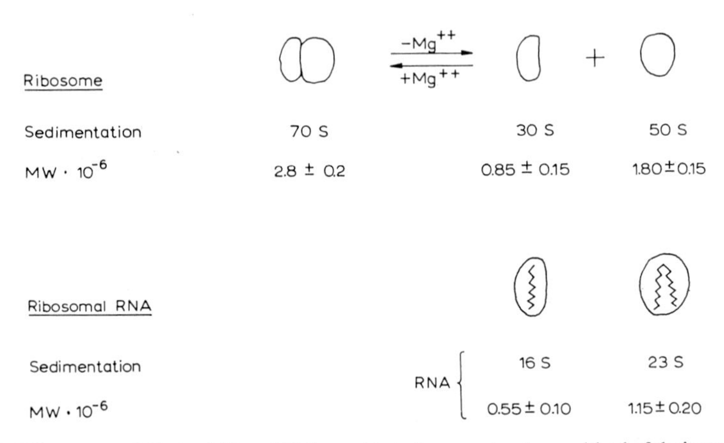

Fig. E–28. Ribosomes of *Escherichia coli* fall apart into fragments when robbed of their magnesium. S = sedimentation constant (Watson, 1963).

Apart from the RNA in the ribosomes, which is sometimes termed ribosomal nucleic acid (r-RNA), there are threads of RNA, to which the individual ribosomes are attached forming the polysomes. Therefore, polysomes isolated in the centrifuge can be broken down by ribonuclease into the individual ribosomes. In contrast to ribosomal RNA, the above-mentioned RNA threads play an important role in protein synthesis.

(*b*) *Function*

There is no doubt that ribosomes fulfil an extremely important function in cell metabolism. They are capable to catalyse the synthesis of polypeptides from amino acids even *in vitro* (Tsugita *et al.*, 1962). RNA plays a decisive part in these syntheses, since under ultraviolet irradiation which disintegrate nucleic acids, the ribosomes lose their protein-synthesising ability. On the basis of such determinations, the ribosomes have been stated to be the centres of protein synthesis in the cell.

The question, therefore, arises whether individual ribosomes, *i.e.* cell particles of the order of magnitude of globular macromolecules, can be validly regarded as organelles, since otherwise this concept could with equal right be extended to all inpividual macromolecular holoenzymes.

In actual fact, it can now be shown that ribosomes themselves are incapable of protein synthesis, but that this requires a more complicated apparatus. Since in the *in vitro* synthesis, polypeptide chains with specific amino acid sequences are formed, the system must receive the genetic information it requires for this action from the nucleus. This was originally explained by assuming that the ribosomes were formed in the nucleolus of the nucleus and then migrated out through the large pores of the nuclear envelope carrying the genetic information into the cytoplasm (Bonner, 1958). It is now known, however, that the nucleolus is composed of ribosome-like particles (Fig. F–6), which stay where they are and presumably build nuclear proteins *in situ*, whilst the transmission of genetic information is achieved by a special RNA thread (Ts'o, 1962) which is called messenger RNA (m-RNA) (Jacob and Monod, 1961).

Messenger RNA obtains its information from the DNA of the chromosomes, in contact with which it is formed. It is therefore equipped with the code of the base triplets, which is necessary for the construction of specific amino acid sequences (see p. 112). In the cytoplasm outside the nucleus, m-RNA is adsorbed by ribosomes, and polypeptide polymerisation then takes place on its surface. Natural m-RNA with its complicated code has been successfully replaced *in vitro* by artificial RNA molecules having a simpler base sequence. If, as was described on p. 112, a mixture of amino acids and polyuridylic acid, *i.e.* an RNA whose nucleosides consist entirely of uridine, is added to a system of ribosomes capable of protein synthesis in the presence of the requisite energy source (ATP), then a polypeptide is formed with only phenylalanine members (Nirenberg and Matthaei, 1961). It is thus clear that the UUU triplet of messenger RNA can seek out phenylalanine molecules out of the amino acid mixture offered for polymerisation.

In studying the mechanism of how a triplet of the code selects the correct amino acid, a third kind of RNA molecule was discovered, which was low molecular and therefore easily soluble. This was consequently termed soluble RNA (s-RNA) or, since it transfers amino acids, transfer RNA (t-RNA). The nucleotide sequences in this transfer RNA have been partly elucidated; we know the terminal members of this relatively short chain with 70–80 nucleotides (Wittmann, 1963). A different transfer RNA is found for each amino acid, which at one end carries the relevant amino acid and at the other end the complementary bases to the corresponding code triplet of messenger RNA. The code triplets for some amino acids according to Nirenberg (1963) and the complementary triplets of transfer RNA are presented in Table E–VI.

A triplet of nucleotides is called a codon, each codon representing a particular amino acid.

Fig. E–29 gives a schematic representation of how, according to the present state of our knowledge, protein synthesis is conducted (Nirenberg, 1963; Ochoa, 1964a). Jacob and Monod (1961) proposed that each gene, or with a new term each DNA cistron, acts as a template for synthesis of messenger RNA molecules to which the cistronic polynucleotide sequence is transcribed. The nascent m-RNA

TABLE E–VI

NUCLEOTIDE TRIPLETS (CODONS) FOR THE SELECTION OF AMINO ACIDS
(Nirenberg, 1963; Ochoa, 1964a)

Amino acids (arranged alphabetically)	Triplet of code along m-RNA	Complement end triplet of t-RNA
Ala	UCG	AGC
Arg	CGC	GCG
Cys	UUG	AAC
Glu	AGU	UCA
Gly	UGG	ACC
Lys	AAU	UUA
Met	UGA	ACU
Phe	UUU	AAA
Pro	CCU	GGA
Try	GGU	CCA
Tyr	AUU	UAA
Val	UGU	ACA

enter into temporary union with ribosomes which synthesise the polypeptide chain. Their amino acid sequence is inscribed as a succession of codons. The amino acids themselves are temporarily attached to the t-RNA which has a codon complementary to those of the m-RNA. They are thus aligned in the manner prescribed by the code next to one another, and are polymerised into the specific polypeptide chain. At the same time, the amino acids free themselves from the t-RNA chain, which for its part detaches itself from the m-RNA. These processes must take place at a surprising rate, since it can be calculated that only 5 sec are needed for the polymerisation of a polypeptide chain to a molecular weight MW of 20,000 (Staehelin *et al.*, 1963).

The establishment of the fact that the ribosomes are held together in the polysomes by the m-RNA, shows that this system serves for the polymerisation of polypeptide molecules. It is also assumed that the 15 Å thick m-RNA thread passes through the contact plane of the 50 *S* and 30 *S* sub-units (Watson, 1963).

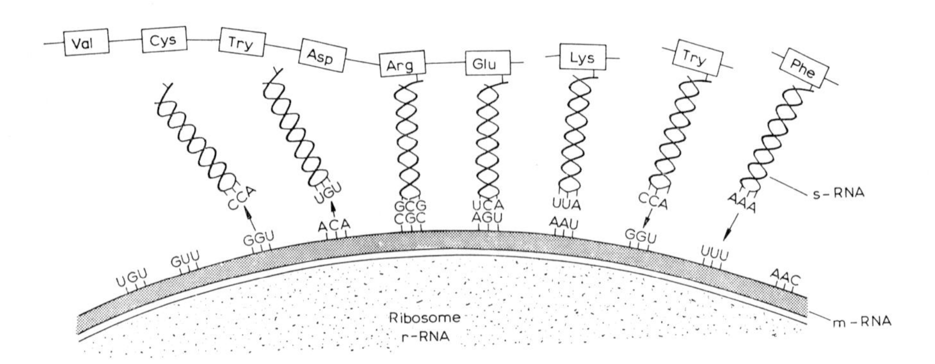

Fig. E–29. Operation of the ribosomes (after Nirenberg, 1963).

The m-RNA thread is considerably longer than the circumference of a ribosome. For a protein of MW 20,000 with about 200 amino acids, 200 codons, *i.e.* a DP of 600 is required; as the length increment per nucleotide in the RNA chain is 2.5 Å, such an m-RNA thread must be at least 1500 Å long, *i.e.* about 10 times longer than the diameter of a ribosome. For this reason, Goodman and Rich (1963) outlined a more dynamic model as compared with the static scheme of Fig. E–29. These authors postulated migration of the ribosome particle along the RNA chain with its information (Fig. E–30). An oncoming ribosome may traverse the entire code over 1000 Å long. The polypeptide chain grows during this process, as indicated in Fig. E–30. The ribosome and the completed polypeptide chain are released at the end of the m-RNA chain. A workable model for the control of this mechanism has been postulated by Jacob and Monod (1961) and by Stent (1964).

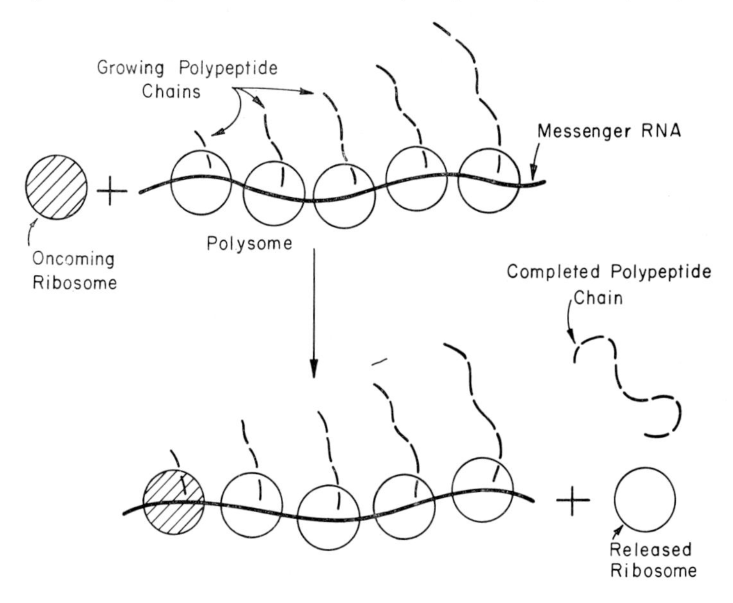

Fig. E–30. Ribosomes moving along an RNA chain to acquire information on the specific amino acid sequence of the polypeptide chain undergoing synthesis (Goodman and Rich, 1963).

This picture reveals the polysome in contrast to the individual macromolecular ribosome as the real organelle. It has been found for the synthesis of haemoglobin (Warner *et al.*, 1962), as well as in reticulocytes, liver cells, HeLa cells, slime moulds (*cf.* Goodman and Rich, 1963), and in coli bacteria (Staehelin *et al.*, 1963). Although polysomes have not yet been isolated from higher plants, the spiral arrangements of ribosomes as reported (p. 175) might indicate the possibility of polysomal interrelations in such agglomerations.

As a rule, a polysome consists of 4 to 6 (mostly 5, and in *Escherichia coli* up to 11) ribosomes with a long messenger RNA thread which links the ribosomes together. In HeLa cells, the polysomes contain 30–40 ribosomes. In cells infected by the poliomyelitis virus, polysomes consisting of 50–70 ribosomes have been observed (Fig. E–31). A broad distribution of polysome sizes implies that a cell contains messenger RNA of many different lengths. Some of the long m-RNA strands associated with polysomes that contain 20 or more ribosomes may carry information for the synthesis of more than one kind of polypeptide chains (Rich, 1963).

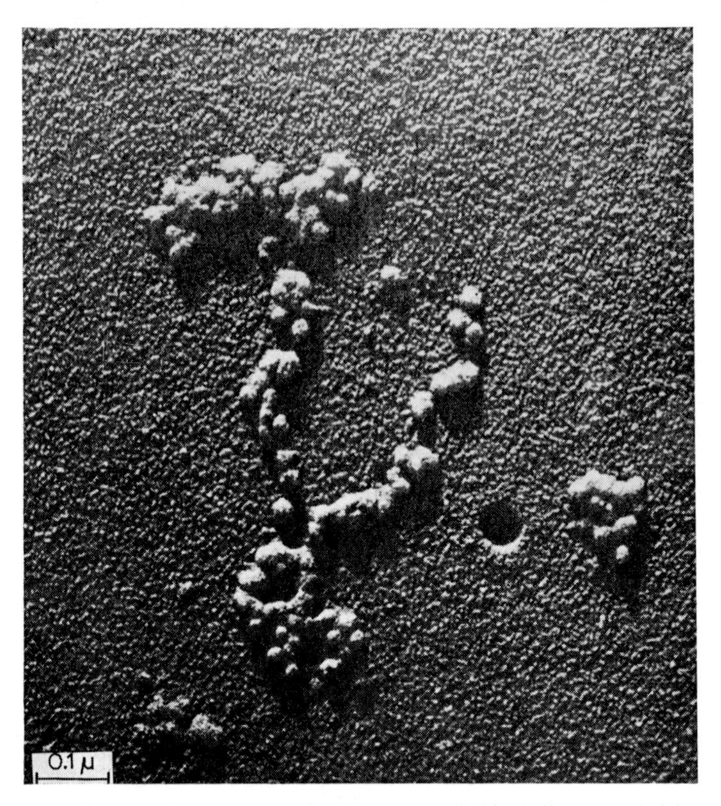

Fig. E–31. Polio virus polysomes with at least 50 individual ribosomes (Rich, 1963).

The m-RNA is a single polynucleotide chain, and its codons are thus free for action. They attract the transfer RNA, which is a rather short paired chain (DP 70–80). Only the ends of the t-RNA double chain are active, one for the attachment of a specific amino acid and the other with a free base triplet for combination with the code triplet of m-RNA.

(*c*) *Origin*

The ribosomes could be formed *de novo* through reconstitution from the groundplasm, by autoreplication, or in the nucleus. Certain facts have been brought forward which speak in favour of nuclear origin, for in contrast to the chromatin the nucleolus contains RNA but not DNA (Caspersson, 1941). Moreover, when viewed in the electron microscope, the nucleolus consists, as already mentioned, of individual particles the size of the ribosomes (Fig. F–6). If cells are fed with labelled uracil, the radioactivity first appears in the nucleolus, indicating the latter to be the site of RNA synthesis. Subsequent appearance of radioactivity in ribosomes outside in the cytoplasm shows that RNA migrates from the nucleus. As has already been stated, whole ribosomes were initially considered to migrate. In actual fact, however, messenger RNA alone leaves the nucleus, after it has taken over the code of the hereditary information from the DNA of the chromosomes, while the particles of the nucleolus, comparable with the ribosomes, remain in the nucleus (Wittmann, 1963).

The structural RNA of cytoplasmic ribosomes in pea seedlings can be hybridised with denatured DNA (p. 111) of their nuclei. The same is true of RNA of the nucleoli. It is therefore concluded that the bulk of the ribosomal RNA is manufactured by non-nucleolar regions of the chromatin, and that ribosomal RNA is then transferred to the nucleolus for final assembly into nucleolar ribosomes (Chipchase and Birnstiel, 1963).

It must therefore be assumed that the ribosomes in the groundplasm, like those of the nucleolus, are formed *in situ*. Another site of protein synthesis has been found in the mitochondria, and there ribosomes also appear to be involved, though the mechanism of their formation is still unknown. Whether they arise *de novo* or through autoreplication remains to be determined. Since their function in protein synthesis is rather unspecific, in that they simply provide the enzyme structure required for amino acid polymerisation, their mode of formation may be less complicated than the code transfer for the building up of specific proteins by the messenger RNA.

F. Nucleus

The cell nucleus was discovered in 1833 in plant material by Robert Brown. Using cells of the stamen filaments of *Tradescantia*, this worker observed, in addition to the chaotic movement of very small particles (now known as Brownian motion), 'areolae' slowly displaced by the plasma stream, and termed them 'nuclei' (see Küster, 1933). A synonymous term for this organelle is the Greek word *karyon*.

In the active nucleus, classical cytology distinguishes, apart from the nucleolus, between the denser regions which can be stained with basic dyes or with Feulgen (p. 107) and which are consequently called chromatin, and the non-staining ground substance, achromatin. In the living nucleus, this is a colloidal liquid and is thus called nuclear sap or karyolymph; it may also be gel-like, in which case it is called karyoplasm. It is probable that a more liquid and a rather solid phase occur together. Since however this cannot be definitely decided at the present time, the entire ground substance is termed nucleoplasm in the present discussion. The contents of the nucleus are separated from the cytoplasm by a nuclear envelope.

The chemical composition of the nucleus was first determined by bulk analysis of fish spermatozoa. Today it is no longer necessary to use sperm nuclei in the study of nuclear biochemistry, since intact nuclei and even nucleoli may be collected in cane sugar solutions and $CaCl_2$ from broken plant cells (Accola, 1960, onion root tips; Johnston *et al.*, 1959, and Birnstiel *et al.*, 1962, pea embryos).

The nuclear components can be separated into proteins and phosphorus-containing nucleic acids. Other compounds, such as lipids, (Hirschle, 1942) are present in insignificant amounts. In 1963, Birnstiel *et al.* reported the following values for isolated nuclei of pea seedlings:

DNA	14.0% by weight	Basic proteins	22.6% by weight
RNA	12.1% ,, ,,	Non-basic proteins	51.3% ,, ,,

The basic proteins are protamines and histones, rich in the basic amino acids such as lysine, histidine, and especially arginine. They combine with the nucleic acids in a salt-like manner, forming nucleoproteins. The non-basic proteins seem not to be related to nucleic acids, and can occur as separate phases in the nucleus or as structural protein in the anucleal segments of the chromosomes (see p. 194).

Both DNA and RNA are present in the nucleus, but they are not evenly distributed. The nucleolus is rich in RNA, whilst the chromatin contains DNA only. The above-mentioned staining properties with Feulgen of the chromatin is due to DNA.

1. NUCLEAR ENVELOPE

The nature of the nuclear envelope has been an object of discussion, since the envelope can only be made visible under a light microscope by external interference, such as plasmolysis, fixation, or staining. Many cytologists considered that it is not a consistent membrane, but merely a phase boundary (Luyet and Ernst, 1934; Pischinger, 1950). Other authors, however, believed it to be a real envelope, the birefringence of which has frequently been found to differ from that of the nucleus itself. In 1939, Schmidt discovered evidence of an optically negative spherite texture in the boundary layer of the nucleus (lamellar birefringence). According to F. O. Schmitt (1938) the sign of the spherite cross is reversed after imbibing with glycerol, urea, or sugar solutions; this procedure would neutralise the form birefringence, and the intrinsic birefringence of radially oriented lipids would become apparent. Baud (1949) emphasized that there is no optical anisotropy in living nuclei, and that a birefringent nuclear membrane surrounded by a birefringent perinuclear zone appears only after fixation; the optical anisotropy of which is that of a negative spherite, indicating a lamellar protein texture.

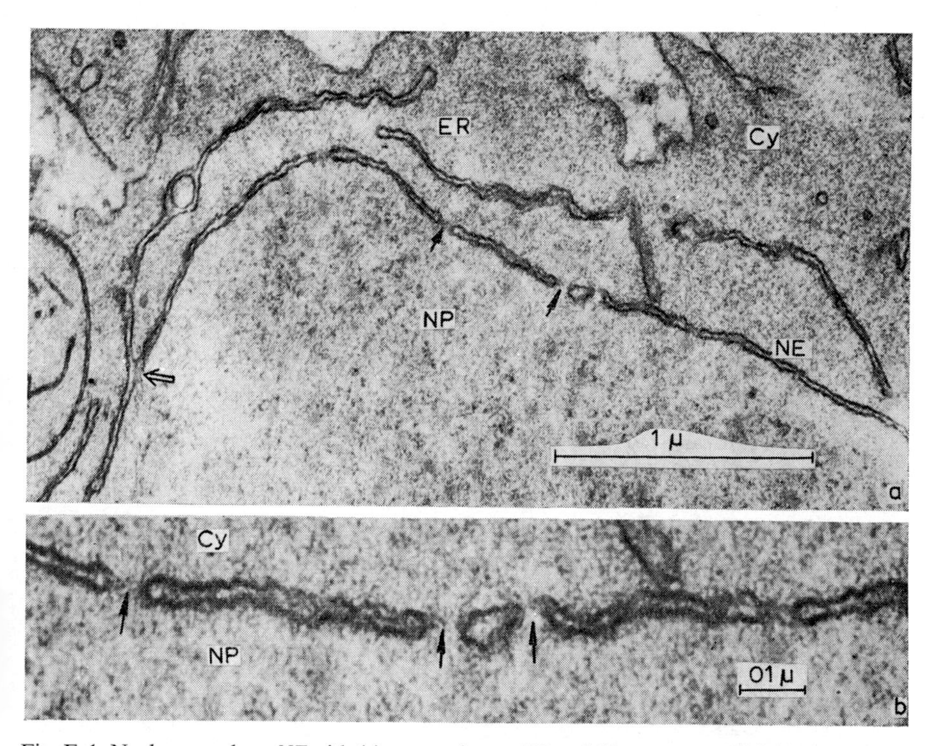

Fig. F–1. Nuclear envelope *NE* with (a) connection to *ER* and (b) pores ← establishing continuity between nucleoplasm *NP* and cytoplasm *Cy*. Cell in root of *Ricinus communis*.

(*a*) *Ultrastructure*

Electron microscopy has now confirmed the presence of an organised nuclear envelope in the living cell; all nuclei studied up to now are found to be surrounded by a double membrane (Fig. F–1). The pair of membranes is separated by a gap of varying width, known as the perinuclear space. Its diameter may vary from 100 to several hundred Å. It is bounded on both inside and outside by unit membranes showing their more contrasted osmiophilic stratum towards the cytoplasm and the nucleoplasm. It may be assumed that the space surrounding the cell nucleus is filled with a low-density liquid, *i.e.* with a kind of serum.

This conjecture is supported by the fact that numerous direct connections with

Fig. F–2. Surface view on the nucleus with pores (*Allium cepa*) (Branton and Moor, 1964). Nuclear envelope intact, on top with removed outer layer.

strands of the ER have been established (Watson, 1955; Marinos, 1960). The gap between the nuclear membranes must thus contain the same serum-like fluid, known as enchylema, as the endoplasmic reticulum. This relationship also explains why inclusions of ergastic and secretory nature occur in the perinuclear space, just as in the ER.

Certain authors (*e.g.* McAlear and Edwards, 1959) have tried to establish a direct connection with the plasmalemma through the ER channels. Since, however, the plasmalemma is a simple unit membrane, this would mean that the perinuclear space is in direct open connection by capillarities with the environmental medium of the cell (Robertson, 1959). On close examination of the photographs concerned, however, no direct continuity can be established between the plasmalemma and the ER, such as is found between the nuclear envelope and ER (Whaley *et al.*, 1959). Since a direct relationship between the endoplasmic reticulum and the plasmalemma is unlikely (p. 157), there is also no possibility of a straightforward ontogenetic connection of the nuclear envelope with the plasmalemma.

A further general property of the nuclear envelope is its porosity. In contrast to the plasmalemma, it is penetrated by large pores, 200–300 Å in diameter. Such pores were first described in the membranes of the large nuclei of amphibian oöcytes (Callan *et al.*, 1949) and amoebae (Bairati and Lehmann, 1952). They are also found in all plant cell nuclei (Figs. F–2 and F–1). The pores provide direct contact between the cytoplasmic groundplasm and the nucleoplasm. These two components of the active cell must be equated with one another, although the nucleoplasm may occur as a liquid sap (karyolymph). Apparently the gel $\rightleftharpoons$ sol transformation described (p. 27) for the matrix of the cytoplasm (groundplasm) is equally valid for the nucleoplasm. It has been shown to be a nucleic acid-free protein colloid in which 12 different amino acids can be demonstrated by paper chromatography (Brown *et al.*, 1950). In contrast to this plasmic contact, there is no open connection between the enchylema of the ER strands or the perinuclear space on the one hand, and the groundplasm or the karyolymphatic nucleoplasm on the other. Whilst macromolecules can be freely translocated through the pores of the nuclear envelope, the exchange of metabolites with the enchylema is only possible by passage through the permeability barrier of a unit membrane. Nucleoplasmic sap (karyolymph) and the sap of the plasmic reticulum (enchylema) are thus cell fluids of quite different character. In any event, only the enchylema may be considered as a serum, whilst the term karyolymph is rather ill-chosen; it should certainly not be considered as an ergastic sap, but as a metabolically and morphogenetically active cell component. The name nucleoplasm (or karyoplasm) should be preferred.

Ringshaped structures called annulus were observed around the nuclear pores in animal objects (oöcytes, Afzelius, 1955; Wischnitzer, 1958; hepatic cells, Watson, 1959). No such differentiation has however as yet been described in the nuclei of plant cells.

(*b*) *Function*

The pores do not show a static pattern but form changing sieve areas. Using yeast cell nuclei, Moor has shown that, judging from marks on the surface of the nuclear envelope, the pores can be closed under certain circumstances (Fig. F–3). Formerly, Dawson *et al.* (1955) had established that the openings of the nuclear

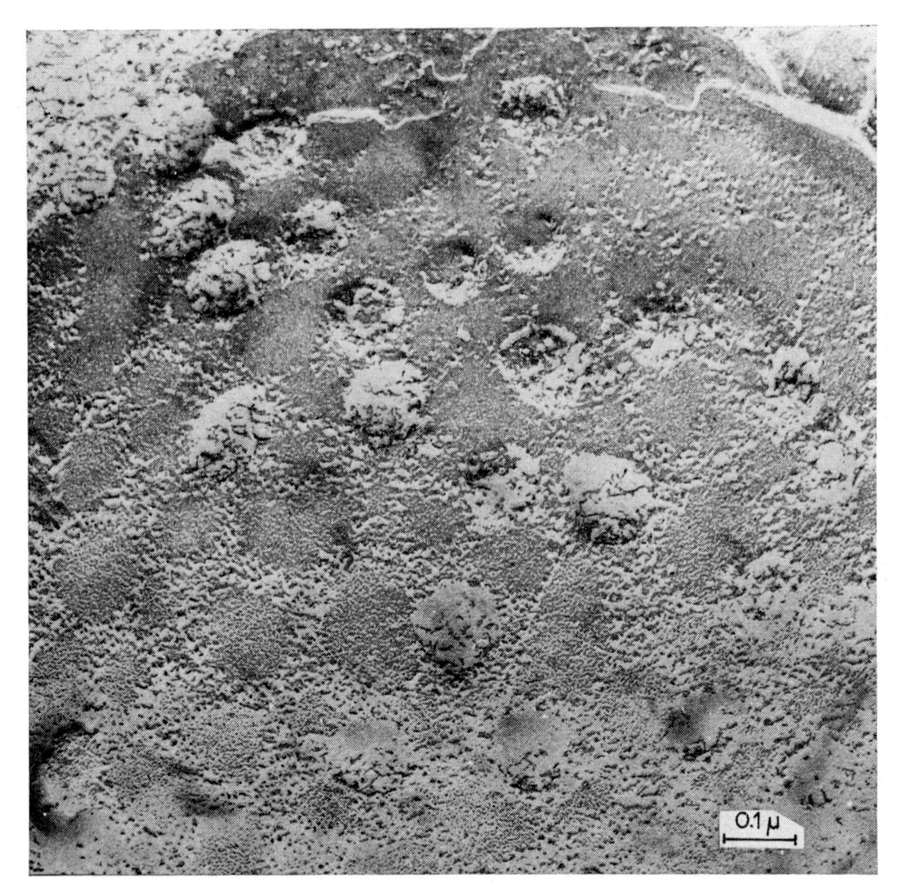

Fig. F–3. Yeast nucleus with open and healed pores. Freeze-etching (Moor and Mühlethaler, 1963).

envelope are not permanent. The physiological factors governing the formation, closing, and as the case may be, reforming of the pores remain at present unknown. Fig. F–3 shows that the active openings are assembled to a pore field, whilst the closed pores are distributed over the whole surface of the nuclear envelope.

It is difficult to attach a definite functional significance to the nuclear envelope, in view of its coarse perforations and the analogous natures of the groundplasm and the nucleoplasm. The fact that it is lacking around the chromatin areas of the nucleoids in Prokaryonta (see p. 196), and that it temporarily vanishes at each mitosis (see p. 188) makes it appear to be merely a mechanical boundary between the cytoplasm and the nuclear region. On the other hand, the active opening and closing of the pores points to a regulation of the escape of specific nuclear substances (messenger RNA, morphogenetic hormones, etc.). If the chromosomes in the active nucleus occupy fairly definite positions, the emission of their genetic code could be hindered, for example, by the closing of the neighbouring pores. Such a consideration may be correlated with the surprising and far too neglected fact that of the many genes present in a nucleus, only a restricted number participate actively at a definite moment in each ontogenetic phase of the cell, the tissue, and the organism, while the others only intervene at a later stage or remain permanently latent in specialised cells.

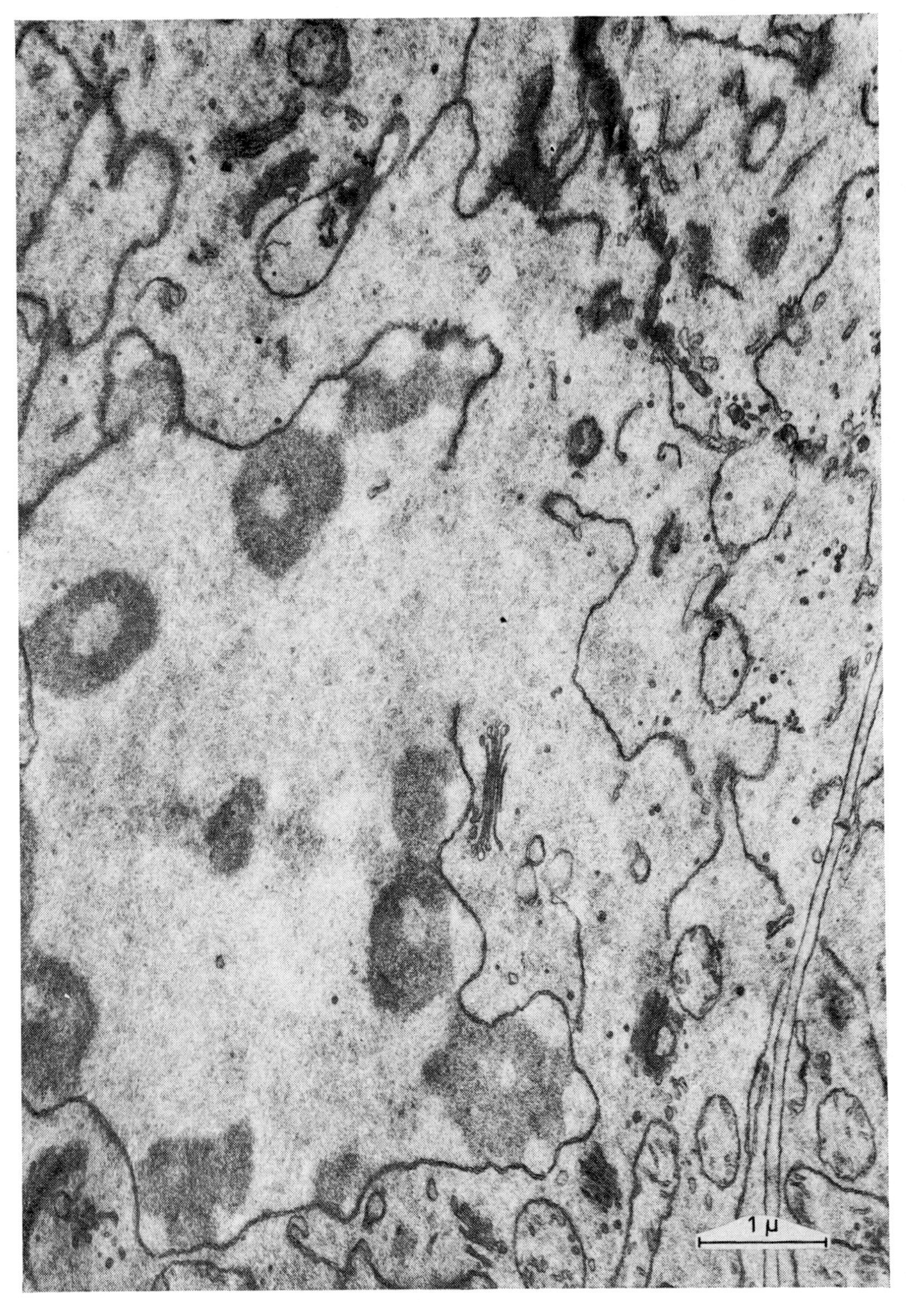

Fig. F–4. Reconstitution of the nuclear envelope during the telophase in mitosis. Cell from root tip of *Phalaris canariensis* (photo J. López-Sáez).

(c) Origin

Formation of the nuclear envelope can be followed at the end of each mitosis. According to classical cytology, the nuclear membrane disappears at the end of prophase and re-appears at the end of telophase. Hassenkamp (1957) observed the disintegration of the double membrane in electron micrographs, even in the pachytene of meiotic prophase, and in 1960 Porter and Machado showed that the membrane does not quite disappear, but merely breakes up into fragments which remain around the chromosome region during the whole period of cell division. Under these circumstances, the elements of the dispersed nuclear envelope cannot be differentiated from the strands of ER with which they mix. Towards the end of the telophase, a new nuclear envelope is reformed from the surrounding elements of ER; it is thus a part of the endoplasmic reticulum (Fig. F–4). This conclusion is also arrived at from the regularly observed evaginations of the outer membrane of the nuclear envelope, and the open communication of the perinuclear space with the cisternae of the ER.

According to whether the evaginations of the nuclear envelope are formed only in the outer membrane or in the double membrane (Fig. F–5), either strands of ER containing enchylema, or double-walled initials of organelles (see p. 228) containing stroma originating from the nucleoplasm, are formed (Gay, 1956; Mühlethaler and Bell, 1962).

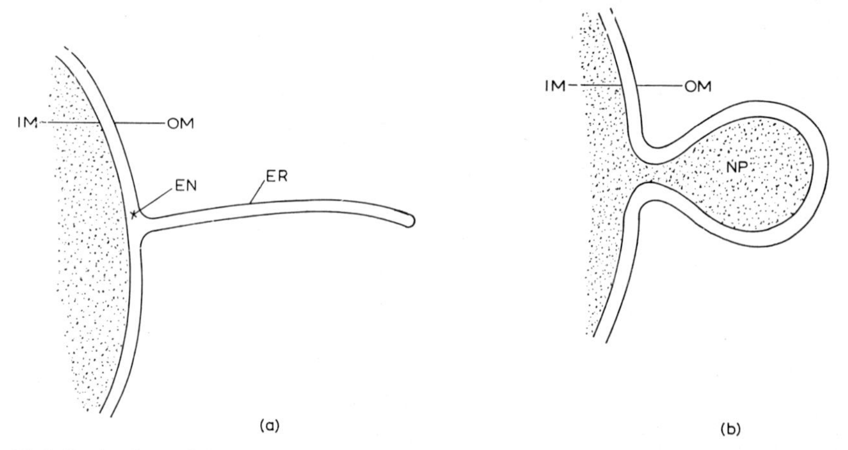

Fig. F–5. Derivatives of the nuclear envelope; (a) evagination of outer membrane *OM* including enchylema sap *EN* → *ER*, (b) evagination of double membrane (outer membrane *OM* and inner membrane *IM*) including nucleoplasm *NP* → organelle initials.

According to Gay, such 'blebs', with a double membrane appear in the salivary gland cell nuclei of *Drosophila*, especially where the nuclear envelope is in contact with the chromosomes. Foldings of the inner membrane into the nucleoplasm have also been observed (Hoshino, 1961), but this may indicate a pathological condition.

The derivation of the ER and the initials from the nuclear envelope would indicate that the plasmic cell organelles are possibly of nuclear origine, a fact which would stress the importance of the nucleus as the central organelle.

2. NUCLEOLUS

(a) *Ultrastructure and origin*

By special staining (*e.g.* with Pb) the nucleolus can be made to appear dark, like the chromatin, under the electron microscope (Fig. F–6). Both show the ultraviolet absorption maximum at 260 nm, characteristic of nucleic acids (Caspersson, 1941). Whilst however the chromatin is characterised by DNA (Feulgen-positive), the nucleolus contains RNA (Feulgen-negative), and can thus be degraded enzymatically with ribonuclease. It shows a specific reaction with the acidic dye, methyl green, allowing differential staining under the light microscope for comparison with the red Feulgen reaction of chromatin (Semmens and Bhaduri, 1939). This double staining technique has become important in the problem of the formation of nucleoli.

Initial work on the ultrastructure of the nucleolus seemed to indicate a coiled filament, which was called a 'nucleonema' and to which a diameter of 90–180 nm

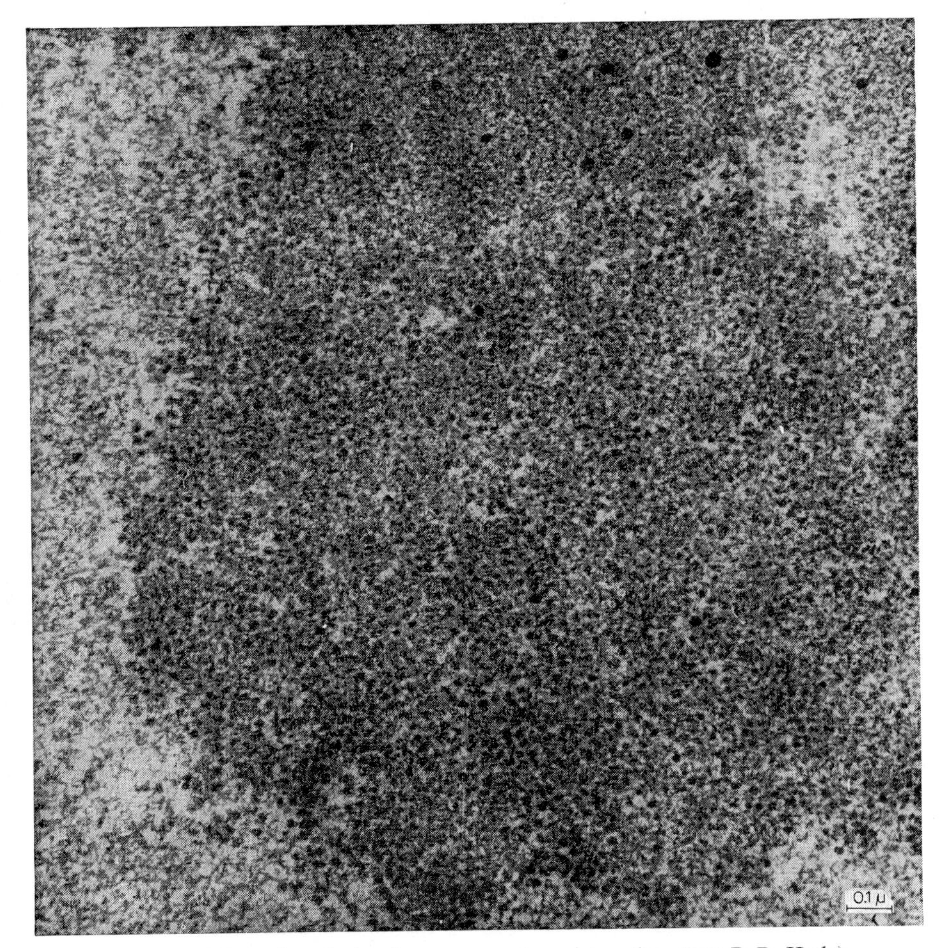

Fig. F–6. Edge of nucleolus from pea root meristem (courtesy B. B. Hyde).

(Bernhard *et al.*, 1955; Peveling, 1961) was assigned. On better fixation, however, the nucleolus was found to consist of coarse granules of the size of ribosomes – diameter about 150 Å (Mühlethaler, 1959/61; Falk, 1962a).

The nucleoli disappear during mitosis and reappear during the telophase at a predetermined place in the nuclear area. Two or even more nucleoli are frequently formed, which condense in contact with special chromosomes (Fig. F–7) with secondary constrictions (Heitz, 1935; Håkansson and Levan, 1942). Under light microscopes, the nucleoli appear to behave initially like vacuoles, for in the presence of several chromosomes capable of condensing nucleolar material, the nucleoli formed can subsequently coalesce into a larger one.

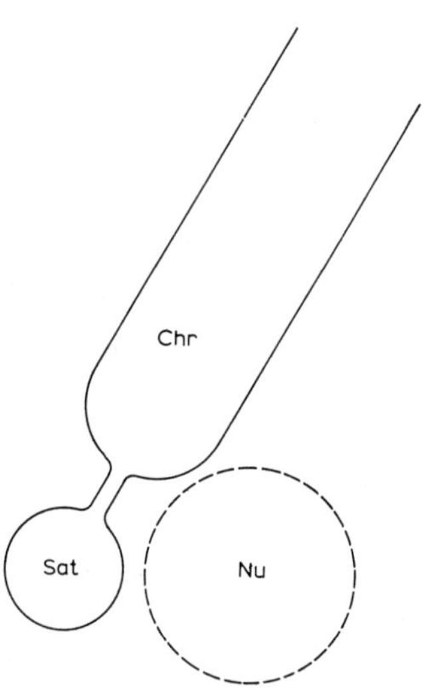

Fig. F–7. Chromosome *Chr* with secondary constriction and satellite *Sat* condensing nucleolus *Nu* (Heitz, 1935).

This means that the nucleolar matrix is a special form of nucleoplasm, which is liquid when the nucleolus is formed. In the giant chromosomes of the *Drosophila* salivary glands, the nucleolus originates not next to an anucleal constriction, but in contact with a DNA region which is thought to produce the nucleolar substances (Stich, 1956). Thus it seems that the nucleolar RNA is produced in cooperation with the DNA of the chromosomes.

(*b*) *Function*

The interpretation of the function of the nucleolus is an old problem. Its protein content and its behaviour during mitosis lead to the conjecture that it stores reserve substances. These views have been reinforced by the fact that nuclei are found in some plant families which contain a crystalline protein (Tischler, 1934, p. 163; Küster, 1951, p. 176). These crystals disappear when the plant suffers from malnutrition. Thus for example, the cell nuclei of *Pinguicula* (Klein, 1882), which carry large crystals in the summer, contain no crystals in winter. Artificial feeding of this insectivorous plant with protein *via* the leaves generates protein vacuoles within the nuclei, which dry up to form crystals (Saurer, 1962). Although, in contrast to the nucleolus, the crystals are not an indispensible nuclear constituent, a reciprocal relationship exists between these two nuclear inclusions, in that the volumes of nucleoli decrease with increasing crystal size and *vice versa* (Fig. F–8). Whilst however, as has already been mentioned, the crystals can completely disappear, the volume of the nucleolus, a necessary nuclear component, cannot fall below 5 μ^3. The same behaviour was observed in the nuclei of *Polypodium punctatum*.

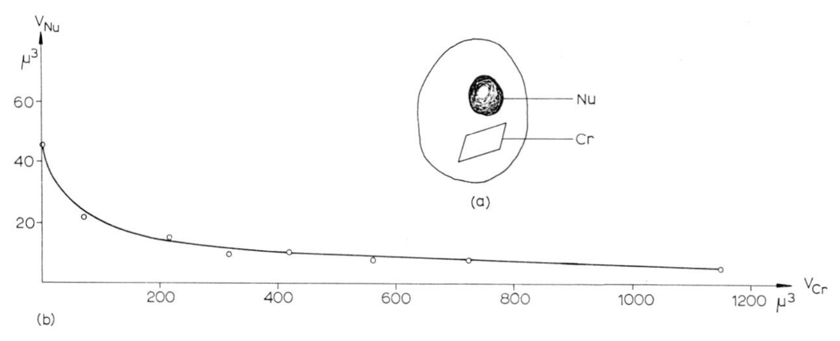

Fig. F–8. Protein crystals in the nucleus in glandular cells of *Pinguicula caudata* (Saurer, 1962); (a) nucleus with nucleolus *Nu* and crystal *Cr*; (b) relation volume of nucleolus V_{Nu} to volume of crystal V_{Cr}.

Since the quantity of matrix of the nucleolus varies so markedly, attention has more recently been turned to the permanent osmiophilic granules visible under the electron microscope. On account of their size and their behaviour towards osmic acid, they have been compared, and even identified, with the ribosomes. The nucleolus would thus seem to be a centre of protein synthesis. Another hypothesis, which has now been abandoned, assumed that the nucleolus produced new ribosomes by autoreproduction, and that these ribosomes were then dispersed into the cytoplasm through the pores in the nuclear envelope (Bonner, 1958). That theory has now been replaced by the view that the nucleolus takes part in the emission of the nucleotides for the messenger RNA, whose sequence is established in contact with the DNA in the chromosomes (p. 177). This RNA would then transmit to the ribosomes of the cytoplasm the code for the synthesis of specific protein molecules (Figs. E–29 and E–30). Using radioactive cytidine, it can be shown that the nucleolus produces RNA not only in interphase, but also up to its disappearance towards the end of the mitotic prophase.

To prove such a function *in vitro*, it is necessary to isolate the nucleoli from the cell nucleus. This has been achieved by a group led by J. Bonner (Birnstiel *et al.*, 1962). The homogenate obtained by squeezing pea embryos in a so-called enucleator was worked up to isolate the nuclei, and the latter were 'broken up' in sodium citrate, which disintegrates the nuclear envelope. High-speed centrifugation in suitable suspension media then yielded four different fractions, which were characterised, on grounds of their light-microscopic structure and total chemical analysis, as respectively the nucleolar apparatus, chromatin, the ribosome fraction, and nuclear sap. None of these fractions was very uniform, but as they were enzymatically active *in vitro*, it was hoped that the reactions of their chief components could

TABLE F–I

FRACTIONATION OF ISOLATED PEA NUCLEI
(Birnstiel *et al.*, 1962)

	RNA	DNA	Protein
Whole nucleus	0.75	1	6
Nucleolar apparatus	2.60	1	12
Chromatin	0.35	1	2.5
Ribosome fraction	1.64	1	3.1
Nuclear sap	1.54	1	78

be elucidated. The protein and nucleic acid contents of the fractions (relative to DNA) are collected in Table F–I.

The fraction which contains the nucleoli is characterised by a high content of RNA, which has already been shown histochemically, and also contains an appreciable amount of DNA, which may in part arise from impurities (*e.g.* remainders of chromatin). In the chromatin fraction, the RNA/DNA ratio is reversed. The ribosomal fraction can be related to the ribosomes solely on the ground of its predominant RNA content. The nuclear sap, which can be regarded as the karyoymphatic nucleoplasm (see p. 182), is composed chiefly of proteins.

3. CHROMOSOMES

It was first discovered by Strasburger in 1875 that thread-like structures appear during nucleus division, and because of their affinity towards basic dyes, these were named chromosomes (Gr. *chroma* = colour). Their strikingly elongated shape was the reason for the creation of the term mitosis (Gr. *mitos* = thread) for this karyokinesis (Flemming, 1882).

(*a*) *Autoreproduction of chromatids*

In all types of Eukaryonta the nucleus contains a definite number of chromosomes of individual size and shape; they are invisible in the active nucleus, but come into view at the beginning of mitosis (prophase). In diploid nuclei, each chromosome has a morphological homologue. The chromosomes arisen during prophase arrange themselves in the equatorial plane of the cell, separate lengthwise into two identical daughter individuals (metaphase), which finally move apart to opposite poles of the cell (anaphase). At each of the poles, a new nucleus is formed in which the chromosomes disappear, whilst the nucleoli and nuclear envelope reappear (telophase). If such a nucleus divides further, and thus at once passes again into the state of prophase, it is called an interphase nucleus. After a final division, the nuclei are referred to as 'resting'; however, in view of their activity in the organisation of the cell, it may be better to call them metabolically active nuclei.

The processes of mitosis, which should be understood and explained in terms of molecular biology, consist of the apparently directed migration of the chromosomes to the centre of the cell where they are held side by side at a definite distance from one another, their lengthwise splitting to form not halves, but two complete, identical individuals (autoreproduction), and the separation (apparent mutual repulsion) of the daughter chromosomes.

Further problems are posed in the reduction division known as meiosis (Gr. *meion* = smaller, less) which reduces the diploid number of chromosomes, 2 *n*, to the haploid (single) number, *n*. In this process the chromosomes change into very long double threads, the chromatids, which consist of one or (according to certain authors) of two chromonemata (Gr. *nema* = yarn, twin). This initial state of meiosis prophase is called leptotene (Gr. *leptos* = thin).

The most noteworthy event in this phase is that the leptonemata of homologous chromosomes are mutually attracted, and pair with one another (zygotene: Gr. *zygon* = yoke). The zygonemata now consist of four lightly twisted chromatids (tetrads), in parallel or crossing over one another. The threads become thicker by swelling of the chromomeres in the third stage of meiotic prophase (Fig. F–9b), pachytene, (Gr. *pachyteros* = thicker), and finally the chromatid tetrads shorten when their twin nature derived from paired chromosomes becomes especially evident (the diplotene stage). As the shortening progresses, the conjugated chromosomes coalesce so that their double structure becomes obscure, and their number of 2 *n* monovalent chromosomes is now reduced to *n* bivalent chromosomes. These tend towards the equatorial plane in the cell, where they arrange themselves at the greatest possible distance from one another (diakinesis). The haploid chromosome number, *n*, of a cell is especially easy to count in this phase. Meta- (Fig. F–9c), ana- and telophases now follow, as in mitosis, in which the four chromatids of the bivalent chromosomes are distributed as independent daughter chromosomes in four daughter cells (tetracytes or gones) by two division steps. Each daughter chromosome contains once again two chromatids. Therefore, an autoreproduction of the chromatids, which become separated from each other in the metaphase of each cell division, must thus take place soon after metaphase.

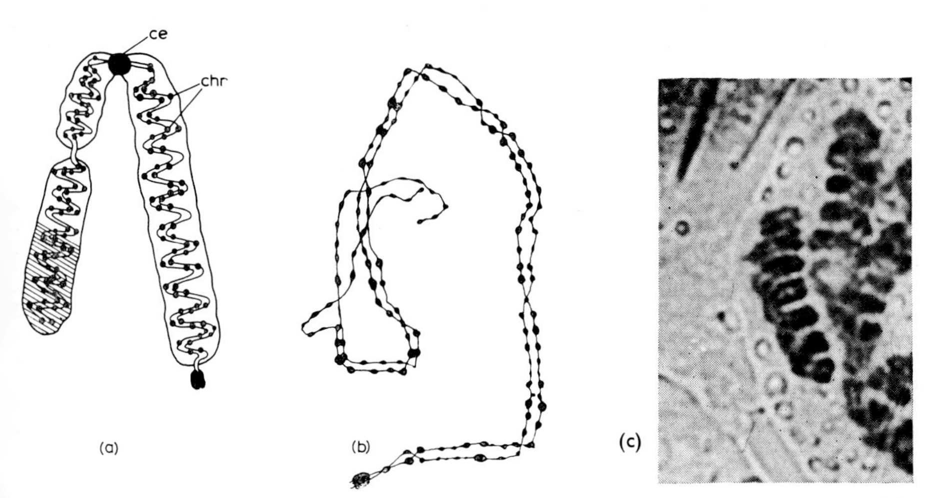

Fig. F–9. Microscopic chromosome structure. (a) Idealised chromosome with helical chromatids: heterochromatic region hatched; in the upper part, the primary (kinetic) constriction; in the lower part at the right, secondary constriction with satellite (from Heitz, 1935, corrected to satisfy Geitler's criticism, 1938, p. 98). (b) Pachynema of meiosis in pollen mother cell of *Trillium errectum*, consisting of two paired leptonemata, each holding two chromatids (after Huskins and Smith, 1935). (c) Metaphase of meiosis in pollen mother cell of *Tradescantia virginica* (Ruch, 1949).

An increase in the chromatids can also occur within a cell (without any visible mitosis). They then manifest themselves in an increased volume and chromatin content of the nucleus, which may be detected by ultraviolet and fluorescence microspectroscopy (Bosshard, 1964). In this way the cell can change from the diploid to a tetraploid, octaploid, etc. up to a 256-ploid state. Such an increase in the number of chromosomes within the nucleus, known as endomitosis, is characteristic of trichocytes, of trichomes, of the mother cells from which latex tubes, vessel members, or collenchyma cells arise, and of other specialised tissue cells such as idioblasts and the like (Geitler, 1940; Tschermak-Woess *et al.*, 1954). The establishment of the polyploid state of such cell nuclei by a chromosome count is only possible when the cells concerned, which normally have lost their capacity for cell division, can be successfully stimulated to further division by external agents or chemicals.

The parallel increase in the chromatin and in the number of chromosomes during endomitosis gives evidence that the chromosomal material is identical with chromatin. It can be shown that the chromosomes of an endomitotic nucleus are normally coiled up in the form of leptonemata. If, however, a considerable autoreproduction of chromatids takes place, the numerous homologous leptonemata formed do no longer contract by individual coiling, but remain extended and paired. Chromatid bundles then arise, whose number corresponds to the haploid chromosome number. Such endomitosis, without chromatid contraction, combined with somatic pairing and without an increase in the number of chromosomes, is called polytenia (Gr. *tainia* = ribbon). Polytenia gives rise to the formation of giant nuclei and giant chromosomes (Beermann, 1962), which are well known in the salivary glands of diptera. They are also occasionally found in the ovula of phanerogames (antipodal nuclei of *Aconitum*, synergid nuclei of *Allium ursinum*, and haustorial nuclei in the chalaza of *Rhinanthus alectorolophus*).

(*b*) *Light microscopic structure*

The structure of the chromosomes (Geitler, 1938) is best disclosed during the leptotene stage of meiosis. The previously coiled chromatin threads are then despiralised to their full length, and exhibit a series of bead-like nodules known as chromomeres (Fig. F–9b). These may be strongly stained with basic nuclear dyes and show the Feulgen nucleal reaction. Their behaviour is thus 'nucleal', whilst the thread connecting the chromomeres is said to be 'anucleal', because it does not react with Feulgen. The chromomeres have an absorption band in the ultraviolet, at 260 nm, which is not shown by the interchromomeric segments, from which we may conclude that the DNA of the chromosomes must be located in the chromomeres. The sizes and the spacings of individual chromomeres are not constant, (Fig, F–9b), so that every leptonema has its particular pattern.

In the zygotene of the meiotic prophase, when the chromosomes pair for conjugation, identical chromomeres of homologous leptonemata are in contact with each other. In Fig. F–9b the double strand of the leptonema is not resolved, so that only two threads instead of the four chromatids in the resulting bivalent chromosome are visible. It can be observed that the two leptonemata cross one another at certain points, giving rise to a chiasma. In polytenic chromosomes the identical chromomeres of the many homologous chromatids fuse together and form transverse cross bands of chromatin (Fig. F–10). In 1951 Engström and Ruch proved that these Feulgen-positive bands contain 2–10 times more mass than the Feulgen-negative ones. Spleen deoxyribonuclease digests the nucleic acid of the chromomeres (Mazia and Jaeger, 1939) without disturbing the ground structure of the chromosomes of the salivary glands. The ability to take the Feulgen stain disappears, whilst the ninhydrin test for proteins becomes positive over the entire length of the chromosome. The chromonema is thus not a chain of alternating protein and nucleic acid links, but a continuous protein thread, in which nucleic acid knots are intercalated (Fig. F–15b) at regular intervals. The nucleoprotein is therefore limited to the chromomeres.

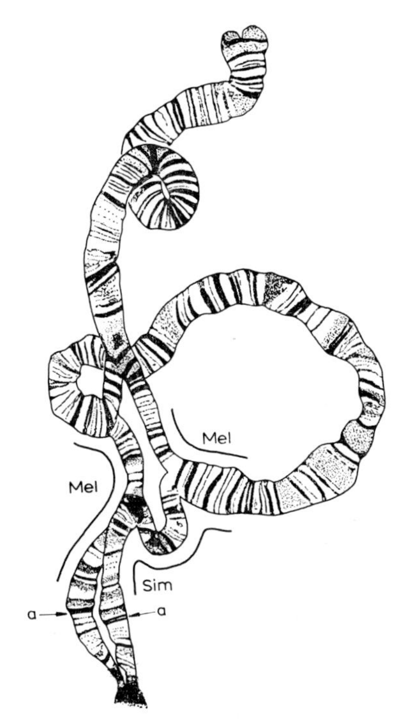

Fig. F–10. Two incompletely conjugated polytenic chromosomes of the nuclei from the salivary gland of a *Drosophila* hybrid, with a chromosome pattern characteristic of the two parental species (from Pätau, 1935); *Mel* from *Drosophila melanogaster*, *Sim* from *Drosophila simulans*, $a \rightarrow$ structural difference.

It can be shown that the genes contained in the chromosome are located in the chromomeric band, so that the chromomeric pattern differs in homologous chromosomes of related species according to their different genetic constitution. In Fig. F–10 a polytenic chromosome of the salivary gland in a *Drosophila* hybrid is split at the arrows because in that section the patterns of the relevant chromatid bundle do not concord (Pätau, 1935).

Following the pairing of the chromosomes in meiosis, the zygonema formed from four chromatids shortens to a more and more compact helix (Fig. F–9c). At first the chromomeres are still visible in pairs; they then simulate an optical section through a smaller helix, which, in contrast to the main (or major) helix, is described as the minor helix (Fujii, 1926; Manton, 1950). In 1949, however, Ruch demonstrated that there is no minor helix which can be resolved under a light microscope (see Fig. F–14).

The condensed helical ribbon is set within the kalymma (Gr. *kalymma* = veil), a strongly staining substance which gives the nucleal reaction and by which the inner structure of the chromosome is obscured.

In the active nucleus the helix expands extensively, whereby some sections may be coated with Feulgen-positive material. These especially strongly staining regions have been termed heterochromatic by Heitz (1935) in contrast to the normal euchromatin. The so-called chromocentres and the smaller chromatin granules probably represent such heterochromatin. Under the electron microscope they appear as thick osmiophilic regions in contact with the nuclear envelope or the nucleolus (Peveling, 1961).

During mitosis the loosely entwined chromosome ribbons are not fully extended as in the leptotene of meiosis, but each of them is wound up in a helix and their double strands are so placed within one another that they can easily come apart sideways in the metaphase, as shown in Fig. F–16b.

As a rule such a chromosome possesses two arms. The point where the two arms meet is the so-called centromere, an anucleal particle to which the spindle complex is attached during the anaphase (Matthey, 1945). The two arms may be of different lengths, and one can even be missing, thus giving rise to a terminal centromere. The sites of centromeres are known as primary constrictions. There are, moreover, secondary constrictions in certain chromosomes onto which the nucleoli are deposited, as was already mentioned. The chromonemata traverse uninterruptedly these constrictions. In certain cases the end of one of the limbs carries a satellite, which is again connected to the chromosome through the chromonemata (Fig. F–9a).

(*c*) *Nucleoids of the Prokaryonta*

Classical cytology could show no presence of cell nucleus in cells of the phyla Schizomycetes (Bacteria) and Schizophyceae (Cyanophyceae). This group of nucleus-less organisms, the Akaryobionta, could thus be contrasted with the cytologically better differentiated nucleated Karyobionta, but it could not be definitely decided whether the Karyobionta should be considered as a higher cytological development arising from the way of life of the Akaryobionta, or whether these latter cells originated through loss of nuclei from nucleated cells.

Since the discovery of the nucleal reaction, however, Feulgen-positive bodies have been found in the cells of bacteria (see Robinow, 1956) and Cyanophyceae; these have been denoted as nuclear equivalents. In electron micrographs, these bodies show up on fixation with 1% OsO_4 at pH 6 as areas of low intensity, which were at first compared with nuclear vacuoles. They contain a coagulated mass of chromatin, or a criss-cross network of extremely fine threads of precipitated DNA (Ris, 1961). However, by an improved method of fixation developed by Ryter and Kellenberger (1958) using 10% tryptone (culture liquid) in the presence of uranyl acetate and Ca^{2+} ions, 1% OsO_4 (at pH 6) indicated a well organised fibrillar bundle of DNA strands. Thus an actual bacterial nucleus does exist, which differs from the nuclei of higher organisms primarily by the lack of a nuclear envelope: it is therefore best called a nucleoid (Fig. F–11). Accordingly the term Akaryobionta has been changed in Prokaryonta and Karyobionta in Eukaryonta.

Giesbrecht (1958/60) explained the fibrillar bundle as a single chromosome, discussed its fine structure (1961), and suggested the three possibilities shown in Fig. F–12. In the nucleus of *Bacillus megatherium* and *Amphidinium elegans* (Dinoflagellata) he distinguished (1958/60) different stages of a helical system. A 20–40 Å thick thread is proposed as the basic structural component, which may correspond to the Watson–Crick double helix (1st order helix). Two such threads are, like the chromatids, paired in parallel and wound in a molecular helix (2nd order helix).

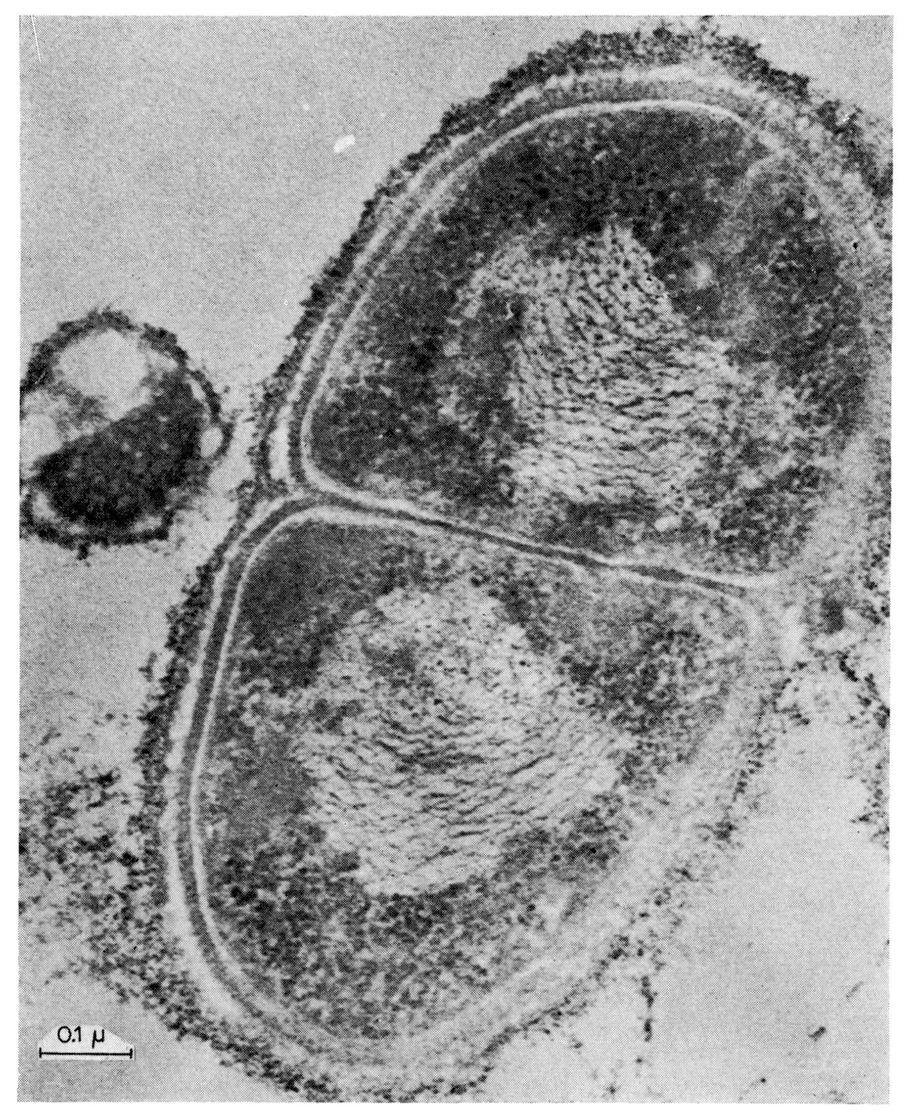

Fig. F–11. DNA in bacterial nucleoid. Fibrillar bundle in a dividing *Coccus* (Van Iterson and Robinow, 1961).

The diameter of this double helical thread appears to be 120 Å. This in turn is rolled up into a 3rd order 500 Å wide helix which acts as cross-striation of a. chromonemal minor helix (4th order helix) (Fig. F–12c; see p. 201 and Fig. F–14).

It is an open question to what extent Giesbrecht's results may be generalised. Many authors, as Schlote (1960), mention only one stage in the coiling of the threads (Fig. F–12b) or even none at all, as *e.g.* Van Iterson and Robinow (1961) (Fig. F–12a).

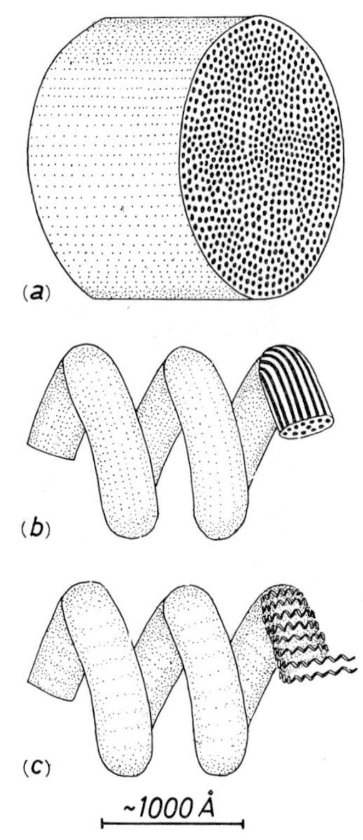

Fig. F–12. Nucleoid of bacteria, considered as a chromosome (from Giesbrecht, 1961). Theories concerning its structure. (a) Polytenic bundle, (b) helical thread consisting of microfibrils, (c) major helix formed by minor helix of a double chain.

In 1962 Kleinschmidt and Lang succeeded in spreading the DNA of the nucleoid into molecular strands by the method developed using bacteriophages. They find only one thread per cell and no ends of the strand are visible. Thus endless coiled DNA threads, closed on themselves, are found in the nucleoid of bacteria.

In *Escherichia coli* fed with tritiated thymidine, Cairns (1963a) has demonstrated microautoradiographically the nucleic acid strand of the nucleoid to be closed to a ring (Fig. F–13a). The thread is over 1 mm (10^7 Å) long! Assuming it to be a DNA double helix, and since a nucleotide member is about 2.5 Å long, the thread must contain something like $10^7/2.5$ *i.e.* 4×10^6 nucleotide pairs (Table F–II).

TABLE F–II

FEATURES OF GENETICALLY ACTIVE DNA THREADS IN PHAGES, NUCLEOIDS AND CHROMOSOMES
(in part after Cairns, 1963b)

DNA thread from	Length (μ)	Time for replication at 37° C (min)	Rate* of replication (μ/min)	Approximate number of nucleotide pairs
T2 bacteriophage	50	1–2	(25–50)	2×10^5
Escherichia coli	1,000	30	33	4×10^6
Human chromosome	*ca.* 20,000	*ca.* 400	(*ca.* 50)	8×10^7

* Rates between brackets are calculated, not observed.

Cairns (1963b) demonstrated also the replication of the DNA thread by feeding with tritiated thymidine for two generations, as shown in Fig. F–13b. At one place, which is called a swivel, the thread splits into its two strands, and each single strand is then completed to a double strand by DNA, which contains radioactive thymidine, so that the resulting thread consists of one cold and one hot strand. The replication of the strands proceeds from the swivel along loops, which are indicated in scheme of Fig. F–13b as circles. One generation after the addition of tritiated thymidine the thread of the DNA with its radioactive (hot) and non-active (cold) strand replicates again. Then a thread with two hot strands and another with only one hot strand are formed. This is shown in Fig. F–13a taken two generations after addition of tritiated thymidine, where loop B is twice as strongly radioactive (two hot strands) as loop A + C (one cold + one hot strand).

(*d*) *Ultrastructure*

There is as yet no convincing picture of the ultrastructure of the chromosomes, for the fixation methods so far developed do not produce sufficient contrast or give rise to some artefact formation. It is possible to discern by conventional methods very fine fibrils, 20–40 Å in diameter, which then appear like the Watson–Crick DNA strands. Whether however this is the natural state of these fibrils before fixation is not certain, for similar fibrils can also be obtained by the precipitation of DNA solutions (Watson, 1962). Similar reservations hold true against the widely-held opinion that any 25 Å strands detected in the cell organelles are DNA strands; in this way Ris (1962) explained the very fine fibrils found in proplastids, albino chloroplasts, and mitochondria as DNA molecules. Although this is theoretically possible, the random pattern of the threads does not indicate a natural structure.

Another handicap is that the exact structure of the protein components of the nucleoprotein in the chromosomes is unknown, as is their relationship to DNA. Even in the leptotene chromosomes does the protein form an essential constituent of the structure besides the DNA, so that the chromosomal ultrastructure cannot be explained without considering the proteins.

Many biologists disregard the vast difference in the orders of size of the Watson–Crick double helix (20 Å) and the chromatid pair (more than 0.2 μ or 2000 Å)

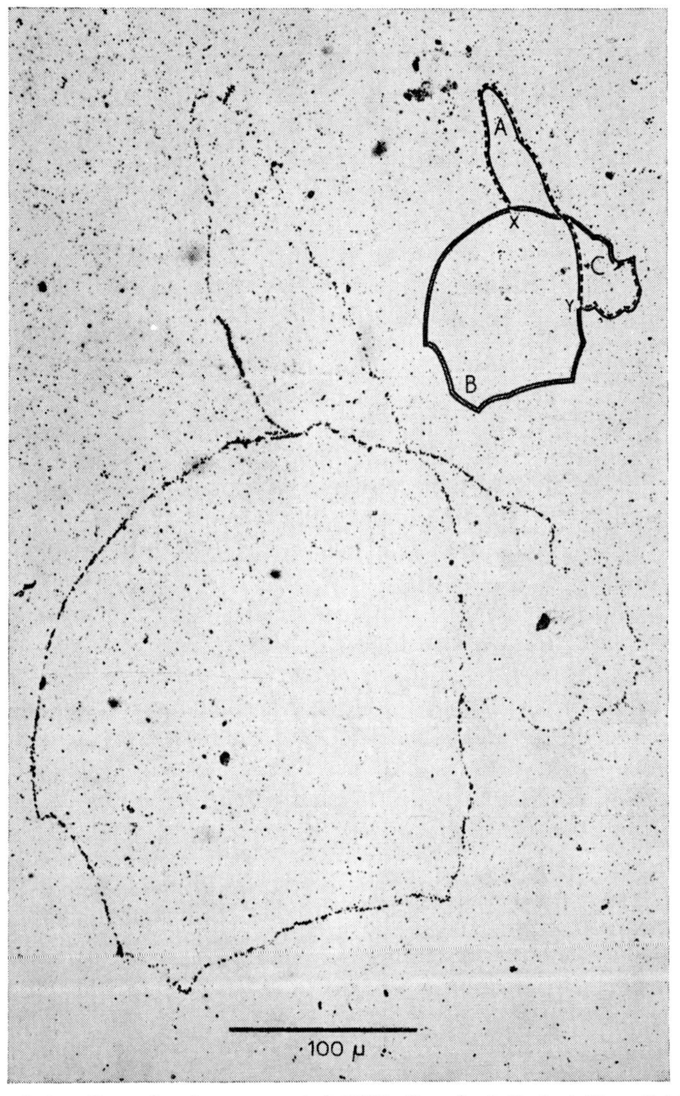

Fig. F–13a. Autoradiograph of an expanded DNA thread of *Escherichia coli* labelled with tritiated thymidine for two generations. Loops A and C only one, loop B both DNA strands radioactive (Cairns, 1963a).

diameters. The morphological analogy, that in both cases genetic threads with linear arrangements of the genes are mutually twisted with crossing over and involvement in chiasma, has lead to the optimistic idea that the genetic event was now clearly explained. But this is at best a molecular biological, and by no means a cytological answer, since segment exchange, required for the recombination of associated genes (coupling), must work at the stage of the chromatid pairs with a distance of $> 0.1\ \mu$ of the chromomomeres as gene centres, quite different to the 100 times smaller distance of < 10 Å of the individual genes (Leupold, 1958) on the DNA double helix.

(see p. 200); furthermore, it does not touch upon the problem of the protein constituents, and does not approach the question of the chromomeres.

In 1956 Wilkins suggested that the basic protein (histone) is built into the groove of the Watson–Crick helix. This could be one reason why the subfibrils known as molecular threads are mostly found to be wider than the 18 Å thick double strand of DNA. The 100 Å thick elementary fibrils observed in zoological specimens would then be due not to coiling, but to pairing of such nucleoprotein strands. Further bundling and a coiling of the elementary fibrils would then lead directly to the leptotene strands of the chromosomes. The chromomeres would arise, according to Ris (1961) by local tangling of these elementary fibrils (Fig. F–15a).

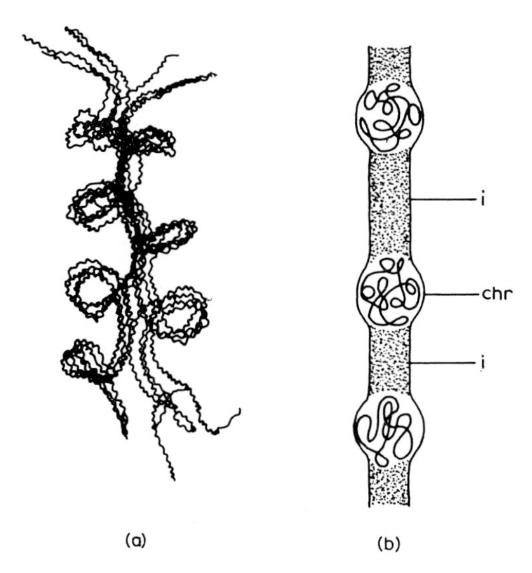

Fig. F–15. Diagram illustrating the molecular organisation of a prophase chromosome; (a) prophase chromosome with 8 continuous DNA threads (after Ris, 1961), (b) chromatid with interrupted DNA thread segments located in the chromomeres (after the present authors). *chr* = chromomere, *i* = interchromomere. The arrangement of the DNA threads is arbitrary; it may be as well a helical one.

If this theory were correct, the Feulgen-positive chromomeres and the anucleal interchromomeral regions should be chemically identical. This is, however, certainly not the case in the giant polytene chromosomes of Diptera, for only the transverse striations formed by the lateral merging of the chromomeres show the ultraviolet absorption maximum at 260 nm characteristic of nucleic acids. On the other hand, there are some observations indicating coiled threads in the chromomeres (Yasuzumi and Ito, 1954); also, in the so-called lampbrush chromosomes, loops of such threads are extruded only from the chromomere segments on treatment with ions (NaCl, KCl) which promote swelling (Callan and Lloyd, 1960). However, every argument for a chemical identity of the anucleal sections and the nucleal chromomere bands has failed. As only the latter are genetically active, it is very unlikely that the intermediate segments contain DNA. If continuous elementary fibrils do exist, they must consist of DNA-containing and DNA-free segments of light microscopic length. Nothing points to a chemical homogeneity of these segments differentiated as chromomeres and interchromomeres. On the contrary, it is known that the basic proteins (histones) are only found in the chromomeres. Thus the interchromomeres probably consist of neutral structural protein with a passive role, for only the chromomeres and not the interchromomeres of homologous leptonemata attract one another during chromosome pairing.

Thus far convincing electron micrographs of chromosomes have only been obtained during meiosis of spermatocytes (amphibians: Moses 1958, 1960; birds: Nebel and Coulon, 1962; insects: Guénin, 1963). These meiotic chromosomes consist of a paired filament having a width of almost 1000 Å. It is perfectly smooth and no helical structure whatsoever is visible! According to enzyme digesting tests (Nebel and Coulon, 1962) the filament seems to be formed of structural protein. It is covered with fringe-like short fibrils of DNA which are oriented perpendicular to the filament axis. In lampbrush chromosomes, the fringes of each chromomere yield two DNA threads upon treatment with appropriate salt solutions. The threads form loops up to 50 μ long (Gall, 1958). As they are covered with a thick sheath of ribonucleoprotein, they are visible in the light microscope, although the two DNA threads together are only 200 Å wide.

These considerations suggest a model of the leptotene chromatids which is presented in Fig. F–15b. According to this conception, the DNA strands which carry the genetic code do not transverse the complete length of the chromosome, but each chromomere shelters in its core a suitably shortened coiled thread. Each single chromomere would be similar to a bacterial nucleoid or a bacteriophage head, containing an independent DNA thread which is perhaps closed on itself. The interchromomeres thus have the function of filing these sites of genetic information in a linear fashion, according to a prescribed pattern, as is indicated by the genetic map (Frey–Wyssling, 1964c).

On the assumption that a chromosome, like the bacterial nucleoid, contains only a single DNA strand, Cairns (1963b) calculated its lengths to be 2 cm (Table F–II). Since the DNA double helix is only about 20 Å thick, this strand, were it not divided into sections independent of one another as indicated in Fig. F–15b, would have a slenderness ratio of 10 million!

Table F–II shows also how much time is needed by the DNA strand to duplicate itself, and how rapidly the untwisting (Fig. F–16a) proceeds (rate of replication). Cairns (1963b) points out that the mechanism for the unwinding of the DNA double helix (p. 111) is energetically no longer possible when the strand reaches 1 mm length.

The most important immediate problem of cytogenetics consists in establishing the spatial relationship between the genes as sections of the DNA strands (called operons) and the loci on the gene map which are discernible with the light microscope. For there are two types of linear arrangements with two different orders of magnitude: on one hand there is the continuous alignment of the amicroscopic nucleotides on the DNA strand, and, on the other hand, the discontinuous sequence of the chromomeres as loci of the genes on the chromatids.

(*e*) *Function*

The function of chromosomes as carriers of heredity is generally known, so that only the question of the autoreplication of the genetic material and the lengthwise splitting of the chromosomes will be discussed in this book.

Of the many helical structures observed and postulated in the chromosomes, only those of the first and last order correspond to reality, and therefore are of special importance to the process of replication:

(1) the double helix of the molecular DNA strands (detected by X-ray analysis and

(2) the major helix of the chromatids during metaphase of mitosis (visible in the light microscope and observed live by phase contrast).

Morphologically speaking, these two double helices are fundamentally different from one another. The two chains of the molecular DNA helix are so coiled together that they cannot be separated from one another without much ado (Fig. F–16a), whilst the two entwined chromatids of the major helix fit into each other in such a way that they can be pulled apart sideways.

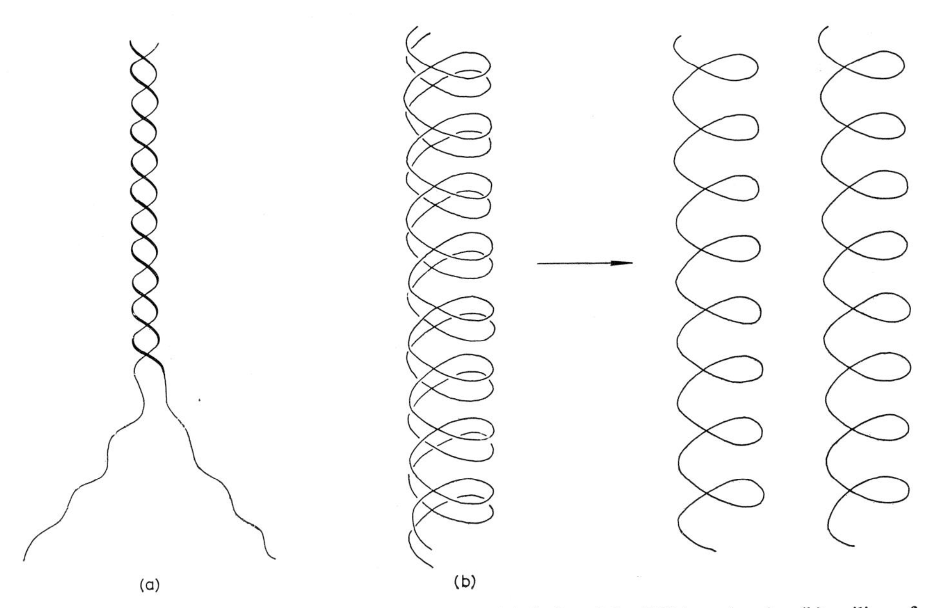

Fig. F–16. (a) Twisting of the two chains in the double helix of the DNA molecule; (b) coiling of the two chromatids in the major helix of the chromosome.

Thus if during mitosis the genetic material of the helical thread is divided equally into two daughter chromatids, there is no morphological obstacle to splitting the chromosome into two daughter chromosomes without unwinding the helix.

On the other hand, in the replication of the DNA chain the two threads must first unwind (Fig. F–16a) before they can serve as templates for new chains (see p. 110). Levinthal and Crane (1956) have shown that the double strand must rotate very quickly, and assume that during the separation each single chain immediately completes itself to a new double helix (Fig. F–16a), according to the principle mentioned on p. 109. Although vibration, translation, and rotation of molecules are well-known random processes, the necessary rotational movement involved in the unwinding of a 25,000 Å long double helix may well stand out in the science of molecular motion. It can only be conceived if, as is assumed, it is supported by chemical reactions. We do not know what role is played in this by the chromosome protein, or how it behaves during the unwinding stage.

This discussion of the morphological interpretation how the genetic material is doubled in the chromosome confirms the view brought forward in the introduction that electron microscopy has up to now been able to do little to elucidate the molecular morphology of the nucleus, in contrast to the successful explanation of the ultrastructures in the remaining cell organelles. Clearer indications about these topics can be obtained by biochemical methods, using microautoradiography.

If plant roots are fed with tritium-labelled thymidine, the DNA molecules, and consequently also the chromosomes, become radioactive. With the help of the microautoradiographs of tissue sections, the H^3-thymidine in the chromosomes can be identified (Taylor, 1958). If the roots are then allowed to grow without further labelling, the two daughter chromosomes of the tritiated chromosomes remain radioactive. After the next cell division, however, only one of the two daughter chromosomes is labelled. This means that thymidine incorporated in the chromatids remains in the strands concerned and is not exchanged during the formation of new strands. Referring to Fig. F–16a, this means the following: the two single chains in the Watson–Crick model behave towards one another as positive and negative (see p. 110). If both are radioactive, one unlabelled positive chain and one unlabelled negative chain are formed at the first splitting, which then at the second splitting lead to a moiety of non-radioactive double helices. This demonstrates the accuracy of the cytologically postulated theory of chromatid splitting during mitosis, and of the formation of chromatid tetrads during meiosis.

After the autoreproduction of the DNA double helices, the question of the transmission of the genetic code from the chromosomes to the active sites in the cell arises. As was explained earlier, a template is transferred from the DNA code to an RNA chain, and this latter molecule then leaves the nucleus as messenger RNA so that an efferent system acts in the cytoplasm in conjunction with the ribosomes (Fig. E–30).

If all messenger RNA from the gene sites were sent out at the same time, no normal differentiation of the cell could take place; this requires that the different efferent impulses do not happen simultaneously, but in carefully controlled succession, according to the forthcoming phases of the ontogenetic development. Two mechanisms, which can be followed karyologically, have been claimed to be correlated with the occurrence of ordered cell and tissue differentiation. On the one hand, the sites on the chromosomes where the transfer of the code takes place appear to swell. This can be observed in the salivary glands of Diptera, where the polytene chromosomes with their linearly arranged gene loci exhibit in certain chromomere bands so-called puffs during the course of ontogenesis in insects. These local swellings temporarily coincide with a definite step in differentiation, and afterwards disappear again (Beerman, 1961; Kroeger, 1963). Thus according to these observations the genes, *i.e.* the fixed operon regions in the DNA code, are actively involved in the transmission of messenger RNA only in the area of a puff.

In certain regions of the chromosomes, additional DNA is so thickly deposited that they respond especially strongly to nuclear stains and to the Feulgen reaction. As mentioned before, such nuclear regions were designated as heterochromatic (Fig. F–9a) or as chromocentres (Heitz, 1935). As a rule they are genetically inactive, and in certain cases it can be demonstrated that in developing organisms their number increases with progressive differentiation of the cell. They show not only a higher nucleic acid concentration than the remaining nuclear constituents, but also a disproportionate increase in the histone content (Rosselet, 1963). Inactivation of the DNA chain can thus be explained by increased combination with basic protein.

With all due reservation we could say in an oversimplified way that during ontogeny, the activity of the genes are switched on by the puffs and switched off by heterochromatin.

4. NUCLEAR SPINDLE

(*a*) *Structure visible in the light microscope*

The microscopic structure of the spindle which becomes apparent in nuclear divisions has long remained an enigma. In fixed preparations, divergent fibrils are visible, some of which stretch from one pole of the cell to its equator, while other, shorter ones, coalesce with the chromosomes at special points of attachment (centromeres). In the living state, however, all this remains invisible; before the phase miscroscope was available the spindles appeared homogeneous, structureless and optically empty. Microsurgical interventions revealed a relatively rigid double cone with distinct cleavability but without a visible structure (Bélar, 1929). The spindle fibres were therefore believed to be artefacts due to the fixing process.

On the other hand, it was possible to elucidate structural details with the aid of polarising (Runnström, 1929) and phase microscopes (Michel, 1943; Jacquez and Biesele, 1954). Schmidt (1937a) found the spindles in living sea urchin eggs to show positive birefringence. The images visible in the fixed material thus proved to be real structures, existing *in vivo* (Wada, 1955). Since the poles of the spindle behave like positive spherites whose rays can be followed throughout the cell (Inoué, 1953), they must consist of optically positive fibrils. Undoubtedly the same fibrils stretch from each pole to the chromosomes. They are sublight-microscopic and should therefore be visible in the electron microscope.

(*b*) *Ultrastructure*

When acid fixation is used (*e.g.* Bouin's solution: Beams *et al.*, 1950, or cadmium chloride: Satô, 1958) coarse strands show up in the electron microscope. Rozsa and Wyckoff (1950) found on the other hand that the spindle region of dividing cells in onion root tips appears to be structureless when fixed with neutral formalin, and the fibrillar elements of the spindle were consequently thought to be amicroscopic. Careful treatment with osmic acid at pH 6.1 in the presence of divalent cations (Ca^{2+}, Mg^{2+}) indicates however the presence of 150 Å thick fibrils (Harris, 1962); these have the appearance of capillaries because they show electron-dense borders with an empty core (Fig. F–17; Roth and Jenkins, 1962), and their cross-sections display an open bore (Roth and Daniels, 1962).

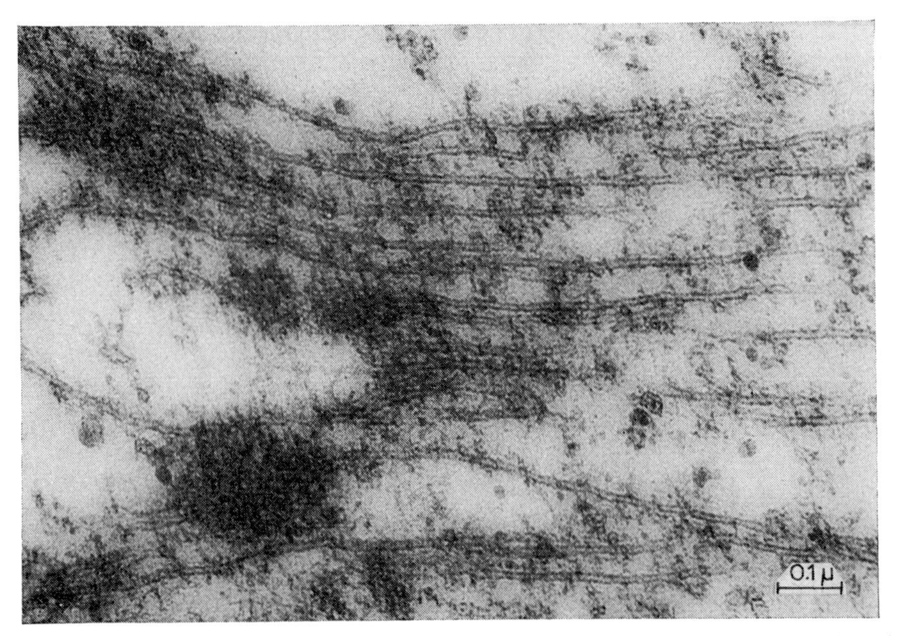

Fig. F–17. Spindle fibrils (OsO_4 + $CaCl_2$) at the metaphase in a giant amoeba (Roth and Jenkins, 1962).

The explanation of this structure, which is remarkably reminiscent of a cross-section through a unit membrane, is not easy. It could be an artifact due to a surface reaction of the fibrils with the metallic fixative, a possibility which must also be taken into consideration with the three-layered unit membrane. In contrast to the case of unit membranes, there is no question of the existence of a protein sheath and a lipid core (Fig. E–10), for in the spindle there is only pure protein and not lipoprotein. Also, there are certainly no capillaries or pores. Since significant cross-striations appear, Harris (1962) suggests that the spindle is a tightly coiled spiral, whose helix can, however, not be clearly resolved.

The attachment of the spindle fibrils to the centromere (or kinetochore) of the chromosomes can be observed in the electron microscope. The centromeres consist of strongly electron-scattering membraneless plates in whose vicinity are found large vesicles and collections of ribosome-like particles.

The so-called 'strain theory' (Ger. Zugfasertheorie) has been applied to the motion of the chromosomes on the basis of phase-microscopic observations (Michel, 1943). Actually, the expected thickening of the spindle fibrils in the anaphase is not observed in the electron microscope; according to Harris (1962) this can be accounted for by the fact that only the pitch of the postulated helical thread, and not its diameter, is actively involved.

(*c*) *Origin*

The spindles are formed primarily in the cytoplasm when the nucleus is still intact, and even enucleated cells are capable of forming spindles (E. B. Harvey, 1936). In special cases the spindle can form inside the nuclear membrane, or it may be observed that both cytoplasmic and nuclear fibrils take part in the construction of the nuclear spindle. This has intrigued the classical cytologists, but since morphologically there is no difference between the cytoplasmic groundplasm and the nucleoplasm, which communicate freely through the open pores of the nuclear envelope, it is no longer of any special importance.

Ever since Mazia and Dan (1952) succeeded in isolating the mitotic apparatus from sea urchin eggs, the chemistry of the spindle has been well established. As could be expected from its light-polarising properties, the spindle consists of a fibrous protein, and amino acids, such as aspartic and glutamic acids (25%), as well as serine and threonine (15%) play a special role in its behaviour (Mazia, 1959). Precursor protein molecules are probably synthesised in the interphase, polymerised into non-oriented chains and then brought into an oriented paracrystalline state by hydrogen bonds. These elongated micellar particles are surrounded by more highly hydrated non-oriented chains. Thus an equilibrium is thought to exist between parallelised and randomly arranged chains that is sensitive to slight changes in the environing conditions. Since the absence of divalent ions prevents the formation of fibrils, fixation without such stabilisers does not reveal the preformed spindle fibrils in the electron microscope (Roth and Daniels, 1962).

Whether the numerous ribosome-like particles visible around the centromeres and the spindle fibrils are related to the precursor protein must be established by further study. On treating pollen mother cells of *Lilium* with $CdCl_2$, Satô (1960) found innumerable granules in the nuclear cavity at the end of prophase which seems to aggregate into short, fine fibrils at the beginning of the metaphase. If it could be proved that these particles are different from the ribosome-like bodies still visible after the fibril formation, they might represent precursor protein whose globular macromolecules have been made visible by the action of cadmium ions.

G. Mitochondria

The cell organelles known as mitochondria have long been overlooked, both on account of their smallness (*e.g.* 0.5 μ in diameter and 2 μ long) and because they are destroyed by the frequently applied fixatives alcohol and acetic acid. There are, indeed, cells in which the mitochondria are especially large, and in these cases they were detected and described quite early. The recognition of their general distribution and fundamental importance in all cells is, however, relatively recent (see Fauré-Fremiet *et al.*, 1909).

Flemming (1882) and Altmann (1890) are regarded as the discoverers of mitochondria. Flemming described thread-like structures in animal cells which he called 'fila', and Altmann granules resembling bacteria of an autonomous kind (so-called plastosomes), which could divide and arrange themselves into rows of particles. These cell inclusions were later identified with the mitochondria. The name (Gr. *mitos* = thread, and *chondrion* = granule) was given by Benda in 1897. Other names such as chondriocontes (small rods), chondriomites (emphasising the granular rather than the threadlike form) and chondriosomes (granules) have rightly disappeared from the literature, for the shape of the mitochondria is changeable, so that the same individual may assume spherical form, or may stretch into a small rod. In the electron micrographs slender thread forms (Fig. G–1) are a rather rare manifestation; instead the mitochondria usually appear as ovoids or short rod-like particles. This is due to the fact that in ultrathin sections elongated forms

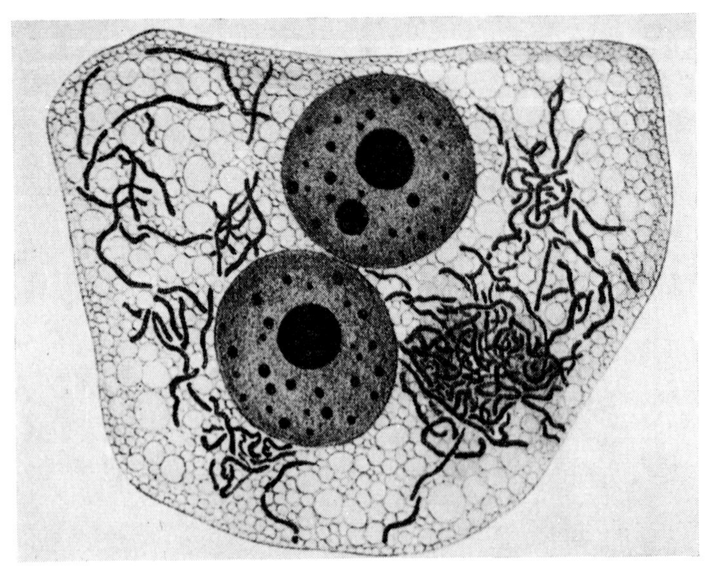

Fig. G–1. Mitochondria in a tapetum cell of *Nymphaea*. First evidence in plant cells (Meves, 1904).

are practically never cut along their whole longitudinal extent, but always obliquely or transversely.

The evidence that mitochondria also occur in plant cells (in tapetum cells of the anthers of *Nymphaea*) first dates from the beginning of this century (Meves, 1904). Since then, mitochondria have been found in the cells of all plant tissues with the help of special methods of mitochondrial fixation (OsO_4, $K_2Cr_2O_7$, Altmann, 1890; H_2CrO_4, OsO_4, Benda, 1901; $K_2Cr_2O_7$, formaldehyde, Regaud, 1910). The mitochondria of a cell are collectively designated by the term chondriome, coined by Meves in 1908 (Guilliermond *et al.*, 1933).

1. CHEMISTRY

Mitochondria can be isolated from animal tissues, especially from the liver and kidney, and also from plant seedlings and leaves, so that it is possible to investigate their chemistry and their enzymatic behaviour *in vitro*.

The mitochondria from tissue homogenates remain relatively unaltered in isotonic salt or sugar solutions. By pressing the tissue brei through a suitable muslin or silk cloth, they can be separated from the coarser cell constituents. As the resulting turbid cell sap contains different kinds of particles, the mitochondria are obtained by fractional or density gradient centrifugation. Chloroplasts sediment in a reasonable time at a centrifugal acceleration of 120 G (Du Buy *et al.*, 1950), while for the mitochondria about 100 times (*e.g.* 12,000 G) and for the ribosomes almost 1000 times greater accelerations (*e.g.* 100,000 G) are necessary (Table E–V). In the liver of the frog *Xenopus* a density of 1.10–1.20 is found for the mitochondria, and of 1.25–1.30 for the ribosomes (Holter *et al.*, 1953), so that a clean separation of the two types of particles is possible by density gradient centrifugation.

The mitochondrial fraction obtained by centrifugation can be made salt-free by dialysis, dried, and analysed chemically (Hogeboom *et al.*, 1947). The mitochondria consist of lipoprotein, with a considerable rate of phospholipids, and small amounts of ribonucleic acid. Some analytical results from the literature are collected in Table G–I.

TABLE G–I

CHEMISTRY OF THE MITOCHONDRIA

	Guinea pig liver (Barnum and Huseby, 1948) (%)	Rat liver (Leuthardt and Exer, 1953) (%)	Pea seedlings (Stafford, 1951) (%)
Protein	70.7	71.7	35–40
Ribonucleic acid	3.7	2.85	0.5–1
Phospholipids	15.5	16.75	30
P-free lipids	11.9	8.55	
Total	101.8	99.85	71

Although the large variations indicate the uncertainty of these values, the figures are sufficiently consistent to lead to the conclusion that the compositions of mitochondria from different types of cells are basically similar, in that they consist of about two thirds protein and one third lipid (of which more than half is phospholipid) and 1–3% of ribonucleic acid; other authors find as little as 0.5% RNA (Lindberg and Ernster, 1954). Thus the mitochondria can hardly be called rich in nucleic acid; their characteristic is rather their high phospholipid content. According to Green and Fleischer (1963) these account for 95% of the lipid components. The latter authors are of the opinion that the lipids and proteins do not occur in separate strata, but that lipoproteins exist in which the acid groups of the phospholipids are bound to basic proteins. Liver mitochondria have also been successfully fractioned into their individual constituents and the composition of their membrane has been ascertained (Table G–II).

The following hypothesis has been proposed for the function of the substances found in the mitochondria: the phospholipids are involved in electron transfer (Green and Fleischer, 1963), while the protein components are divided into the physiologically passive structural proteins and the highly active enzymes.

2. MORPHOLOGY

(*a*) *Vital staining*

The mitochondria are hyaline particles, of the order of 0.5–2 μ, and are not always easy to detect. They are instable structures, which disintegrate in lipophilic liquids and swell up in water. The difference between the light refraction of mitochondria and the groundplasm is smaller than that between the refractivities of the groundplasm and fat droplets or other lipid inclusions. This makes it difficult to recognise mitochondria in the living state. They are however very easily visible in the phase contrast microscope (v. Albertini, 1947).

Their differentiation by supravital staining with Janus green or dahlia violet can be taken as fairly reliable. Brenner (1953) has investigated such staining with phenosafranine (the basal compound of Janus green B) and six of its derivatives, as supravital dyes. Accumulation of the dyestuff by lymphocyte mitochondria is increased with increasing methylation, and especially ethylation, of the two amino groups of phenosafranine. The strongest staining is however obtained with Janus green B (diethylphenosafranine-azo-dimethylaniline); clearly it plays the role of a specific mitochondrial stain, not only as a basic and simultaneously strongly lipophilic substance, but also as an azo dye. Janus green is moreover a redox indicator (rH), in that it is reduced to a colourless leuco-form at low oxygen concentrations (low redox potential): the mitochondrial coloration is thus invisible under anaerobic conditions (Brenner, 1949). According to Lazarow and Cooperstein (1953), Janus green B is taken up not only by the mitochondria, but quite generally by the protein components of the cell, being reduced in the latter to the leuco-form. The cytochrome oxidase content of the mitochondria is therefore responsible for the appearance of the dye in its oxidised green form. Thus the specific supravital dye Janus green B is to some extent an indicator of the localisation of cytochrome oxidase in the cell. On observing the mitochondria under a cover glass the coloration fades after a while owing to lack of oxygen; this reaction can be taken as an indication of active mitochondria. Other lipophilic cell particles, as e.g. the spherosomes, which are also stained green with Janus green B, do not subsequently reduce the dye, for they lack the necessary enzymes for electron and hydrogen transfer (Bautz, 1955).

The capacity of the mitochondria to oxidise α-naphthol to the quinone form of indophenol blue (Nadi reaction, Pearse, 1961) can likewise be ascribed to cytochrome oxidase (see p. 98).

In accordance with their low RNA content, the mitochondria show no marked ultraviolet absorption and are not coloured by pyronine (Monné, 1948).

Especially notable is the osmotic activity of the mitochondria. In hypotonic solutions they absorb water and swell (Ruska, 1962) increasing their volume many times (ballooning), while in hypertonic media they shrink so as to be unrecognisable. They must thus be handled as much as possible in isotonic fluids, e.g. in 0.5 *M* saccharose.

(*b*) *Ultrastructure*

The mitochondria are surrounded by a double envelope consisting of an inner and an outer membrane, each of which, at high resolution shows the structure of a unit membrane. The outer membrane separates the mitochondrion from the groundplasm, whilst the inner membrane encloses the mitochondrial matrix, which is known as stroma or chondrioplasm. The two membranes are separated by a narrow ($\sim$ 100 Å) gap, the perimitochondrial space, which, like the perinuclear space, is probably filled with a serum-like fluid. Thus for the function of the mitochondria three aqueous phases must be taken into consideration, which are separated from one another by two different membrane systems: the surrounding groundplasm, the serum-like fluid in the space around the entire mitochondrion, and the stroma (Fig. G–4a).

No direct relationship has ever been observed between the membranes of fully grown mitochondria and the membranes of the endoplasmic reticulum or the nuclear envelope. Thus no direct connection exists between the perimitochondrial space and the enchylematic spaces of the ER or the nuclear envelope. In yeast mitochondria the outer membrane shows fine invaginations which perhaps form cracks (Moor and Mühlethaler, 1963), so that possibly a direct contact does exist between the groundplasm and the perimitochondrial space (Fig. G–5a).

Surface area of the inner mitochondrial membrane is very much greater than that of the outer owing to numerous invaginations into the stroma. These are almost perpendicular to the mitochondrial surface, and are shaped like fingers (tubuli, microvilli or sacculi) or septa (cristae mitochondriales, Palade, 1953) (Fig. G–3). On ultrathin sectioning the scattered tubuli are cut obliquely, and then appear as vesicles.

Mitochondria with cristae are found in animal cells with especially intensive respiratory metabolism (liver, pancreas, kidney) and are thus not common in plant cells; here, as a rule, one finds tubuli. Heitz (1959) believed that he had found a special type of mitochondrial organisation in fungi, with an accumulation of tubuli at both poles of the elliptical mitochondria. Moore and McAlear (1963a), who investigated 50 different species of fungi, found a wide variety of forms with

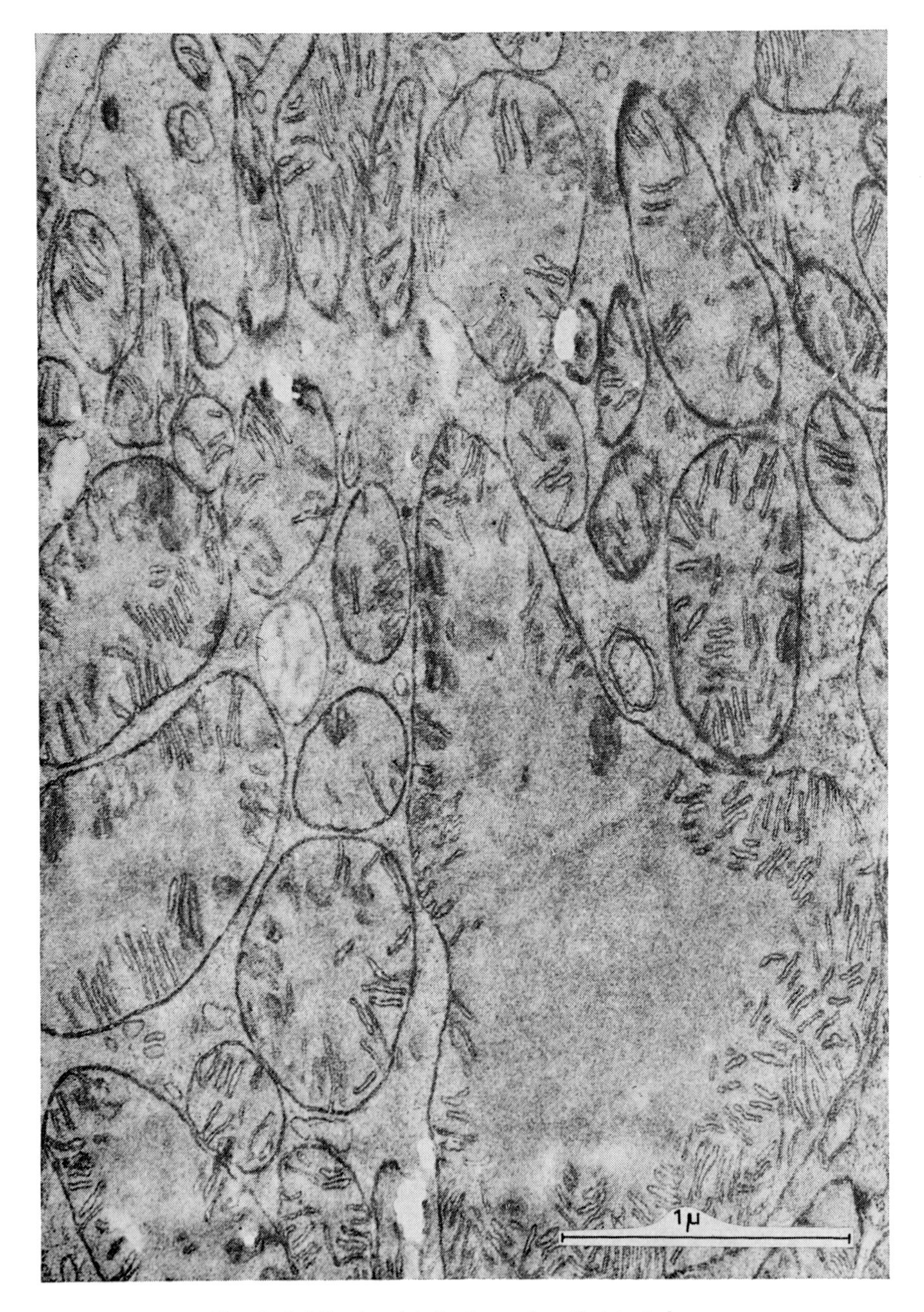

Fig. G–2. Mitochondria in the euglenoid *Astasia longa*.

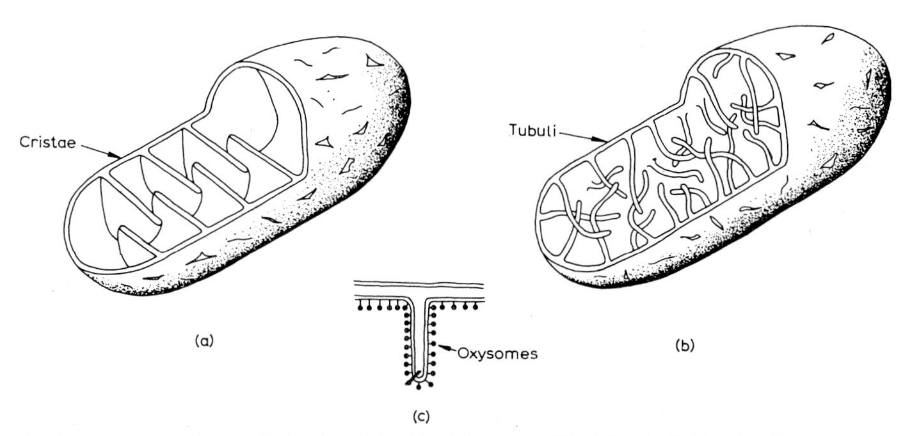

Fig. G–3. Structural types of mitochondria, (a) with cristae, (b) with tubuli, (c) sub-units (oxysomes) associated with the inner membranes or tubuli.

tubuli, sacculi and cristae, so that no taxonomic qualification as to the type of formation exists, but rather the intensity of function determines the organisation. Since this can alter during the course of the development of a cell, not only the external form of the mitochondrion, but also its internal structure may take different and changing forms.

Stalked particles are attached to the inner mitochondrial membrane on the side of the stroma (Fig. G–3c), which can be rendered visible by negative contrast staining (Fig. G–5b) (Fernández-Morán, 1962; Parsons, 1963; Stoeckenius, 1963). These were called oxysomes (Chance *et al.*, 1963), because they contain cytochrome (Chance and Parsons, 1963).

Like the development of the inner mitochondrial surface, the number of mitochondria per cell depends on the intensity of metabolism. Glandular cells contain as a rule more such organelles than do cells with a passive function. Thus, for example, the glandular tissue of plant nectaries is very richly supplied with mitochondria in comparison to the surrounding ground tissue (Eggmann, 1962). In view of their inner structure, a small number of large mitochondria can of course be as effective as a large number of small mitochondria.

Although in general the mitochondria are scattered throughout the entire cell, they may sometimes accumulate near the nucleus or other active centres. The euglenoid alga *Astasia* may serve as an example (Frey-Wyssling and Mühlethaler, 1960) in which the mitochondria are concentrated peripherally in the ectoplasm, while the paramylon granules are accumulated inside the cell (Fig. G–2).

(*c*) *Function*

The outer membrane of the mitochondrion may be interpreted as the boundary of the cytoplasm and the inner membrane as the boundary of the chondrioplasm. It was initially impossible to isolate mitochondria by ultracentrifugation from tissue homogenates with their double membranes intact, and only in later preparations in buffered, isotonic and isoionic suspension media did the outer membranes remain undamaged. Failing this, the simple inner membranes become distended by surface tension and inflated like a balloon by considerable expansion of the stroma, so that frequently they break up. In consequence, the mitochondrial fractions prepared in early days showed in the electron microscope a lamentable picture of membrane fragments and vesicles. It is amazing that such extensively disintegrated organelles still retain their enzymatic activity. There is no essential difference between the *in vitro* reactions of mitochondria obtained under optimal conditions, with their envelopes intact, and of disintegrated fractions consisting of membranes in shreds. This indicates that the enzymes of the mitochondria are not localised in the stroma, but rather on the inner membrane, which makes the remarkable increase of its surface area understandable. It can in fact be shown that the activity of enzymatic succinate dehydrogenation parallels the increase in area of the tubuli (Simon and Chapman, 1960).

In vitro experiments show that the enzymes of the tricarboxylic (citric) acid cycle as well as flavoprotein and cytochrome (Estabrook and Holowinsky, 1961) occur in mitochondria (see e.g. Leuthardt, 1949). Succinic acid dehydrogenase may be detected most readily. The mitochondria are thus the respiratory centres of the cell. This explains their necessary presence in all aerobic cells and the accepted fact that enucleated cells or nucleus-free cell components obtained by microsurgery are still able to respire; in other words, although with loss of the nucleus the capacity to grow and all regenerative abilities of the cell are lost, the ability to carry on the katabolism is not.

When the oxidative tricarboxylic acid cycle ceases with the transition from an aerobic to an anaerobic way of life, the mitochondria disappear (Hirano and Lindegren, 1963) and simultaneously a remarkable ER lamella system develops, as has been demonstrated in yeast cells by Linnane *et al.* (1962). A similar proliferation of the ER has been observed by Mercer and Rathgeber (1962) in the nectar tissue of *Abutilon* sepals under anaerobic conditions (nitrogen atmosphere).

These observations are in agreement with the biochemical findings that glycolysis of the respiratory substrate (conversion of the carbohydrates to pyruvic acid) occurs outside the mitochondria, in the groundplasm. Anaerobiosis is thus not only characterised chemically by the absence of oxygen, but also morphologically by the lack of mitochondria.

Consideration of these facts leads to the following hypothesis concerning the mechanism of action of the mitochondria: The substrate of the tricarboxylic acid cycle, pyruvic acid, must be considered to be in the perimitochondrial space, while the enzymes are anchored in series in the inner membrane (Frey-Wyssling, 1960). Such an assumption is necessary, for the different enzymes of the respiratory chain work at different redox potentials. The necessary stepwise increase in potential is made possible by the serial arrangement of the enzymes, whilst in a mixture of freely mobile enzymes a mean potential would occur as a result of all redox buffer systems acting on one another. Fig. G–4a shows how a micromolecule of the respiratory substrate would migrate over the enzymes of the respiratory chain. When it arrives at the final stage in the cycle, the series can start again from the beginning. Thus the biochemical cycle would be in fact morphologically performed by innumerable repeating linear enzyme series.

One problem is posed by the question of how the substrate molecule reaches the perimitochondrial space from the groundplasm (Ruska, 1962). According to Fig. G–4a, it must pass through the outer mitochondrial membrane. By the

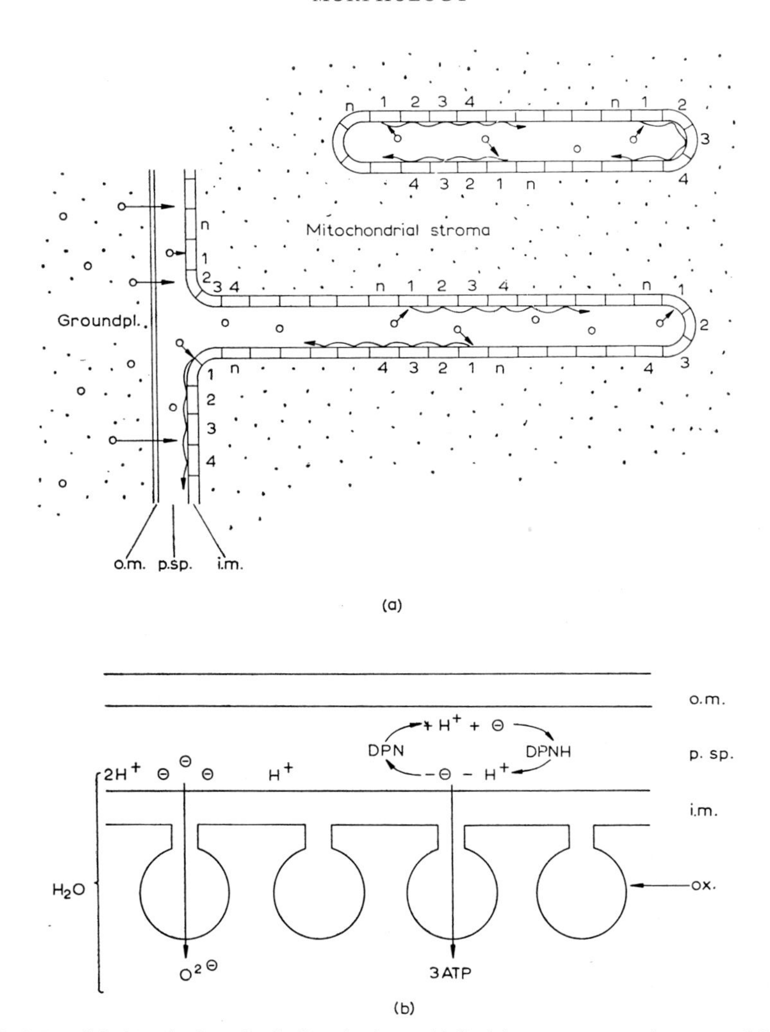

Fig. G–4. Possible functioning of tubuli and cristae. (a) Serial arrangement of enzymes 1,2,3......*n* on the inner mitochondrial membrane (*i.m.*) and substrate molecules ○ in the perimitochondrial space (*p. sp.*) translocated across the outer membrane (*o.m.*) (Frey-Wyssling, 1960); (b) electron transport mechanism in oxysomes (*ox.*) (D. E. Green, 1964).

discovery of microcracks in the outer membrane of yeast mitochondria (Moor and Mühlethaler, 1963) the possibility arises, however, of a direct contact between the groundplasm, and the content of the perimitochondrial space (Fig. G–5a). Furthermore, Feldherr (1962) has discovered that ferritin molecules introduced into the cytoplasm of amoebae by micro-injection find their way rapidly into the perimitochondrial space, so that a direct route must exist, by which the macromolecules penetrate.

The question now arises of how the molecules are translocated in such narrow spaces whose width is of the order of 100 Å. A diffusion gradient results from the disappearance of the pyruvic acid molecules in the perimitochondrial space, which is similar to that gradient between the CO_2-content of the outer atmosphere

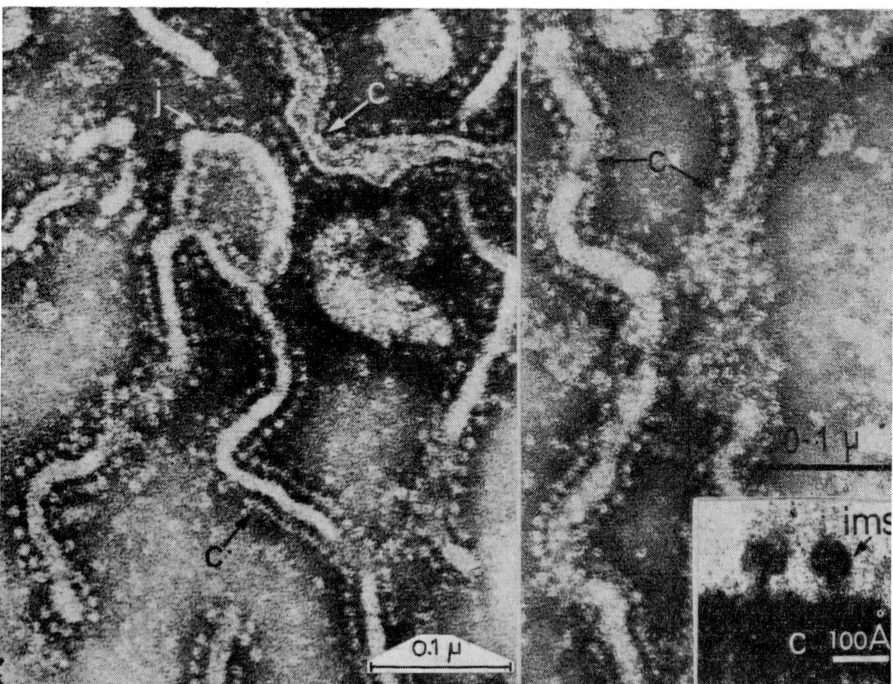

Fig. G–5. (a) Surface view of the outer mitochondrial membrane showing cracks or pores (freeze-etching; Moor and Mühlethaler, 1963); (b) sub-unites (oxysomes) on the inner mitochondrial membrane of rat liver; *c* = cristae, *ims* = inner membrane sub-unit. Negative staining (Parsons, 1963).

(0.03%) and the intercellular atmosphere of the assimilation tissue (ca. zero), by which the stomata of the leaves function as a perfect sink for CO_2 molecules. Whether however this transfer mechanism established for a gas passing through light microscopic pores can be applied to dissolved molecules passing through micropores of molecular dimensions remains rather questionable, in view of the mutual influence of the molecules of the pore membrane and of the migrating substrate. As indicated in Fig. G–4a, in this size range chemical potentials must replace the diffusion potentials. If there exists a discontinuous chemical gradient between locus 1 and locus *n*, the stepwise changed substrate molecules will be passed from locus to locus by chemical forces. Thus, in addition to the classical streaming and diffusion potentials, a further principle governing the migration of molecules emerges in the region of micromolecular distances.

The postulated scheme gives no information regarding the site of the combination with oxygen of the hydrogen transferred by the different dehydrogenases acting as acceptors, nor does it tell us where the energy transfer to ATP takes place. These processes can be imagined to occur on the inside of the mitochondrial inner wall. In this way a special function would be found for the stroma, but it is then difficult to see how the energy-rich ATP molecules could be sent out from the stroma into the cytoplasm for use in the energy-consuming processes (Fig. G–4b).

The mitochondrial membranes have been successfully separated into three fractions, whose compositions are given in Table G–II. The lipid-free fraction consists of dehydrogenase enzymes, which are thought to be localised in the outer membrane. The bulk of the wall is formed from phospholipoproteins, to which is ascribed the capacity for electron transfer.

TABLE G–II

RELATIVE PROPORTIONS OF PROTEIN AND LIPID IN THE MITOCHONDRIAL MEMBRANE
(Green and Fleischer, 1963)

	protein (%)	lipid (%)
Dehydrogenase complexes	20	—
Structural lipoprotein	60	75
Oxysomes	20	25
	100	100

It has been established that in mitochondria *in vitro*, from which the lipids are extracted with aqueous acetone, the transfer is interrupted. The activity may be restored by adding back both phospholipid and the enzyme ubiquinone Q. The inner mitochondrial membrane must be made up of this lipoprotein. The third fraction holds the oxysomes, which contain cytochromes.

(*d*) *Mitochondrial equivalents*

The Prokaryonta are not only exceptional in their lack of normal cell nuclei, but they also possess no typical mitochondria. This is very surprising in aerobic organisms, since mitochondria in the Eukaryonta are known to be obligatory centres of dehydrogenation and oxygen transfer. Thus the idea of mitochondrial equivalents in these cells has been considered.

In several instances granules visible in the light microscope have been found, which can be stained with Janus green B and which are able to reduce tetrazolium compounds to formazan or oxidise α-naphthol to indophenol (Nadi reaction). In view of these observations there is some tendency to ascribe mitochondrial function to these particles. Drews and Niklowitz (1956) have described such granules on the newly formed diaphragm of the blue alga *Phormidium*, which reduce stilbene tetrazolium chloride to a blue formazan and Janus green B to rose red diethylsafranine. However, in the electron microscope these particles show only a granular structure, without a double membrane, vesicles, or tubuli, so that no morphological similarity between these bodies and mitochondria can be proposed. Moreover, nothing is known of the general distribution of such granules in the Prokaryonta, so that the sites of respiratory centres in the cells of these organisms remain an open question. In certain Prokaryonta respiration seems to be performed by the plasmalemma.

A special case is presented by *Bacillus megatherium*, the cells of which contain large organelles known as mesosomes. These are enclosed in a unit membrane, and inside contain a helically wound cavity system, which Giesbrecht (1960)

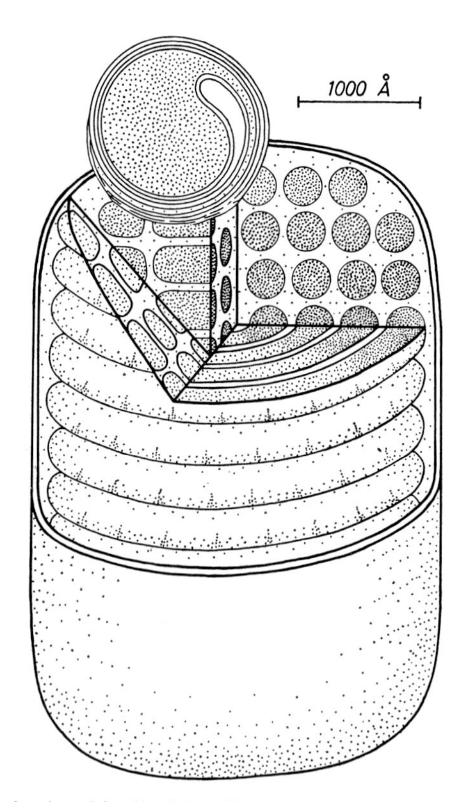

Fig. G–6. Mesosome (mitochondrion?) of *Bacillus megatherium* with a circular system of tubuli (Giesbrecht, 1960). Above: supplementary body with a spirally wound membrane system.

explained as mitochondrial tubuli. This highly organised 'mitochondrion' is further connected to a drop-shaped supplementary body formed of spirally wound membranes (Fig. G–6). According to Fitz-James (1960) the mesosomes in the related *Bacillus medusa* arise by a folding of the plasma membrane. Thus in contrast to true mitochondria they possess only a single membrane in place of the double membrane. Under these circumstances, only the future can decide whether these mesosomes with their complex helical structure and, at present, unknown function, can be considered as mitochondria.

3. ONTOGENY

(*a*) *Chondriome*

The observation that mitochondria divide, and that, in the large elongated mitochondria of locust spermatocytes, this takes place simultaneously with nuclear division, has lead to the idea that these plasma components represent autonomous organelles which multiply by autoreproduction. This opinion was also especially put forward in 1947 by Lehmann, who has coined the term 'biosomes' for such autonomous cell constituents.

Strugger (1954) was an inspired propounder of the idea of autonomous systems in cells. He turned against the classical karyogenetically formulated theories of cells and proposed a pangenetic scheme, according to which not only does the cell nucleus, or its chromosomes, display autonomous elementary duplication, but also all cell organelles, the mitochondria, plastids, spherosomes, and perhaps the ribosomes as well, would replicate themselves independently; they continually would divide spontaneously, being in this way passed from cell to cell, and never would arise *de novo*. Correspondingly the cell physiology would not be centrally controlled by the genom of the nucleus, but the groundplasm would shelter at least three further systems, the chondriome, the plastidome, and the spherome, as organisers of the genetic plasmon.

Cell physiology, on the other hand, reports cases where mitochondria are reformed, *de novo*, after their complete destruction in the living cell. Thus Hirsch (1931) injured the chondriome of the pancreas with X-rays, whereupon reconstruction independent of the damaged mitochondria took place. Similarly, in 1951, Dangeard destroyed the mitochondrial system in the root apex of seedlings with acetic acid, and after all mitochondria had disappeared in this way, a new chondriome re-established itself in the surviving cells.

It is of great theoretical importance whether the mitochondria are autonomous organelles with independent hereditary action, which, so to say, would live in symbiosis with the cell, or whether they depend on the nucleus for their formation and development, and are thus under genetical control of this central organ. It was not possible to decide this question unequivocally by light microscopy, for in spite of the above-mentioned case in which after the mitochondria were destroyed new individuals arose again from 'nothing', it could always be considered that ultramicroscopic pieces of the destroyed mitochondria have remained and regenerated themselves anew. It is therefore very interesting to see whether the electron microscope could decide these fundamental questions.

(*b*) *Initials and Promitochondria*

The hypothesis that mitochondria are autonomous elementary replicators has been contradicted by various authors, (*e.g.* Brachet, 1949; Eichenberger, 1953; Danneel, 1959) who observed that the mitochondria were formed from smaller particles. These particles were sometimes confused with 'microsomes', and were sometimes termed promitochondria. The promitochondria are preceded by even smaller particles, which we call initials. The latter possess a double membrane (consisting of two unit membranes) and are thus easy to distinguish from other particles, *e.g.* young spherosome vesicles. Their diameter is of the order of 500 Å and their contents are somewhat more electron dense than the surrounding groundplasm. These initials are found in considerable numbers in newly formed meristematic cells. Paralleling the growth of these cells, the initials rapidly increase in volume and their inner membranes begin quite early to form small folds lying perpendicular to the surface of the particle. At this stage the initials can be regarded as promitochondria, for now their future destiny can undoubtedly be recognised, which is not always the case with the initials themselves, since plastid initials have a similar appearance. The promitochondria show a vigorous growth in volume, during which the little membrane folds become differentiated into tubuli or cristae. The whole process is shown schematically in Fig. G–7.

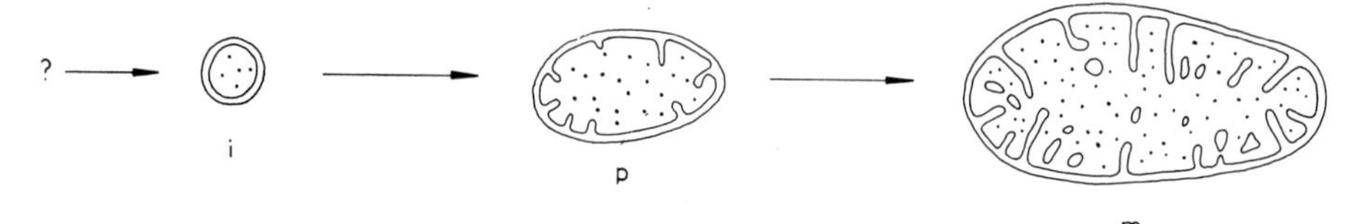

Fig. G–7. Ontogeny of mitochondria. Initial *i* → promitochondrion *p* → mitochondrion *m*.

The origin of the initials was at first unclear. It could be argued that the autonomy of the chondriome is expressed not in the mitochondria themselves but in these sublight-microscopic precursors, by their transfer from cell to cell. On the other hand, we should then be able to find the dividing state of such particles, which has not been the case, whilst the division of grown mitochondria is easy to observe. In the liver Lee (1964) has observed the formation of initials from aggregations of plasmatic granules smaller than ribosomes which form first a single and later a double limiting membrane.

Without autoreproduction the initials cannot however represent an independent chondriome. In contradiction of older views they can arise *de novo*.

The development of the mitochondria involves considerable growth. Concerning the differentiation of the tubuli and cristae, emphasis has been placed on the remarkable increase in the surface of the inner membrane. On the other hand it is often overlooked that the stroma must grow rapidly as well. The possibilities that the mitochondria increase their volume by mere uptake of water, or that the surface of the tubuli grows without a change in volume of the stroma, must be ruled out. If there was a constant volume, the transverse and longitudinal diameters of the mitochondria had to change in opposite directions, through an increase in the degree of slenderness. During the development of the initial into a grown mito-

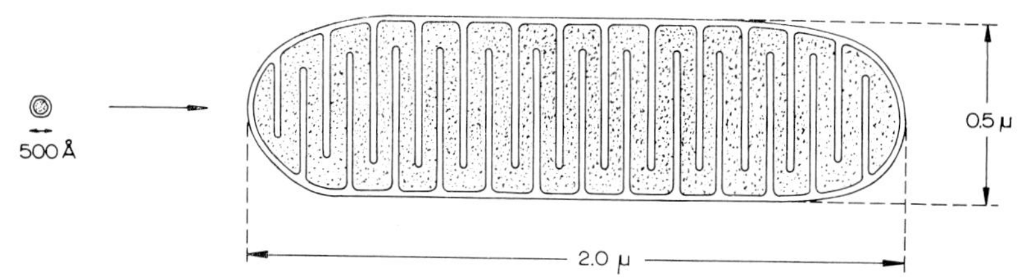

Fig. G–8. Mitochondria. Growth in volume and in surface.

chondrion, however, the dimensions grow in all spatial directions so vastly, that not only the inner membrane but also the stroma must undergo a considerable increase in volume.

These circumstances are represented schematically, and to scale, in Fig. G–8. The initial measures 500 Å in diameter, and grows into a mitochondrion 0.5 μ in diameter and 2 μ long. In the course of this process, twenty-five 300 Å wide cristae are formed, with 500 Å of intermediary space for the stroma. On the basis of estimated measurements, the inner surface is then calculated at 50 times the mitochondrial cross-section, and the stroma volume at 5/8 of the total volume. Following these assumptions, and considering the mitochondrion as a cylinder, the increase in the stroma volume is:

$$\frac{5/8 \times \pi\,(2500)^2 \times 20{,}000\ \mathrm{A}^3}{1/6 \times \pi\,(500)^3\ \mathrm{A}^3} = 3750\text{-fold},$$

and the increase in the inner surface area is:

$$\frac{50 \times \pi\,(2500)^2}{\pi\,(500)^2} = 1250\text{-fold}.$$

Thus in spite of the enormous development of the inner surface, the stroma volume increases three times as much. We cannot therefore regard mitochondrial differentiation merely as convolution of the inner membrane, while the stroma behaves passively, for this must actively grow with it.

For the sake of comparison, let us consider invaginations which can be observed by light microscopic histology. Fig. G–10 shows an incipient invagination on the underside of an embryonic oleander leaf, which in the course of ontogeny will develop into a large hole, into which all the stomata open. The growth in the surface

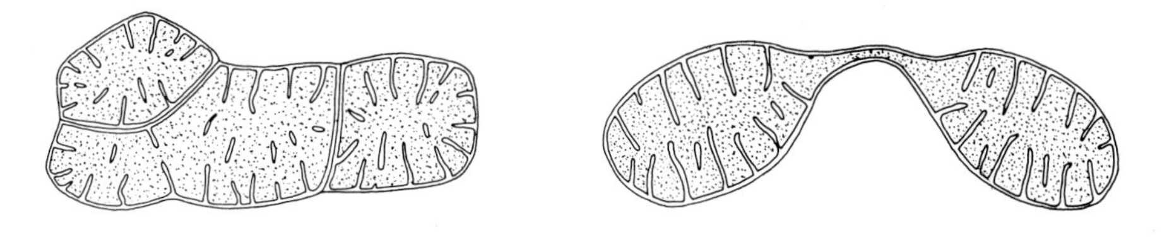

Fig. G–9. Division of mitochondria (Mühlethaler, 1959); (a) by septa, (b) by constriction.

area of the underside of the leaf occurs by division of the epidermal cells which is intensified where the epiderm invaginates; but the mesophyllic cells must also divide simultaneously for the invagination can be deepened only by a synchronised increase in the leaf thickness. The growth of a tubule penetrating into the mitochondrial stroma can be imagined in a similar manner. The bulk of the stroma must correspondingly increase, simultaneously with the increase in the molecular components of the inner mitochondrial membrane.

Two different possibilities must be considered for the growth of the membrane surface. Either the molecular structural components increase by autoreproduction in the membrane itself (this would be comparable to the cell divisions shown in Fig. G–10), or they are synthesised in the adjacent stroma and subsequently fitted into the wall. By analogy with the situation found in the groundplasm, the ribosomes might be considered responsible for the increase in stroma molecules. Such particles have been observed in the stroma, it is true, but we do not understand how the amino acids necessary for a protein synthesis are forwarded across the double membrane. These comments indicate how great are the difficulties involved in the understanding of molecular biological processes, when we turn from static descriptions of the ultrastructure to dynamic considerations of the ontogenetic and metabolic processes.

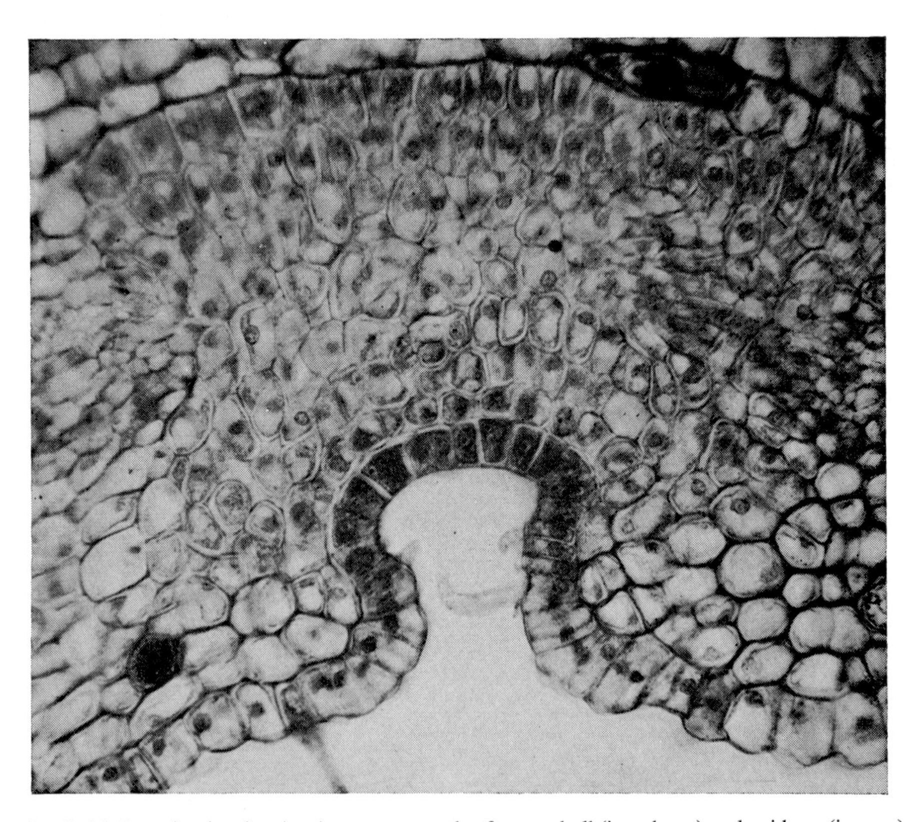

Fig. G–10. Invagination by simultaneous growth of mesophyll (in volume) and epiderm (in area) in the leaf of *Nerium oleander*.

(*c*) *Division and budding*

Differentiated mitochondria can undergo active division. Two different modes of division have been observed, of which, however, only one occurs in all cells of a given tissue.

Usually the division takes place by diaphragm-like foldings of the inner membrane, without the outer membrane being drawn in. The cross septa close little by little, like the shutter of the iris, forming mitochondria with two or more compartments (Figs. G–9a and 11b, c; Mühlethaler, 1959). The resulting conglomerate can split up into its components, resulting in an increase in mitochondria. This form of mitochondrial division can be observed in meristems (Fig. G–9a). However, it can also happen that the central part of the mitochondrion extends to form a long neck, and finally becomes constricted right through (Fig. G–9b). Such a picture is found, for example, in the mitochondria of yeast cells.

A special form of mitochondrial division, which is more accurately termed budding, was discovered during the regeneration of moss plants from cut-off moss leaflets (v. Maltzahn and Mühlethaler, 1962a). The new formation is initiated by a de-differentiation of the green cells, for the specialised assimilation tissue must revert to the meristematic state. The initials for both chloroplasts (see p. 239) and mitochondria are then cut off by budding (Figs. G–11a and G–12). Thus not only the cell as a whole, but also its organelles become rejuvenated. In this way differentiated mitochondria produce from themselves the starting stage for a new onto-

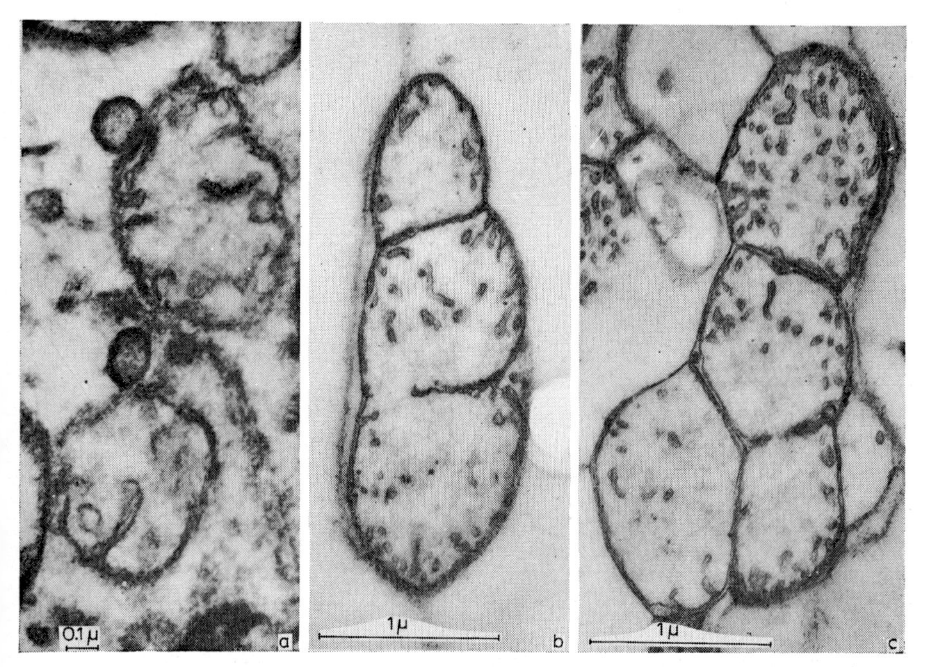

Fig. G–11. Division stages of mitochondria; (a) formation of initials by budding in de-differentiating moss leaves (v. Maltzahn and Mühlethaler, 1962a), (b) formation of a transverse septum prior to division (*Elodea*), (c) later stage before separation (Mühlethaler, 1959).

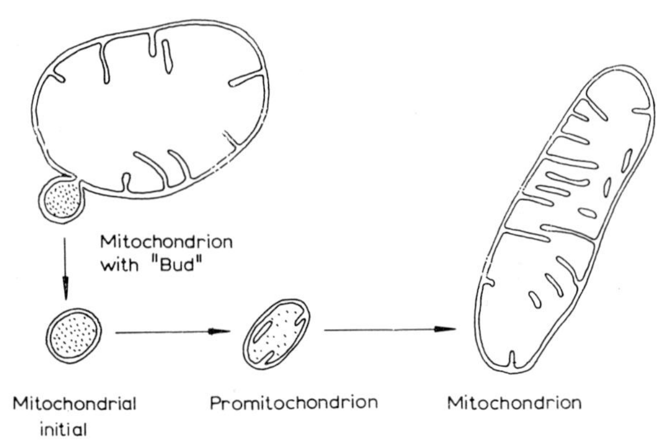

Fig. G–12. Diagram of Fig. G–11a.

genetic development series (Fig. G–7). Thus the mitochondrial stroma must contain the information for the morphogenesis of these organelles.

It should be noted, however, that such new formation of initials from existing mitochondria occurs only after de-differentiation of previously specialised cells which reassume growth, and is never observed in normal meristematic cells or in ova.

(*d*) *Nuclear origin*

As the mitochondria are passed from cell to cell during mitosis, the question of whether in fact there is a permanent chondriome must be determined by investigation of the gametes. The readily accessible ova of fern archegonia, or those of Cycadaceae (*Zamia, Dioon*) which are known to reach especially large dimensions, serve for this purpose.

In 1961 Bell demonstrated during the oögenesis of the eagle fern (*Pteridium aquilinum*) that DNA labelled with radioactive thymidine migrated out of the nucleus before fertilisation. In the electron microscope it now becomes evident that this process takes place by the formation of vesicles, which detach themselves from the nucleus and disperse into the cytoplasm. The evaginations embrace both membranes of the nuclear envelope (Fig. F–5b); they contain homogeneous finely granular nucleoplasm within a double membrane (Fig. H–12), and often display an electron dense inclusion which, according to autoradiographic evidence in archegonia fed with labelled thymidine, is DNA (Bell and Mühlethaler, 1964b). Later on these bodies function as initials.

Preliminary to this general budding from the nuclear surface, the organelles which exist in the oösphere degenerate (Mühlethaler and Bell, 1962a). The chloroplasts and mitochondria lose their characteristic inner structure and shrivel to pyknotic residues, which are finally eliminated from the cell (see Fig. H–11). This stage is then followed by an active development of the newly formed initials to proplastids and promitochondria (Bell and Mühlethaler, 1962b, 1964a).

The establishment of a breakdown of the cell organisation of the maturing ova and the formation of a neoplasm with the *de novo* development of mitochondria and plastids from nucleoplasm given off by the nucleus is incompatible with the theory of an autonomous chondriome and an autonomous plastidome. These organelles are not, therefore, independent gene carriers, but draw their genetic information from the nucleoplasm. From the standpoint of a consistent cell theory, this behaviour is not only comprehensible but even necessary, since it is more probable that the cell events are governed by a single central organelle, rather than by three different autonomous systems operating as it were in symbiosis with one another, and mutating independently of one another. A mitochondrial mutation could for example enforce the transition from the aerobic to an obligatory anaerobic way of life, which is hardly conceivable without the cooperation of the nucleus. According to the new concept the acceptance of genetic information takes place through the nucleoplasm.

Geneticists who deal with plasma genes and plastid inheritance (see p. 240) are of the opinion that the concept of the genetic continuity of plastids which they present could not possibly be false, on the grounds of their genetical experiments (*e.g.* Stubbe, 1962). As, however, the morphological continuity is obviously interrupted in ova, customary theories of extrachromosomal heredity must be reconsidered and brought into harmony with electron microscopic observations. Because the stroma of initials cut off from the ovum nucleus consists of maternal nucleoplasm that cannot be differentiated from the maternal groundplasm with which it is in open communication, it is possible on the basis of the new observations to attribute to it plasma genes and maternal inheritence.

In view of the importance of the questions raised, it is of particular significance that other authors have also observed a neoformation of mitochondria in the electron miscroscope. Buvat (1959) described the *de novo* appearance of mitochondria in the cell plate during mitosis of meristematic cells in the root apex. Here, therefore, a new formation of mitochondria is even postulated in a meristem. On the other hand, the reports of Camefort (1962), in which the disappearance of the old cell organelles of the zygotes of *Pinus laricio* and *Zamia*, and the neoformation of mitochondria and proplastids by vesicles evaginated from the nucleus were established, are also important. A difference to be further clarified, consists in that in the fern archegonia mentioned earlier the reconstruction of cell organelles takes place prior to fertilisation, whilst in the investigated gymnosperm ova it occurs in the fertilised zygote nucleus. As, however, the male gamete brings with it practically no nucleoplasm, this difference is perhaps not decisive.

4. INTERRELATION OF CYTOMEMBRANES

Double membranes should be referred to as envelopes, and single membranes as unit membranes or simply membranes.

The formation of mitochondria from the plasmalemma was considered by Robertson in 1959. As however the plasma membrane is a simple unit membrane, and not a double envelope as in the mitochondria, a complicated assumption must be made of a curved evagination which sinks into a nearby invagination. Observations of this kind have not however been reported.

The different cytomembranes observed in the electron microscope (Sjöstrand, 1956), and their mutual derivation from one another (insofar as this can objectively be established) are briefly reviewed here. It is assumed that the unit membrane has a

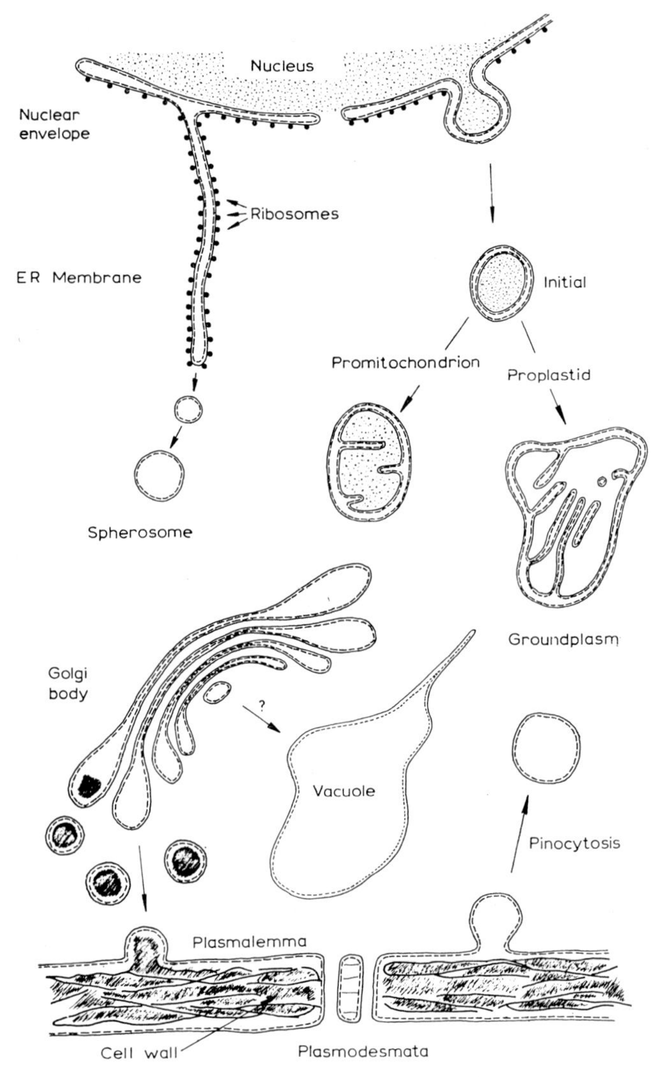

Fig. G–13. Interrelations of cytomembranes.

polar structure, *i.e.* its outer side is constituted somewhat differently from its inner side. Fig. G–13 illustrates these relationships.

So far, the observations allow us to distinguish three, more or less independent, types of membrane:

(1) The double membrane of the nuclear envelope

(a) The cavernous system of the endoplasmic reticulum is formed by evagination of the outer nuclear membrane; this membrane may become vestured with ribosomes. The ER is able to isolate vesicles, which can eventually grow into spherosomes. The single membrane of grown spherosomes is thus indirectly derived from the nuclear membrane.

(b) Evaginations of both membranes of the nuclear envelope lead to initials, which can subsequently differentiate into mitochondria or plastids. The double membrane of these organelles is therefore homologous with the nuclear envelope.

The above discussion shows that the nuclear envelope is of outstanding importance in the formation of organelle membranes and that, from the genetic point of view, it deserves very special attention.

(2) The plasma membrane (plasmalemma)

The plasmalemma is capable of evagination and invagination, the latter process being involved in pinocytosis. Pinocytotic vesicles, like the ER vesicles and Golgi vesicles, are bounded by single unit membranes, so that if their source is unknown, these structures can hardly be distinguished morphologically. A direct relationship between nuclear and plasma membranes has not as yet been established.

(3) Golgi membranes

As in mitosis the new cell wall arises from Golgi vesicles and their membrane becomes new plasmalemma (see p. 278), an ontogenetic relationship exists between the Golgi membrane and the plasmalemma. Furthermore, the Golgi membrane fuses with plasmalemma when the material of the cell wall matrix contained in Golgi vesicles is incorporated into the primary cell wall (see p. 301), which likewise indicates an homology between Golgi and plasma membranes.

These ideas agree with the measurements carried out by Sjöstrand in 1963, according to which the reticulum membrane is only 50–60 Å thick, whereas the Golgi and plasma membranes are substantially thicker, 70–100 Å. According to Yamamoto (1963), however, only the plasmalemma is 100 Å thick, whilst the Golgi and ER membranes measure on average 87 Å.

Three possibilities are discussed concerning the formation of the vacuoles and their membranes (tonoplasts): (1) from the groundplasm (Mühlethaler, 1958/60), (2) by expansion of the endoplasmic reticulum (Buvat, 1962), and (3) from Golgi vesicles (Marinos, 1963).

H. Plastids

With the discovery of the light microscope, natural scientists came to know the green chromatophores in leaves now referred to as chloroplasts. Since Ingen–Housz established as early as 1779 that light and the green pigment, chlorophyll, are necessary for the assimilation of CO_2, it was not difficult to assign the correct function to the chloroplasts. Particular cytological interest was, however, first stimulated by the investigations of A. Meyer (1883), Schmitz (1884) and Schimper (1885), according to which, and in conflict with the opinions prevailing at that time, the green chromatophores found in plant cells could not be formed from the cytoplasm, but always only from existing chloroplasts, *i.e.* they never arise *de novo*. This behaviour is striking in certain algae, whose oögametes are green; it can then be demonstrated that the chromatophores also divide at the cell division of the zygote, so that all chloroplasts of the grown plant descend ultimately from that of the oögamete. Schimper extended these ideas to higher plants. He observed that yellow and orange, and even 'colourless' chromatophores occur, as well as the green. This discovery caused Schimper to abandon the term 'chromatophores' in favour of the now used 'plastids' (Gr. *plastikos* = formed, moulded). Corresponding to the lack or existence of pigmentation, this worker subdivided the plastids into leucoplasts, chloroplasts, and chromoplasts (Gr. *leukos* = white, *chloros* = green, *chroma* = colour), and found that the leucoplasts (as amyloplasts) are able to synthesise starch, the chloroplasts chlorophyll, and the chromoplasts the pigments now known as carotenoids. To Schimper it seemed most important that the different plastids can transform into one another; for example, he observed that the leucoplasts can turn green, and the chloroplasts can again become yellow. Schimper's theory of convertability of plastids can be represented by the following scheme:

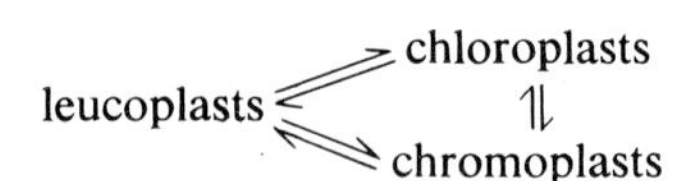

This became of particular significance, for as leucoplasts have been detected repeatedly in the ova of higher plants, their presence in the female gametes formed the main argument for the doctrine of autoreplication, independence, and continuity of the plastid system. These views were later supported genetically, because it was established in transmission experiments that the properties of the plastids, particularly deficiencies in their pigmentation, are inherited purely maternally. Geneticists thus speak of a nuclear independent plastid inheritance, and the term plastidome has been introduced for the plastid system as a carrier of the genes concerned.

1. ONTOGENY

(*a*) *Monotropic development*

According to Schimper's theory, the plastidome is autonomous. Only division and metamorphosis occur, but never new formation of the plastids, so that the same plastid is attributed with the ability to change itself repeatedly in a definite way, and then to change back again. On re-examination of these cyclic events, it was however established that chloroplasts etiolated by lack of light can, if they are sufficiently strongly yellowed, not become green again; they degenerate, and must be replaced by new chloroplasts. Similarly, the transition from chloroplast to chromoplast has proved to be irreversible after complete disappearance of the chlorophyll. From this, it must be concluded that the capacity for change in the plastids is not a cyclic process, but follows a monotropically directed course (Frey–Wyssling *et al.*, 1955), from the leucoplasts, via the chloroplasts, to the chromoplasts; the latter are to be considered as senile and degenerating forms of the plastids.

As this conclusion represents an ontogenetic course, we must consider whether the leucoplasts observed by Schimper and others in ova have any juvenile forms as precursors. The so-called proplastids are described as such by Strugger (1950). The proplastids are small amoeboid particles, little over the limit of resolution of the light microscope, which were originally confused with mitochondria since they exhibit fixation and staining properties similar to those of the latter organelles. This led temporarily to the view that the plastids arise from mitochondria. In the example of the liverwort *Anthoceros*, it could however be shown that the chloroplasts only arise from their like, but never from mitochondria (Scherrer, 1915), and Guilliermond (1922) observed two types of 'mitochondria' in developing leaves of *Elodea*, one of which remained mitochondrial on cell differentiation, while the other grew into plastids. This second type has proved to be identical with the proplastids, so that today the ontogenetic independence of the plastidome from the chondriome is fully established.

We must still decide whether the proplastids are to be held responsible for the postulated continuity of plastids, as Strugger imagined. In the electron microscope, however, smaller particles without any internal lamellar structure have been discovered, which are termed initials (Mühlethaler and Frey–Wyssling, 1959). The following series can thus be drawn up for the monotropic development of the plastids:

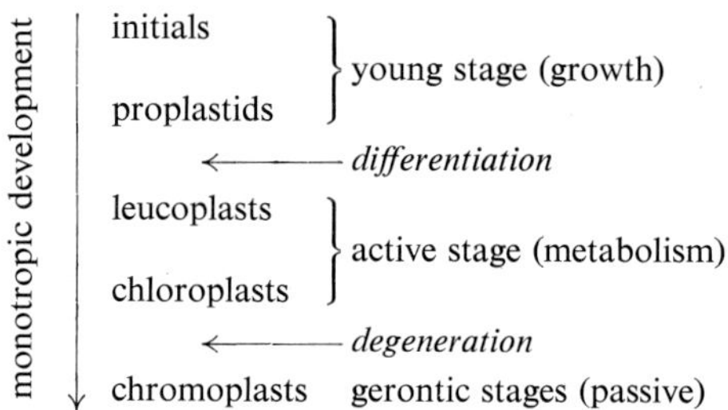

In this way the plastids, like all biological objects and cell organelles (with the exception of the nucleus), follow the undeviating events of unilaterally directed development, which leads from the starting forms (zygotes, initials), by metabolism and differentiation, to fully developed organisms or organelles, which subsequently die or undergo dedifferentiation by aging or decaying processes.

As may be seen from the example of plastid development, the morphogenetic processes of monotropically directed irreversible processes are in principle different from the biochemical cyclic operations (citric acid cycle, ribulose cycle of CO_2 assimilation), with their reversible equilibria, which regenerate again the necessary acceptor substances after a number of coordinated enzyme reactions (Fig. K–7).

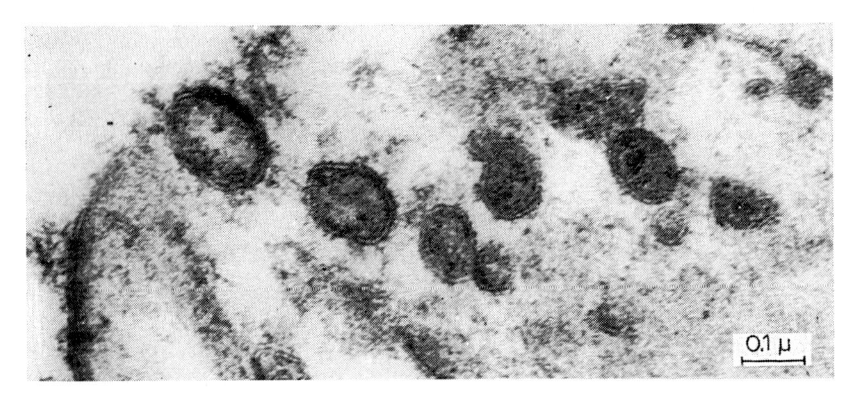

Fig. H–1. Plastid initials in the apical meristem of *Elodea canadensis* (Mühlethaler and Frey-Wyssling, 1959).

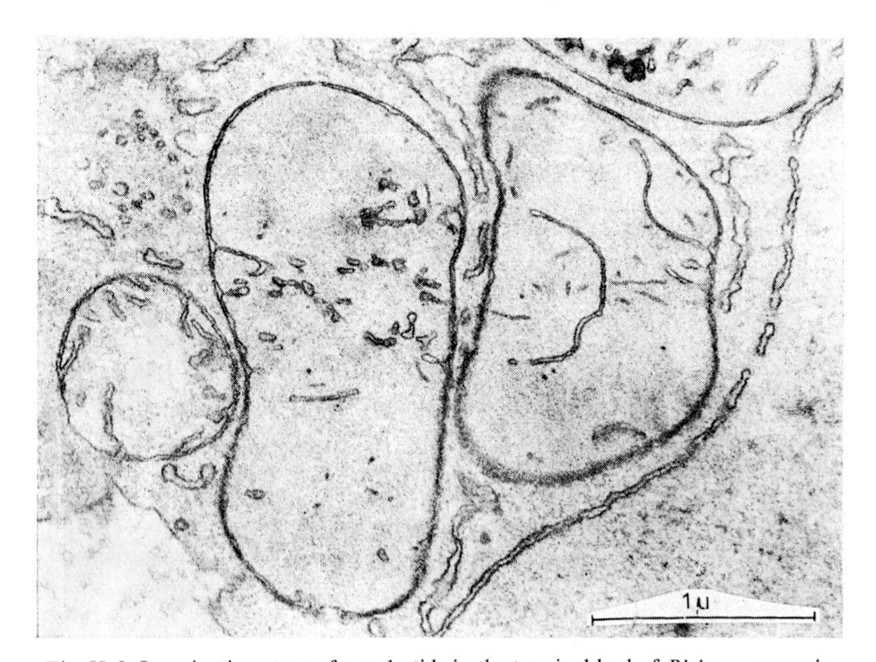

Fig. H–2. Invagination stage of proplastids in the terminal bud of *Ricinus communis*.

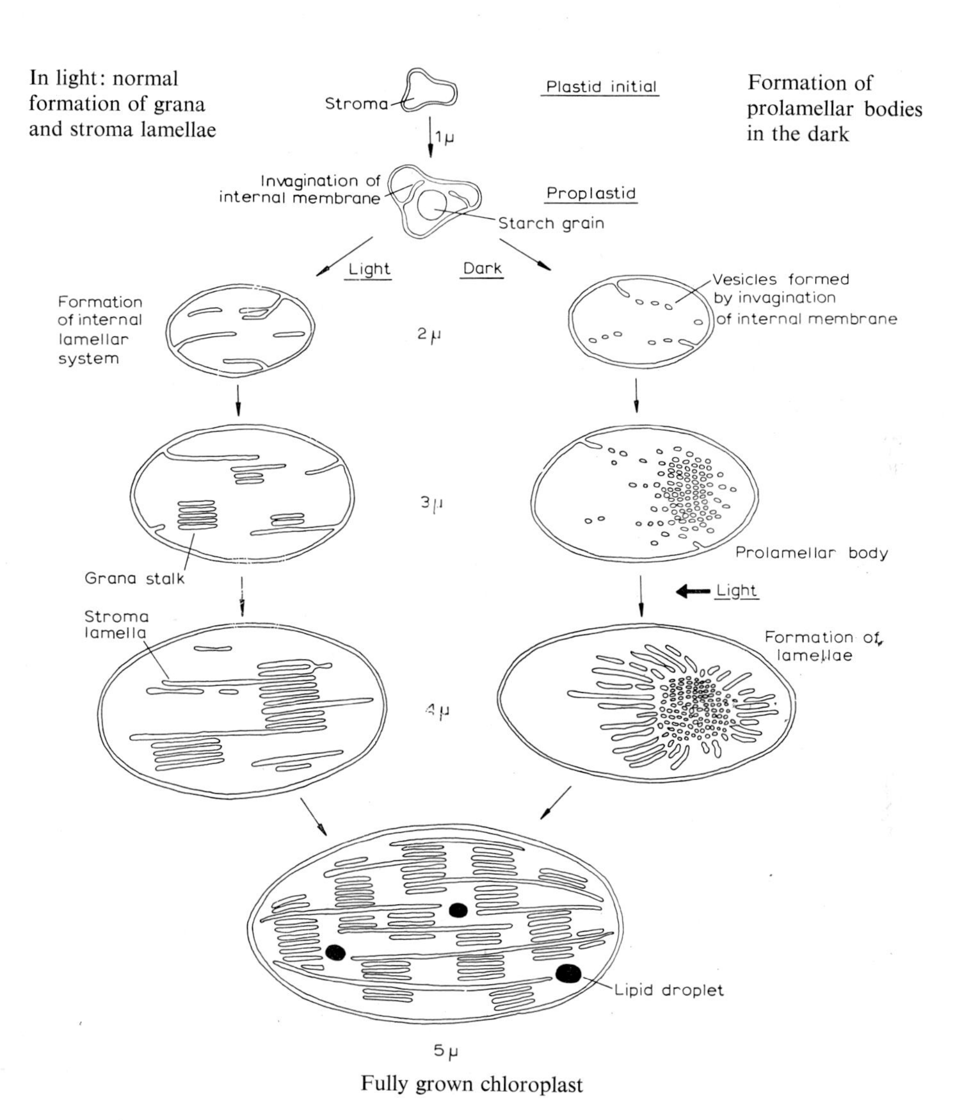

Fig. H–3. Ontogeny of chloroplasts (Mühlethaler and Frey-Wyssling, 1959). Left side: in light, lamella formation; right side: in darkness, formation of prolamellar body.

(b) Initials and proplastids

The earliest stage of plastid development is represented by globular particles. These possess a stroma substantially denser than the surrounding groundplasm, and already have a double membrane. Fig. H–1 shows such initials from a bud of *Elodea canadensis*. They can only be distinguished from mitochondrial initials, if their development is followed. It is then observed that after adequate growth of the initials to oblate ellipsoids, their inner walls begin to fold inwards (Fig. H–2). In contrast to the case of mitochondria, these folds do not run perpendicular but rather parallel to the surface of the particle, *i.e.* the folding occurs more in a tangential than a radial direction (Fig. H–5). This form of invagination is undoubtedly related to the later lamination of the plastids, which is oriented parallel to the longer axis of the ellipsoid. The folds may also show rather flattened formations, instead of linear tubuli. Flat vesicles are cut off from these pockets, so that the stroma, which increases rapidly as in the mitochondria (Fig. H–5a, b), but nevertheless always remains optically more electron-dense than the groundplasm, is permeated by a system of flat, parallel bladders of varying diameters. At this stage, in which the plastids have reached light microscopic dimensions (diameter $>0.5\mu$), and their outline can change in amoeboid fashion, they are known as proplastids.

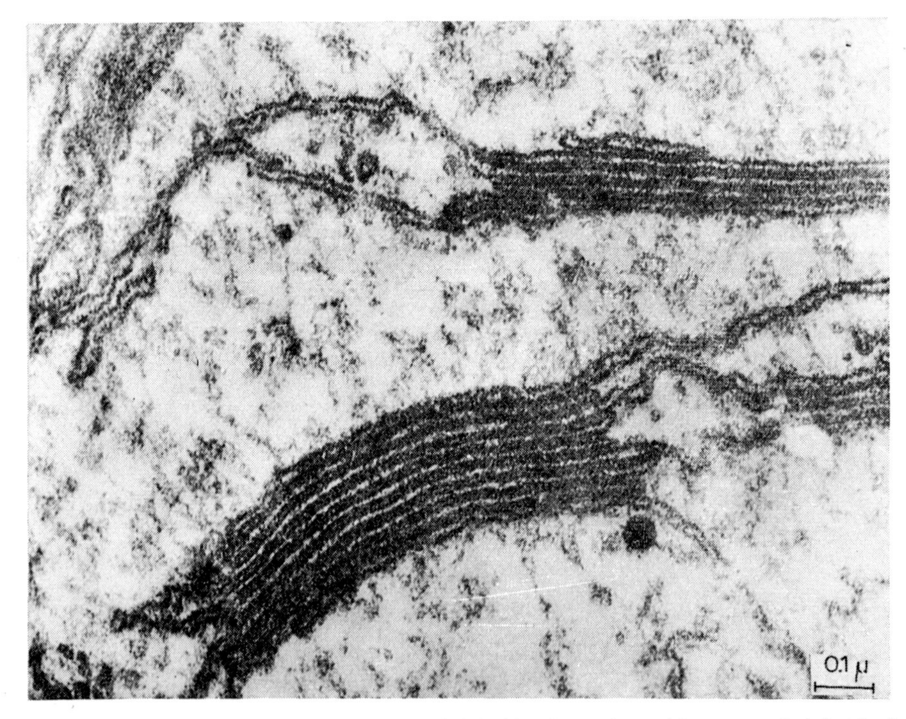

Fig. H–4. Granum stack showing granum thylakoids alternating with stroma thylakoids, in *Elodea canadensis* (Frey-Wyssling, 1960).

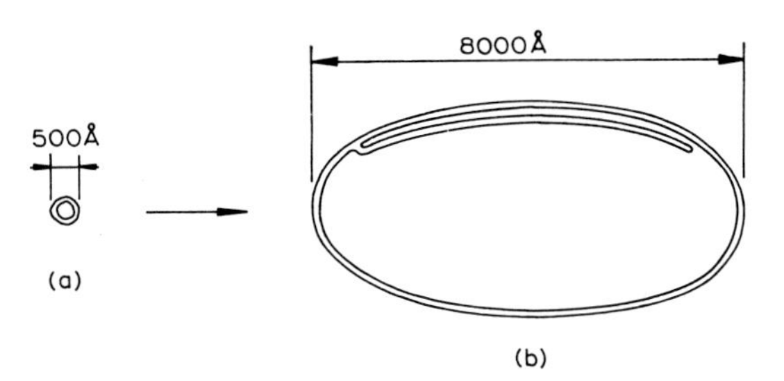

Fig. H–5. Diagrammatic representation and size of chloroplast precursors, (a) initial, (b) proplastid, with tangential invagination (Frey-Wyssling, 1959/61).

Since the flat vesicles play a special role in further development of the plastid, Menke (1962) coined the special term thylakoids for them (Gr. *thylakoides* = sack-like). Two different types of thylakoids develop under the influence of light, as for example in the transparent buds of *Elodea*: those that extend over the whole longitudinal section of the plastids and are thus called stroma lamellae (see p. 256), and much shorter ones, which arrange themselves on top of one another in stacks, and are termed granum lamellae (Fig. H–4). This is expressed schematically in Fig. H–25c (Frey-Wyssling, 1959/61).

No lamellation takes place in the absence of light, which is the case in most buds (*Agapanthus, Aspidistra, Chlorophytum* and *Hordeum*), or when the transparent buds of *Elodea* are darkened (Fig. H–3, right hand side). Thus not only chlorophyll formation but also differentiation to lamellated chloroplasts is suppressed in the dark (Mühlethaler and Frey-Wyssling, 1959; Wettstein, 1959). Instead, countless small particles gather together, which can assume spherical close packing and form quasicrystalline bodies. These attain a magnitude just over the resolving power of the light microscope, and can be stained with rhodamine B. Strugger (1950), who discovered these bodies, named them primary grana. Later, Heitz (1954) and Leyon (1954) thought that they consist of spherical macromolecules in close packing. On better resolution, however, the spheres proved to be vesicles, and finally Menke (1961) showed that the hexagonally arranged vesicles represent sections across helical tubules, whose pitch angle in the proplastids of *Chlorophytum* is 64°, so that only an approximate hexagonal symmetry is realised.

If such proplastids are subsequently exposed to light, thylakoid lamellae grow from the organised lattice (Fig. H–6), which is today termed prolamellar body (Hodge *et al.*, 1956). The thylakoids differentiate themselves into stroma- and granum lamellae (Fig. H–7), so that eventually exactly the same chloroplast structure results by the roundabout route via the leucoplasts as by direct differentiation in the presence of light. The unique formation of prolamellar bodies in the dark can be explained by the assumption that certain substances necessary for lamella formation cannot be synthesised in the dark. Other constituents (probably lipoproteins on the ground of their ability to stain with rhodamine B) can however be formed in the absence of light, and these are thus accumulated in the prolamellar bodies, for as long as they cannot be incorporated into the ultrastructural ladstidal lamellae.

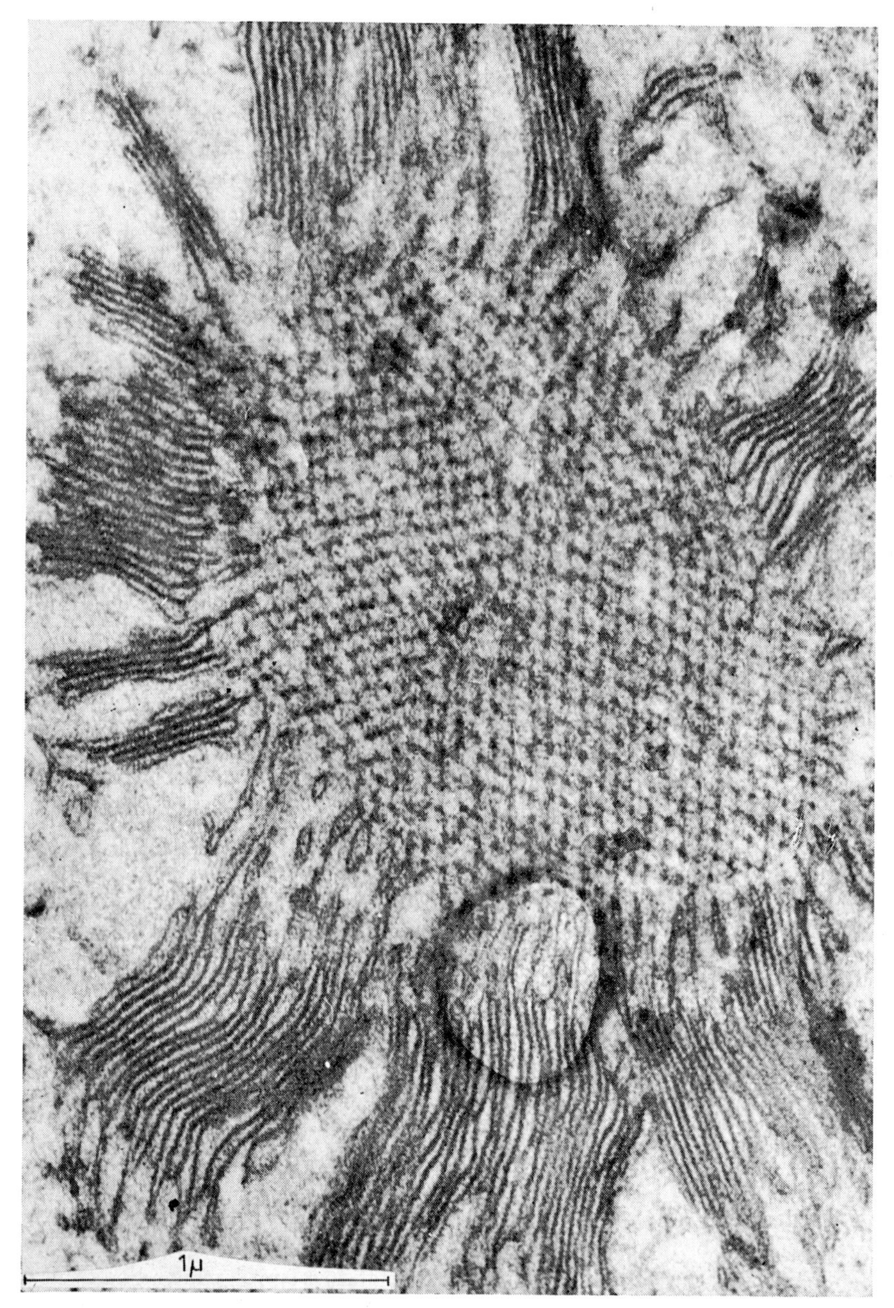

Fig. H–6. Prolamellar body in proplastid; bud of *Chlorophytum Sternbergianum* (Mühlethaler and Frey-Wyssling, 1959).

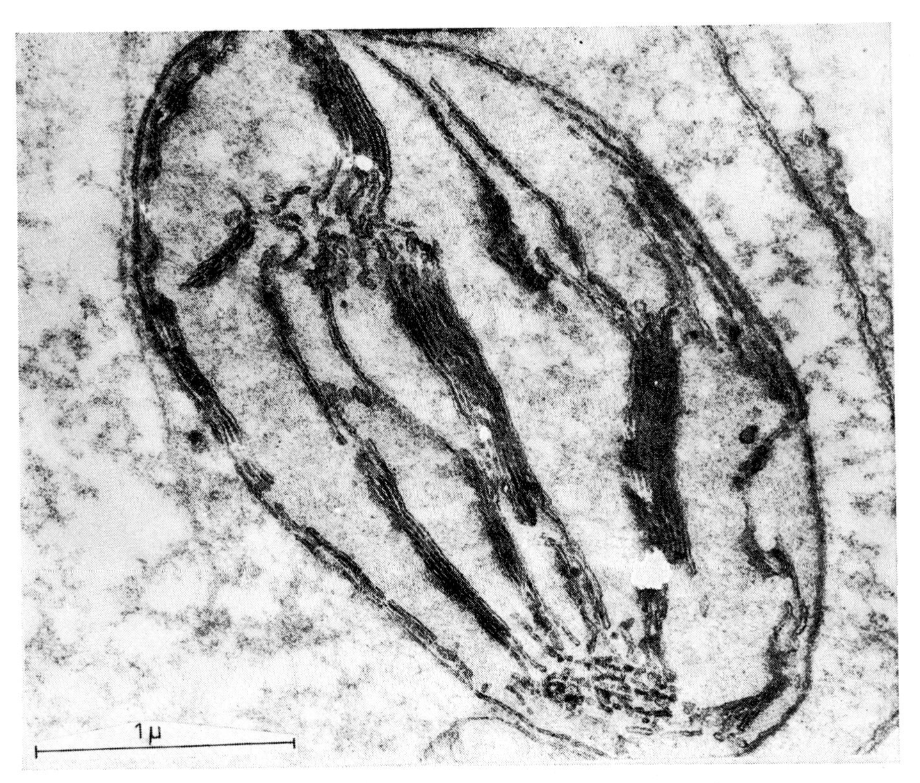

Fig. H–7. Formation of granum and stroma thylakoids from two prolamellar bodies in *Chlorophytum Sternbergianum* (Mühlethaler and Frey-Wyssling, 1959).

(*c*) *Division and budding*

The plastids increase in number by division; but this occurs only after differentiation of these organelles has begun or is concluded. In other words, no autoreproduction of the initials has ever been observed. Detection of a division state of the initials would be of the greatest importance for the theory of the autonomy of the plastidome, but no such observations have been made. On the other hand, the division of leucoplasts with two prolamellar bodies, and of differentiated chloroplasts, can be easily demonstrated.

The transverse constrictions observed during chloroplast division occur without regard for the ultrastructural lamellar system; the latter is simply sectioned (Fig. H–8). Constrictions proceed by a folding of the inner membrane of the plastid, which advances faster from one side than from the other (Mühlethaler, 1960b). Until the transverse separation is complete, the outer membrane remains steady and holds the two daughter individuals together (Fig. H–9). It is not clear whether it is subsequently also folded, or whether it tears and allows the dividing region of the plastid with its single membrane to form an new outer membrane from the periplastidal gap or from the groundplasm.

As in the case of mitochondria, budding of the chloroplasts can occur when differentiated cells are compelled to de-differentiate. Such an event takes place

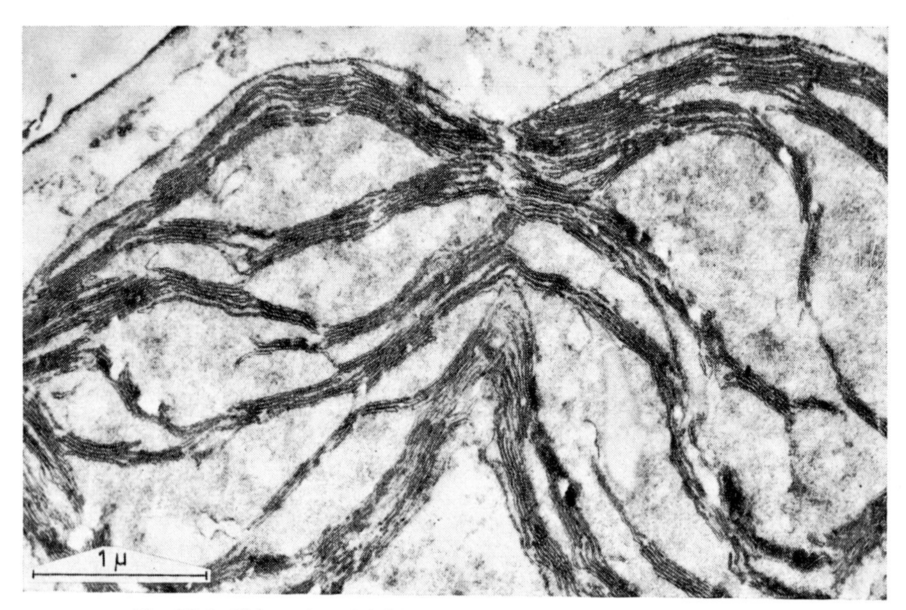

Fig. H–8. Chloroplast. Division by constriction (Mühlethaler, 1960b).

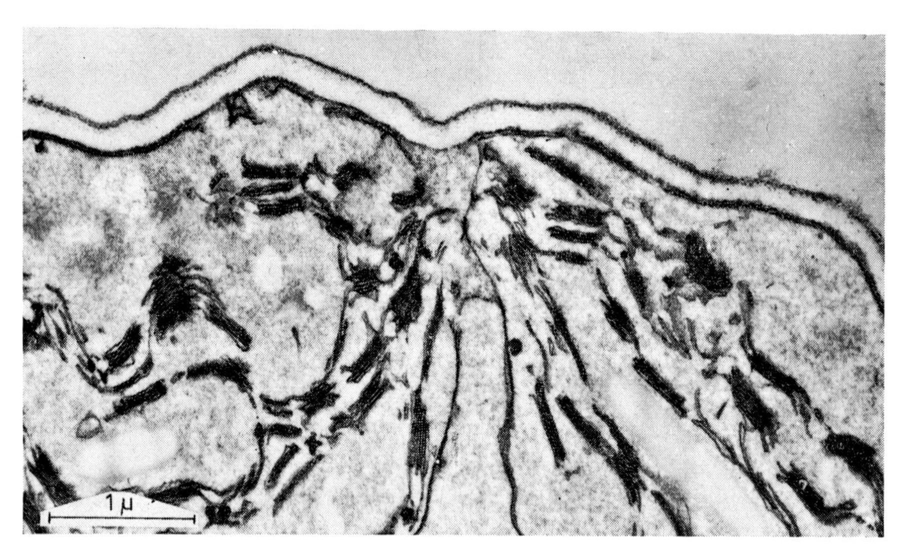

Fig. H–9. Dividing chloroplast in a leaf of *Solanum nigrum*. The outer membrane of the peri-plastidal space is not involved in the invagination of the septum (Mühlethaler, 1960b).

when dissected moss leaves develop by regeneration to a new small moss plant. Their differentiated cells then revert to the meristematic state, the existing organelles disappear, and during this process new cell organelles are produced. It has been observed that small bodies are detached (Fig. H–10) from the degenerating mitochondria and plastids (Maltzahn and Mühlethaler, 1962b). The detached particles show a double membrane and dense stroma and thus possess all the outward characteristics of plastid initials. They are also undoubtedly carriers of the morphogenetic information for the ontogenesis of plastids. The question thus arises whether the continuity principle is maintained with the help of such sprouted initials. This interpretation must however be opposed on the grounds that no such budding has been observed in normal plastid ontogenesis, so that this neoformation must be regarded as an exceptional phenomenon. The problem of where the initials originate in the ova is therefore of paramount importance.

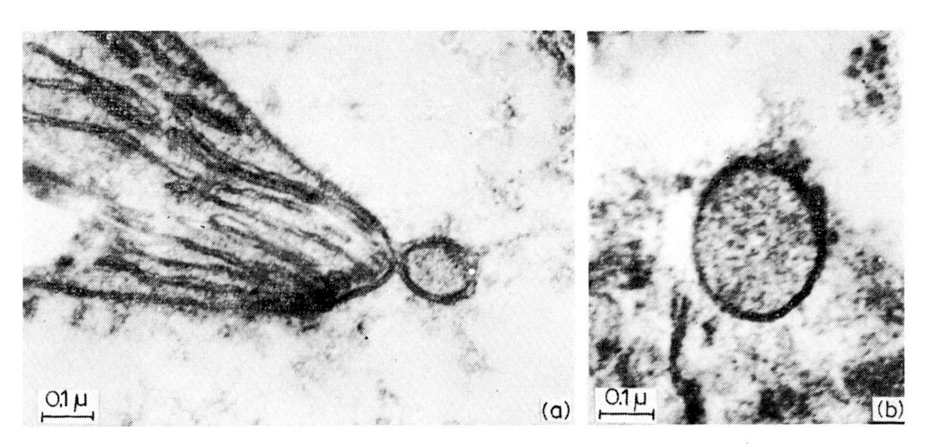

Fig. H–10. Budding chloroplast in a de-differentiating moss leaf; (a) formation of an initial, (b) detached initial (v. Maltzahn and Mühlethaler, 1962b).

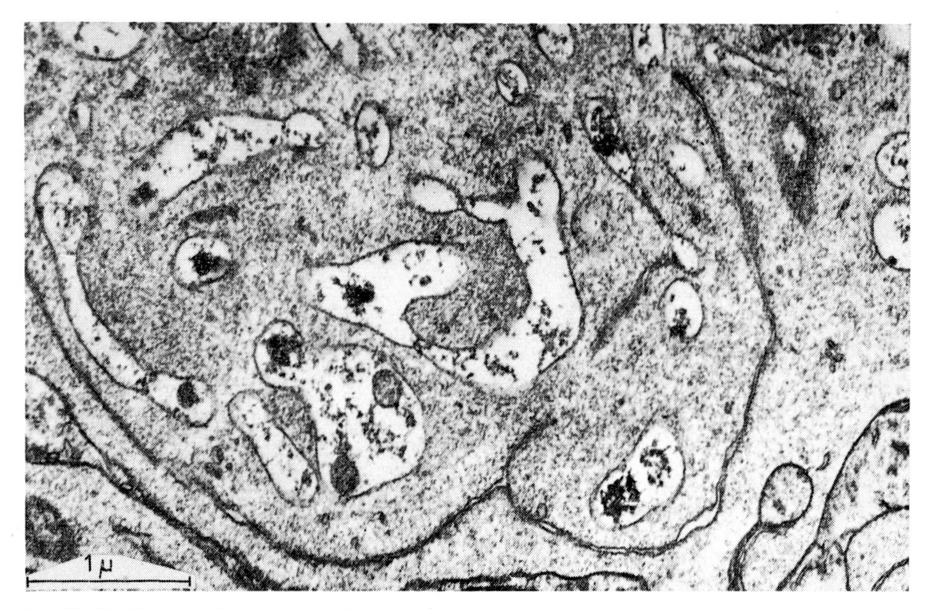

Fig. H–11. Completely degenerated chloroplasts in the egg cell of *Pteridium aquilinum* (Mühlethaler and Bell, 1962).

(*d*) *Nuclear origin*

As was explained on p. 226, the system of cell organelles breaks down in the ovum of the eagle fern, before this is fertilisable. Fig. H–11 shows plastids at this stage, degenerated so far as to be no longer recognisable. While the cell organelles disappear in this way in the outer plasma layer, new initials form in large numbers around the nucleus (Mühlethaler and Bell, 1962; Bell and Mühlethaler, 1962b), arising by evagination of the double nuclear membrane and containing dense nucleoplasm (Fig. H–12).

Mitochondria and proplastids are later formed from these initials which arise *de novo*.

As the double wall and the stroma of such plastid initials come from the still unfertilised ovum nucleus, it is understandable that the first subsequent generation shows pure maternal inheritance of plastid characteristics. Transmission experiments show, however, that all following generations are also purely matroclinous

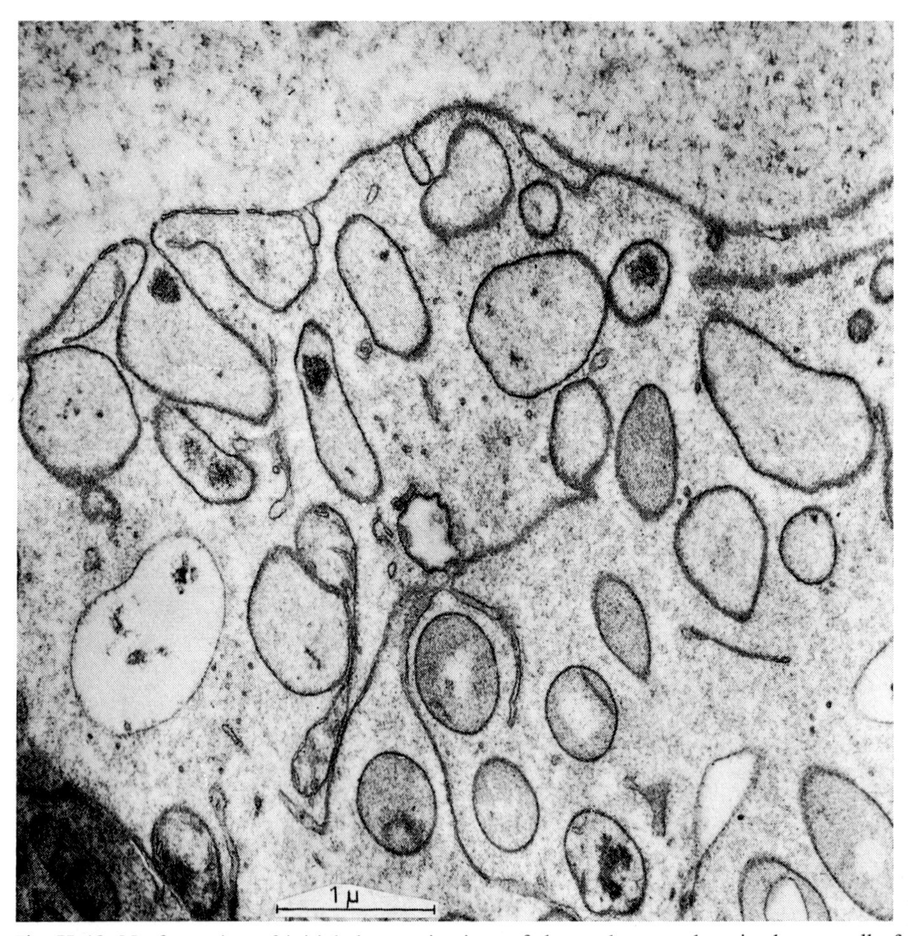

Fig. H–12. Neoformation of initials by evaginations of the nuclear envelope in the egg cell of *Pteridium aquilinum* (Mühlethaler and Bell, 1962).

in this respect, although in fact an influence of the paternal genetic material would be expected after the second generation (Stubbe, 1962).

There is a marked contradiction here between the ideas worked out on the basis of extensive inheritance experiments and the electron microscopic observations. As however the theory of plastid inheritance was based originally on Schimper's assumption that all plastids can arise only by division of their like, and as the electron microscope has now thrown doubt on this proposition, we must try to bring the genetic results into harmony with the factual electron microscopic observations (Arnold, 1963).

In this connection, it seems important that the nucleoplasm is likely the genetic carrier for plastidome inheritance. As shown, an open contact exists between nucleoplasm and groundplasm, so that, if there is a plasma inheritance through the plasmone, such properties can just as well be attributed to the nucleoplasm.

It may be too early, however, to follow up such trains of thought before the relationships outlined are confirmed in examples other than in the ova of fern archegonium. As has already been mentioned (p. 227), the cell organelles do also break down in the ova of Cycadaceae and conifers (Camefort, 1962), so that they must likely be replaced by new formations.

2. LEUCOPLASTS AND STARCH GRANULES

(*a*) *Amyloplasts*

It is difficult to give a concise definition of the leucoplasts. In the ontogenetic developmental series, they join the proplastids from which they differ only by their size. Like proplastids, they contain only a few lamellae, but are able to build up a well organized thylakoid structure (under the influence of light) and to become green. In the dark, they store ergastic materials for the thylakoids in the prolamellar bodies, as well as starch in the stroma. The ability to polymerise glucose to starch, which is considered the chief characteristic of the leucoplasts, is not however a diagnostically decisive indication, for both their precursors, the proplastids, and also their successors the chloroplasts, similarly possess the capacity to build up starch and to store it, mostly in the form of several small granules in the stroma (Fig. H–3). In the chloroplasts these are known as transitory starch, because they appear only temporarily during the assimilation stage of the green plastids, and eventually disappear again; the sugars which thus become free are transported to the consuming centres, or to the storage tissues. There are however leucoplasts whose sole function is the synthesis of large starch granules – these are known as amyloplasts. The starch produced by the amyloplasts remains stored for occasional later use, and is termed reserve starch.

The starch granules are readily separated from an homogenate on account of their particularly high specific gravity (more than 1.6) in relation to the density of the remaining cell components. Starch meals can thus be produced without centrifugation, by simple sedimentation in a gravitational field (*e.g.* potato meal). In the cereals the endosperm of the caryopses is however so densely filled with starch that grinding of the tissue without subsequent washing leads directly to the meal. The latter also contains of course all the other cell components, from which the starch granules must be separated if we want to investigate their chemistry.

As was explained on p. 41, purified starch consists of two components, amylose and amylopectin, the proportion of amylose being as a rule 1/5 to 1/4. It can also be completely absent (waxy maize, ketan), or rise to 1/3 and exceptionally even to $\frac{1}{2}$. These chemical differences have no effect on the morphology of the starch particles. The starch granules of waxy maize, which stain reddish with iodine, appear in the microscope to be the same as those from normal maize, which blackens with iodine. The various specific formations of the starch granules of different plant species and families, as portrayed in classical text-books (*e.g.* Nägeli, 1858), are thus limited to morphogenetic factors independent of starch synthesis.

Of the many kinds of starch, only the two economically especially important groups, distinguished as cereal starch and tuber starch, shall be considered in this book. The meals of the two groups yield, as mentioned earlier, different X-ray spectra, cereal starch the so-called A-spectrum and tuber starch the B-spectrum. Moreover, the granules of the first group are spherical (simple, *e.g.* wheat or composite, *e.g.* oats), whereas those of the second group (*e.g.* potato) are ovoid; these also show an excentric stratification detectable in the light microscope, whilst the concentric layering of cereal starch generally remains invisible without maceration.

(*b*) *Ontogeny of starch granules*

In both cases starch development begins with the formation of a particle, which appears white in the electron microscope, in the stroma of an amyloplast containing numerous often curved, tubuli and thylakoids (Fig. H–13). Occasionally, open or closed rings of this system surround the centrally situated starch particle, so that a vacuole is simulated (Badenhuizen, 1961). It must be emphasised that the starch only appears white in the electron microscope with the usual short fixation times; if the material is exposed longer to the fixing metal oxides, the starch granules become blackened. It is probable that the high density of the starch prevents a rapid penetration of the fixing agents.

Further starch is laid down on all sides of the original granule appearing in the stroma until a centre visible in the light microscope is produced, which is known as a hilum (Latin for navel). This quickly grows further by apposition, until the entire amyloplast is filled with starch. The tubuli of the stroma are thus pressed against the periphery of the plastid. Finally, the granule exceeds the original size of the amyloplast, so that the plastidal envelope (consisting of double wall + stroma residue) expands, becomes overdistended, and withers.

(*c*) *Cereal starch*

In the case of cereal meals we must distinguish between the composite (oats, rice) and the simple (wheat, rye, barley) starch granules.

As Fig. H–13 shows, in the first instance numerous initiatory granules arise in the stroma, which then grow until they mutually touch and flatten. The residual stroma remains in the slits between the component granules (Buttrose, 1960, 1962a). The composite starch grains thus dissociate easily into their component granules

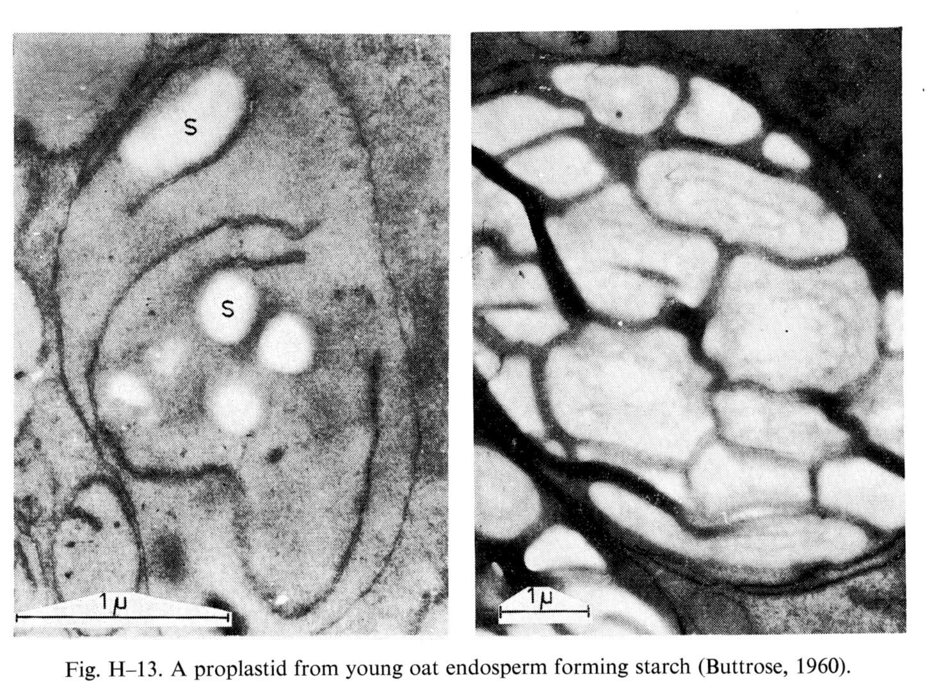

Fig. H–13. A proplastid from young oat endosperm forming starch (Buttrose, 1960).

Fig. H–14. Maturing compound starch granule from oat endosperm. Individual granules show rings after $KMnO_4$ treatment (Buttrose, 1960).

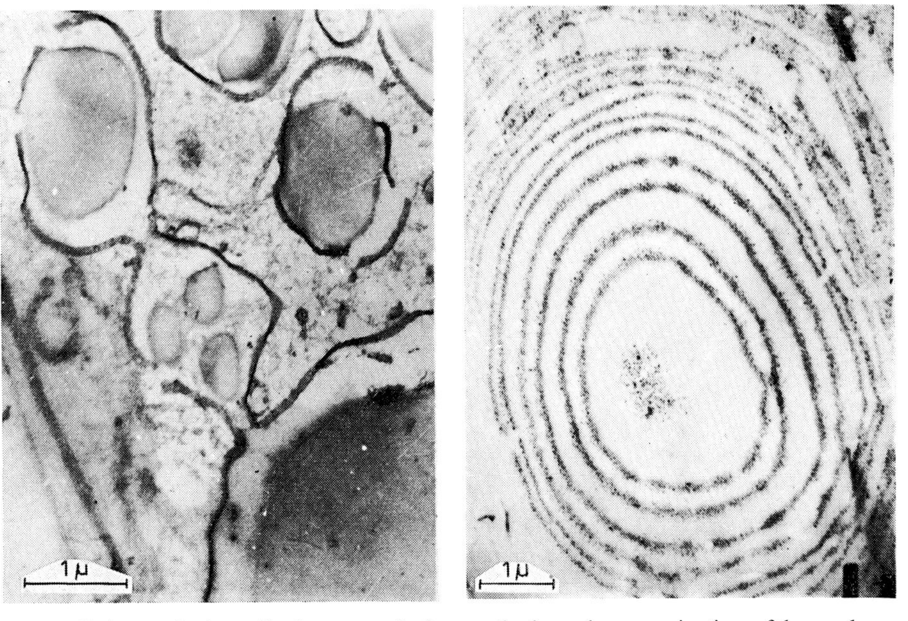

Fig. H–15. An amyloplast of barley two weeks from anthesis produces evaginations of the envelope, with the formation of small starch granules in the processes (Buttrose, 1960).

Fig. H–16. Waxy maize starch granule after acid treatment (7.5% HCl for one month) (Buttrose, 1960).

which in rice are particularly small, so that rice meal can be used as face powder. Each single component granule shows an ultrastructural layered formation.

In wheat and barley endosperm, the amyloplasts produce only one primary granule, which in the course of development reaches a diameter of 20–30 μ. These large starch granules are not spherical but shaped like flattened discs. A further generation of very numerous smaller granules, with a maximum diameter of 10 μ, is also formed. As the bodies producing the latter looked like mitochondria to the classical cytologists, the small granules were occasionally called 'mitochondrial starch', in distinction from the larger 'plastidal starch' in the leucoplasts. Today the genesis of these two forms of starch granules is known precisely.

May and Buttrose (1959) established that in barley the large starch granules begin to grow on about the 7th day after anthesis, their growth is concluded on the 24th day. For these granules, with a 10 μ radius, the average daily growth rate is therefore 0.6 μ. By daily measurements and statistical evaluation of the results, it was found, moreover, that the growth is rapid at first ($> 1\ \mu$/day), slowing down towards the end (see Fig. H–17). The smaller granules make an appearance three days later. Their number is about ten times as large as that of the larger original particles, although, on account of their size, the latter constitutes 90% of the total granule volume in the mature endosperm! The small granules also finish their growth on the 24th day after anthesis, and reach a radius of 5.5 μ at the most. Their average daily growth thus amounts to an about $5.5\ \mu/14 = 0.4\ \mu$ thick layer. Here also the growth is fast initially and then slows down asymptotically.

Electron microscopy has shown that the small granules are formed in an

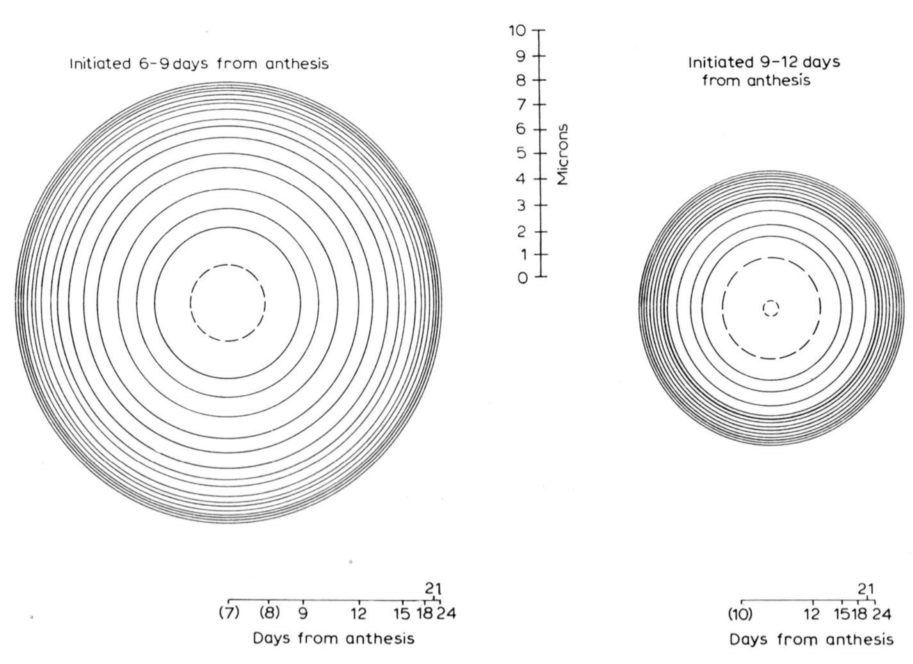

Fig. H–17. Diagrammatic representation of the calculated daily growth in cross-section of an average starch granule initiated either 6–9 or 9–12 days from anthesis (May and Buttrose, 1959).

unexpected way from the same amyloplasts as the large ones. Three days after the primordial particles are observed in them, the plastidal envelope begins by evaginations, to fill up the whole of the space in the wedges between the spherical leucoplasts with vesicles (Fig. H–15). These either remain connected to the original plastid by a neck, or are completely cut off. A small starch granule is formed in each vesicle, and in this way the entire volume of the endosperm cell becomes filled with starch.

After fixing with $KMnO_4$ (Fig. H–14), and particularly after lintnerisation, *i.e.* after treatment of the starch granules with 7.5% HCl for one to two months (Fig. H–16), the concentric lamination becomes plainly visible. The number of layers corresponds to the number of days of growth. No stratification is observed in cereal grains ripened under continuous illumination and constant temperature (Buttrose, 1962b). The layers of cereal starch thus represent daily rings. At the beginning of the development of a granule, these daily growth rings are clearly wider than those formed towards the end (Fig. H–17).

These observations, made with the help of the electron microscope, had already been made earlier with the light microscope by Bakhuizen (1925). They were however disputed, because, in the case of potato starch, culture under constant conditions did not succeed in causing the layers to disappear.

(*d*) *Tuber starch*

Potato starch will be discussed as the representative of tuber starch. In these eccentrically formed granules there are numerous shells, which are easily distinguishable in the light microscope. They are of course only visible in aqueous media, and disappear if the starch granules are dehydrated. Owing to their complete dehydration, the starch granules as a rule appear in the electron microscope to be homogeneous and structureless, although the shells can be made visible by treatment with acid.

TABLE H–I

DIFFERENCES BETWEEN CEREAL AND TUBER STARCH GRANULES (with about 25% amylose)

	Cereal starch (wheat)	Tuber starch (potato)
Structure	concentric	eccentric
X-ray pattern	A	B
Solubilisation by enzymes	more rapid	less rapid
Iodine colouration after acid degradation	still blue	red-violet
Growth periodicity of granule	aitiogenic	endogenous
Rhythm of lamella formation	24 h	ca. 2 h
Cause of periodicity by block in	production of carbohydrate	carbohydrate translocation (?)

Fig. H–18a shows potato starch lintnerised for 4 days in 15% (4 *N*) HCl. We can readily recognise the coarse layers, known since Nägeli's (1858) classic sketches. On stronger resolution, these 1 to 2 μ thick layers appear to be subdivided in

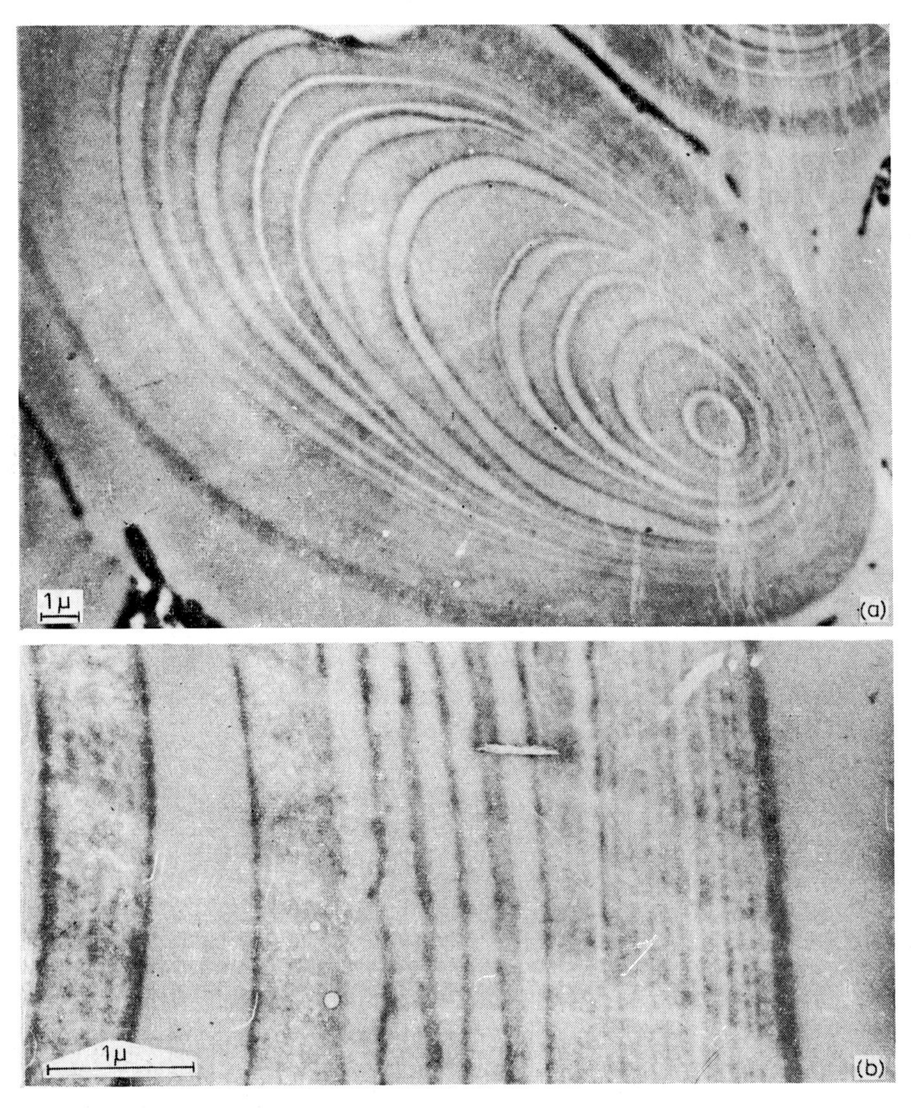

Fig. H–18. (a) Granule of potato starch after acid treatment, (b) resolution of the layers into ultrathin lamellae (Frey-Wyssling and Buttrose, 1961).

periods of about 1000 Å, about 400–500 Å of which is accounted for by lamellae, the remainder being empty space (Frey-Wyssling and Buttrose, 1961). The peculiarity of this lamellation is that it is not subject to day periods. Buttrose (1962b) found that the wide light microscopic shells are formed in a cycle of 18.5 h, and the very fine lamellae in one of 2 h. On eliminating the daily rhythm (continuous lighting and constant temperature), this periodicity is, however, not influenced in any way! In contrast to the case of cereal starch, with an aitiogenic layering, the periodicity in potato starch must thus be endogenous, which is independent of external influences. Thus, here too, the results of classical cytology remain valid. While in cereal starch the assimilatory daily rhythm is responsible for the periodic formation of the starch granules, another principle must be responsible for the 2 and 18.5-h lamellisation cycles of the tuber starch. It is conceivable that the effect is due to a rhythmic block in the translocation of sugars as starch precursors.

With the determination of the endogenous rhythm in the formation of potato starch, a further difference between cereal and tuber starch is found. These differences are summarised in Table H–I (see Buttrose, 1963b).

(*e*) *Macromolecular structure*

It is the task of the molecular biologist to explain the molecular structure of single lamellae of starch granules.

In this respect, considerable preliminary work was accomplished by indirect methods. As the starch granules behave in the polarising microscope as optically positive spherites and show the effect of a rodlet composite body (Speich, 1942), radially directed glucosan chains must be present (Frey-Wyssling, 1948a). These interpretations were confirmed by X-ray analysis, which indicated the existence of a chain lattice with radial axes (Kreger, 1951).

Such observations would lead easily to a simple explanation of the molecular structure if the starch granule consisted exclusively of amylose chains. As Table B–IV (p. 44) indicates, however, amylopectin forms the main substance of the starch granules. The question thus arises of how the two components are divided. It was long believed that in the interior of the granule, or of a layer, the growth began with amylose, which later became replaced by amylopectin in the outer region (K. H. Meyer, 1940; Badenhuizen, 1959/61). This view is however incorrect on various grounds. Firstly, the amylose chains are so long (see below) that they traverse the entire layer and therefore cannot occur only in the inner parts; secondly, the starch granules which consist exclusively of amylopectin (waxy maize) are exactly the same with regard to morphology and lamination as those containing amylose (Fig. H–16). The two compounds must thus occur mixed with one another in such a way that the layers can be regarded as chemically uniform.

A valid explanation for the observed lamination is the idea that at the beginning of growth each single lamella has abundantly available molecules of starch precursors (glucose, glucose phosphate), so that the layer is initially laid down thickly. Gradually, however, the concentration of the precursors, which cannot be procured in sufficient quantity, declines, so that a gradual impoverishment of solid material sets in and is made noticeable by the greater hydration of the outer part of the layer. In consequence of this behaviour, the density of each layer decreases gradually in the radial direction from the inside to the outside (Fig. H–19a), and

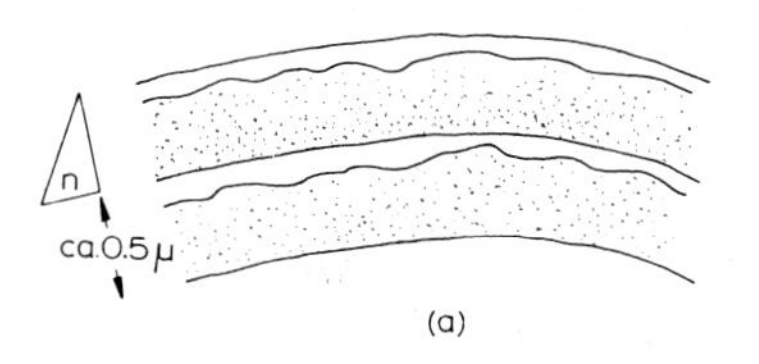

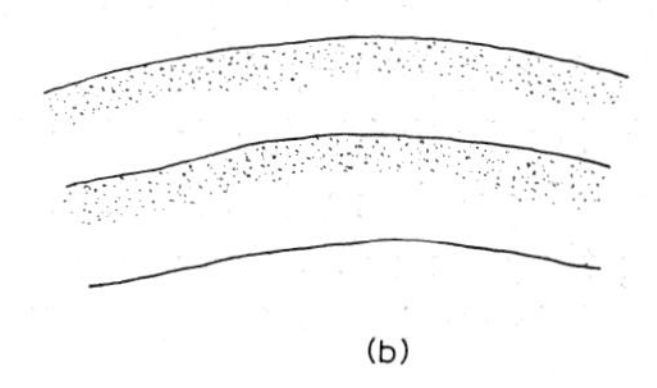

Fig. H–19. Lamellae of cereal starch granules with centripetally increasing refractivity n; (a) enzymatic corrosion (α-amylase), (b) acid corrosion (lintnerisation), see Fig. H–16.

then starts up again suddenly at a higher value at the layer boundary (Frey-Wyssling, 1938). This finding has recently been confirmed by Buttrose (1962b). In germinating seeds, or on treatment of the starch granule with α-amylase, the outer parts of the layers are rapidly corroded, because the enzyme molecule has a better access to the loosely formed structure and the amount of substance to be degraded is smaller. Unexpectedly, treatment with acid (lintnerisation) leads to the converse picture. In this case, a discontinuity of the density gradient at the outer limit of the layers is again observed, but the density decreases from outside to inside (Fig. H–19b). In order to explain this observation, we must bear in mind that, during the lintnerisation of wheat starch, 74% of the starch granule goes into solution, so that only about $\frac{1}{4}$ of the mass remains (Buttrose, 1962b). Obviously, on chemical dehydration with concentrated hydrochloric acid, the outer parts of the layer are converted to insoluble compounds, which appear to be carbonised in the electron microscope (Fig. H–16), whereas the denser inner regions disintegrate completely in the course of one or two months' hydrolysis.

Armed with the knowledge that the carbohydrate chains run radially in the layer, and that they are more densely packed in the inside of the layer than in its outer region, we can now attempt to trace out a picture of the molecular structure. For this purpose, we shall first consider layers of cereal starch, which are 0.5 μ (5000 Å) wide on the average.

A degree of polymerisation (DP) of 940–1300 (average M.W. 180,000) is indicated for the amylose of maize starch. This corresponds to a chain length of $(1120/3) \times 10.6 = 4000$ Å. In contrast, the amylopectin in maize can reach a degree of polymerisation of $\frac{1}{2}$ million (M.W. 80–100 $\times$ 10^6) (Aspinall and Greenwood, 1962). According to Table B–IIIb and p. 43, such molecules are however only about 1000 Å long, and about half as wide. They must be arranged radially in the layer of the starch granule.

This is illustrated in Fig. H–20a. It is recognised that amylose chains of DP 1100 and with length 4000 Å, are almost as long as the width of an average layer. For wheat amylose, however, DP values of 2100 are indicated, which corresponds to a length of 7500 Å. As such a length exceeds the average width of the layers, folded amylose chains could perhaps exist; in order to comply with the results of polarised light studies and X-ray analysis, however, the straight chain segments must run radially, as indicated in Fig. H–20c and e.

In comparison, the amylopectin molecules can be accommodated easily in the layer. They are symbolised in Fig. H–20 as triangles, with the aldehyde groups as vertex. Here, meanwhile, we are faced with the problem of a complete filling of space. In order to attain this, the amylopectin molecules must be set in an anti-parallel manner (Fig. H–20d and f). Their polarity is signified by arrows, which indicate the direction of growth of the branched molecule, proceeding from the aldehyde end group.

When both kinds of molecule are present, the difficulty is that, in contrast to the amylopectin molecules, the amylose chains traverse the whole width of the layer; they thus run through several amylopectin molecules which would not be impossible sterically (see Frey-Wyssling, 1957). With anti-parallel close packing (Fig. H–20f) radial insertion of the amylose chains would, however, be impractica-ble on steric grounds. A further possibility of shortening the amylose molecules to

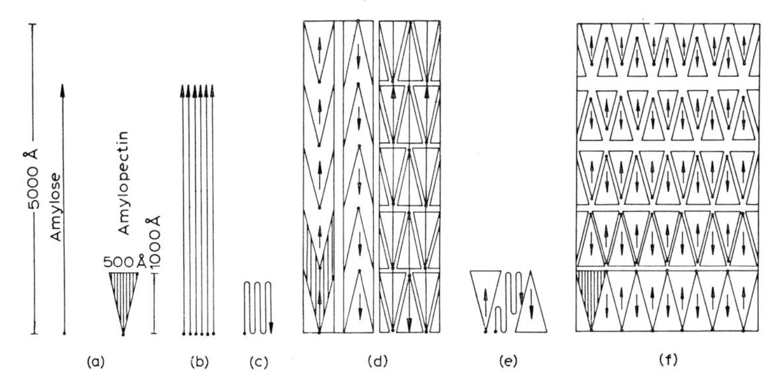

Fig. H–20. Hypothetical molecular structures of a 0.5 μ wide lamella in the starch granule; (a) individual amylose (n = 1100) and amylopectin ($n = 5 \times 10^5$) molecules; • indicates aldehydic pole, arrows → point to the non-aldehydic ends of the molecular chains; (b) crystallised amylose; (c) folded amylose chain; (d) overlapping molecules of amylopectin in anti-parallel array or shifted in such a way as to permit insertion of amylose chains; (e) amylose chain shortened to the length of amylopectin by folding; (f) anti-parallel orientation of amylopectin, accounting for the radially decreasing optical density within the lamella.

the length of the branched amylopectin molecules, consists in the assumption of folding (Hirai *et al.*, 1963). As mentioned before, to comply with the anisotropy and the crystalline spherite structure of the starch granules, the straight line members of such a folded chain must be aligned radially (Fig. H–20c and e). Thus, at present, it is not possible to picture the bimolecular structure of the starch granule in a manner free from contradictions.

In potato starch, in which the layers consist of ten times thinner lamellae with a thickness of the order of magnitude of 500 Å, one must expect half as long amylopectin molecules of DP 5000–10,000 (M.W. of about one million) and amylose chains of DP 150 (M.W. about 25,000) (Frey-Wyssling and Buttrose, 1961).

3. CHLOROPLASTS

(*a*) *Chemistry*

In comparison with the leucoplasts, the chemistry of the chloroplasts has been substantially better explained. Table H–II shows the chemical composition of spinach chloroplasts, as given by Lichtenthaler and Park (1963). It is assumed that the isolated chloroplast material consists only of thylakoids, and its manganese content is taken as the reference point. The molecular contents and molecular weights are then as listed in Table H–II.

If all the constituents were combined in a lipoprotein macromolecule, the total molecular weight would be 960,000. As globular protein particles of this weight have a diameter of about 130 Å (see Table B–IX), the volume of the assimilatory units called quantasomes (see p. 263), whose dimensions are something like 100 × 200 × 200 Å would be twice as large. Consequently, Park and Biggins (1964) postulate a molecular weight of 1,920,000 with two atoms of Mn for the

TABLE H–II

CHEMICAL COMPOSITION OF THE THYLAKOIDS IN SPINACH CHLOROPLASTS
(Lichtenthaler and Park, 1963)

Moles		Mol. weight	
115	Chlorophylls (80 chl a + 35 chl b)	103,200	
24	Carotenoids (7 β-carotene, 11 lutein etc.)	13,700	
23	Quinones (14 plastoquinone, 7 tocopherol, 2 vitamin K)	15,900	
58	Phospholipids (26 phosphoglycerol, 21 lecithin etc.)	45,400	
72	Digalactosyl diglycerides	67,000	
173	Monogalactosyl diglycerides	134,000	
24	Sulpholipids	20,500	
?	Sterols	7,500	
	Unidentified lipids	87,800	
	LIPIDS		495,000
4690	N atoms as protein	464,000	
1	Mn	55	
6	Fe	336	
3	Cu	159	
	PROTEIN		465,000
	Minimum molecular weight per atom of Mn		960,000

quantasomes. As a result the number of moles indicated in Table H–II must be doubled for obtaining the molecular composition of a quantasome.

According to Table H–II, the lamellae in the chloroplasts consist half and half of protein and lipid. Earlier analyses of the total chloroplast material showed 50% protein and only about 33% lipid by weight (Menke, 1938; Comar, 1942). The discrepancy is due to the fact that the specific gravity of proteins is > 1.3 and that of lipids < 1.0.

Stäubli (1957) investigated the nucleic acid content, finding in young spinach chloroplasts 150–200 μg and in grown chloroplasts 3–7 μg of nucleic acid phosphorus per mg of N. With a protein content of 50%, this corresponds in young plastids to a value of over 10% and in grown plastids to only 0.5% of nucleic acids. Similar results are obtained with blood erythrocytes whose nucleic acid content falls in the course of the differentiation, as a consequence of growth, to quite insignificant amounts. New formation of nucleic acid obviously does not take place, so that its relative content is considerably decreased.

The question is forcefully debated of whether RNA alone, or both RNA and DNA are found in the chloroplasts. By electrophoresis on paper, Stäubli (1957) was able to detect in grown spinach chloroplasts the four bases cytosine, uracil, guanine, and adenine, but not thymine. Similarly as in the nucleolus, however, small amounts of DNA can be found histochemically or after feeding with radioactive thymidine autoradiographically (Bell and Mühlethaler, 1964b), which evidently lie outside the limits of detection of the method used for the electrophoretic separation. According to Wollgiehn and Mothes (1963) only growing chloroplasts in tobacco leaves incorporate infiltrated tritiated thymidine, whereas grown ones do not. Jagendorf and Wildman (1954) found 1.3–3.3 μg of DNA-phosphorus/mg N, *i.e.* about half that of RNA. Gibor and Izawa (1963) determined in single chloroplasts of *Acetabularia* 10^{-16} g of DNA a piece (see Ruppel, 1964).

The following energy carriers and active enzymes have also been demonstrated in chloroplasts: ATP (Arnon, 1955), DPN (see p. 265), cytochrome f (Davenport, 1952; Lundegårdh, 1961), quinones (vitamin K; Arnon, 1961), etc.

In contrast to these active substances, which are only present in small amounts, the pigment content reaches considerable levels. Thus 6–8% chlorophyll has been found in spinach chloroplasts (over 10% based on the thylakoids, see Table H–II). Of this, as a rule $\frac{3}{4}$ is chlorophyll *a* (chl a) and $\frac{1}{4}$ chlorophyll *b* (chl b). A relationship with the yellow carotenoid pigments is often present; for example, in tobacco leaves (Heierle, 1935) there is approximately one such molecule per 2 chl; about 2/3 of the carotenoids are xanthophylls (x) and $\frac{1}{3}$ is carotene (c). There are thus the following approximate molecular proportions:

chl a : chl b : x : c = 9 : 3 : 4 : 2 (tobacco leaves)

Seybold (1941) showed that considerable deviations from these ratios may occur. They are thus only an approximate reference point. From Table H–II the ratio is 11:5:2:1 (spinach leaves).

In the chloroplasts, the pigments are associated with proteins. The corresponding chromoproteins could not, up to now, be isolated with certainty (see the porphyrin proteins in Table B–XVIII). The amino acid inventory of the chloroplast proteins hardly varies from that of the cytoplasmic proteins (Noack and Timm, 1942; Weber, 1962).

Chlorophyll can only be synthesised by the action of light. A precursor substance, protochlorophyll, which is bound to protein, forms in etiolated tissues; even though this chromoprotein or holochrome possesses no noticeable colour, it can be spectroscopically detected and characterised. Its molecular weight is estimated at 400,000 (Smith, 1948; Smith and Kupke, 1956).

Virgin *et al.* (1962) have studied the conversion of protochlorophyll to chlorophyll in connection with chloroplast ontogenesis (see Fig. H–3). This follows a two-stage course. Chlorophyll formation sets in with the growing of the prolamellar bodies, but then enters a lag phase during the organisation of the thylakoids, and the discontinued pigment synthesis only carries on with the differentiation of the stacks of grana.

(*b*) *Ultrastructure suggested by indirect methods*

The ultrastructure of the chloroplasts as lamellar packets was known before the discovery of the electron microscope, from indirect methods of investigation. A brief retrospective look at the results of research with the polarising microscope should thus be taken. The optical anisotropy of the chloroplasts of the cormophytes is small, so that the much larger plastids of lower plants were used for quantitative measurements.

In the algae classes of the flagellates and the conjugates, as well as in the moss genus *Anthoceros*, the green cells contain only one or a few large chloroplasts, mostly developed like plates. Such a plate behaves in the polarisation microscope as a negative uniaxial body, whose optical axis is perpendicular to the plate surface. These chloroplasts prove, with the help of form birefringence analysis in the polarising microscope (Menke, 1934; Frey-Wyssling, 1937; Frey-Wyssling and

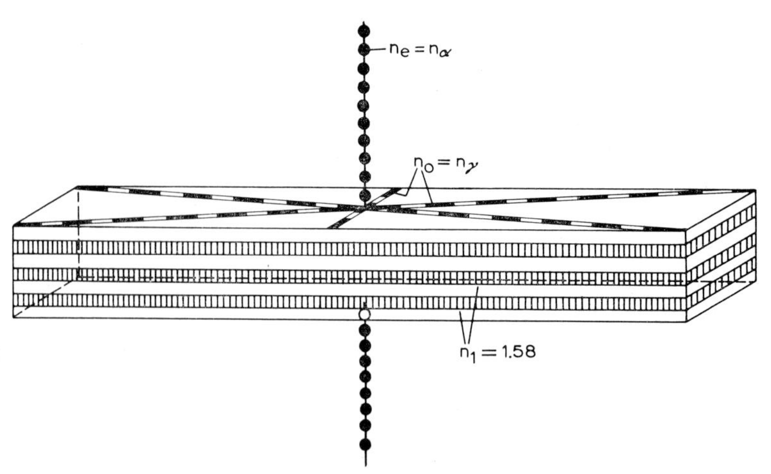

Fig. H–21. Ultrastructure of a *Mougeotia* chloroplast, derived from the form birefringence of its uniaxial anisotropy (Frey-Wyssling and Steinmann, 1948). The smaller refractivity n_α corresponds to the extraordinary beam n_e, and the stronger refractivity n_γ to the ordinary beam n_o. $n_1 = 1.58$ is the refractive index of the protein lamellae.

Steinmann, 1948), to be layered bundles of sublight-microscopic lamellae (Fig. H–21).

While the intrinsic birefringence of crystals is constant for the given material, the birefringence Δn of biological objects proves as a rule to be dependent on the refractive index n_2 of the embedding medium [$\Delta n = f(n_2)$], because the embedding fluid penetrates the object and alters its optics by imbibition. If the object consists of isotropic material of refractive index n_1, it should show no birefringence. If, however, n_2 is different from n_1 ($n_2 \neq n_1$), one observes, in spite of the isotropic constituents, an anisotropy, which is known as 'form birefringence'. The term arises from the fact that the resulting birefringence in a so-called rodlet composite body has a positive sign, whereas in a layer composite body (as in Fig. H–21) it has a negative sign. The form of the sublight-microscopic structural elements can thus be deduced from the character of this birefringence.

In the chloroplasts the form birefringence is negative, so that a layered bundle, not visible in the light microscope, must be present. The dependence of the layer birefringence on the refractive index n_2 of the imbibition medium is hyperbolic, according to Wiener's (1912) theory of composite bodies.

The equation of these hyperbolas is

$$n_e^2 - n_o^2 = -\frac{\delta_1\delta_2(n_1^2 - n_2^2)^2}{\delta_1 n_2^2 + \delta_2 n_1^2},$$

where δ_1 and δ_2 designate the relative volumes ($\delta_1 + \delta_2 = 1$) of the layers and the empty space between them, n_1 is the refractive index of the solid layer, and n_2 that of the imbibition fluid. As it is difficult to evaluate n_e and n_o (Fig. H–21) of the chloroplast in the polarising microscope, the so-called retardation Γ is measured. The latter quantity is proportional to the birefringence:

$$\Gamma = (n_e - n_o)d,$$

in which d is the thickness of the birefringent object. Therefore

$$\Gamma = -\frac{\delta_1\delta_2(n_1{}^2 - n_2{}^2)^2}{\delta_1 n_2{}^2 + \delta_2 n_1{}^2} \times \frac{d}{n_e + n_o}.$$

Since the variable difference $n_e - n_o$ is small (n_e and n_o differing by only about 0.1%), to a first approximation $n_e + n_o$ can be considered as constant. The same is true of d, if we work with fixed chloroplasts whose swelling capacity is reduced to the minimum.

A discussion of the above equation shows that, in the first place, Γ depends on the difference between the indices of the layers n_1 and that of the imbibition fluid n_2; if $n_2 = n_1$, form birefringence disappears, and, irrespective of whether n_2 is greater or smaller than n_1, it is negative; it increases with the square of the difference $n_1{}^2 - n_2{}^2$, forming a hyperbola (Frey-Wyssling, 1940). In the second place, Γ is a function of the product $\delta_1\delta_2$; assuming all other magnitudes to be constant, Γ passes through a maximum if $\delta_1 = \delta_2 = 0.5$, *i.e.* if the layers and the empty spaces are of equal thickness.

The curves of Fig. H–22 indicate further that $n_2 = n_1$ if $n_2 = 1.58$. This means that the submicroscopic protein layers of the *Mougeotia* chloroplast have a refractive index of 1.58. This is rather high for a protein, but it coincides with the refractive index of muscle myosin or of neurokeratin, and indicates that the chloroplast

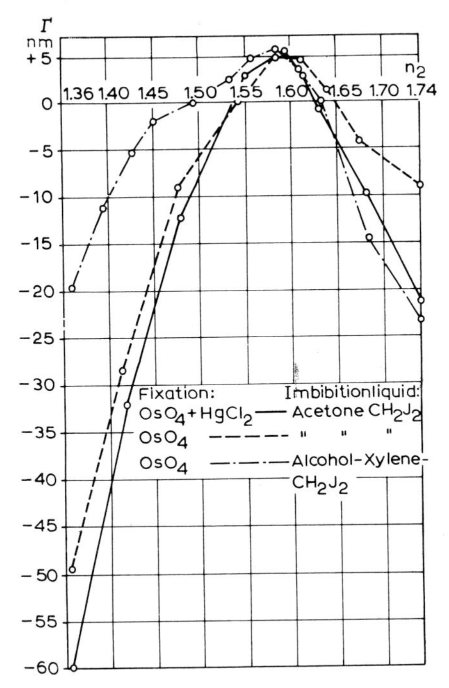

Fig. H–22. Form birefringence of *Mougeotia* chloroplasts (Frey-Wyssling and Steinmann, 1948). Abscissa: refractive index n_2 of the imbibition liquid. Ordinate: retardation (caused by birefringence) Γ, in nm.

protein must contain aromatic amino acids, such as phenylalanine, tyrosine, histidine, or tryptophan. 2.6% of tyrosine, 4.6% of histidine, and 2.7% of tryptophan have in fact been found in *Chlorella*, and similar values were obtained in higher plants (P. Weber, 1962).

It is further seen that imbibition with mixtures of alcohol, xylene, and methylene iodide gives somewhat flatter and less regular curves than that with mixtures of acetone and methylene iodide. This is due to the fact that alcohol and xylene cause some shrinkage of the fixed chloroplasts, so that the submicroscopic space accessible to the imbibition fluid is reduced, with the result that the product $\delta_1\delta_2$ is decreased. In this way, several features bearing on the sublight-microscopic structure can be read from carefully measured curves of form birefringence.

Only after extraction of the lipidic compounds does Γ disappear completely when $n_2 = n_1$. Fixation with OsO_4, preserving the lipids, yields curves such as those reproduced in Fig. H–22, where the vertex of the curve penetrates into the positive region. This indicates that there is a component of positive intrinsic birefringence ($\Gamma = 5$ nm) combined with the form birefringence. The observed intrinsic birefringence must therefore be due to oriented lipid molecules.

The lipids must have a polar structure, for on mild hydrolysis with dilute ammonia they can be caused to shift and to form myelin forms (F. Weber, 1933).

On the basis of these observations, a basic scheme of chloroplast ultrastructure was formulated as early as 1937. According to Hubert (1935), it shows bimolecular layers of phospholipids (formed like tuning forks), which are associated with the chlorophyll (Fig. H–23). For each two of the stamper-shaped chlorophyll molecules, there is one rod-shaped carotenoid molecule. The pigments are adsorbed on a protein layer. The layer thickness could be calculated from the chlorophyll content of the chloroplasts, and the periodicity was almost correctly predicted as about 300 Å (Fig. H–24) (Frey-Wyssling, 1937).

When the lamellar structure was first confirmed in the electron microscope (Steinmann, 1952), the periodicity of the lamination was subsequently measured with the help of X-ray diffraction. This method has the advantage that the object need not be completely dehydrated, as for measurements with the electron micro-

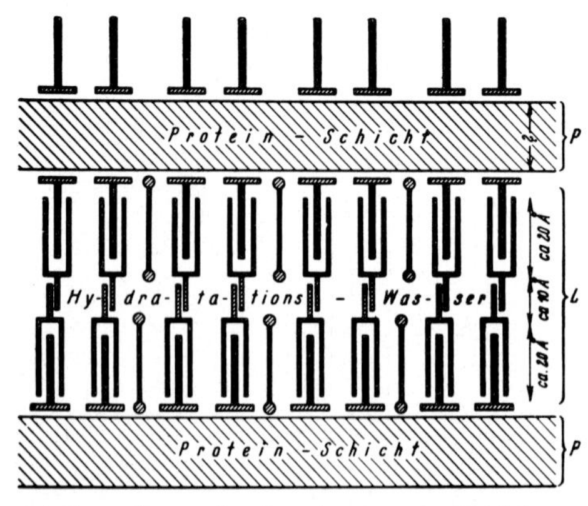

Fig. H–23. First representation of a molecular structure in lipid layers of chloroplasts (Frey-Wyssling, 1937): phospholipid (shape of tuning fork), chlorophyll (T), carotenoid (I).

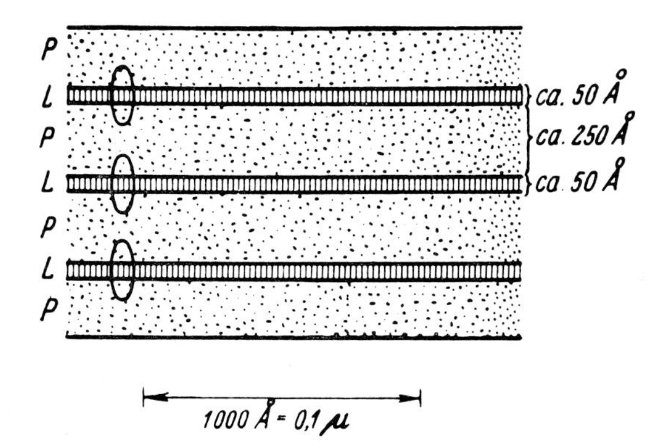

Fig. H–24. Calculated thickness of protein (*P*) and lipid (*L*) layers in chloroplasts (Frey-Wyssling, 1937).

scope, but can also be investigated in the wet state. A layer period of 250 Å was found in chloroplasts both of *Aspidistra* (Finean *et al.*, 1953) and also of *Allium porrum* (Kratky *et al.*, 1959).

The scheme in Fig. H–23 could not withstand subsequent criticism. Only two discrepancies will be briefly mentioned in the present work. The allotment of one molecule of chlorophyll to each molecule of phospholipid presupposes a stoichiometric ratio of 1:1 to lipid phosphorus and chlorophyll in the chloroplasts. Much too little lipid phosphorus is however present, (Bot, 1939), so that no such chromolipid can exist (see Table H–II).

Moreover, the arrangement of all the porphin rings of the chlorophyll molecules in one plane would cause a strong intrinsic dichroism. Dichroism can in fact be detected, especially in red light. For the *Mougeotia* chloroplasts Goedheer (1955) found at 680 nm an extinction ratio E_o/E_e of 1.15 and Ruch (1957), $E_o/E_e = 1.36$. E_o is the extinction parallel (ordinary ray) and E_e that perpendicular to the plane of the layer (extraordinary ray; see n_e and n_o in Fig. H–21). Menke (1943, 1958) indicated that in myelin figures artificially coloured with chlorophyll, the extraordinary ray is more strongly absorbed than the ordinary ray ($E_e > E_o$). In the chloroplasts the reverse effect is observed ($E_o > E_e$). In other words, the green-coloured myelin tubes are positively, and the chloroplasts negatively dichroic. From this, it follows that no intrinsic dichroism of the chlorophyll can exist in the chloroplasts. Actually, Ruch (1957) was able to show by imbibition experiments that form dichroism is mainly present, *i.e.* the absorption anisotropy is produced not so much by aligned orientation of chlorophyll molecules, but mainly by the layered ultrastructure of the chloroplasts. The intrinsic dichroism of the existing chromoprotein is minimal. The same applies to the difluorescence of the chloroplasts, *i.e.* the property that the fluorescence excited by shortwave light is emitted with different intensities parallel and perpendicular to the plane of lamination. This anisotropy can also be reduced by appropriate imbibition of the chloroplasts.

Finally, the dichroism in ultraviolet light was also tested (Ruch, 1957). After extraction of all the pigments, the above method may be used to see whether the ultraviolet-absorbing amino acids (Phe, Tyr, His and Try) are arranged in any way in the plastid protein as it is *e.g.* the case for tyrosine in silk fibroin. As no such effect appeared, it must be concluded that the plastid protein is organised rather as globular macromolecules than as extended polypeptide chains.

(*c*) *Ultrastructure as seen in the electron microscope*

The chloroplasts are enveloped in a distinct double membrane, the periplastidal gap between outer and inner membranes being *e.g.* 100–300 Å in *Nicotiana* (Düvel, 1963). A peristromium, which was described by Senn (1908) for the green chromatophores of moss and algae, is not present in the fixed state. Living spinach chloroplasts were however found by cinephotomicrographic studies to be surrounded by a colourless jacket in motion (Wildman *et al.*, 1962). A jacket of structureless stroma appears also in fixed chloroplasts, as a pathological phenomenon in manganese-deficient spinach (Possingham *et al.*, 1964).

Electron microscopy has revealed the lamination of the chloroplasts in higher plants to be more complicated than shown in Fig. H–21, since stacked thylakoids (p. 235) with a diameter of the order of 0.5 μ (granum lamellae), and others which traverse the wide extent of the stroma (stroma lamellae), are present (Figs. H–4, H–25c).

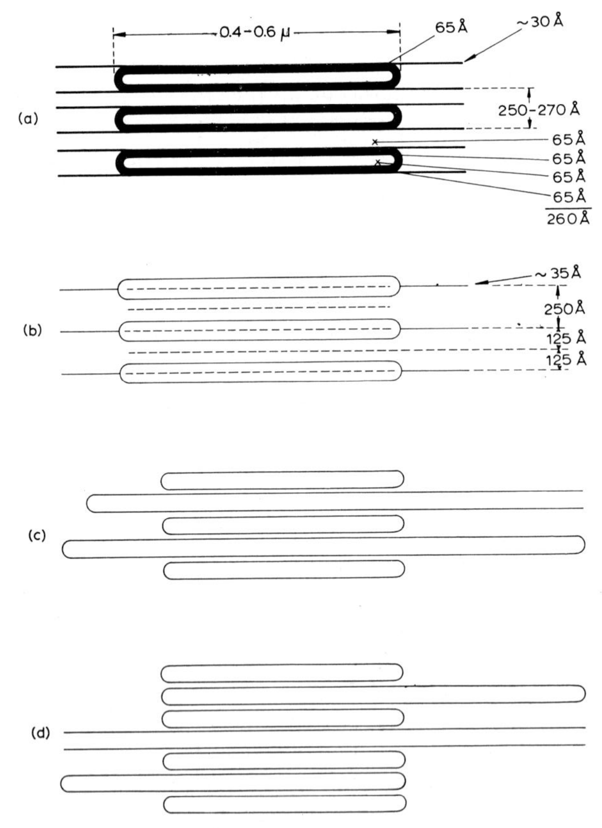

Fig. H–25. The interrelation between the stroma and the granum lamellae, (a) *Aspidistra* (Steinmann and Sjöstrand, 1955), (b) *Zea mays* (Hodge *et al.*, 1955), (c) *Elodea canadensis* (Frey-Wyssling, 1959/61 and Mühlethaler, 1960b), *see* also Fig. H–28, (d) *Spinacia oleracea* (Wehrmeyer and Perner, 1962).

Differing views have been expressed concerning the relationship of the granum and stroma lamellae. Steinmann and Sjöstrand (1955) found in *Aspidistra* chloroplasts just as many lamellae of both types, and thought that the stroma lamellae grow from those of the grana, as indicated in Fig. H–25a. On the other hand, Hodge *et al.* (1955) observed only half as many stroma lamellae in maize chloroplasts, and thus published the scheme of Fig. H–25b, according to which the granum discs are locally swollen thylakoids. On the basis of their research on barley chloroplasts, Wettstein (1957), as well as Erikson *et al.* (1961), likewise presented the opinion that such carrier lamellae for the grana are present in the stroma. The ontogeny of the lamellae shows however that the stroma and the granum thylakoids have the same origine (Figs. H–4, H–5b). Pictures like those in Fig. H–25 a and b result from aggregation of the stroma lamellae with the granum lamellae, or by the stroma thylakoids becoming compressed between the granum thylakoids, so that their lumen disappears locally; outside the granum stacks however, their lumen may be inflated to double thickness, so that adjacent stroma lamellae touch and aggregate.

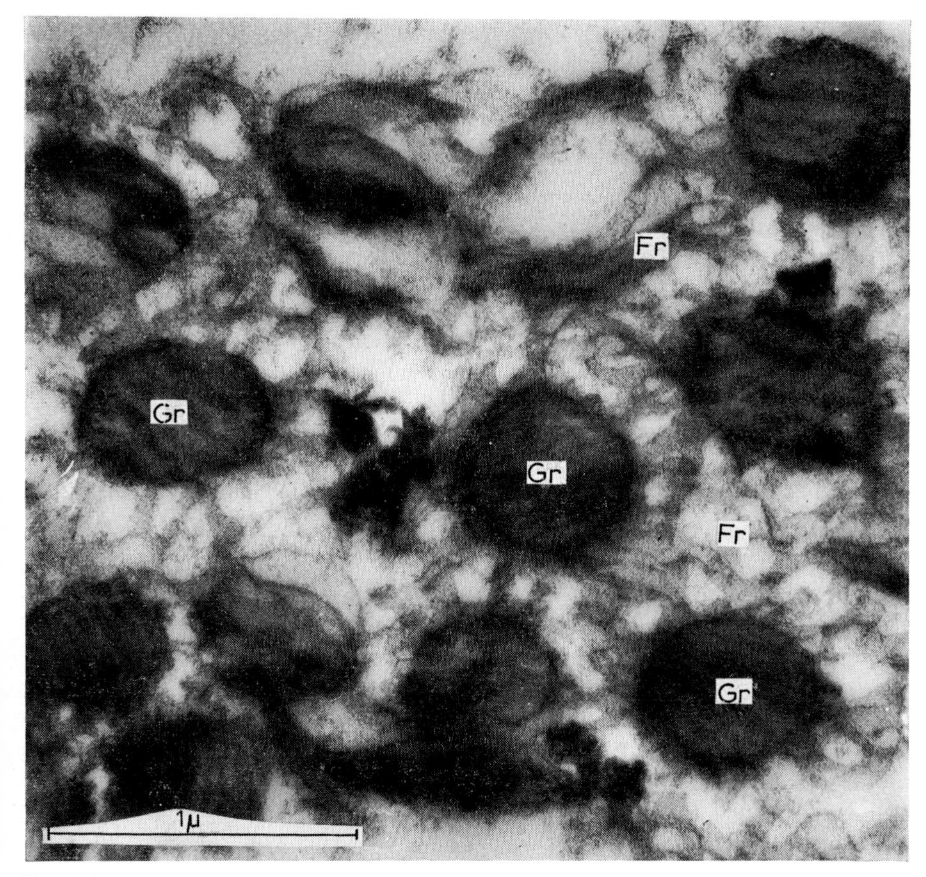

Fig. H–26. The stroma lamellae are perforated by large pores forming a fretwork (*Fr*) between the granum stacks (*Gr*). *Nicotiana rustica* (Weier and Thomson, 1962).

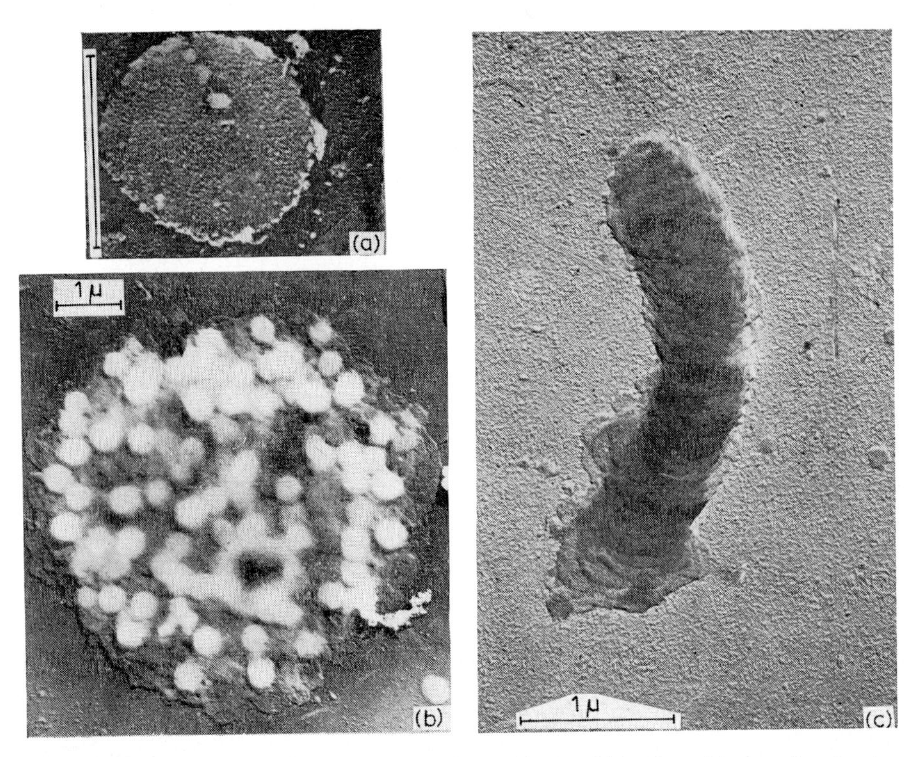

Fig. H–27. Isolated grana; (a) by ultrasonics, first evidence of lamellae with granular structure and organised as thylakoid structure indicated by rim and broken upper membrane, *Aspidistra* (Frey-Wyssling and Steinmann, 1953); (b) by swelling of stroma in 0.5 *M* sucrose, *Spinacia oleracea* (ibid.). (c) Granum stack isolated in phosphate buffer, *Aspidistra* (courtesy of E. Steinmann).

The stroma thylakoids frequently terminate, however, in ultrathin sections, in the granum stacks, so that one can distinguish continuous and discontinuous stroma lamellae (Fig. H–25d). This means that the stroma thylakoids do not represent uninterrupted layers, but they display large perforations (Fig. H–26, Weier, 1961). Weier *et al.* (1962, 1963) and Heslop-Harrison (1963) showed how the granum thylakoids mutually grow together by anastomosis, so that a two- or even three-dimensional fretwork results. This posed the question of whether there is a difference between granum and the stroma lamellae at all. No difference can actually be established on ultrathin sectioning, apart from that of their longitudinal extent (Wehrmeyer, 1963). In order to judge their identity, however, we must bear in mind the results of earlier experiments (Frey-Wyssling and Mühlethaler, 1949; Frey-Wyssling and Steinmann, 1953), when it was not yet possible to obtain ultrathin sections. At that time, preparations of chloroplasts were obtained with the help of the centrifuge, ultrasonics, and considerate swelling. In this way, it became evident that the stroma lamellae were completely destroyed in the centrifuge, while the grana remained as stacks or single discs (Fig. H–27). The same was true of the treatment with ultrasonics. The stroma lamellae were disintegrated, while the granum thylakoids showed a resistance to such conditions, even when parts of the membrane broke down (Fig. H–27a). From such pictures it was, moreover, first recognized that the granum discs consisted of flat vesicles. During the isolation of spinach chloroplasts in 0.5 *M* sucrose, the stroma disintegrates (Fig. H–27b), while the grana are preserved. In an aqueous phosphate buffer (pH 6.4) the chloroplast of *Aspidistra* breaks down, and only the granum stacks remain, slightly swollen (Fig. H–27c). These can be denatured to dense protein membranes, while the stroma lamellae are lost (Frey-Wyssling and Steinmann, 1953). All these results make the granum lamellae appear to be more robust than the stroma lamellae, for they exhibit greater resistance against mechanical force (centrifugal shear, ultrasonics) and against swelling.

On freeze-etching, the granum surfaces appear to be much more thickly charged with quantasomes (see p. 263) than the stroma lamellae (Fig. H–31). This

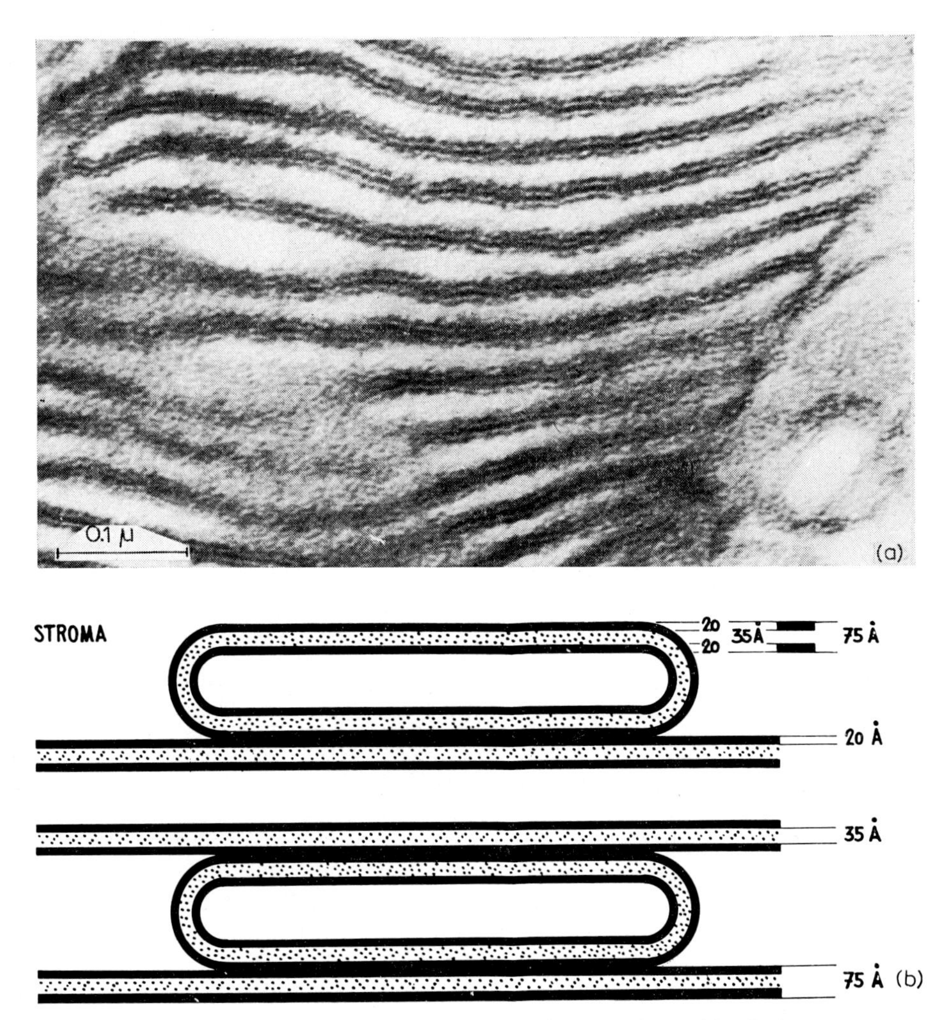

Fig. H–28. Evidence of unit membranes in granum and stroma thylakoids of *Elodea canadensis* (Mühlethaler, 1960b); (a) granum stack, (b) thickness of strata in unit membranes.

could thus mean that these undergo a greater surface development than the granum lamellae. In any case, the grana prove to be differentiated functionally by their wealth of quantasomes, whilst the stroma lamellae seem to possess a rather passive mechanical role for the spatial stabilisation of the grana.

Various reports have been made regarding the thickness of the lamellae. Steinmann and Sjöstrand (1955) found the stroma lamellae to be 30 Å and the 'granum lamellae' 65 Å thick. Since the latter (as has been shown on the basis of Fig. H–25c) represent, however, double layers, of one stroma and one granum membrane, the thickness of the granum lamellae is reduced to 35 Å. Hodge *et al.* (1955) likewise gave 35 Å for the layer blackened by osmium, but added, on the basis of the intermediate lines discovered (Fig. H–25b), 45 Å thick lipid strata on both sides of the layer regarded as protein. The total thickness is thus 125 Å. Mühlethaler (1960b) established that the lamellae have the character of a unit membrane (Fig. H–28a), from which the scheme of Fig. H–28b results. According to this, the unit membranes measure 75 Å; no morphological difference may be perceived between the stroma and the granum lamellae. Thicknesses of the unit membrane, found by different authors, are collected in Table H–III. As the X-ray period of lamination is greater

TABLE H–III

THICKNESS OF UNIT MEMBRANES AND X-RAY PERIODS IN CHLOROPLASTS

Unit membrane:		
Frey-Wyssling (1957)	Proposal	65 Å
Hodge (1959)	*Zea*	65 Å
Wolken (1959)	*Euglena*	100 Å
Mühlethaler (1960b)	*Aspidistra*	75 Å
Park and Pon (1961)	*Spinacia*	80 Å (–100 Å)
Kreutz and Menke (1962)	*Chlorella*	61 Å
X-ray period:		
Finean *et al.* (1953)	*Aspidistra*	250 Å
Kreutz and Menke (1962)	*Chlorella*	177 Å

than the thickness of the lamellae, the latter must be arranged in polar fashion, including a lumen between them; and this thickness of a thylakoid accounts for the measured X-ray period (see also Fig. H–28b).

The problem now arises of explaining the macromolecular structure of the lamellar system which has been demonstrated.

(*d*) *Macromolecular structure*

Hodge *et al.* (1955) discovered polarity of the granum lamellae by the detection of intermediate lines in granum stacks (Fig. H–25b). Nevertheless, the above authors believed that a protein layer about 35 Å thick, and of unknown fine structure, is symmetrically covered on both sides by a film of lipid (Fig. H–29a). In comparison with the original scheme reproduced in Fig. H–23, this representation brought no real improvement. In both cases a molecular 1:1 ratio of chlorophyll to the lipid phosphorus is assumed, which did not prove to be correct (Bot, 1939).

Moreover, it is not borne in mind, that an intrinsic dichroism stipulated by the chlorophyll in the chloroplasts and which, according to the suggested orientation of the porphin rings, should be strong, is in fact absent (Ruch, 1957). These shortcomings were overcome by Calvin (1959), and by Hodge (1959), who reversed the mutual positions of the lipid and the protein and assumed the existence of a polar lipid layer sandwiched between two protein films in line with the postulated structure of the unit membrane. Further club-shaped chlorophyll molecules are introduced, whose chromogenic rings no longer stand parallel, but which are

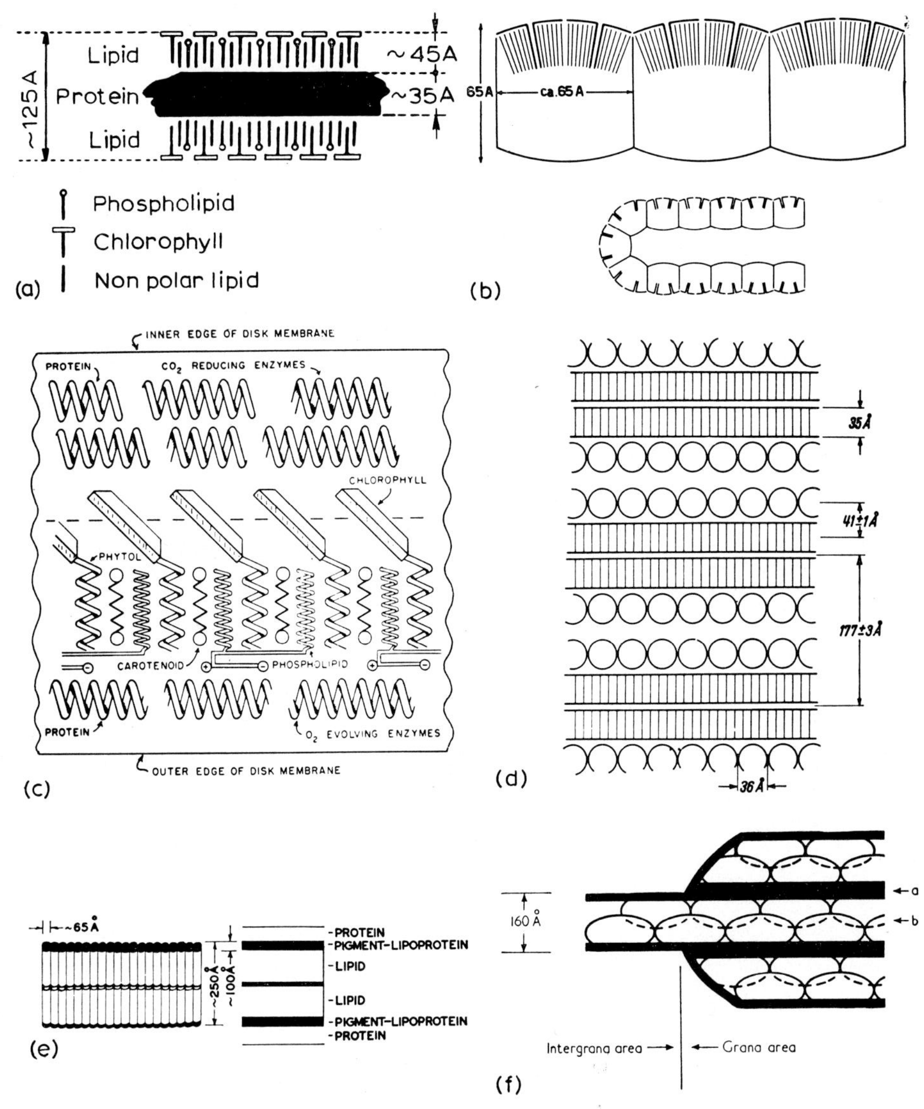

Fig. H–29. Models of the macromolecular structure in thylakoid membranes, after (a) Hodge *et al.*, 1955 (see Fig. H–25), (b) Frey-Wyssling, 1957, (c) Calvin, 1959, (d) Kreutz and Menke, 1962, (e) Wolken, 1959, (f) Park and Pon, 1961.

inclined in different directions (statistical isotropy) to the lamellar surface. The protein molecules are drawn as α-helices and the lipid chains of the phytol, the carotenoids, and the phospholipids are collected in a polar stratum, covered by the porphin rings serving as photoreceptors (Fig. H–29c).

On the basis of his researches with *Euglena* and on the retinal rods of vertebrate eyes, Wolken (1956a, 1959) supposed that rod-shaped 125 Å long macromolecules of lipids and proteins, which were covered with the pigment, serve as the photoreceptors (Fig. H–29e). Although lipoproteins are spoken of in the text, this is not expressed in the scheme; as in most representations, the fact that the weight ratio of protein to lipid amounts to 1:1 in the chloroplast lamellae (Table H–II) is disregarded.

In contrast to the above scheme, Frey-Wyssling and Steinmann (1953)

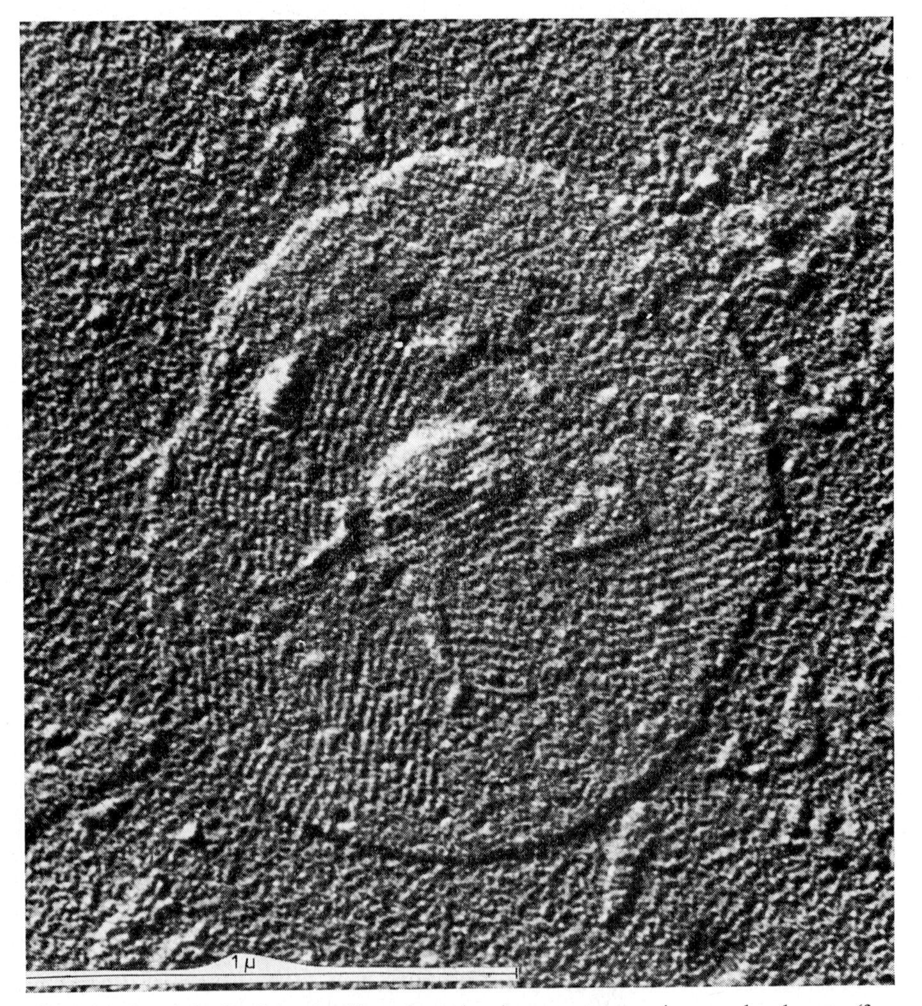

Fig. H–30. Granum of a *Spinacia* chloroplast, showing quantasomes in an ordered array (from Park and Pon, in Calvin, 1962).

postulated, on the basis of the granular surface of the granum discs, a structure of globular macromolecules, which would then represent chromolipoproteins. Corresponding to the lamella thickness, these particles would be of the order of 65 Å. These ideas are refined in Fig. H–29b (Frey-Wyssling, 1957), by assuming a chromolipid layer anchored in the globular protein molecule. The one-sided arrangement of the chlorophyll on the outer side of the granum thylakoids is chosen because their surface corresponds approximately to the total surface of the chlorophyll molecules present arranged in a monomolecular film (Frey-Wyssling, 1937; Wolken, 1956a), and because the chlorophyll can be extracted very easily.

The macromolecular blocks of Fig. H–29b show a pronounced polarity, since their curved outer side is lipophilic, whilst the flat inner side is hydrophilic. The sandwich model of the unit membrane is abandoned on account of the globular texture of the granum lamellae.

With the help of X-ray diffraction, Kreutz and Menke (1962) likewise found in the chloroplasts of *Chlorella pyrenoidosa* only two strata in the unit membrane, namely, a compact lipid layer of 30 Å in thickness, and on its inner side globular protein molecules 36 Å in diameter (Fig. H–29d). The period of the stacking amounts to only 177 Å, in contrast to the situation in the chloroplasts of *Aspidistra* (250 Å).

Recently, globular elements have been taken into general consideration for the formation of the grana. Park and Pon (in Calvin, 1962) published electron micrographs of plastid lamellae with a regular array of particles which they named quantasomes (Fig. H–30). The quantasomes protrayed have diameters of 100 × 200 Å. Such particles could hardly be fitted in the unit membrane of 75 Å thickness. They project inside the lumen of the thylakoids (Fig. H–29f), and are notched in with the particles of the opposite lamellae.

Park and Pon (1961) succeeded in isolating the quantasomes, as a green particulate fraction, from spinach leaves. The quantasomes are oblate ellipsoids, with axes of about 100 Å and 200 Å; they consist of four or more subunits (Park and Biggins, 1964). Their form allows an orientation of the particles in an electric field. Their intrinsic dichroism can then be investigated, and is not interfered with by form dichroism governed by the lamellar system of the intact chloroplasts. Whilst values of 1.15–1.36 were found for the dichroic ratio parallel and perpendicular to the longitudinal extent of intact chloroplasts (see p. 255), the quantasomes give the considerably smaller values of only 1.03–1.10 (Sauer and Calvin, 1962). From this, it must be concluded that the chlorophyll molecules are embedded in the quantasomes in such a way that their intrinsic anisotropy is in practice mutually cancelled out (statistical isotropy).

The lumen in the thylakoids shows up very differently according to the method of preparation. Steinmann and Sjöstrand (1955) found it to be 65 Å wide, (Fig. H–25a) and in Fig. H–28a it is over 150 Å wide. In contrast to this, in the electron micrographs which are the basis for the schemes of Figs. H–25a, H–29d, e and f, it is regarded as non-existent. One must assume that in the instances where no lumen appears, the thylakoids are collapsed by dehydration, which supposes a very dilute solution to be their content between the quantasomes *in vivo.*

The reality of a lumen can be demonstrated with the help of freeze-etching. As Fig. H–31 shows, the grana membranes have a flat outer and a granular inner

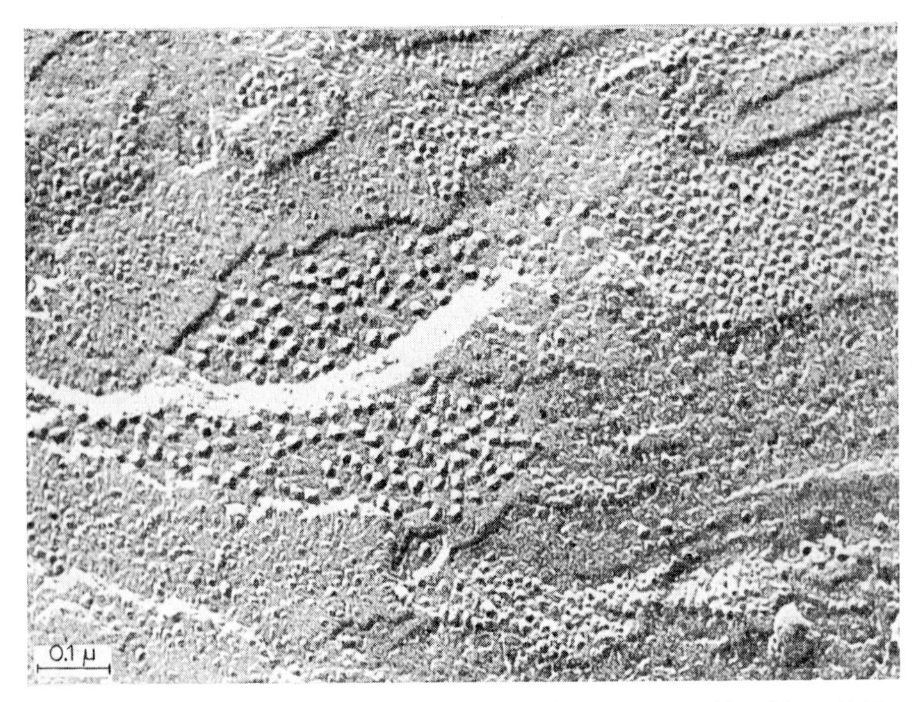

Fig. H–31. Quantasomes on membranes of *Nicotiana tabacum*. Freeze-etching (Moor, 1964).

side. This corresponds to the findings of the schemes in Fig. H–29d and f. Two types of granules can be observed, the first of the order of 150 Å, and another type with an average diameter of 35 Å, which is in accordance with the statement of Kreutz and Menke (1962).

Broken freeze-etched thylakoids show a white lumen and a bilateral flat unit membrane. This demonstrates most impressively that a lumen exists *in vivo* and that granular surfaces cannot actually be recognised in sections across membranes whose particulate elements are smaller than the thickness of the fine section.

With this, we return to the question of the molecular structure of the unit membrane (see p. 92). As Fig. H–28a shows, the thylakoid lamellae appear in fixed specimens as typical three-strata unit membranes. Their molecular construction is however by no means a sandwich structure, but a polar membrane with a flat outer surface (chromolipid layer) and a granular inner surface (layer of globular protein macromolecules), in which both layers, as indicated in Figs. H–29b, d and f, are bound up chemically with one another. Similar values for the tickness of these complicated membranes (Table H–III) may appear in very different chloroplasts.

In spite of this polar construction, both sides of the membrane are similarly blackened on fixation with permanganate. This gives the impression that the blackening, at least as regards the chloroplast thylakoid lamellae, is rather a non-specific surface reaction than a chemical interaction with specific lipids or proteins.

Although no consistent explanation has so far been proposed for the relationship of the chromolipids to the protein and the size of the blocks termed quantasomes, Figs. H–29b, d and f do show clearly that the unit membrane of the chloroplasts possesses a completely different structure to, say, the myelin lamellae of nerves. The term 'unit membrane' can thus have only a morphological, but by no means a molecular-structural, significance.

(*e*) *Function*

The photosynthesis of chloroplasts is divided into the following partial reactions:

(1) Absorption of light energy (photons) by the leaf pigments and its transfer to ADP, which is thus raised to ATP (photosynthetic phosphorylation – Arnon, 1955).

(2) Photolysis of water $H_2O \rightarrow 2\,H^+ + O^{2-}$, transfer of hydrogen to the system TPN $\rightarrow$ TPNH by ferredoxin (Tagawa *et al.*, 1963), and liberation of oxygen, O_2. This process is called the Hill reaction. It is associated with light and the presence of a hydrogen acceptor, which *in vitro* can be replaced by ferricyanide or quinone; added phosphopyridine nucleotides (DPN, TPN) can also be hydrogenated.

(3) Transfer of CO_2 from the air to a C_5 sugar (ribulose 1,5-diphosphate) and cleavage of the resulting C_6 compound to two molecules of phosphoglyceric acid (Calvin, 1962).

(4) Hydrogenation of the phosphoglyceric acid by the action of TPNH to phosphoglyceroaldehyde $C_3H_5O_3$-P, which, as a triose, can polymerise to hexoses (CO_2 assimilation) and regenerate ribulose (Calvin cycle).

Reactions 1 and 2 are both light reactions. They proceed according to the following stoichiometric proportions (P_i = inorganic phosphate):

$$2\,\text{ADP} + 2\,\text{Pi} + 2\,\text{TPN} + 2\,H_2O \xrightarrow[-2\,H_2O]{\text{light}} 2\,\text{ATP} + 2\,\text{TPNH} + O_2 + 2\,H^+$$

In experiments *in vitro*, only half as much ATP was liberated under certain circumstances (Jakobi, 1963).

Reactions 3 and 4 are independent of light, and are termed as Blackman's dark reaction. The energy necessary for this is produced from the ATP formed with the aid of light, and the hydrogen for the hydrogenation from TPNH.

It is conceivable that the light and the dark reactions do not take place at the same site within the chloroplast. As the light reaction only proceeds in the presence of chlorophyll, its locality can be deduced from the distribution of the above pigment in the chloroplast. In cytology, chlorophyll is detected by means of its characteristic red fluorescence. This test is very sensitive, for in the fluorescence microscope the smallest amounts of chlorophyll, which produce no measureable green colour in transmitted light, glow deep red.

On the basis of this test, Strugger (1950) believed to have demonstrated that in the proplastids the chlorophyll is localised in the prolamellar bodies, which he at that time called primary grana. In 1953, Heitz and Maly found, however, that the fluorescence was homogeneously distributed throughout the whole proplastid, and that the chlorophyll diffusely formed in the stroma is later absorbed into the granum thylakoids. Thus by the time the chloroplasts are grown, the grana glow red in the fluorescence microscope while the stroma appears black (Düvel and Mevius, 1952). This makes it difficult to understand the equivalence of grana and stroma lamellae, established by recent electron microscopy. Since, as a rule, there are about half as many stroma as granum thylakoids (see Figs. H–25b, c), and as the concentration of quantasomes containing the chlorophyll, is much lower than in granum thylakoids (Mühlethaler and Moor, 1964), it would be difficult to detect their feeble fluorescence.

The chloroplasts can accomplish all four steps of the CO_2 assimilation mentioned above as long as they are intact. If, however, the plastids are centrifuged in a not completely isotonically and iso-ionically standardised medium, they burst, and separate into two fractions which are differentiated as 'broken chloroplasts' and 'chloroplast extract' or 'stroma' (Arnon *et al.*, 1956). The broken chloroplasts consist essentially of granum thylakoids. This fraction can accomplish only the photophosphorylation and the Hill reaction; CO_2 reduction may be induced by the addition of the 'chloroplast extract' which contains DPN. This can be considered as an indication of the localisation of the hydrogen transfer system in the stroma.

It has been known for a long time that the stroma possesses a stronger capacity for reduction than the grana. It is able to reduce $AgNO_3$ in acid solution (Molisch reaction), which has led to the assumption that the stroma would contain ascorbic acid (Weber, 1937). Today, it can be established in the electron microscope that the silver particles arising by reduction are not free in the stroma, but remain associated with the stroma lamellae (Brown *et al.*, 1962). If TPNH is responsible for the silver reduction, then the TPNH system may be localised in the stroma lamellae.

Such observations must be drawn upon if the division of biochemical activity between the grana and the stroma is to be explained. Phylogenetic points of view must also be considered.

In the blue-green algae (Cyanophyceae) and photosynthetically active bacteria (purple bacteria) series of clearly equivalent thylakoids are found free in the cytoplasm. Neither a plastid membrane nor grana are present. As the thylakoids cannot be resolved in the light microscope, these homogeneous green-coloured areas were termed chromoplasm. Other algae, like the conjugates *Spirogyra*, *Closterium*, *Mougeotia*, the flagellate *Euglena* and the liverwort *Anthoceros*, are distinguished by especially large chloroplasts (megaplastids), which have a plastid envelope, but no grana. On the other hand, they possess special unpigmented organelles known as pyrenoids (Gr. *pyrèn* = stone of stone fruits). These are spherical structures surrounded by condensed starch granules. A division of labour occurs in the chloroplasts with pyrenoids, in that definite regions do not assimilate CO_2 but convert the assimilated sugar to higher polymeric carbohydrates.

As we established that in higher plants a differentiation of the chloroplasts into granum thylakoids, stroma thylakoids, and ground stroma takes place, this must, as in the case of the differentiation of the pyrenoids, have a particular functional significance. It is known that the starch forms steadily between the stroma lamellae, so that polymerisation to glucosan chains takes place in the stroma. Moreover, we can ascribe a mechanical function to the stroma lamellae for the spatial fixation of the grana stacks (see Fig. H–26). Whether and to what extent they also differ biochemically from the granum thylakoids, has yet to be elucidated.

The theory according to which all four of the above-mentioned partial reactions of CO_2 assimilation take place exclusively in the quantasomes, meets spatial difficulties. This is because, as in the oxysomes of the mitochondria (see p. 215), it is difficult to imagine the enzymes, necessary for the many single steps of the light reactions and the Calvin cycle, fitted into a particle only about 150 Å in diameter.

4. CHROMOPLASTS

(*a*) *Chemistry*

In contrast to the abundant literature on the chemistry of the chloroplasts, few investigations have been carried out on the chemical composition of isolated chromoplasts. In carrot chromatophores, Straus (1954) found 20–56% of carotene, and in the pigment-free plastids

lipids	58% by weight
proteins	22%
RNA	3.3%

These compositions may apply generally as characteristic of the chromoplasts. Compared with the chloroplasts (Table H–II), in which the lipids make up about $\frac{1}{3}$ and the proteins $\frac{1}{2}$ of the weight, the lipid content is in this case well over onehalf, and the protein content falls to 1/5 of the weight. We may speak of an intrinsic lipophanerosis. The nucleic acid present is RNA; it has not been possible to detect DNA. The important β-carotene of the chloroplasts is converted to epoxides in the chromoplasts, so that it falls off considerably, or is completely lacking. It is replaced by other carotenes, as by α-carotene in the carrot root, or by lycopene [which is not found in green organs (Goodwin, 1958)] in the tomato and other solanaceous fruits. On yellowing of the leaves in autumn, the chlorophyll is degraded and the nitrogen-containing pyrrole compounds withdraw from the leaves. On the other hand, the carotenoids remain behind, in the oxidised form in the plastids or as water-soluble esters in the cell sap (Goodwin, 1960).

(*b*) *Ultrastructure*

The chromoplasts arise as a rule from chloroplasts, and rarely from leucoplasts as in the root of carrots. The discoloration of the chloroplasts can be readily followed in the development of petals or in the ripening of fruit. From this, two different routes can be followed: either yellow-coloured droplets arise in the plastids, which were termed globuli (Wettstein, 1957) by electron microscopists, or the pigment appears in the plastids in the form of spindle-shaped bodies or little crystals. These crystal-like structures show intrinsic birefringence and dichroism; elongated forms are optically positive with reference to their long axes.

The petals of *Ranunculus* (buttercup) are involved in the first type of chromoplast formation (Frey-Wyssling and Kreutzer, 1958a). The chromoplasts arise from pale green chloroplasts which contain starch granules. In the course of development, the chlorophyll and the starch gradually decrease quantitatively, while the proportion of the yellow pigment is rapidly increased. It appears in spherules, the globuli, which in the light microscope have been mistaken for grana. In the electron microscope, however, it is recognised that these are strongly

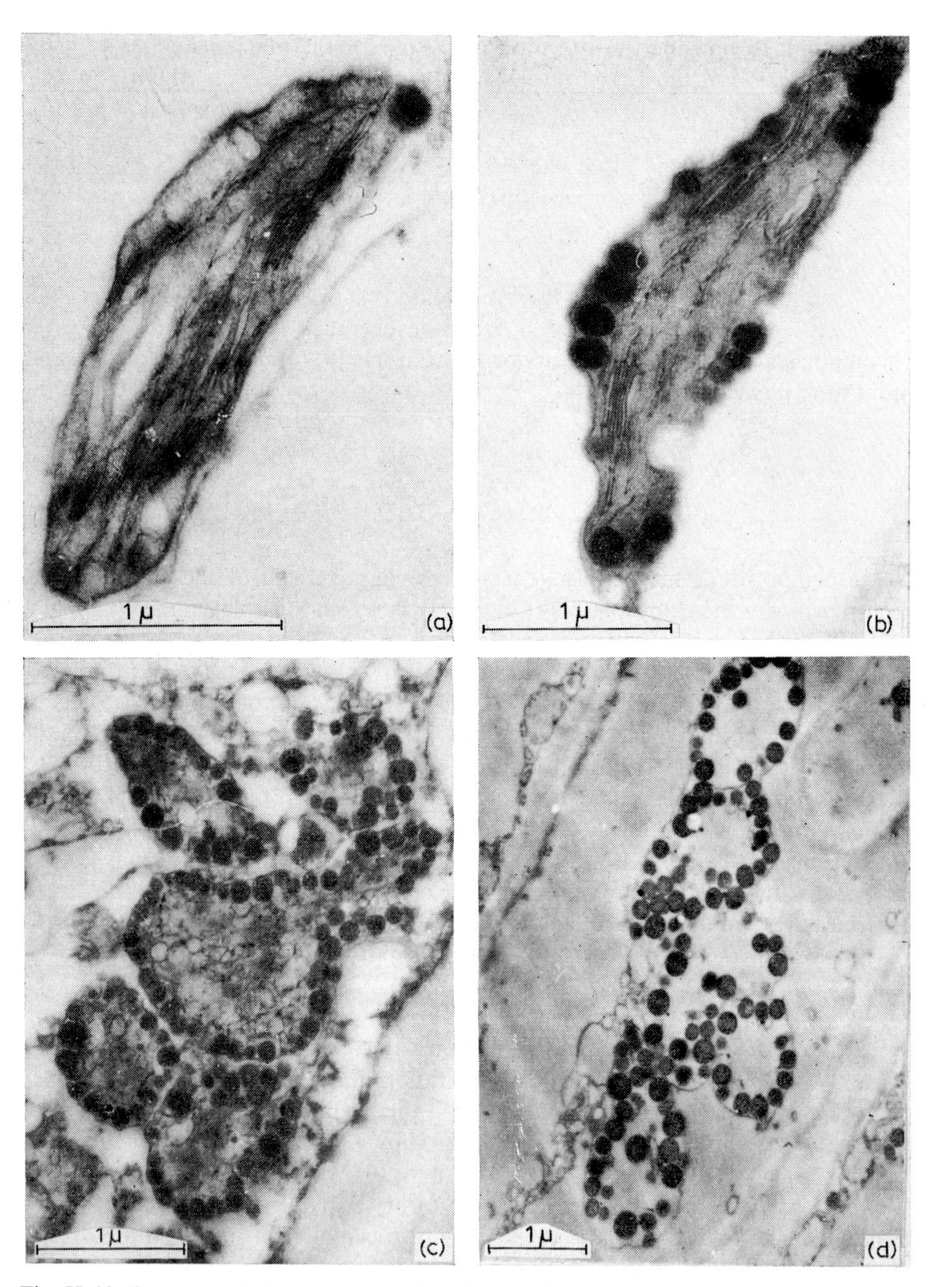

Fig. H–32. Ontogeny of chromoplasts in the adaxial epiderm of the honey-leaf of *Ranunculus repens* (Frey-Wyssling and Kreutzer, 1958a); (a) pale chloroplast with grana and starch in flower bud, (b) production of large globuli before anthesis, (c) disintegration of thylakoids at the beginning of anthesis, (d) complete lysis of plastid contents, except globuli, at anthesis.

Fig. H–33. Chloroplasts of *Elodea canadensis*, with osmiophilic globuli (Mühlethaler and Frey-Wyssling, 1959).

osmiophilic homogeneous particles. While large globuli are formed in considerable number, the lamellar structure of the chloroplasts breaks down (Fig. H–32a, b), and the stroma degenerates (Fig. H–32c). The globuli arrange themselves along, and finally completely occupy the inner plastid membrane. Meanwhile, the stroma disappears from the plastid, so that its centre appears optically empty (Fig. H–32d). The plastid envelope however remains, and individual, bright yellow pigment carriers, formerly termed chromatophores, are formed.

The most remarkable phenomenon in this development, which consists in a degeneration of the structural arrangement in the chloroplasts, is the formation of globuli, which can be resolved in the light microscope. Such bodies sometimes also appear, however, as sublight-microscopic particles in the chloroplasts (Fig. H–33). They are found there sporadically, or in a hexagonal arrangement (Sitte, 1963a) between the stroma lamellae. It is a formation found in older chloroplasts; no such particles have been observed in young chloroplasts, which may contain starch instead. They give the impression of disintegration forms of the very labile chloroplast structure. Isolation of the globuli from spinach chloroplasts has been achieved (Murakami and Takamiya, 1962); they show the absorption spectrum of a β-carotene-lipoprotein complex. Globuli appear in large quantities in the xantha mutants of barley, which through a genetic block have lost the capacity to form chlorophyll, and whose plastids can build up no thylakoids (Wettstein, 1957). We can hence conclude that they contain building materials for the differentiation of the lamellar ultrastructure which, however, cannot be used owing to damage of the corresponding morphogenetic gene. Conversely, we may regard the origin of the globuli during the chromoplast formation in yellow flowers, as separation of the lipids from the lipoproteins, when the structure of the chloroplasts collapses. Simultaneously, in the xantha mutation instead of chlorophyll, and in the chromoplasts perhaps at the cost of the phytol of the disappearing chlorophyll, abundant quantities of xanthophyll are formed, whose molecules dissolve in the lipid droplets. The situation described for the petals of *Ranunculus* is also valid for *Aloë* flowers (Steffen and Walter, 1958), and surely also for many other objects.

The genesis of the spindle-shaped chromoplasts follows an essentially different course, as is found in the fruits of *Capsicum* (cayenne), *Sorbus* (mountain ash) and *Rosa* (hips). The chromoplast development in the pericarp of *Capsicum annuum* will be described as an example (Frey-Wyssling and Kreutzer, 1958b).

Soon after fertilisation, young green fruits contain typical chloroplasts with numerous green grana in all cells. The plastids develop normally, until, when the fruit has almost reached its full size, those in the inner cell rows begin to store starch in the form of numerous small granules. Parallel with this, the chloroplasts become paler green and the grana less distinct, until they can no longer be discerned in the homogeneous, pale yellowish-green plastids. Starch and chlorophyll gradually disappear with increasing intensity of the yellow coloration. The coloured mass of the slowly elongating plastid often gets drawn to the side and is concentrated into striated, elongated shapes or crescents which leave the rest of the plastid colourless. This is accompanied by a change in colour, from a pale yellow to an increasingly intense orange-red, along with an ever greater distortion of the originally oval plastid to the characteristically pointed shapes in the mature fruit.

In the electron microscope, we can first see during this transformation the

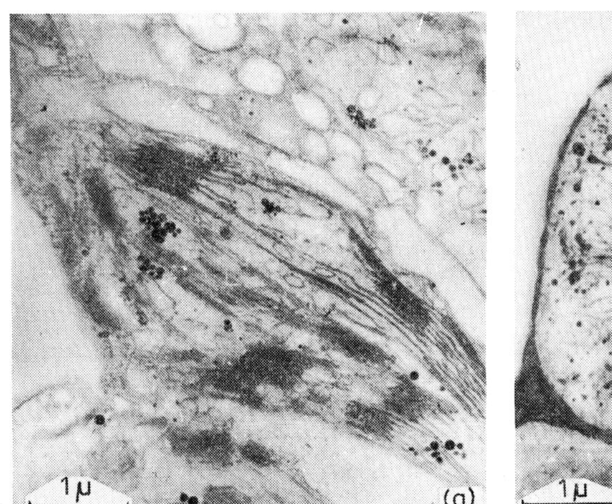

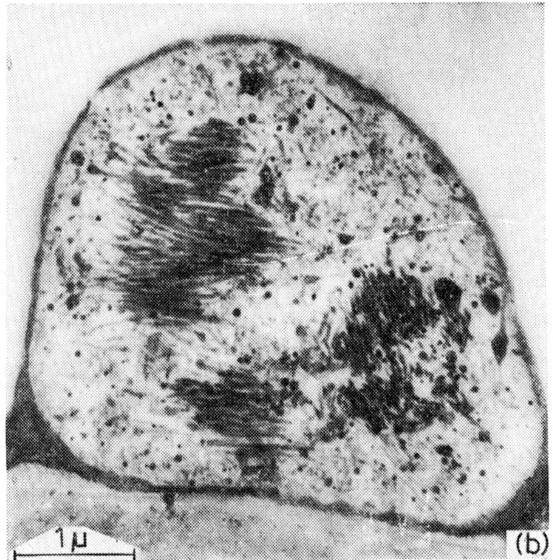

Fig. H–34. Ontogeny of chromoplasts in the fruit of *Capsicum annuum* (Frey-Wyssling and Kreutzer, 1958b); (a) plastid in pale-green fruit; lamination, small globuli; (b) plastid in ripe red fruit; stroma disintegrated, bundles of filaments.

appearance of small globuli (Fig. H–34a); the chloroplast structure then collapses and filament packets are formed (Steffen and Walter, 1955), which in Fig. H–34b are found in long, oblique, and transverse sections. As a result of this differentiation, the stroma becomes completely disorganised.

In Fig. H–35, the filaments appear as extended capillaries, resembling the myelin tubules. We obtain the impression that here also a degeneration in the form of a lipophanerosis occurs. The lipid tubules accumulate large quantities of the orange-red carotenoids, and in the light microscope simulate acicular carotenoid crystals.

In the chromoplasts of carrot roots, rhombic crystals are formed at the degeneration of the leucoplasts, which finally fill up the whole plastids.

According to Straus (1950) the crystal platelets are formed from fibrils laid down in parallel, and in consequence there would be not three, but only two different types of chromoplast differentiation, namely the formation of globuli and fibrils (Straus, 1961). It can, however, be shown that the longitudinal striations of the crystal-like chromoplasts seen in the light microscope depend on diffraction phenomena (Steffen and Walter, 1958). Ultrathin sections across such chromoplasts disclose laminated platelets embedded in a disorganised watery stroma. These sheetlike platelets, which appear orange in the light microscope, are thinner than 0.1 μ so that their specific absorption power for short-wave light must be considerable (Schwegler, 1964).

Summarising, it can be said that the chromoplasts are to be interpreted as plastids degenerated by lipophanerosis. In contrast to metabolically active plastids (CO_2-assimilation; starch synthesis) their protein content declines; on the other hand their lipid components increase, and large amounts of carotenoids are formed. Breakdown of the plastid structure leads (a) to the formation of lipid droplets (globuli), in which the carotenoids are dissolved – example: petals of *Ranunculus* –, (b) to the formation of longitudinally extended ultramicroscopic myelin tubules which are termed filaments – example: fruits of *Capsicum* –, and (c) to the formation of crystalline sheets piled up to thin platelets which contain large quantities of carotenoids. The fluctuating pigment content (20–56% α-carotene in carrot chromoplasts) indicates that the chromoplasts can contain arbitrary amounts of these terpenoids.

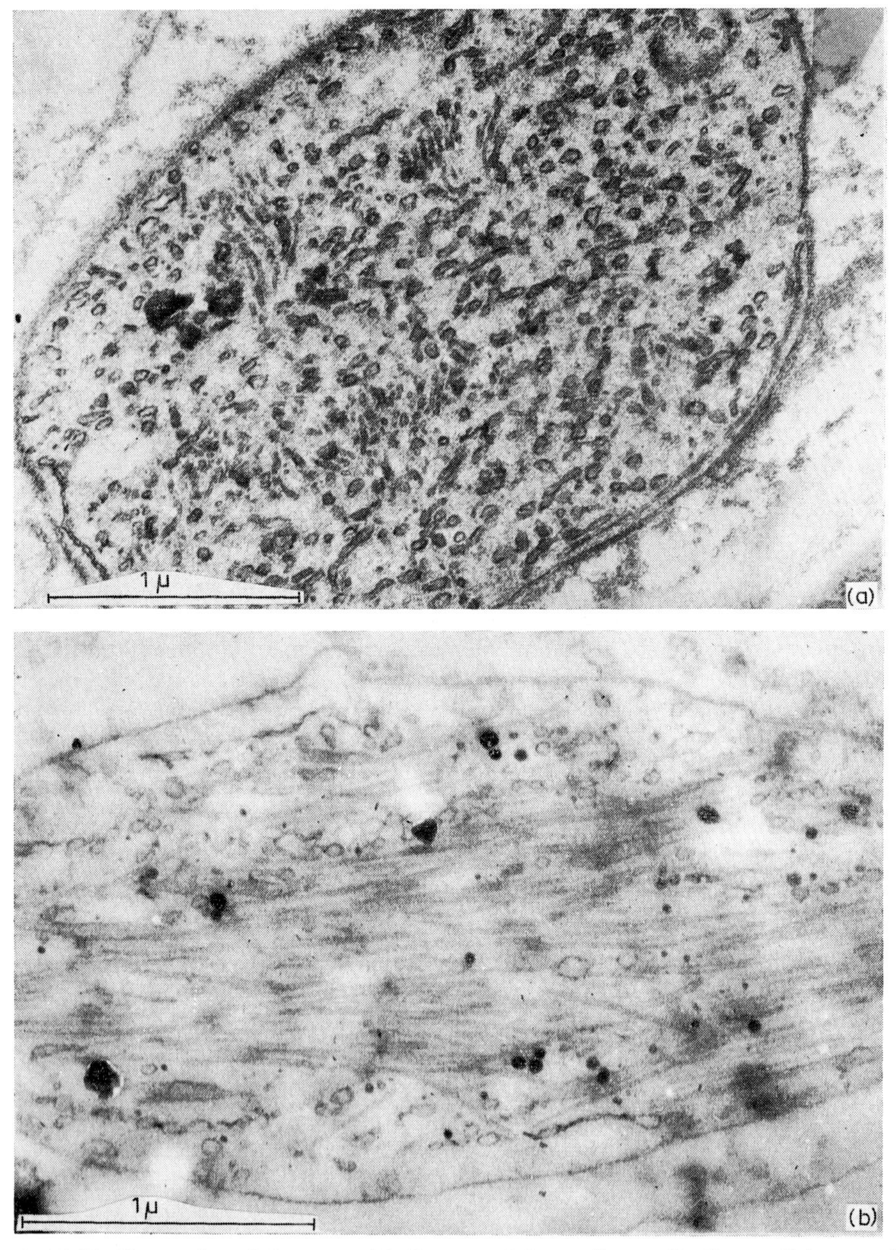

Fig. H–35. Chromoplast of *Capsicum* with filaments of chromo-lipomyelin. Stroma disintegrated by lipophanerosis (?); (a) transverse section (photo K. Mühlethaler), (b) longitudinal section (Frey-Wyssling and Kreutzer, 1958b).

(*c*) *Function*

The chromoplasts have been interpreted in this monograph as gerontic end products of the monotropic plastid development. The fine structure of the chloroplasts collapses, and valuable compounds like chlorophyll-building materials and other nitrogen-containing micromolecules are withdrawn; the easily replaced nitrogen-free compounds like the lipids and the phytol from chlorophyll remain. The physiologically active carotenoids are inactivated, and additional inactive carotenoids are formed (possibly from phytol) as secondary plant substances.

On the other hand, we cannot definitely conclude from this that the chromoplasts are functionless, like senile cell organelles, for they may have a passive function, similarly to the conducting tracheary elements which arise in the xylem by the degeneration and drainage of the plasma content. In fact, the flower biologists and carpologists have demonstrated that insects and birds respond to organs brightly coloured with carotenoids. We can thus assume that the chromoplasts, like the originally extrafloral nectar glands and the sugar content of fleshy fruits play an indirect part in the propagation and the dissemination of seeds.

The fact that no function of any kind has as yet been found for the chromoplasts of the root speaks in favour of such an interpretation.

5. PHYLOGENY

The interpretation of the plastids as organelles of photosynthesis, which belong only to the autotrophic plants and are missing in all heterotrophic organisms (including the fungi), justifies a brief mention of their phylogenetic origin. In general, it is rather pointless to speculate on the phylogeny of single cell organelles, for the cell lives only as a gestalt, through the interplay of the nucleus (or its equivalent) with the groundplasm, the plasmalemma, the reticulum, the mitochondria, etc. It is thus difficult to imagine how the precursors of the modern cell could exist with a simpler outfit of cell organelles. We cannot forsake, for example, the nucleus as the organiser of reproduction, or the mitochondria as the centres of respiration, without excluding what we today understand by life. From this impossibility to explain the origin and development of terrestrial life by cytological considerations, the question of plastid phylogeny forms an exception, for here nature exceptionally demonstrates before us cells which live with or without these organelles.

Geochemists and biochemists hold that at the time of the origin of life on Earth, the atmosphere contained not oxygen but hydrogen (Oparin, 1957). It has been shown that amino acids can be formed under these conditions from CO_2 and NH_3 (Miller and Urey, 1959). There are also hypotheses concerning the formation of optically active compounds (Keosian, 1960). Above all, however, the enigma worth solving is of how the nucleic acids developed from polyphosphates and became catalysts, energy stores, and gene carriers. Nothing is known as regards the chronology of these developments, for which over two thousand million years were available (from the solidification of the Earth's crust about three billion years ago to the appearance of cells about one billion years ago).

We must thus proceed from the first cells. Their conjectured precursors arose about 2 billion years ago, in an atmosphere containing hydrogen. It is interesting that the existence of plastids is theoretically conceivable under such conditions, for their activity is not bound to the presence of oxygen; on the contrary, they can liberate oxygen from water. There are therefore theories according to which oxygen of the present day atmosphere is completely, or to a large extent, attributed to the CO_2 assimilation of plants. The action of the mitochondria, which need free oxygen as an acceptor for their liberated hydrogen, was hardly possible at the time when eobiontic life first arose on Earth, since the atmosphere was then devoid of free oxygen. Thus in all probability the plastids should be phylogenetically older than the mitochondria.

If, in addition to the morphological similarity of these two types of organelles, we further consider their inventory of biochemically active compounds, we arrive at an astonishing conformity. With the exception of the assimilatory pigments, both organelles contain the ATP system for energy transfer, the TPN system as the hydrogen carrier, cytochrome with its relations to oxygen, a few per cent of RNA and small amounts, difficult to detect, of DNA, about 30% of lipids, and over 50% of protein.

Such a conformity leads naturally to the assumption of relationships in the developmental history. In fact, an ontogenetic relationship was already discussed in the first half of this century (Guilliermond *et al.*, 1933). Since however it could be demonstrated that the presumed mitochondria from which the plastids were supposed to arise are morphologically distinct proplastids, we now believe that the plastids and the mitochondria belong to independent systems, and that consequently the initials of the plastids and of the mitochondria are genetically determined as different bodies, in spite of their morphological similarity.

Phylogenetically, however, the situation could be different. As the mitochondria probably appeared later in the Earth's history than the plastids, and as biochemically both organelles display unexpectedly close relationships, it is not impossible that the system of mitochondria is derived from that of the plastids. In principle, the same reactions (though in reverse directions) are carried out in respiration as in CO_2-assimilation, so that the light energy originally stored in the assimilates by ATP, is in the end again set free, and is in turn passed on by ATP to the endergonic life processes.

Even though the mitochondria are found today as necessary organelles in all aerobic living cells, while the plastids are only indispensible in the autotrophic plant cells, the plastidome seems to represent the older organelle system in cytology.

J. Cell Wall

The cells of plants are surrounded by a strong cell wall lined throughout with the plasmalemma. They are thus coated, in contrast to animal cells which are termed naked. The cell wall imparts protection and firmness to the protoplasts. Its presence or absence serves as a criterion as to whether a given genus of the Protobionta is to be allotted to the plant or to the animal kingdom, (*e.g.* Flagellates with a cell wall are counted as Protophyta, and naked ones as Protozoa), and this characteristic is more reliable than the presence or absence of chlorophyll, for plant cells can change by mutation to the heterotrophic way of life. The representatives of the phylum of slime moulds (Myxomycetes), which consist of naked plasmodia, develop typical cell walls when they proceed to the formation of sporangia; on the other hand, the reproductive cells of the lower plants are often naked (zoospores, isogametes, spermatozoa). In the higher plants, however, even the female gamete (ovum) and the pollen tube, the transmitter of the male gamete, are walled.

The cell wall, often termed the cell membrane, is distinguished chemically and physically from the plasma membrane (plasmalemma) which is occasionally also called the 'cell membrane'. It does not consist of lipoproteins but largely of carbohydrates, and, in contrast to the semi-permeable plasmalemma, is as a rule holopermeable. It is thus soaked by extracellular liquids like a sponge, and all molecules and ions of a nutrient solution can advance without resistance to the plasma surface. Special measures such as cutinisation or incrustation must be taken if the free diffusion in the cell wall should be limited.

Historically, the plant cell walls played a special role insofar as they gave rise to the origin of the name 'cell' itself. When Robert Hooke (1667) discovered cell walls in the tissue of bottle cork with the aid of a light microscope, he compared the observed texture with a honeycomb and called the observed compartments cells. Had the cytological unit not been discovered in a plant tissue but in the form of a creeping amoeba, it would scarcely have been termed 'cell', and indeed the name cytology, which is derived from the Greek *kytos,* vase, container or envelope, would hardly have been adopted for our science.

1. ONTOGENY

(*a*) *Cell plate*

Formation of a new wall takes place by the so-called phragmoplast (Gr. *phragma* = fence, separation), which appears as a defined plasma body in the region of the equatorial plane of the dividing mother cell, after the chromosomes have receded from one another in the anaphase. The spindle fibres, which run from pole to pole (Fig. J–1a) traverse the phragmoplast (Strasburger, 1888). At first its thick plasma mass is spindle-shaped, but it soon flattens and assumes the form of a biconvex lens. In this way the spindle fibres, which shorten simultaneously (Timber-

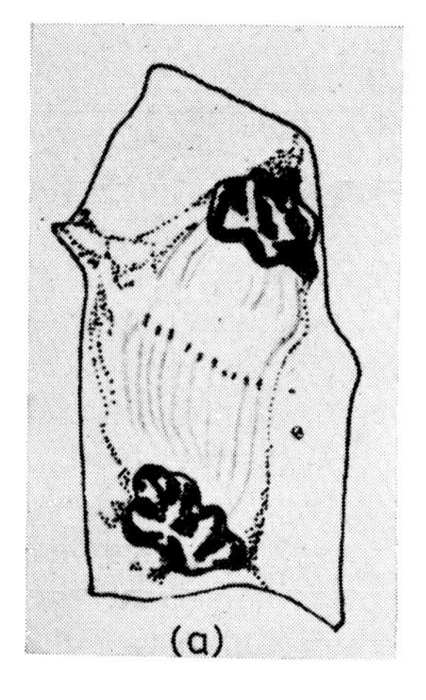

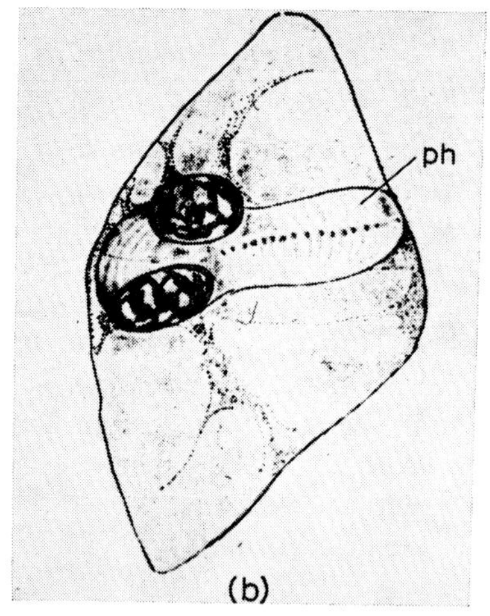

Fig. J–1. Formation of a new wall across the cell during mitosis (Chodat, 1922); (a) telophase, (b) formation of the phragmoplast *ph* by contraction of the spindle fibres.

lake, 1900), become pressed apart. As a result, the two daughter nuclei are drawn to the middle of the mother cell and rest finally on both sides of the flattened phragmoplast (Fig. J–1b).

In anaphase, the first indication of a new transverse wall is the appearance of stainable nodules in the equatorial plane (Fig. J–1a). It was originally believed that these were thickened spindle fibres but Becker (1934) showed that the 'nodules' are small coacervate droplets, which can be vitally stained with vacuole dyes (neutral red, cresyl blue). In fixed preparations they appear as granules agglutinated with the spindle fibres, whilst on observation in the living state e.g. in the stamen filament hairs of *Tradescantia virginica*, they are recognised as semi-liquid differentiations of the phragmoplasts. Their number increases until the droplets finally coalesce laterally to a semi-solid layer, the so-called cell plate.

The droplets are clearly perceptible in the electron microscope (Buvat and Puissant, 1958; Porter and Caulfield, 1958/60), and prove to be vesicles (Fig. J–2) arising from Golgi bodies accumulated at the edge of the phragmoplast (Whaley and Mollenhauer, 1963; Frey-Wyssling *et al.*, 1964).

The cell plate may be stained vitally with the basic dyes methylene blue and ruthenium red (Becker, 1934), indicating that it contains acid cell wall materials like pectins and other uronides, which assume the function of the matrix (see p. 291) in the future primary wall. The cell plate now grows peripherally by the addition of further Golgi vesicles, until it reaches the longitudinal wall of the mother cell and fuses with it.

Before achieving contact with the longitudinal wall, the cell plate is already weakly birefringent, *i.e.* two narrow bright layers can be observed on both sides of an isotropic (and in the polarising microscope dark) central lamella. The growing cell plate thus already consists of three sheets, the central one being known as the middle lamella, on both sides of which the two future daughter cells establish their so-called primary walls. Birefringence of the two primary wall lamellae is

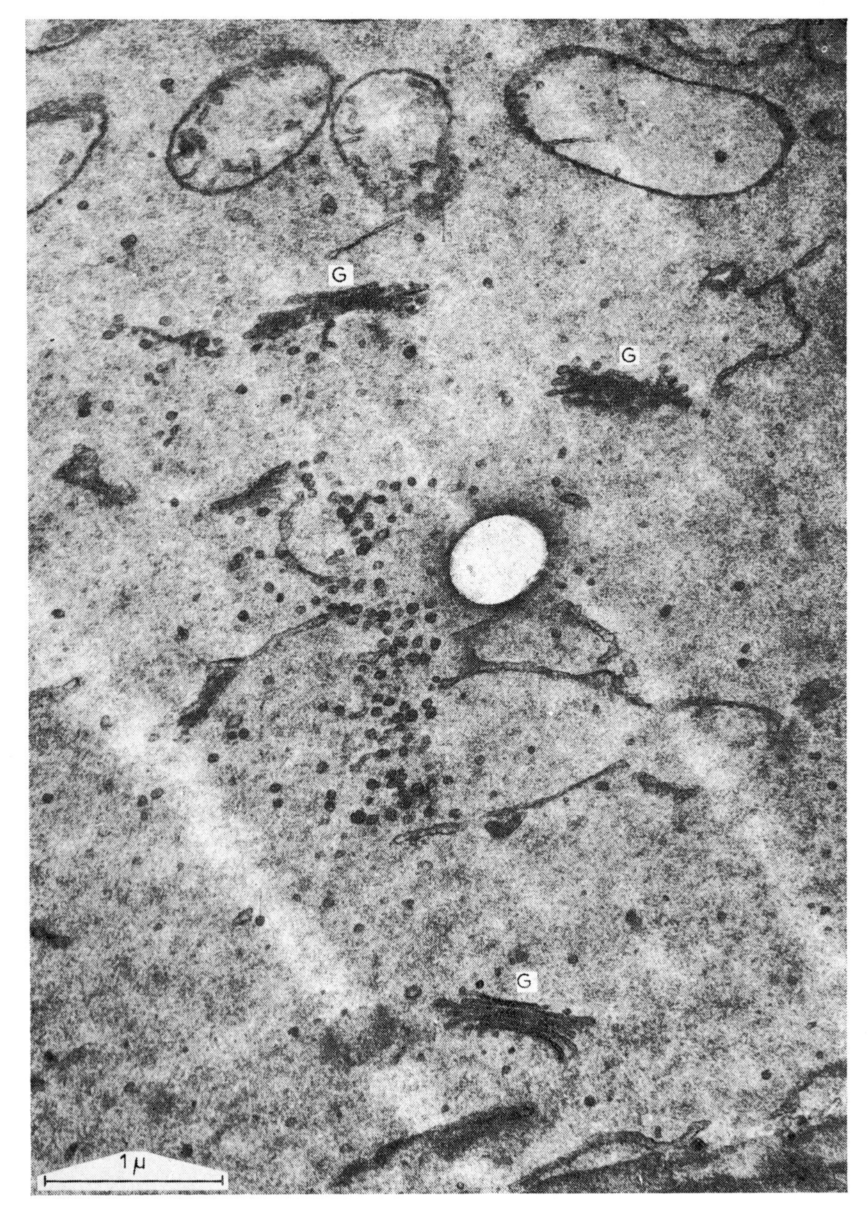

Fig. J–2. Future cell plate with Golgi vesicles in late anaphase. *G* = Golgi (photo J. López-Sáez).

caused by the fact that they already contain small amounts of elementary cellulose fibrils.

It is particularly significant, that after the coalescence of the Golgi vesicles to the cell plate, the unit membrane which surrounds them becomes part of the plasmalemma of the two now separate protoplasts, so that this important organelle may be derived from the Golgi apparatus.

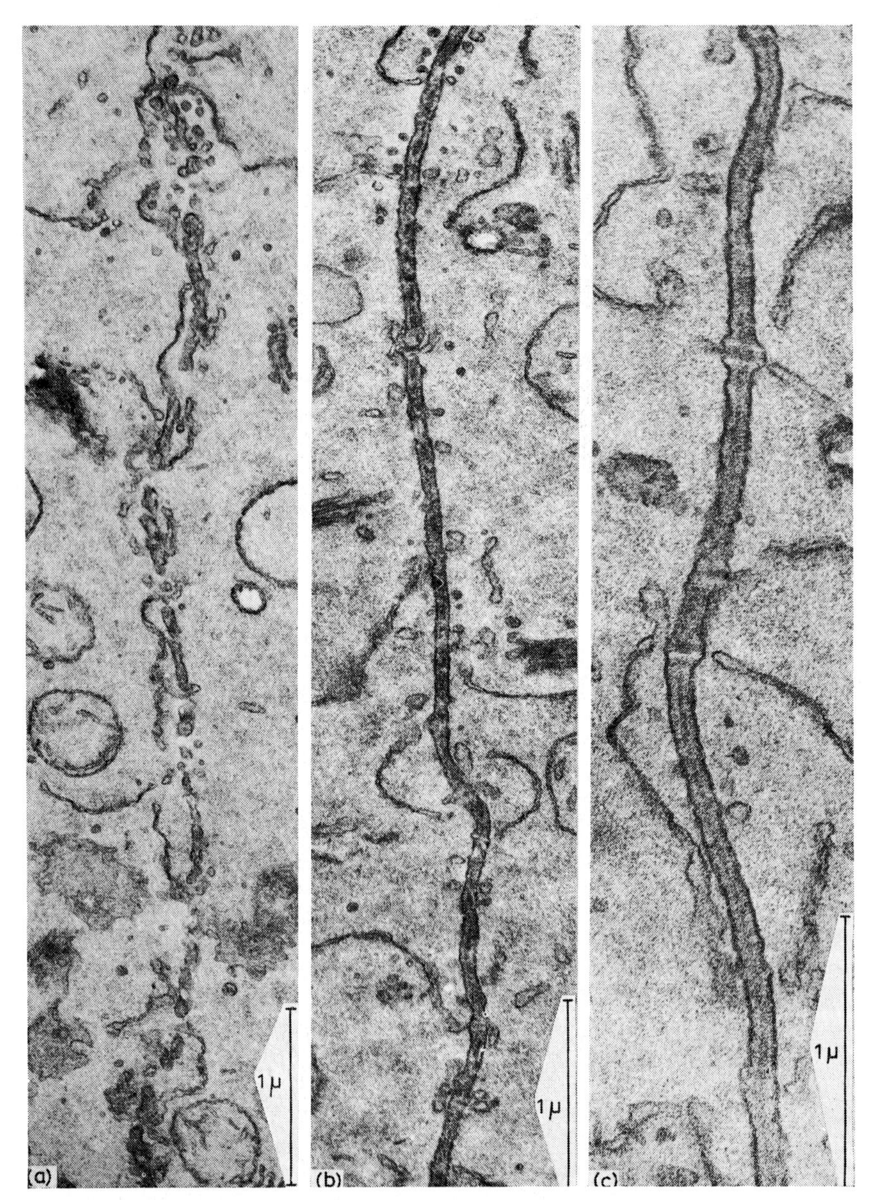

Fig. J–3. Developmental stages of new cell walls with future plasmodesms (photo J. López-Sáez); (a) coalescence of Golgi vesicles, (b) growth in thickness by lateral incorporation of additional vesicles, (c) differentiated primary wall with plasmodesms.

Fig. J–2 shows the first stage of a developing cell wall . The vesicles given off by the Golgi complex migrate to the cell equator, and fuse there to the middle lamella. Individual plasmatic linkages known as plasmodesms (Fig. J–3 a–c) persist between the Golgi vacuoles. The cell plate grows not only centripetally in surface area but also in thickness, by lateral deposition and incorporation of additional Golgi vesicles (Fig. J–3a and b).

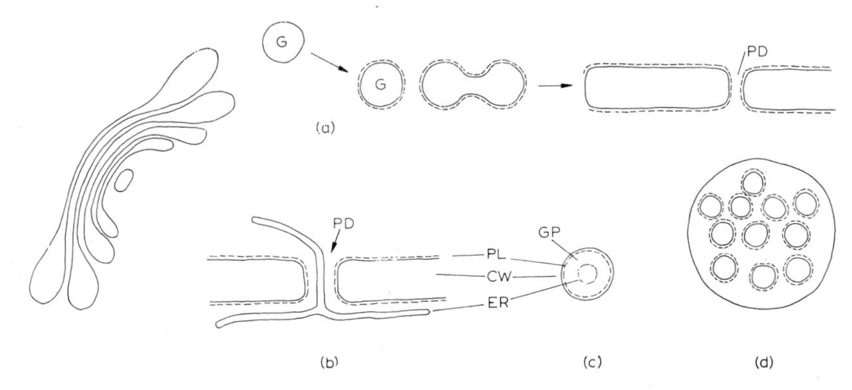

Fig. J–4. Plasmodesms; (a) Golgi vesicles *G* fuse to the cell plate, leaving pores (plasmodesms *PD*). Cell wall *CW* coated by plasmalemma *PL* (originally a Golgi membrane); (b) branch of ER grows across a plasmodesm *PD*; (c) section across a plasmodesm whose lumen contains an ER membrane, minute film of groundplasm *GP*, and plasmalemma *PL*; (d) plasmatic strand in a sieve pore holds numerous ER capillaries (see p. 324).

Plasmodesms (Gr. *desmos* = ribbon, ligament) are extremely fine plasma strands which pass through the cell wall and bridge adjacent cells. They were discovered by Tangl (1879) and named by Strasburger (1882). Their diameter lies at the limit of resolution of the light microscope, and they can thus be rendered visible only by suitable staining or by impregnation with heavy metal salts such as silver or mercury (Lambertz, 1954). They can traverse thick cell walls, branched or unbranched. In the latter case, their formation in the extremely thick walled endosperm cells of seeds of *Strychnos*, *Areca* and *Phytelephas* shows a picture similar to the spindle figure in mitosis. The idea was therefore advanced that they are residues of the spindle fibres embedded in the cell wall (Jungers, 1930/33). However, Figs. J–3a and b show clearly that the plasmodesms are of plasmatic origin.

In cross-section, the plasmodesms show, on sufficient resolution, a double membrane surrounding a thin lumen. The outer unit membrane corresponds to the plasmalemma which arises from the membrane of the Golgi vesicles, and the inner unit membrane from the ER. Between the two membranes is found a thin layer of groundplasm, and the lumen corresponds to the space containing enchylema in the ER (Fig. J–4).

During and after the formation of the cell plate osmiophilic particles become evident in the phragmoplast, which can reach light microscopic dimensions and which are known as phragmosomes. On fixation with permanganate, these appear to be delimited from the groundplasm by a membrane. No clear insight as to their origin and function can yet be obtained, but they could perhaps be lysosomes.

(*b*) *Primary wall*

As has been mentioned, the phragmoplast deposits wall lamellae, termed primary walls, against both sides of the middle lamella. The deposition of these lamellae is not however confined to the cell plate, and the whole cell becomes lined with new wall material by the deposition of Golgi vesicles. Even so, the old walls of the mother cell do not become thicker, for powerful cell growth occurs at mitosis, in conjunction with considerable extension of the existing longitudinal wall.

With the usual periods of fixation, the cell wall takes up neither osmium nor manganese, and thus appears white in electron micrographs, similarly to the starch granules. The middle lamella may be an exception to this, showing up as a grey zone. It has also been possible to convert the pectic substances of the middle lamella into an insoluble iron complex, which in ultrathin sections appears in the form of black particles (Albersheim *et al.*, 1960).

In meristematic tissues the cells assume a polyhedral shape and appear, on an average, as tetrakaidecahedra (Matzke, 1946), like the polyhedral bubbles within foams. This indicates a semi-solid plastic consistency of the young primary wall, so that, as in the bubbles of foam, surface tension is a major factor governing the shape of the meristematic cells (Frey-Wyssling, 1959, p. 8).

Later the corners of the polyhedra become rounded off, giving rise to intercellular spaces. Cleavage of the cell wall, necessary for this phenomenon takes place along the middle lamella. In the region of the intercellular spaces, the young cell walls thus consist only of a primary wall lamella, which becomes thickened, however, by the apposition of additional lamellae which form the secondary wall.

A distinction between primary and secondary walls is necessary because the wall frequently becomes unusually thick and then behaves quite differently to the slender envelope of growing cells. It is however difficult to define the two layers with certainty, and to distinguish one from the other, as there are certain transition lamellae; that is, no simple morphological distinction is possible, and the two layers may be defined only on the basis of their ontogenetic development.

According to Bailey, the primary wall is the slender envelope which surrounds the protoplasts of the cambium cells before they begin to differentiate by the deposition of secondary thickening layers (Kerr and Bailey, 1934). As in these cell walls the growth in area is in general concluded, it was assumed that in the first phase of cell wall development the primary cell wall grows by intussusception and later the secondary wall by apposition, whereby the long-debated question of which mode of substance incorporation into the growing wall is employed by the cell seems to be decided. When we further established that the primary walls showed disperse textures (see p. 293) whilst the secondary walls, in contrast, were characterised by parallel textures, we thought we had found a morphological distinction between the two types of wall (Frey-Wyssling and Mühlethaler, 1951a). Later, however, it was found that parts of the primary wall can also show a parallel texture (Fig. J–12b) and a proposal was thus made to speak of the primary wall as long as the wall grows in area (Beer and Setterfield, 1958; Wardrop, 1962). But since walls can also grow in the surface after the deposition of many apposition layers (*Valonia*, collenchyma cells), this definition departs too far from Bailey's original concept to which we would like to adhere. We shall therefore define the primary wall as the lamella first laid down, which wholly or partly shows a disperse texture, and whose microfibrils are individually displaced from one another in a multi-net growth (see p. 302).

It should be noted that formation of the primary wall is not preceded by the formation of a cell plate in all cell divisions. In the algal threads of *Spirogyra*

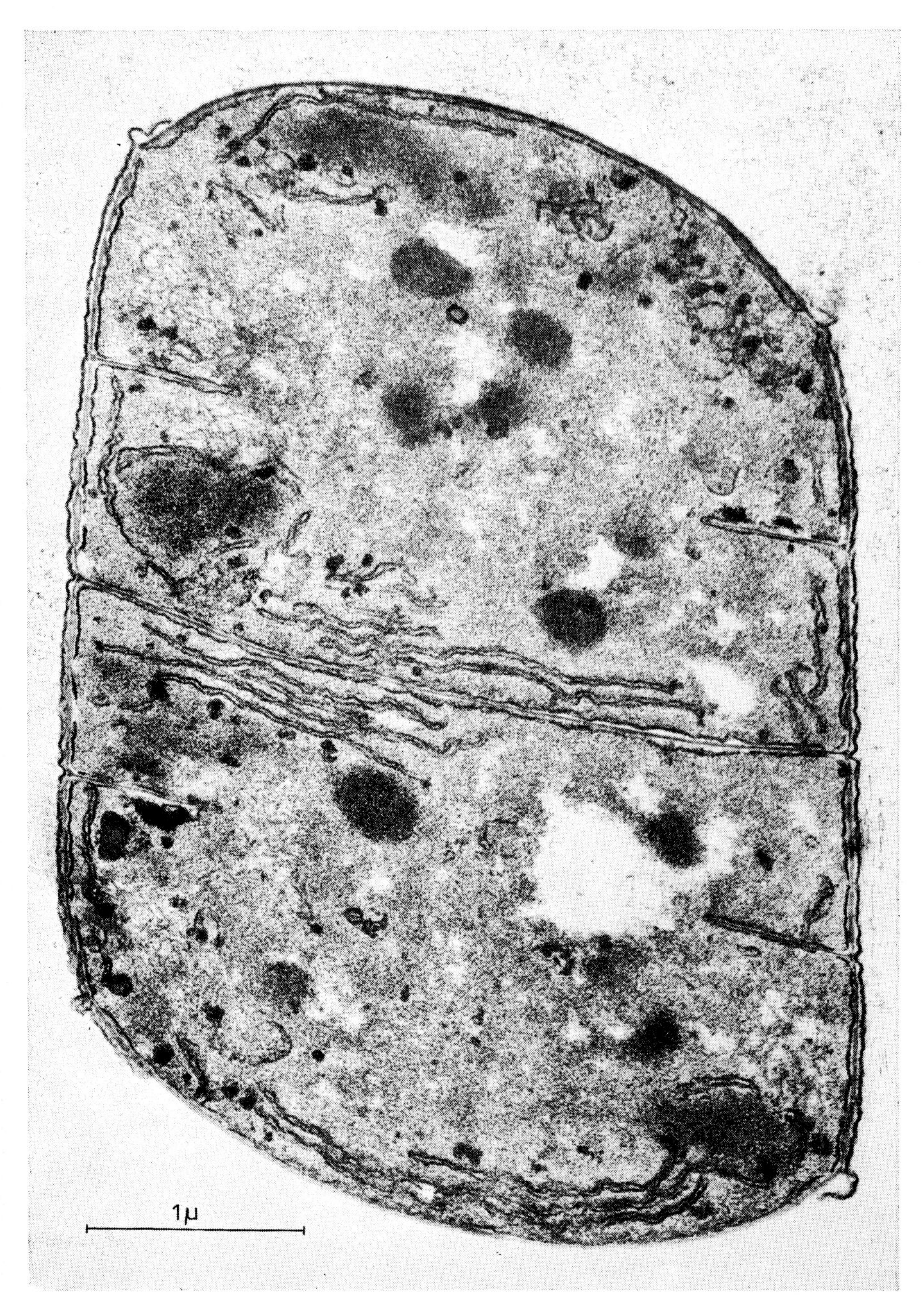

Fig. J–5. Formation of the transversal cell wall in *Oscillatoria rubescens* (photo M. Jost).

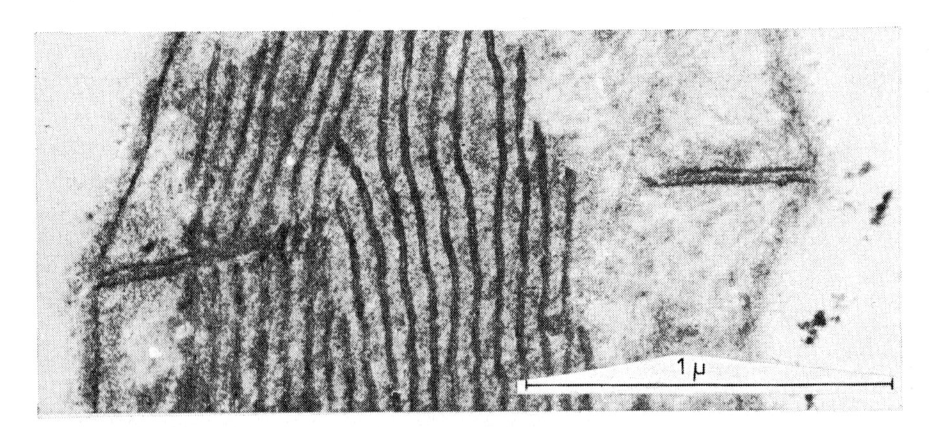

Fig. J–6. Thylakoids of chromatoplasm in *Oscillatoria*, intersected by a transversal cell wall (photo M. Jost).

(Conjugatae) or *Oscillatoria* (Cyanophyceae), for example, new transverse walls arise by the deposition of a ring wall, which gradually closes like an iris shutter so that the mother cell is partitioned by this diaphragm. The situation may be briefly described on the example of cell division in *Oscillatoria rubescens* (Figs. J–5 and J–6).

Normally, the cell wall and the plasmalemma can be separated by plasmolysis. In the Prokaryonta (Bacteria and Cyanophyceae) this is not the case. Thus in these organisms the transition layer between living cytoplasm and the cell wall with its passive function is termed the pellicle (from Lat. *pellis* = skin, hide). Meanwhile, electron microscopy has established that a typical plasmalemma is present as a unit membrane, which however cannot be separated plasmolytically from the cell wall.

As Fig. J–5 shows, the transverse wall arises by invagination of the primary longitudinal wall of the mother cell. Where both the inner surface of the folds come together, a grey-tinted 'middle lamella' is formed which should not however be considered homologous, from the point of ontogenetical view, with the middle lamella arising from a cell plate. The fold grows centripetally, by the action of the plasmalemma, which is clearly visible as a unit membrane. It is especially remarkable how the cell content becomes intersectioned by this growth. The closing diaphragm cuts the existing thylakoids of the chromatoplasm into two halves, without these becoming bent or pushed aside by the advancing edge of the cell wall (Fig. J–6).

(*c*) *Secondary wall*

The bulk of a full-grown cell wall consists of secondary wall. It gives the cell its final shape, and, in the fibre cells, forms the basis of the plant's mechanical supporting tissue. It has received close technological attention, because it is responsible for the special characteristics of the raw materials, wood, textile fibres, paper fibres, cellulose, straw, and cork. As a consequence of the formation of secondary layers, the plastic properties of the cell wall recede in favour of pronounced elastic properties.

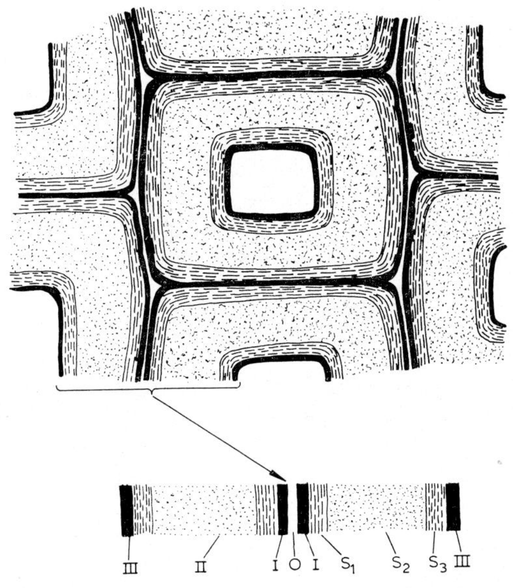

Fig. J–7. Cell wall with a three-layered secondary wall. O middle lamella; I primary wall; II secondary wall with layers S_1, S_2, S_3 (see Fig. J–14); III tertiary wall.

In the tracheids of conifers, the secondary wall appears to be divided into three layers (Kerr and Bailey, 1934), which are termed S_1, S_2, and S_3 (Fig. J–7). The primary wall is so thin that it cannot be easily differentiated from the middle lamella in the light microscope. In the polarising microscope, the middle lamella is isotropic, whilst the weakly birefringent primary wall is eclipsed by the bright S_1 layer (Fig. J–8). The very thick S_2 layer appears dark in the polarising microscope while the S_3 layer is again bright. The secondary wall is frequently coated on the inner side by a terminal lamella, which may be explained as a dried up residue of a degenerated membranogenic plasma lining. It is debated (see Frey-Wyssling, 1959, p. 32) whether this lamella, which really no longer belongs to the wall, should be called a tertiary wall. The tertiary lamella possesses different staining properties (Bucher, 1957) and, like the primary wall, different swelling properties from the secondary wall (Fig. J–9). In the electron microscope it appears warty on the interior surface. In longitudinal section the warts show particles coated by a membrane (Wardrop *et al.*, 1959), and it can thus be assumed that they are dried up residues of plasma organelles, whilst the membrane covering them is explained as the dried up tonoplast. A similar electron-dense membrane derived from degenerated organelles has been found by Bell and Mühlethaler (1962b) around the mature egg cell of the fern *Pteridium*.

The pattern of subdivision of the secondary wall into three layers (S_1, S_2, S_3), which is found in the coniferous tracheids (Fig. J–7) may not be generalised, for in the woody fibres of angiosperms the S_3 layer may be missing, and in the fibres of the so-called tension wood, which show in the interior a thick gelatinous layer *G*, *G* may appear instead of S_3 or indeed be deposited on S_3 (Wardrop and Dadswell, 1955). In other cell walls, as for example in the laticifers of *Euphorbia splendens*, the

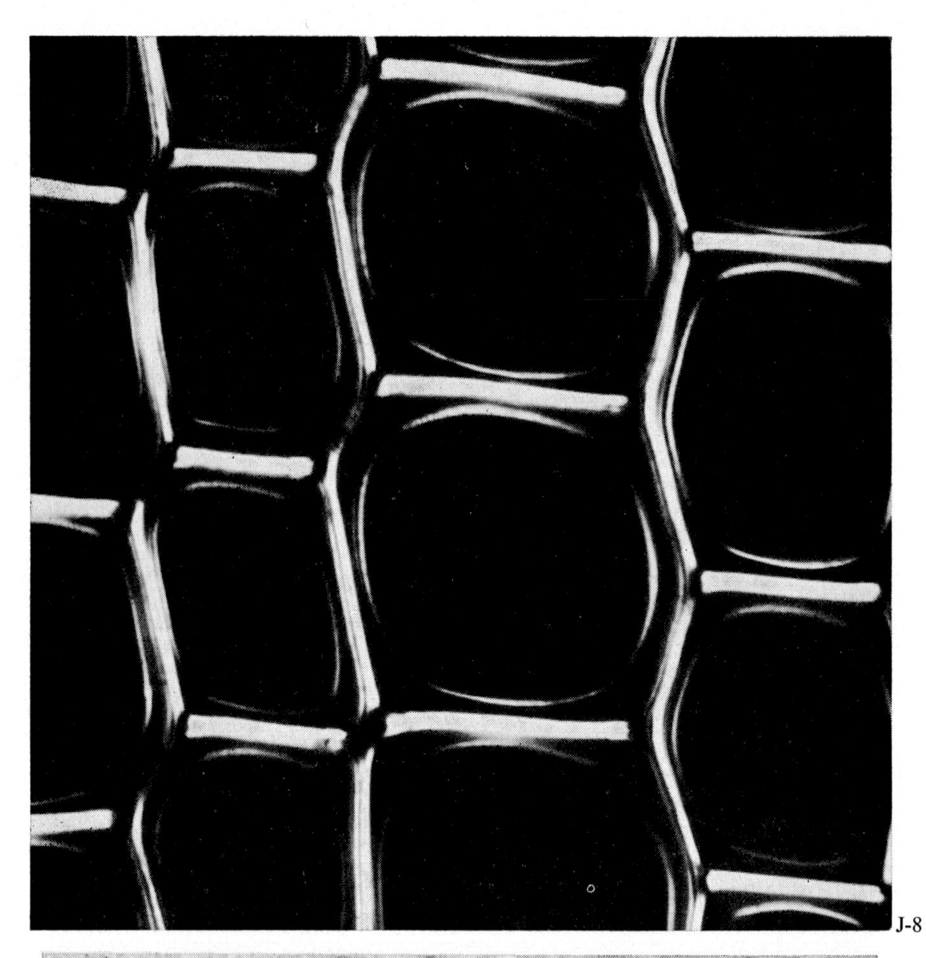

J-8

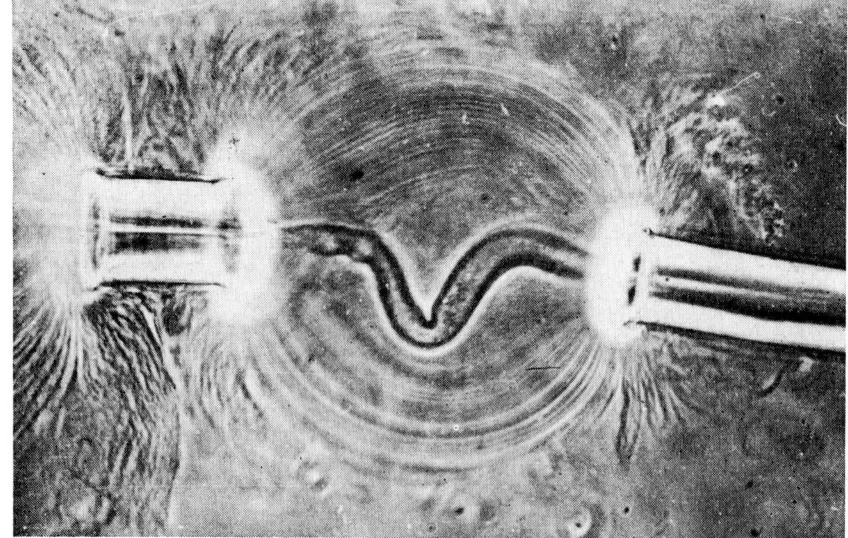

J-9

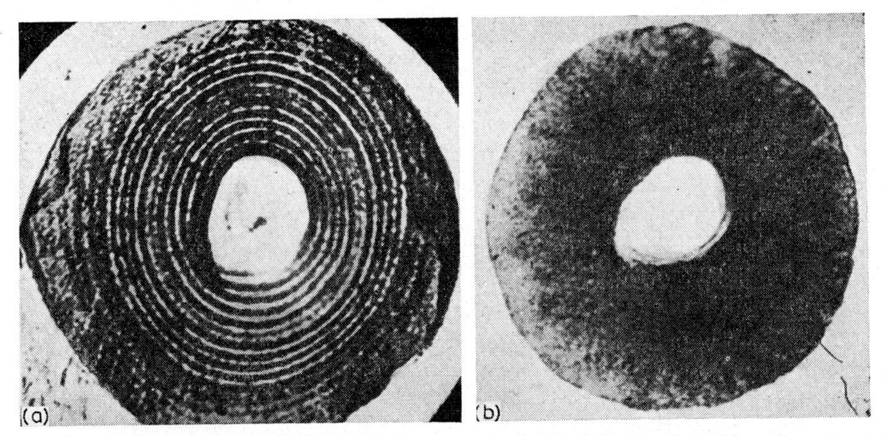

Fig. J–10. Sections across cotton hairs (Anderson and Kerr, 1938); (a) normally cultivated, with daily growth rings, (b) cultivated under constant illumination, without rings.

secondary wall is subdivided into many sublight-microscopic lamellae (Moor, 1959). The same holds for cotton hair, whose secondary wall may consist of 25 lamellae 0.4 μ in width (Balls, 1919). As the growth in thickness of the cell wall of cotton hair continues for 25 days after the end of longitudinal growth (16 days), the apposition rings are explained as day rings. When cotton is cultured under constant conditions (constant temperature, moisture, and continuous illumination) the secondary wall remains free from lamellae (Anderson and Kerr, 1938; Fig. J–10).

2. CHEMISTRY

Plant cell walls contain several groups of compounds which differ by their behaviour towards chemicals used in extraction analysis. This procedure, applied in textile technology, consists of successive treatments with different solvents. Extraction of washed cell walls with lipophilic liquids yields a fraction which is considered to be the soluble part of the cutinised epidermal cells, consisting of the so-called adcrusting substances. Hot water and alternating extraction with dilute alkali and dilute acids remove pectic material and hemicelluloses, the ground substance (matrix) of the cell walls. A fraction called crude fibre is obtained after such treatment, consisting of fibrous frame substances (cellulose, chitin) and incrusting substances such as lignin which is left over when the fibrous substances are destroyed by hydrolysis with strong mineral acids.

(*a*) *Fibrous frame substances*

The extraordinary mechanical properties of the plant cell wall and the

←——

Fig. J–8. Birefringence in the cross-section of a three-layered secondary wall. S_1 strongly, S_3 weakly birefringent; S_2 isotropic (Wardrop and Dadswell, 1953).

Fig. J–9. Tertiary wall of a birch libriform fibre after dissolution of the secondary wall with copper ethylenediamine (Meier, 1955).

technological value of plant fibres (textiles, paper, wood) are due to the presence of fibrous material. In general, this important wall component is the high polymeric polyglucan cellulose, whose molecular structure is treated on p. 34. Cellulose is a very stable compound. The enzyme cellulase produced by cellulose-destroying bacteria and fungi, is in general not found in the cells of higher plants and it is not even produced by herbivorous vertebrates which must rely on symbiosis with cellulose-hydrolysing bacteria.

Crystalline cellulose is often accompanied by chains of glucomannan or xylan. Some authors have even thought that such chains were included in the cellulose lattice, but since xylan turned out to be not a homo- but a heteropolymer, 4-*O*-methyl-D-glucuronoxylan, with X-ray features strikingly different from those of crystalline cellulose (Marchessault and Liang, 1962), such intimate association does not seem possible.

In the cell wall of most fungi, cellulose is replaced by chitin (see p. 47). Although these high polymeric carbohydrate chains contain nitrogen in the side groups, chitin is even more insoluble and more resistant to hydrolysis than cellulose. There are however some microorganisms which can attack chitin by producing chitinase.

In some cases the fibrous component is neither cellulose nor chitin. In the extremely thick cell wall of the endosperm in certain palm seeds, which consist of polymannans with the function of carbohydrate reserve substances, Meier (1956) found the fraction mannan B to consist of microfibrils. In the yeast cell wall the fibrous component is a polyglucan. Although the hydrolysate of this yeast glucan yields exclusively glucose, it differs from cellulose by its X-ray diffraction pattern (Kreger, 1954) and by its insolubility in copper tetrammine (cuoxam).

Different types of microfibrils occur in the diversified phylum of green algae, from which the higher plants are derived. Besides cellulose (in *Valonia*), microfibrils of a β-1,3-linked polyxylan (see p. 34) have been found in *Bryopsis*, *Caulerpa*, etc. (Preston, 1962), a glucomannan in *Hydrodictyon* (Kreger, 1960) and short rods of crystalline mannan A in *Codium* and *Acetabularia* (Preston, 1962). It seems that nature has experimented with many types of polysaccharides before selection showed that microfibrils of the β-1,4-polyglucan cellulose chain are the most appropriate tool for the strengthening of the cell wall in plants striving for a terrestrial life.

(*b*) *Ground substances* (*Matrix*)

Carbohydrates which in distinction from cellulose are soluble in concentrated alkali (17.5% NaOH, 24% KOH), are known as hemicelluloses. Even though this term seems very unsatisfactory chemically, it is very useful for our purpose because in addition to the pectic substances it embraces all amorphous carbohydrates which form the ground substance of the plant cell wall. Only a few hemicelluloses like special xylans, glucans, and mannans (see above) are known in fibrillar form.

On hydrolysis, the hemicelluloses lead to different hexoses (in addition to glucose, mannose, and galactose), pentoses (xylose, arabinose), and uronic acids (galacturonic acid, glucuronic acid). These components appear in widely differing relative amounts and in numerous combinations. The compounds of the series glucose (Gl)—glucuronic acid (Glur)—xylose (Xy), and of the series galactose (Ga)—galacturonic acid (Gaur)—arabinose (Ar), are sterically related to one another. In the uronic acids, the primary alcohol group —CH_2OH of the corresponding hexose (see p. 34) is oxidised to an acid group —COOH, which can then be lost by decarboxylation to give the homologous pentose. One may thus expect the mixed polymerisation Gl—Glur—Xy or Ga—Gaur—Ar. Actually, there are various galacto-arabans, glucurono-xylans, etc., but much more frequently representatives appear of the two series mixed among one another, or with the hexose sugar mannose (galacto-mannan, glucuronoarabino-xylan, etc.) (see Frey-Wyssling, 1959, p. 140 ff.).

There are various reasons why such polymeric chains do not as a rule crystallise and form elementary fibrils. Frequently the chain members following one upon another do not alternate with the rigorous regularity necessary for a chain lattice. Secondly, the 1,4-bridge between the monomers (see p. 33), leading to straight fibres, is often replaced by 1,3-linkages which give rise to helical or branching chains (see callose, p. 290). Thirdly, the inclusion of uronic acids with their polar -COOH groups results in such strong hydration that crystallisation in an aqueous medium is impossible.

The uronic acid-containing hemicelluloses known as uronides lead to the pectic substances, in which the commonest combination is Gaur—Ar. They are differentiated from the galacturonides merely by the esterification of a considerable number of their -COOH groups with methyl alcohol, giving -$COOCH_3$. Estimation of methoxy groups is thus used for the quantitative detection of pectic substances in cell wall analysis. Histochemically, however, the uronides cannot be distinguished from the pectic material, as both groups of substance exhibit the same coloration with basic dyes (methylene blue, ruthenium red) and a similar swelling ability. The content of pectic material in the ground substance has thus often been overrated, and frequently uronide-containing plant mucilage has been erroneously termed pectic mucilage (see p. 326).

(*c*) *Composition of primary and secondary walls*

The analyses of primary walls of coleoptiles, stems, leaves, and hairs, compiled by Setterfield and Bayley (1961), yield the following average values (Table J–I).

TABLE J-I

COMPOSITION OF PRIMARY CELL WALLS

(Setterfield and Bayley, 1961)

	dry (%)	fresh (%)	
Water		60	83.2 (Water, Hemicelluloses, Pectic substances)
Hemicelluloses	53	21.2	
Pectic substances	5	2	
Cellulose	30	12	
Protein	5	2	
Lipids	7	2.8	
	100	100	

Over 50% of the dry wall is the ground substance, and if the pectic substances are included (because of their similarity to the hemicelluloses) the matrix amounts to almost 60%, against only 30% of the skeletal substance cellulose. If it is further considered that in fresh cell walls water is almost quantitatively bound to the hemicelluloses, because hydration of the crystalline elementary fibrils of cellulose is small, the swollen ground substance amounts to over 80% of the whole, and the fibrous material to only 12%.

The protein and lipid contents vary considerably depending on the pretreatment and the origin of the cell walls. If they are extensively washed or taken from older tissues, where the intimate contact of wall and plasmalemma (Fig. J–19a) is broken, the nitrogen content of the young wall may be almost zero. As to the lipids, it is found that hairs (cotton hairs) and epidermal cells which become cutinised contain considerably more lipidic substances than other cell walls.

The alkali extract from the secondary wall of cotton hairs amounts to less than 10%; moreover, the wall consists of pure cellulose. This must however be regarded as an exceptional case. The cellulose content is usually very much lower and frequently additional wall substances are incrusted (see p. 289).

The question whether the three layers S_1, S_2, and S_3 of the triple secondary wall of the tracheids (Fig. J–14a) behave similarly or differently from the chemical point of view has all along intrigued the wood anatomists. The fact that somewhat different reactions to staining are often observed seems to indicate differences, but no one has yet managed to separate the three layers quantitatively from one another in order to subject them to precise analysis. Meier (1961) has therefore examined the cell walls of wood fibres and tracheids for their carbohydrate content throughout their growth, and in this way obtained the composition of the primary wall with the intermediate middle lamella ($M + P$), somewhat later, the wall with the first apposition layer ($M + P + S_1$), even later $M + P + S_1 + S_2$, and finally the grown wall, $M + P + S_1 + S_2 + S_3$. The contents of single layers can be calculated from these analyses (Table J–II). As only the polysaccharides were determined, we still have no complete picture of the composition of the wall. In the primary wall $M + P$, the polyuronides (pectic substances) are not allowed for. Even though, it is evident, how the ground substance (hemicelluloses) predominates in the primary wall and recedes in the secondary wall. Conversely, the cellu-

TABLE J–II

POLYSACCHARIDES IN DIFFERENT WALL LAYERS OF SPRUCE TRACHEIDS

(Meier, 1961)

	$M+P$	S_1	S_2	S_3
Galactan	16.4	8.0	0.0	0.0
Arabinan	29.3	1.1	0.8	0.0
Glucurono-arabino-xylan	13.0	17.6	10.7	12.7
Gluco-mannan	7.9	18.1	24.2	23.7
Total hemicelluloses	66.6	44.8	35.7	36.4
Cellulose	33.4	55.2	64.3	63.6

lose content rises from 1/3 to 2/3. Of the hemicelluloses, galactan and araban can only be found in the primary wall ($M + P$) in considerable amounts. In S_2 and S_3 they are lacking, and S_1 is found to be a typical transition layer for these substances. Xylan is fairly uniformly represented, whilst gluco-mannan, which is associated with the cellulose, increases systematically, together with that skeletal material, from the outside to the inside. The differences between S_2 and S_3 are immaterial. This may be due to the fact that the period of growth of S_3 is so short that parts of the inner S_2 layer have been included in the analysis.

(*d*) *Incrusting substances*

The so-called incrusting substances can appear in older cell walls. The most important of these is lignin, which arises in the cell as a macromolecular substance by polymerisation of coniferylic alcohol (Freudenberg, 1954):

HO—C_6H_3(OCH₃)—CH=CH—CH_2OH

coniferylic alcohol

As can be seen, this is a methylated C_9-compound in which a diphenol bears an unsaturated C_3-chain. The methyl group substitutes one of the two phenol groups. Owing to its double bond in the side chain, this compound can polymerise in different ways, so that in contrast to the linear polymerisation of polysaccharide and polyuronide chains, cross-linking can occur in different directions in space by the so-called reticular polymerisation.

Lignin retains some phenolic properties of the monomer, in the polymeric state. Since free phenol groups are slightly acid, it can be stained with basic dyes, and owing to its aromatic nature, gives typical colour reactions with a number of reagents, the best known being the cherry red coloration with phloroglucinol and hydrochloric acid. In the ultraviolet spectrum, it shows a pronounced maximum at 282 nm, so that lignification of the cell wall may be followed quantitatively in the UV microscope with the aid of the 280 nm line of a mercury lamp.

In wood, the cell wall contains close to 50% cellulose, 20% (soft wood) to 30% (hard wood) hemicellulose, and 20–30% lignin.

In late wood and heartwood, additional incrusting substances occur, which belong to the class of tannins. On polymerisation, these are converted to characteristic insoluble dyestuffs. The *Acacia* tannin, catechol, may be mentioned as an example. This is a colourless C_{15}-compound related to the anthocyanidines, which assumes a red-brown colour on polymerisation.

HO, HO (on benzo ring) — O — CH — C_6H_3(OH)(OH); CHOH; C H_2

catechol

Besides these organic compounds, the ash constituents of the cell wall (SiO_2, $CaCO_3$) are also to be counted as incrusting substances. Their incorporation is a passive process, since they are simply the mineral substances left behind by water evaporated during transpiration.

(*e*) *Adcrusting substances*

In addition to the incrusting materials embedded in the cell wall layers, there are also substances which accumulate on the cell wall surface; these are known as adcrusting substances.

A membranous material may be deposited on the inner side of the walls of various types of cell (sieve tubes, pollen tubes, fungus hyphae), which was called callose by Mangin (1890) because it was the main component of calli, which close the sieve tubes at the end of their activity. Aniline blue and resorcinol blue endow callose with a conspicuous green fluorescence, so that this wall substance can be easily localised in the fluorescence microscope. Similarly to cellulose, callose is a polysaccharide which yields exclusively glucose on hydrolysis, but its solubility in 5% alkali classes the substance among the hemicelluloses. It is further distinguished from cellulose by its isotropic character. Chemical structural analysis shows that, as in lamillarin, the glucose units are connected to one another by β-1,3-glucoside linkages (Kessler, 1957/58). There does not appear to be any chain branching, for only 2,4,6-trimethylglucose and no dimethylglucose is found on methylation. Furthermore, the chain must be of considerable length as no end groups (2,3,4,6-tetramethylglucose) can be detected. The chains are probably wound helically, which makes the isotropy of callose easy to understand.

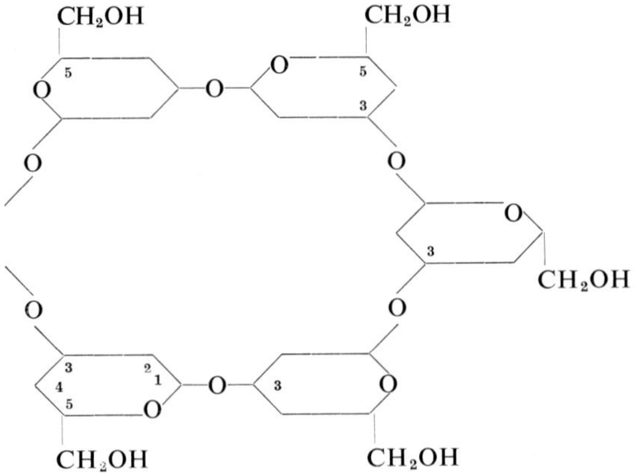

callose

Mucilaginous substances are deposited on the outer side of the epidermal cell walls of most aquatic plants. In terrestrial plants, however, all cell wall surfaces in direct contact with air are covered with lipophilic cell wall substances, which serve to limit transpiration. The lipidic wall substances must be differentiated into extractable, low molecular weight (wax group) and high polymeric (suberin – cutin group) compounds.

The waxes are found in the outer wall of the epidermal cells, or they are deposited as exudations, which cover the epidermis, in the form of granules, rodlets or platelets. Chemically, they consist of esters or mixtures of the aliphatic wax alcohols, $CH_3(CH_2)_nCH_2OH$, and the corresponding fatty acids, in which n can assume values between 22 and 32. In addition, paraffins or ketones may be mixed in as further aliphatic chains with a similar number of carbon atoms.

The cutins arise by net polymerisation of hydroxycarboxylic acids with several esterifiable groups, such as phloionolic acid, $OH \cdot C_{17}H_{32}(OH)_2 \cdot COOH$. Completely insoluble high polymeric substances are formed in this way. These macromolecular esters can be saponified, but an unsaponifiable residue always remains (*e.g.* in bottle cork, 12%; in *Ilex* cutin, 38%). The unsaponifiable components may be due to polymerisation of unsaturated hydroxy-fatty acids which probably harden on exposure to atmospheric oxygen, like the varnish-forming drying oils (*e.g.* linoleic acid $C_{18}H_{34}O_2$, linolenic acid $C_{18}H_{32}O_2$).

According to the resistance which these membrane substances show to saponification, they are divided into suberins (saponified in 3% aqueous NaOH), cutins (saponified in 5% methanolic KOH) and sporopollenins (saponified in alkaline melts). The sporopollenins are probably the most resistant organic compounds in existence; they have preserved pollen grains and fungal spores from decomposition in peat for hundreds of centuries. A similar situation is found in the cutin of leaves, which on decay in the ground withstands decomposition for a long time. It was long believed that there were no saponifying cutinases, until such an enzyme was discovered in certain fungi (Heinen, 1960, 1961).

3. ULTRASTRUCTURE

No ultrastructure can be recognised in fine sections of the cell wall, because its constituents (hemicellulose, cellulose, lignin) diffract electrons in a similar manner in view of their similar make up from C, O, and H. To render the ultrastructure visible in the electron microscope, recourse must be made to the method of maceration. By varying treatment with dilute acids and alkalis, the hemicellulosic and pectic materials can be removed from the wall. The sublight-microscopic fibrils then remain, which were detected in 1948 in the cell walls of algae (Preston *et al.*, 1948) and of higher plants (Frey-Wyssling *et al.*, 1948).

The ground substance or matrix in which the fibrils are embedded consists, as mentioned previously, of hemicelluloses and mostly contains pectin only in small proportions. It forms a soft plastic mass reinforced by the fibrils. In this way, the same principle is followed as in the reinforcement of hyaluronic acid by collagen fibrils in connective tissue, or of gelatinous gels by siliceous needles in siliceous sponges. The science of engineering shows that the strengthening of a plastic mass increases with the square of the length of the embedded rods or piles (see Frey-Wyssling, 1962).

In contrast to the macrofibrils visible in the light microscope, the fibrils of the cell wall are termed microfibrils. As explained on p. 40, they consist of crystalline cellulose, and rarely of chitin (fungal cell walls, Frey-Wyssling and Mühlethaler, 1950) or of mannan.

The cellulose microfibrils are mostly bands 100 (to a maximum of 250) Å wide.

They are often very flat and then appear fasciated (Frey-Wyssling, 1951). Bands of 100 Å wide and about 30 Å thick are found in ramie fibres (Vogel, 1953). The finest fibres as yet resolved are isodiametric, and have a diameter of about 35 Å (Mühlethaler, 1960a). They possess a homogeneous chain lattice and are brittle like crystalline needles (Frey-Wyssling and Mühlethaler, 1963). They are known as elementary fibrils (see p. 39). The flat 100 Å fibrils described by Vogel may be interpreted as the association of three such elementary fibrils. Single elementary fibrils can be observed in primary walls, whilst in secondary walls they are mostly aggregated to thicker microfibrils.

(*a*) *Ultratextures*

The relative amounts of matrix and fibrils vary widely, from almost no cellulose in the cell plate of mitosis to over 95% of cellulose in bast fibres or cotton hairs.

In the primary wall, which displays a conspicuous growth in area during cell extension and differentiation, the cellulose fraction in the hydrated wall reaches only about 12% (see Table J–I), whilst in the thickened secondary wall of fully grown cells, cellulose largely prevails over the other wall constituents. In consequence, there is space for a dispersed arrangement of the cellulose microfibrils in the primary wall; but in the secondary wall they must be closely packed because only a parallel arrangement can yield a sufficiently high cellulose content. In other words, although primary and secondary walls have the same ultrastructure consisting of a matrix and reinforcing fibrils, the pattern of these microfibrils is subject to considerable variation. We can therefore speak of different ultratextures.

On the basis of the orientation of microfibrils with respect to the axis of elongated cells, as determined with the polarising microscope (Frey-Wyssling, 1930), the ultratextures as specified in Table J–III have been distinguished.

TABLE J–III

ULTRATEXTURES OF PLANT CELL WALLS

Orientation of microfibrils with respect to the cell axis	Term	
parallel	fibrous texture	parallel textures
oblique	helical texture	
perpendicular	annular texture	
± longitudinal	fibroid texture	disperse textures
± transverse	tubular texture	
random	foliate texture	

These terms are also useful for the description of ultratextures as seen in the electron microscope. If a sufficiently thick wall layer has the same texture throughout its width, the electron microscope confirms the textures derived from observations with the polarising microscope. If however there are ultrathin lamellae with parallel textures criss-crossing each other, a disperse texture can be simulated in the polarising microscope.

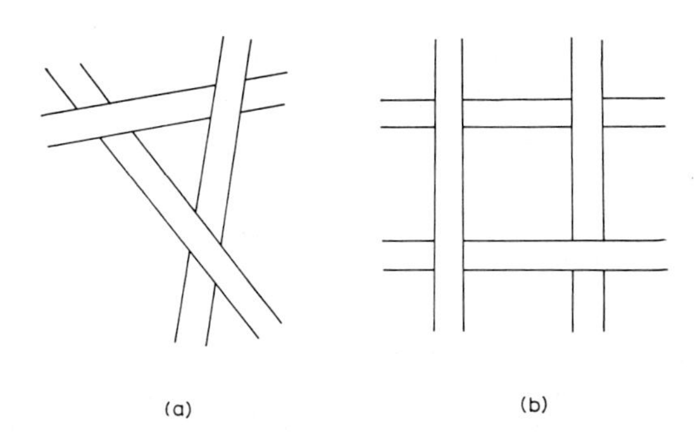

Fig. J–11. Texture of microfibrils in the primary cell wall; (a) interwoven, (b) superposed (Frey-Wyssling, 1957).

(*b*) *Disperse textures*

In isodiametric cells, which later grow on all sides so that they remain isodiametric, the cellulose fibrils in the primary wall are embedded at random (Fig. J–15a) in the matrix. Their longitudinal axes point in arbitrary directions within the lamellae. As however the length of the fibrils ($> 1\ \mu$) considerably exceeds the thickness of the lamellae ($< 0.1\ \mu$) they run crosswise approximately parallel to the cell surface, and, moreover, they are not only laid down dispersedly upon one another (Fig. J–11b), but they can also be interwoven (Fig. J–11a); i.e. the fibrils are arranged in such a way that it is impossible to take one away without disturbing the course of the other. Such cell walls appear optically isotropic in surface view, but show birefringence in cross-section, which is positive with reference to the tangential direction. Their fibrillar arrangement is called a foliate texture.

If these cells are destined later to assume cylindrical shape by cell extension, the fibrils are deposited more or less perpendicular to the future cell axis (Fig. J–12a). This arrangement (tubular texture) shows up in the polarising microscope in that the cell wall, referred to the cell axis, appears optically positive in radial section and optically negative in tangential section (Frey-Wyssling, 1942). In the direction of future growth, the tubular texture induces for the stress of the wall minimal, and for the cell strain maximal values. Cell extension caused by turgidity pressure leads therefore to a cylindrical form.

Attempts have been made to explain the fibril arrangement of the tubular texture mechanistically, by assuming that the fibrils were deposited parallel to the maximal stress (Roelofsen, 1959) or perpendicularly to the maximal strain (Green and Chen, 1960). Such considerations are however invalid, because the tubular texture is laid down before the cell assumes a cylindrical shape; if, therefore, we presupposed this form for the explanation of the ultratexture, this would amount to an inversion of cause and effect.

Ledbetter and Porter (1963) detected microtubules under the plasmalemma of cylindrical cells fixed with glutaraldehyde, which run in the cortical groundplasm and look like spindle fibres (see p. 208). As these microtubules are oriented like the elementary cellulose fibrils of the primary wall, a relationship to these wall fibrils was considered. Similar differentiations, interpreted as contractile plasma filaments, have been found in the plasma gel of *Amoeba proteus* (Danneel, 1964).

The original assumption that the primary wall is generally characterised by a disperse texture of its microfibrils has been found to be false. Already during its surface growth, the cell edges may be strengthened by strands with parallel texture (Fig. J–12b, Mühlethaler, 1950). Also, for this type of fibril arrangement parallel to the axis, which contradicts the rule of the transverse arrangement in cylindrical cells, possible mechanistic explanations were considered with the aid of surface

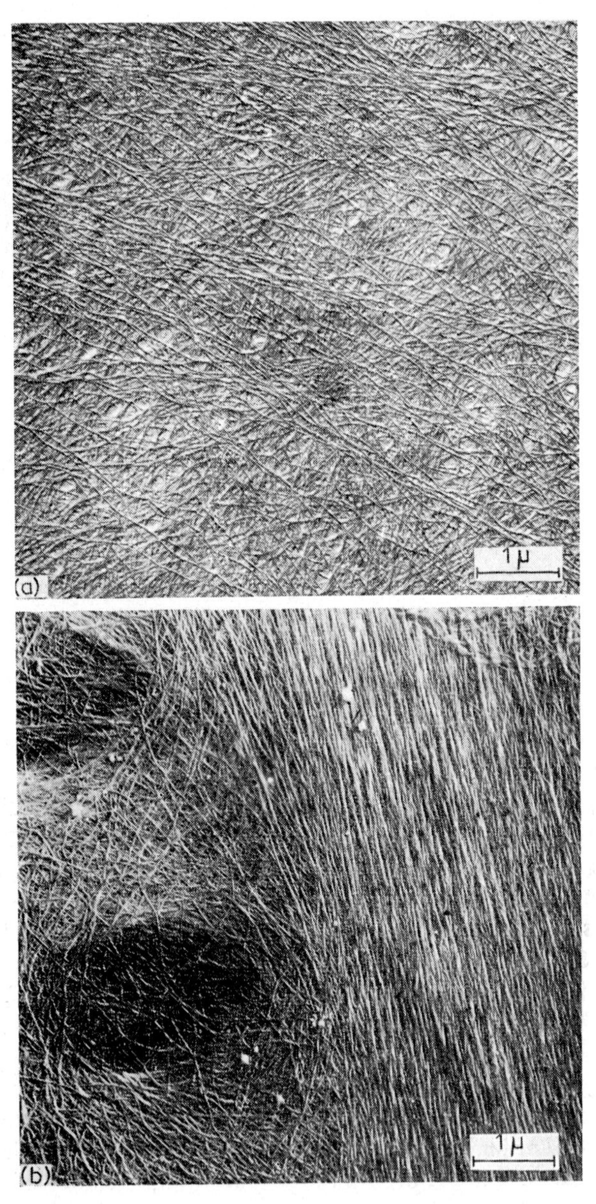

Fig. J–12. Disperse texture of the primary cell wall (Mühlethaler, 1950c); (a) tubular texture, pith cell of *Ricinus*; (b) strengthening of cell edge and pit fields, coleoptile of *Avena*.

tension which differs in magnitude in the faces and along the edges of foam polyhedra (Roelofsen, 1959). Meanwhile, since the stress theory proved to be incorrect, the formation of different textures must be ascribed to the morphogenetic creative power of the plasmalemma, which organises the cell wall with regard to its future function by means of genetic information.

(*c*) *Parallel textures*

As has already been mentioned, the cellulose content of the secondary walls is so great that the fibrils cannot be other than parallel. In rare cases the orientation is strictly parallel to the cell axis. Such a texture is termed fibrous, because it is characteristic of a number of bast fibres (ramie, linen, hemp). This kind of cell wall shows maximum anisotropy (birefringence, dichroism, tensile strength). On X-ray diffraction the cell walls give the classical fibre diagram, which led to an elucidation of the crystalline lattice of cellulose. The refractive indices, the tensile strength, and the bending strength of this chain lattice have also been determined quantitatively.

In such secondary walls the elementary fibrils are bundled into thicker microfibrils, so that the fibres display a heterogeneous ultracapillary system; on the one side the finest hollow spaces (< 10 Å) between the elementary fibrils, and on the other, coarser capillaries (ca. 100 Å) between their microfibrillar bundles. Only micromolecules like H_2O (swelling) and I_2 (zinc chloride-iodine staining) are able to penetrate into the first, while colloidal particles like Congo red or colloidal silver and gold can be infiltrated into the coarser ones.

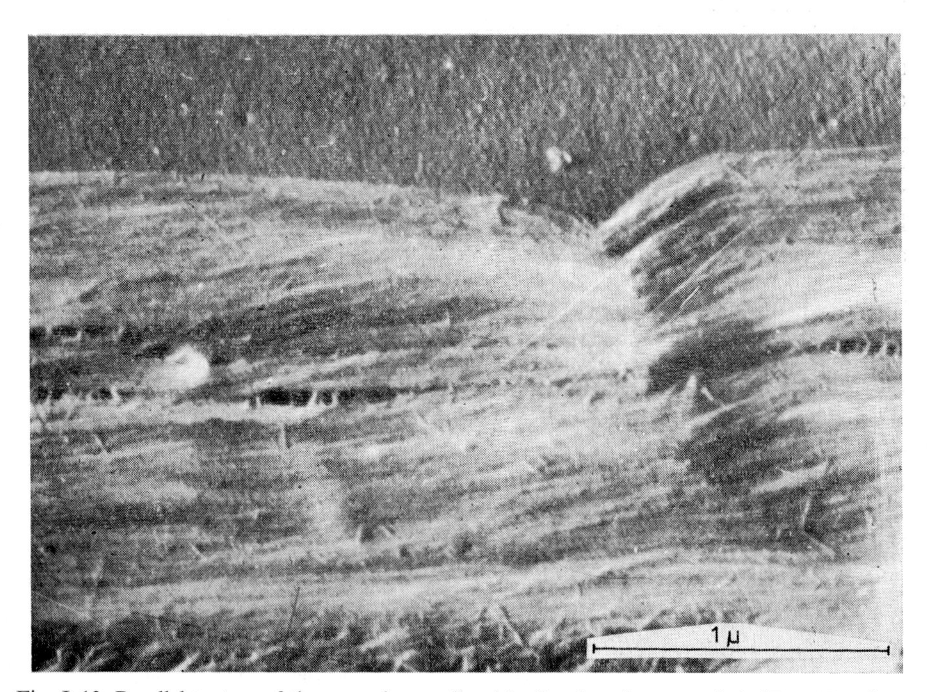

Fig. J–13. Parallel texture of the secondary wall, with slip plane in cotton hair (Frey-Wyssling, *et al.*, 1948).

The so-called slip planes or displacement lines are another characteristic feature of cell walls with fibrous texture. They originate when the walls are subjected to high pressure in direction of the cell axis (Fig. J–13), and are produced by a local displacement in the parallel texture. The spacing of the microfibrils somewhat increases locally, whereby the previously circular fibre cross-section becomes an ellipse; as an elliptical section can only be placed obliquely through the cylinder, the slip planes always run through the fibre at a fixed angle, whose magnitude depends on the local loosening of the texture caused by compression (see Frey-Wyssling, 1959).

Ideal fibrous textures are also found in the so-called gelatinous layer of the tension fibres (optics: Jaccard and Frey, 1928; X-ray diagram: Wardrop and Dadswell, 1955) in the tension wood of deciduous trees (Fig. J–14). We must assume that these cells become shortened, when a shoot is bent geotropically by such reaction wood. How this is possible by a complete longitudinal orientation of the stressed fibrils is still an open question.

As a rule, the parallel texture in the cells of mechanical tissue runs at an angle to the cell axis, resulting in a helical texture. It can be recognised in the light microscope in the oblique arrangement of slit-shaped pits and in the occasional appearance of cleavages running obliquely to the cell axis. In the polarising microscope, cells with helical texture show no complete extinction, and the X-ray diagram shows characteristic crescents, whose arc angle is a direct measure of the pitch of the helix. In the electron microscope, the helical structure is less conspicuous, as the field of vision is so minute that only a small part of a cell can be surveyed (Fig. J–13).

A classical example of such texture is the secondary wall of cotton hair. Each single lamella in Fig. J–10a possesses a helical texture. The helical angle, *i.e.* the angle between the inclination of the pitch and the cell axis is about 30°.

In the tracheids of conifers, all three layers S_1, S_2, and S_3 are coiled. The inclination of the pitch of S_1 and S_3 is small, so that, on a cross-section, they appear bright in the polarising microscope (Fig. J–8), whilst the helix of the thick S_2 layer runs very steeply, so that the microfibrils are cut almost transversely to their axis in which direction no optical anisotropy is manifest

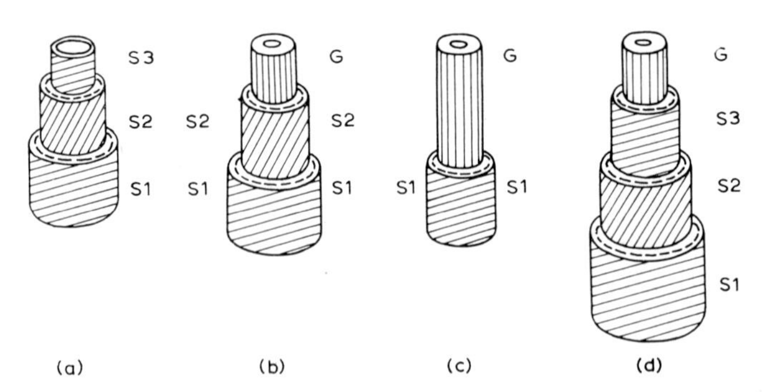

Fig. J–14. Strengthening of libriform fibres in tension wood. Layer diagram (see Fig. J–7) S_1 outer, S_2 central, S_3 inner, G gelatinous layer. Orientation of microfibrils is indicated by striation; (a) normal wood fibre, (b, c, d) different types of tension wood fibres; only the S layers are lignified (Wardrop and Dadswell, 1955).

(*d*) *Crossed textures*

The helices of the three *S*-layers cross mutually over one another. Layers S_1 and S_3 are, however, so thin that the helical texture of the S_2 layer is predominant. This is not the case when numerous equally thin lamellae with oppositely-running helical pitches cross over one another alternately, and we then speak of a crossed texture.

This system has been thoroughly investigated in the wall of the marine alga *Valonia* (Preston, 1952; Steward and Mühlethaler, 1953). The cell wall of germinating sporlings shows a disperse texture (Fig. J–15a), which after some time becomes thickened by apposition lamellae having a parallel texture. The direction of the fibrils changes on average by 60° in successive lamellae (Fig. J–15b). This angle can vary markedly – values of 50° to 70° have been found. The fibrils in a set of three layers crossing over one another are dissimilarly developed, for the X-ray diagram shows only two principal orientations, so that it was originally conjectured that the fibrils were crossed over almost perpendicularly, with a crossing angle of 78° (Astbury *et al.*, 1932).

The wall of the grown *Valonia* cell, which can reach the macroscopic dimensions of 1 cm in diameter, is about 40 μ thick, and the individual lamellae, which each consist of a microfibril stratum about 250 Å thick and a similarly thick stratum of matrix, measure 500 Å. The entire wall is thus built of about 800 lamellae. The surprising thing is that, in spite of its great thickness, this wall shows an extraordinary growth in area while the microscopic sporling grows up to be a macroscopic cell. It owes this ability to its lamination. The innermost apposition lamella is continually laid down somewhat larger than that deposited earlier, so that in consequence the parallel texture of the older lamellae is forced open resulting in cracks and clefts parallel to the texture axis (Fig. J–15b). The crossing fibrils must then slide over one another, which is made possible by interstratification of the plastic matrix material.

According to the proposal to define the primary wall as a cell envelope exhibiting growth in area, the entire cell wall of *Valonia*, in spite of its thickness and its pronounced layering, had to be considered as primary; this is avoided in our definition of the primary wall (see p. 280).

Numerous cases are also known in the higher plants of secondary walls growing strongly in thickness and simultaneously increasing their length in a surprising way. The walls of collenchyma cells, certain sclerenchyma cells and thick-walled laticifers may be mentioned as examples. In all these cases the cell walls show a system of parallel textured lamellae separated from one another by intercalated plastic matrix, so that ultrastructural lamination seems to be a prerequisite for the growth in area of thickened cell walls.

The collenchyma cells of celery petioles lengthen by a factor of about 30, whilst the leaf stalk grows from 1 cm to 30 cm in length (Beer and Setterfield, 1958). The apposition strata of parallel textured cellulose fibrils oriented parallel to the cell axis (Preston and Duckworth, 1946) are separated from one another in the angular thickening by wide intermediary strata of matrix (Fig. J–16), so that lamellae of light microscopic thickness are formed (Anderson, 1927). An analogous situation exists in the unilaterally thickened epidermal cells (so-called lamellar collenchyma). In the epidermis of *Avena* coleoptiles, the epidermal outer wall lengthens in 4 days

Fig. J–15. Cell wall of *Valonia*; (a) primary wall, disperse texture, (b) secondary wall, crossed parallel texture (Steward and Mühlethaler, 1953).

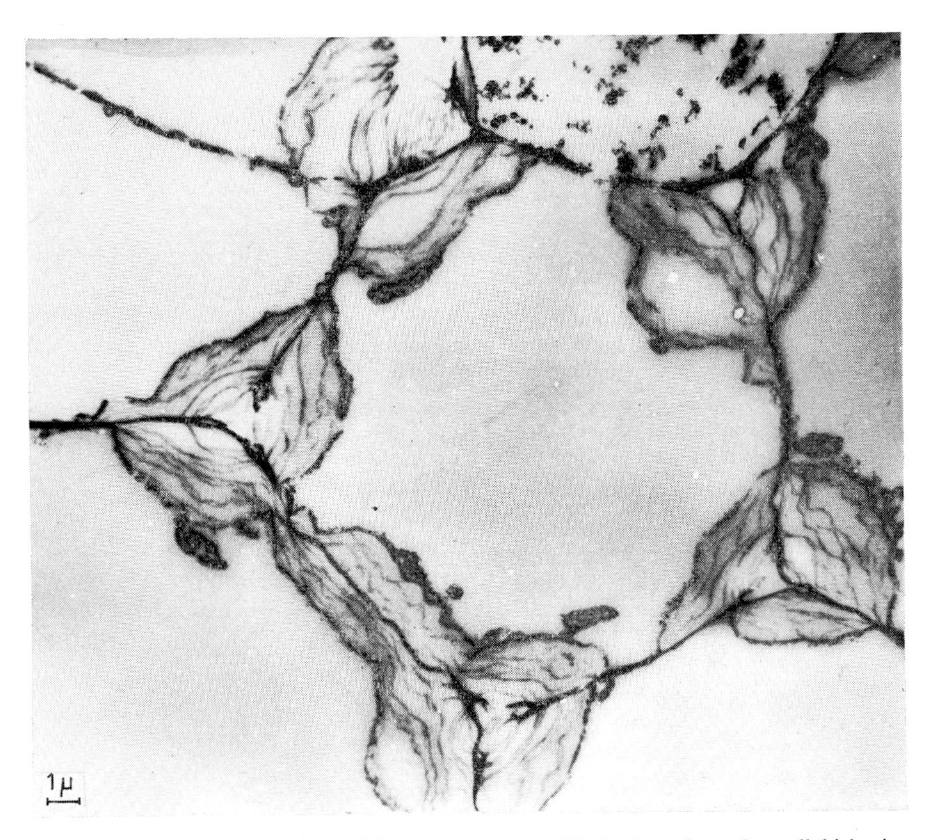

Fig. J–16. Macerised collenchyma cell from a *Ricinus* stem. The laminated angular wall thickenings are capable of longitudinal extension growth (Mühlethaler, 1961).

from 13 µ to 2 mm, corresponding to a 150-fold extension (see Frey-Wyssling, 1948b).

Extending laminated secondary walls are also found in the subcortical sclerenchyma fibres of *Asparagus* shoots. The sclerenchyma elements then grow from 15–30 µ long and 4–7 µ wide cells into 700–900 µ × 10–15 µ fibres. The lamellae growing in area show a parallel texture whose fibril directions cross over one another (Sterling and Spit, 1957). The longitudinal extension follows the same principle as in the collenchyma walls, with the difference that in this case we have a crossed texture and not a fibrous texture. In contrast to the collenchyma cells, these cells show growth not only in length but also in width. Thus the crosswise arrangement of parallel textured lamellae appears to be a characteristic feature of growth in diameter of cells with secondary cell walls.

This growth in width is especially conspicuous in the laticifers of *Euphorbia splendens* (Moor, 1959). Cells which show a diameter of only 3 µ in the region of the vegetative apex grow to 60 µ wide tubes, with (in the hydrated state) walls over 10 µ thick. The outermost lamellae (original primary wall) shows at first a disperse texture (Fig. J–17a), but later the microfibrils become pushed apart and aligned parallel to the axis. An apposition layer follows upon this, which Moor (1959)

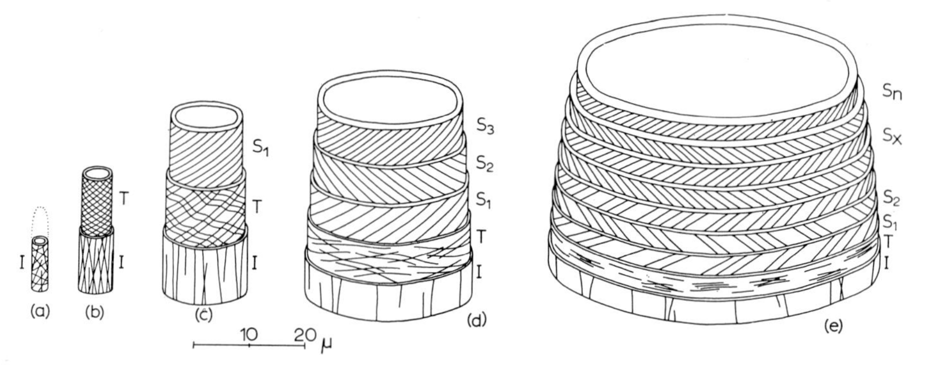

Fig. J–17. Fine structure of latex tubes of *Euphorbia splendens* (Moor, 1959). (a–e) Texture of cell wall during growth from the meristematic (a) to the fully differentiated stage (e). I primary wall, *T* transition layer, $S_1 \ldots S_n$ layers of secondary wall II.

called the transition lamella. This contains obliquely deposited fibrils (Fig. J–17b), which are later displaced by the growth in width and appear to be oriented horizontally. Finally, the deposition of lamellae with helical texture follows (Fig. J–17c), whose direction of coiling in successive layers alternately varies from counterclockwise to clockwise (Fig. J–17d). The texture of the outer lamellae is strikingly loosened. Obviously this lamellar system allows an almost unlimited growth in the length and the width of the cell.

4. GROWTH

Growth of the plant cell wall presents special problems, insofar as this organelle is considered to be a metaplasmatic formation of the living protoplast. It is secreted outside the plasmalemma and it is therefore difficult to visualise active morphogenetic behaviour in this envelope with its passive function.

(*a*) *Production of the matrix*

Classical cytology distinguished between tip growth (of hyphae, root hairs, cotton hairs, pollen tubes), extension growth (cylindrical cells) and isodiametric growth of plant cell walls. This classification assumed that local parts of the wall (tip, longitudinal faces) were capable of independent growth, which certainly appears true when *e.g.* a root hair sprouts from the basal part of an elongated epidermal cell. Tracer experiments showed however that secretion of wall material occurs along the entire surface of the protoplast (cotton hairs, O'Kelley, 1953; coleoptiles, Wardrop, 1956; cells of root cortex, Setterfield and Bayley, 1957). In the thick walls of *Nitella*, Green (1958) found with tritiated water that deposition of new wall material is restricted to the inside, whilst according to Setterfield and Bayley (1958) the lamellar collenchyma of epidermal cells fed with ^{14}C becomes radioactive throughout the total depth of the thickend wall.

Since as long as the cell grows the cell wall formation takes place over the whole surface of the protoplast, local thickening (angular and lamellar collenchyma) must come about by locally increased production of cell wall substances. The question now arises, whether the intensified secretion of material in increased quantity concerns the matrix, the fibrils, or both in a similar way.

The secretion of amorphous carbohydrates (hemicelluloses and pectin) precedes ontogenetically the formation of supporting fibrils. The cell plate arises from pectic vesicles (Figs. J–2, J–3) which are formed from the Golgi apparatus (Whaley and Mollenhauer, 1963; Frey-Wyssling *et al.*, 1964). The same process is involved in the growth of the primary wall. Mollenhauer and Whaley (1962) have demonstrated how Golgi vesicles are sluiced outwards through the plasmalemma, and Sievers (1963) described how such vesicles are incorpoıated into the cell wall of root hairs (Fig. J–18). The vesicles show contrast after prolonged fixation with manganese; the primary wall therefore appears black after fixation with potassium permanganate.

At the time of maximal cell wall growth, the plasmalemma loses its usual appearance of a smooth boundary of the protoplast, and assumes a strange surface configuration (Fig. J–19a), giving the impression of vast dynamic processes (Frey-Wyssling, 1962). Large numbers of vesicles are obviously simultaneously pushed out of the cell, and imbibition fluids of the wall are drawn into the cell by pinocytosis. In this phase the plasma cannot be separated from the cell wall by attempts of plasmolysis, so that it appears coalesced with it. Boysen-Jensen (1954) had already correctly recognised this situation before its elucidation with the electron micro-

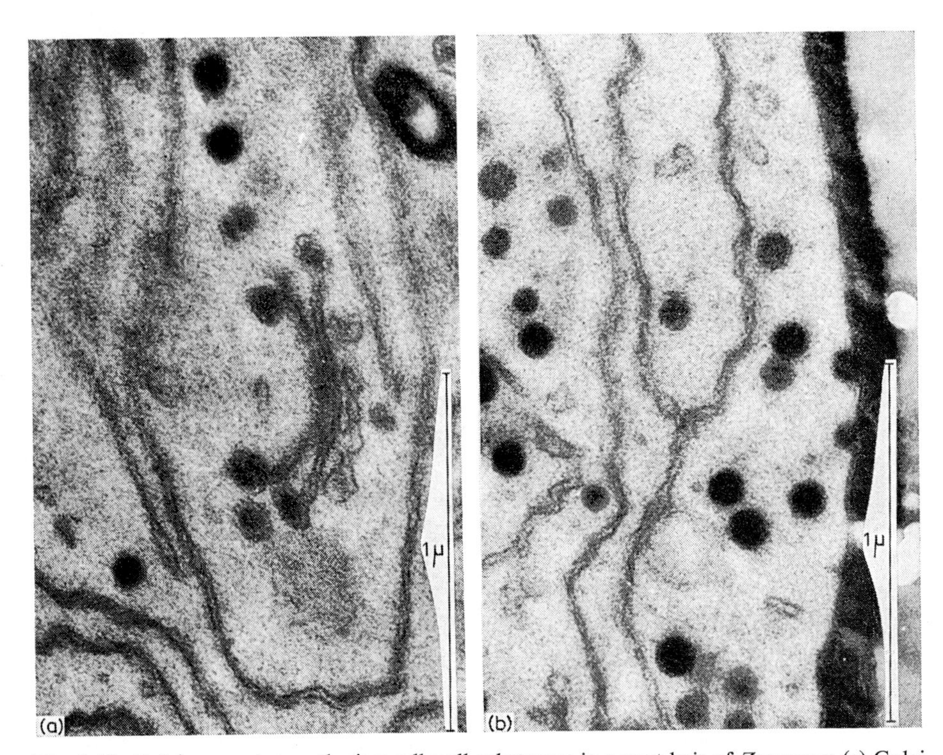

Fig. J–18. Golgi apparatus synthesises cell wall substances in a root hair of *Zea mays*; (a) Golgi apparatus producing vesicles with dense content, (b) incorporation of Golgi vesicles into the cell wall (Sievers, 1963).

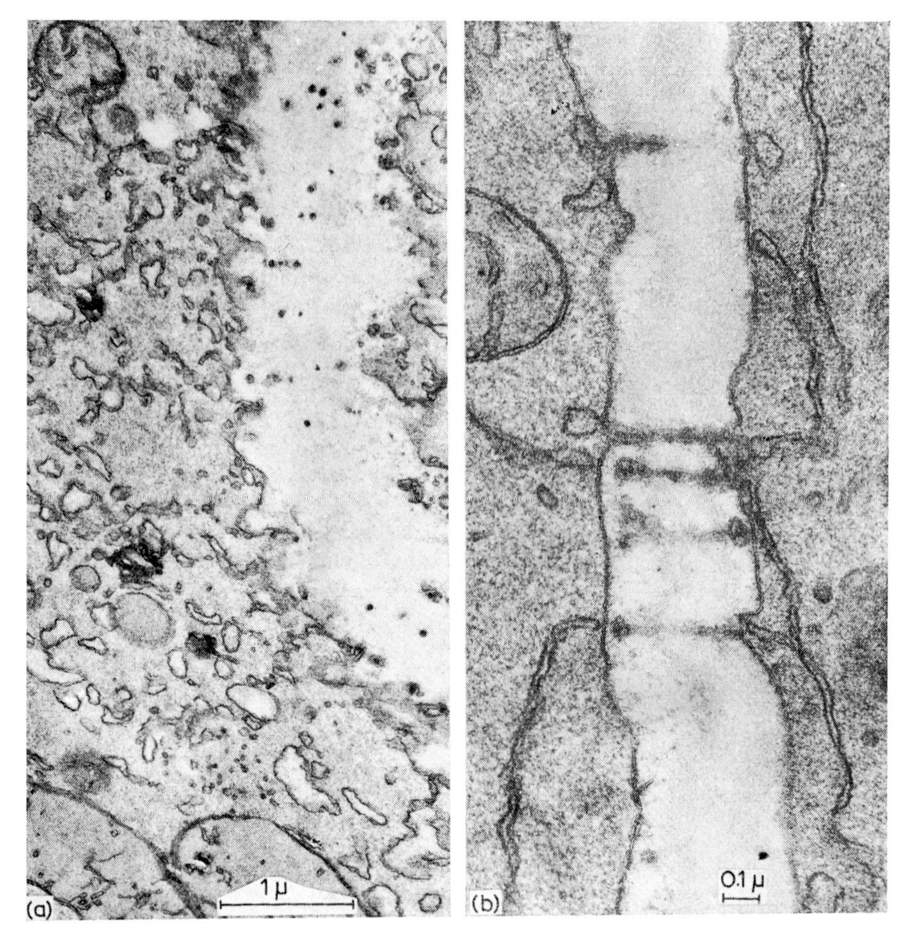

Fig. J–19. Contact plasmalemma/cell wall in the root cortex of *Ricinus*; (a) entangled during formation of the primary wall which is pierced by plasmodesms (black dots), oblique section; (b) smooth in the full-grown primary wall. Strands of ER penetrate into plasmodesms and traverse the wall (Frey-Wyssling, 1962).

scope, as he showed that the hyaloplasm is probably linked with the growing wall by papillae.

The picture of intense cell wall secretion in Fig. J–19a is exagerated because the primary wall is cut obliquely, so that it appears disproportionately wide and shows the plasmodesms in oblique cross-section. The period of this growth phase is of course rather brief. Later, the turbulent occurrences quieten down, and the plasmalemma again becomes smooth as in Fig. J–19b. At that stage a surface organisation of the plasmalemma as in Fig. E–9 may be expected.

(*b*) *Multi-net growth*

It is certain that at the time of the turbulent cell wall formation the reinforcement of the matrix by microfibrils is already in progress. As has already been mentioned, their arrangement takes place at random in cells which show isodiametric growth; in those which assume a cylindrical shape it is more or less perpendicular to the future cell axis.

Later the orientation of these transverse running fibrils is passively changed during extension growth. When the cell elongates, their disperse texture, which is predominantly perpendicular to the extending axis, alters so that after some time the fibrils cross over at right angles, and finally become aligned more or less parallel to the longitudinal axis of the cell. Simultaneously, new fibrils are laid down on the inner side of the plastic lamellae following the pattern of a tubular texture. This type of growth with a passive shift of the fibrils in the outer region and adherence to the original constructional plan by the formation of new fibrils in the innermost lamella was described by Houwink and Roelofsen (1954) (Fig. J–20), and was called multi-net growth by Roelofsen (1959).

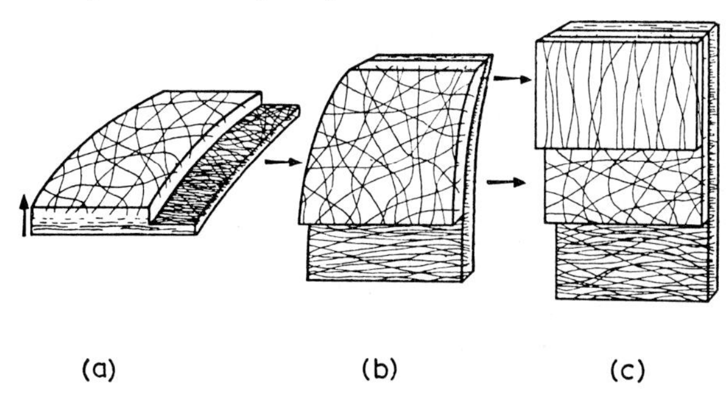

Fig. J–20. Diagram of multi-net growth. (a), (b), (c) passive changes of fibril orientation in the outer part of the primary wall (Houwink and Roelofsen, 1954). Arrows indicate corresponding layers with their textural changes.

Figs. J–20 and J–21 show how the fibrils in the outer regions of the wall are shifted and reoriented by the growth in area. The sketch comes from the tip of a growing cotton hair. It is seen how during growth the cell wall formation on the inner side of the wall advances unperturbed, whilst the outer region becomes extended and plastically moulded. Multi-net growth has been described in various types of cells (stellate pith cells, Houwink and Roelofsen, 1954; laticifers, Moor, 1959 (Figs. J–17 and J–21); *Nitella* cells, Green, 1960), so that this form of extension of the primary wall appears to be fairly general.

The lamellae indicated in Fig. J–20 are not delimited from each other; there is a gradual transition from the inner to the outer texture. It must be realised that plastic matrix is continuously secreted not only into the inner, but also into the outer regions. In terms of classical cytology, the matrix thus grows by intussusception and the fibrillar system by apposition! From the moment, however, at which secondary lamellae appear, the primary wall stops growing, becomes stretched, and can finally tear.

Surface growth of the primary wall can be compared with the growth in area of the laminated secondary wall with crossed parallel textures (see p. 299). The matrix strata between the fibril strata serve as a 'lubricant', when the superimposed parallel textures slide over one another. As we may see in Fig. J–15b, the fibrillar lamellae burst open, preserving however the parallel arrangement of the fibrils; only their crossing-over angle is altered. This is the principal distinction from the

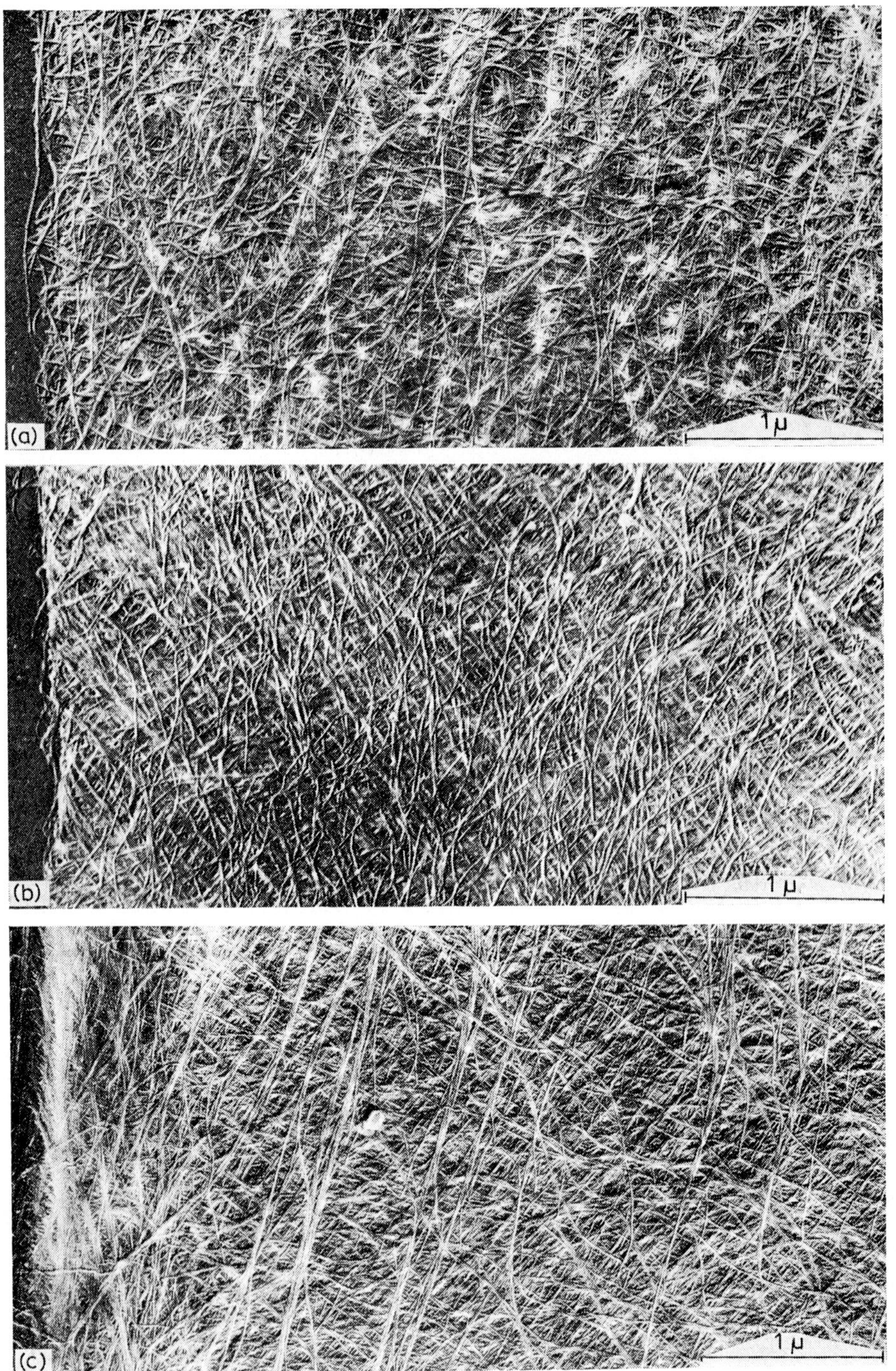

Fig. J–21. Passive behaviour of fibrils in the outer part of the primary wall. Latex-tube wall of *Euphorbia splendens*; (a) primary wall before expansion; (b) the original primary wall is stretched in the direction of growth and its constituent microfibrils are reoriented in the same direction; (c) mature wall, with longitudinally scattered microfibrils on the outer wall surface (Moor, 1959).

primary walls in which the fibrils move individually with respect to one another in multi-net growth; in secondary walls they maintain their mutual positions within a lamella. Not only is an extension perpendicular to the parallel texture necessary for an isodiametric growth of spherical cells, or for the simultaneous growth in length and width of cylindrical cells, but the fibrillar system must also be elongated.

Two mechanisms can be proposed. Either the texture 'grows' in length by the fibrils sliding parallel to the axis, or the individual fibrils must show growth at the tips. The sliding mechanism is easy to understand because the matrix is plastic and because the gaps between the fibril bands (Fig. J–15b) permit a mutual displacement of the microfibrils also in the longitudinal direction. In the outermost secondary wall lamellae of the *Valonia* wall, the microfibrils lie far removed from one another (see also Fig. J–17d); nevertheless the lamellae all appear of similar thickness in dehydrated preparations, because the fibril thickness remains constant and the extremely shrunken matrix does not contribute perceptibly to the thickness of an absolutely dry lamella. Thus it cannot be ascertained whether the amount of matrix has decreased corresponding to the increase in surface of the lamellae.

Setterfield and Bayley (1957) considered a third possibility. According to these authors, cellulose fibrils can grow not only in length, but also arise *de novo* in the outer lamellae of the secondary wall, far removed from contact with the biochemically and morphogenetically active plasmalemma. They substantiate this opinion with the discovery that the outermost lamellae themselves do not become thinner; that in secondary walls grown in the presence of isotopes, not only the innermost but also the outermost apposition lamellae show radioactivity; and that the extracellular microfibrils of *Acetobacter xylinum* also arise and grow without direct contact with the bacterial plasma (Colvin *et al.*, 1957). The reason why no reduction in the thickness of the outer lamellae can be observed in the electron microscope has already been mentioned. As regards the radioactivity of the entire cell wall due to tracer supplied during growth, this effect can just as well be an indication of a continuous permeation of the whole wall with fresh matrix, for on the secretion of mucilages (cutin, see p. 315) through the epidermis, such materials must migrate across the wall. In any case, there is as yet no evidence that the induced radioactivity of the outer cell wall is caused by a neoformation of microfibrils in the peripheral lamellae. Comparison with the formation of extracellular bacterial cellulose is also unconvincing, for those microfibrils undergo no sort of orientation, but grow completely at random on the surface of the nutrient solution. The most important characteristic of fibrillar textures in the plant cell wall is that, in view of the future function of the wall, they become oriented accordingly in the zone of contact with the plasmalemma. This happens, however, only in the innermost wall lamella, while the secreted microfibrils in the primary wall and in the outer secondary wall lamellae behave completely passively. They are inertly displaced and re-oriented in those lamellae according to the laws of flow of viscous fluids, and in this way completely miss any active morphogenetic impetus. It is hard to see how a morphogenetic process observed in the vicinity of the plasma could continue after the plasma had completely lost control over the shaping of the texture. It had to be assumed that the cellulose precursor transfer mechanism discovered by Colvin (1964) in *Acetobacter xylinum* cultures also acts in the cell wall, and that new fibrils are then aligned by the texture of the already existing microfibrils. A whole series of hypotheses must thus be proved before we can raise the new formation of cellulose fibrils in the outer cell wall lamellae to the status of a principle of the growth of the wall.

According to Frei and Preston (1961), the outer plasma layer is responsible for the formation of the cellulose fibrils. By their freeze-etching method, Moor and Mühlethaler (1963) have found globular macromolecules in the plasmalemma of yeast (Fig. E–9), which appear to be somehow involved in the production of glucan fibrils and which show a hexagonal arrangement similar to that in the interior of the bacterial membrane of *Spirillum* (Houwink, 1953). Green (1962, 1963) established that the active fibrillar orientation can be cancelled by the cell poison colchicine; he thus assumed that a protein, resembling the spindle protein, is responsible for the alignment of the cellulose microfibrils in contact with the plasmalemma.

5. INCRUSTATION AND ADCRUSTATION

(*a*) *Lignification*

The structure of full grown cell walls can be stabilized by incrustation with lignin. Their strength is improved by this process. In addition to their tensile strength due to the cellulose microfibrils, they gain considerable compressive strength. It is an important phylogenetic fact that lignin appeared for the first time in connection with the transition from aquatic to terrestrial plant life, when higher mechanical requirements were demanded from the supporting system.

Lignified cell walls display an ultrastructure comparable to the structure of

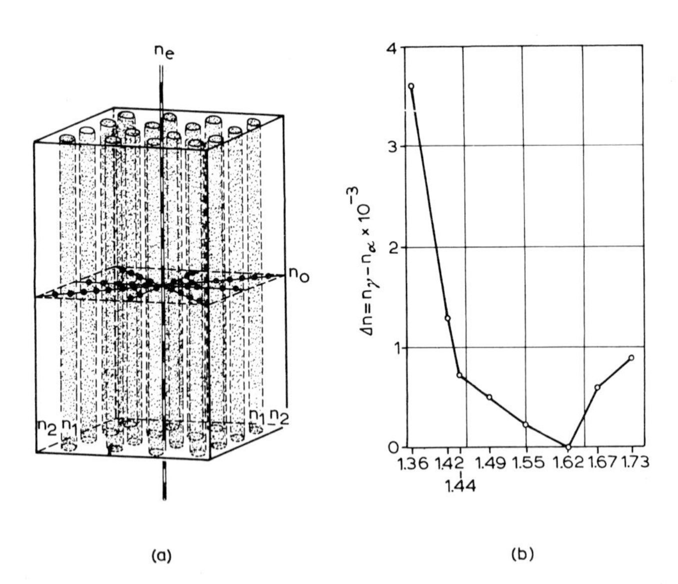

Fig. J–22. Rodlet form birefringence; (a) rodlet composite body (Ambronn and Frey, 1926): n_o ordinary, n_e extraordinary beam of composite body, n_1 refractive index of rods, n_2 of imbibition liquid; (b) experimental form birefringence curve on radial sections of lignified ghosts from spruce tracheids; ordinate: birefringence Δn, abscissa: refractive index n_2 of imbibition liquid.

reinforced concrete (see Fig. J–22a), the microfibrils having properties similar to the iron rods whilst the lignin, which replaces the original highly plastic matrix, serves as an elastic ground substance with a high compressive strength. The two components cannot be distinguished in the electron microscope because their powers to diffract electrons are equal. The lignified cell wall therefore appears homogeneous, as it was in the unlignified state when the microfibrils embedded in the matrix were also invisible.

Nevertheless, if one of the two components is destroyed, the other becomes visible: the fibrillar frame after decomposition of the lignin with chlorine, and the porous ground substance after dissolution of the cellulose with cuoxam (Mühlethaler, 1949). The porous lignin mass is however shrunk by the destruction of the major component, so that it is impossible to recognise details of its original structure. There is an old controversy whether lignin is inserted between the microfibrils in an oriented or in an amorphous state. Since no direct evidence can be obtained in the electron microscope, this question has been solved by indirect methods (Frey-Wyssling, 1964a, b).

The lignin ghost left over when the cellulose is removed from a lignified cell wall is birefringent. Since it is a template of a rodlet composite body (Fig. J–22a), the empty spaces occupied previously by microfibrils can be filled with different imbibition liquids (Freudenberg *et al.*, 1929). If such a liquid has the same refractive index as lignin ($n_D = 1.61$), so that (in Fig. J–22a) $n_1 = n_2$, the birefringence disappears (Fig. J–22b), proving the absence of intrinsic anisotropy caused by some molecular orientation. The birefringence of the system is due only to the special shape and orientation of its capillaries with bores smaller than the wavelengths of light rays; we are therefore dealing with rodlet form birefringence (see p. 252). Since lignin ghosts display mere form anisotropy, it was concluded that the lignin incrusting cell walls is in an amorphous state.

In 1944, however, Lange discovered striking UV dichroism in lignified cell walls. Since cellulose and other wall carbohydrates do not absorb ultraviolet, he attributed this dichroic effect to lignin and assumed an oriented adsorption of its molecules by the cellulose microfibrils, similar to the dichroism induced in cellulose fibres (ramie, linen, cotton) by iodine or direct cotton dyes. Such dichroic staining is due to the intrinsic absorption anisotropy of the dye molecules lined up along the surface of cellulose. By analogy, an intrinsic UV dichroism was ascribed to the lignin molecule.

It can however be shown by imbibition experiments that most of the observed UV dichroism in lignified cell walls is form dichroism, *i.e.* it is not caused by molecular orientation but by the arrangement of the microfibrils (H. P. Frey, 1959).

There is a third anisotropic effect in lignified cell walls, called difluorescence. The blue fluorescence emitted by lignin, when lignified cell walls are illuminated with ultraviolet, is slightly polarised, the intensity being greater parallel than perpendicular to the cell axis. This difluorescence may again be intrinsic or caused by the ultratexture. On the basis of imbibition experiments and their interpretation by the theory of composite bodies, it can be proved to be a form effect (Hengartner, 1961).

The absence of a striking orientation of lignin in the interfibrillar spaces of the cell wall is thus demonstrated by various items of indirect optical evidence. This also explains why lignin does not give any clear X-ray diffraction spots. The result is remarkable insofar as there are possibly chemical links between the oriented cellulose chains and the amorphous lignin. The three-dimensional polymerisation of lignin, urging its molecules to grow in all possible directions of the space available in the rodlet composite body, probably leads to a statistical isotropy of the anisotropic monomer, coniferyl alcohol.

A further question concerns the distribution of lignin in the different wall layers. With qualitative staining reactions, an accumulation of lignin in the primary wall is observed, whilst the secondary wall is less lignified. Quantitative photometric analysis in the UV microscope yields absorption curves of the type shown in Fig. J–23a. From this, it was concluded that the lignin content decreases from a maximum in the primary to a minimum in the outer secondary wall (Lange, 1954). This would characterise S_1 as a transition layer not only in a morphological but also in a chemical sense.

Similar maxima curves are obtained when the intensity of the lignin fluorescence is measured (Fig. J–23a). In this case it can be shown that the slope of the curve is caused by diffraction along the non-homogeneity of the phase boundary between the primary and the secondary wall. A model of the lignin distribution as seen in Fig. J–23b yields a calculated intensity curve exactly as measured in the densitometer (Ruch and Hengartner, 1960). The lignin content thus seems to be constant throughout the whole secondary wall of the investigated object (jute fibre).

If the lignin distribution is as indicated in Fig. J–23b, its content can be calculated from the known UV extinction coefficient of dissolved lignin at 280 nm. For jute fibres, which are only slightly lignified, extinction measurements in the UV microscope yield a value of 3.4% lignin throughout the secondary wall and 7.3% lignin in the primary wall (Ruch and Hengartner, 1960).

We may ask why the primary wall becomes more than twice as heavily lignified

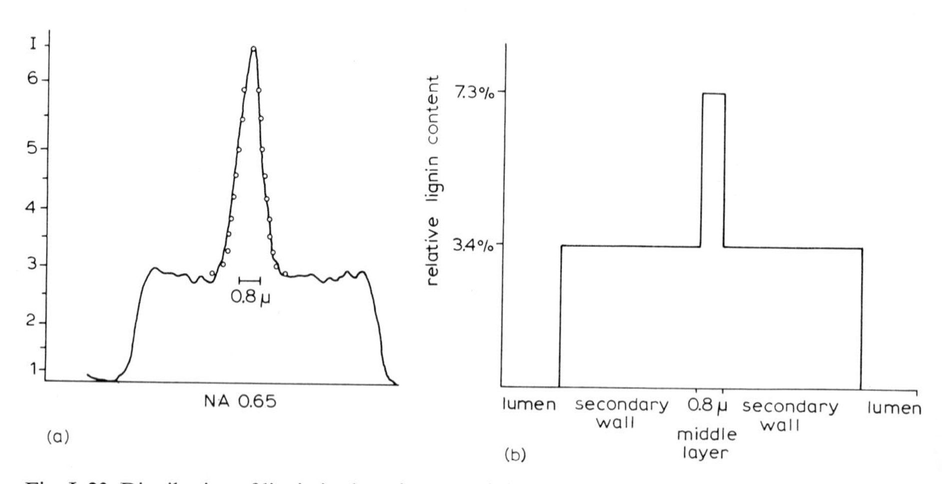

Fig. J–23. Distribution of lignin in the primary and the secondary wall of jute fibres; (a) intensity (I) of lignin fluorescence shown by densitometer curve across the wall of adjacent cells, ○ calculated values based on diffraction at the phase boundary primary/secondary wall; (b) actual lignin distribution, without obscuration from diffraction effects (Ruch and Hengartner, 1960).

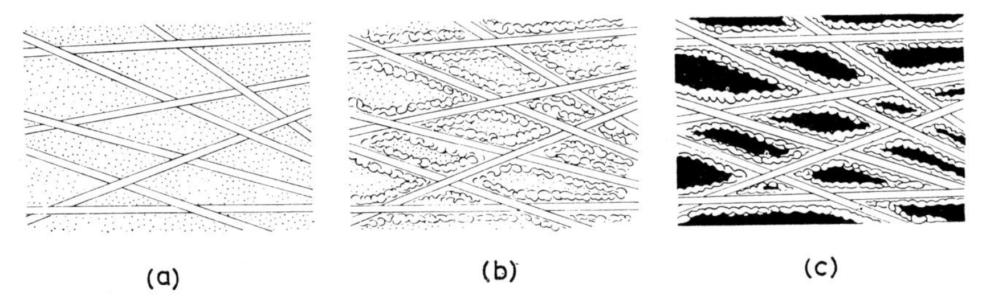

Fig. J–24. Incrustation of the cell wall; (a) fibrillar frame work with interfibrillar matrix, (b) incrustation of lignin, some matrix is left (loss of extensibility), (c) further incrustation by phenols and/or mineral salts (hardening) (Frey-Wyssling, 1959).

than the secondary wall. A space problem is probably involved. Since the primary wall is poor in cellulose, a much greater amount of lignin is necessary to produce in it a solid mass of fibrillar and interfibrillar material, than in the secondary wall with its high percentage of parallel cellulose microfibrils. Since lignification starts in the middle lamella and the primary wall, and because the lignin formed replaces the original matrix, some wood chemists considered the possibility of a transformation of hemicelluloses or pectin into lignin. This does not take place. The volume of the matrix substances is reduced by dehydration and their amount by chemical degradation (Fig. J–24b). If the space requirement of the polymerising lignin is higher than the shrinkage of the matrix, the wall can swell up irreversibly during lignification.

Lignification proceeds for some time. When it comes to an end, there is still some space left containing certain residues of the matrix and of water of hydration. This space can give rise to further incrustations with polymerising phenols (p. 289) in the cell walls of heartwood (Fig. J–24c) or with mineral substances, especially in epidermal cells. Both these types of incrustation can be combined, resulting in a hardened cell wall with a high specific gravity and a reduced stainability.

(*b*) *Silicification*

Heavily mineralised cell walls leave behind ghosts of ash on suitable incineration. The mineral residues obtained are known as spodograms. These are especially stable if the tissue had been silicified. In this way the cell walls of barley awns furnish a siliceous aerogel which shows the form birefringence of a rodlet composite body (Frey, 1926).

Siliceous earth represents similar mineral residues of diatoms, which display a variety of beautiful porous microstructures in their cell walls. No microfibrils have as yet been detected in those walls so that silica must be considered as their skeletal substance. As a result, diatoms cannot be cultivated without silica, which is possible with higher silicophilic plants whose cell walls are only facultatively silicified.

The matrix consists, as in other cell walls, of a gel containing pectin and carbohydrates, which is reinforced by the deposition of silicic acid. In contrast to cellulose and chitin, silicic acid does not polymerise linearly to a chain lattice, bu to a three-dimensional amorphous system which causes a foam structure visible in the electron microscope (Helmcke, 1954). The foam capillaries have a diameter of under 100 Å. A chemical combination between silicic acid and the carbohydrates of the matrix is suspected (Jørgensen, 1953) but we know as little of its nature as of the postulated linkage between cellulose and lignin.

The cell walls of siliceous algae are of special importance for the investigation of the resolving power of light microscopes. The pore pattern can today be resolved

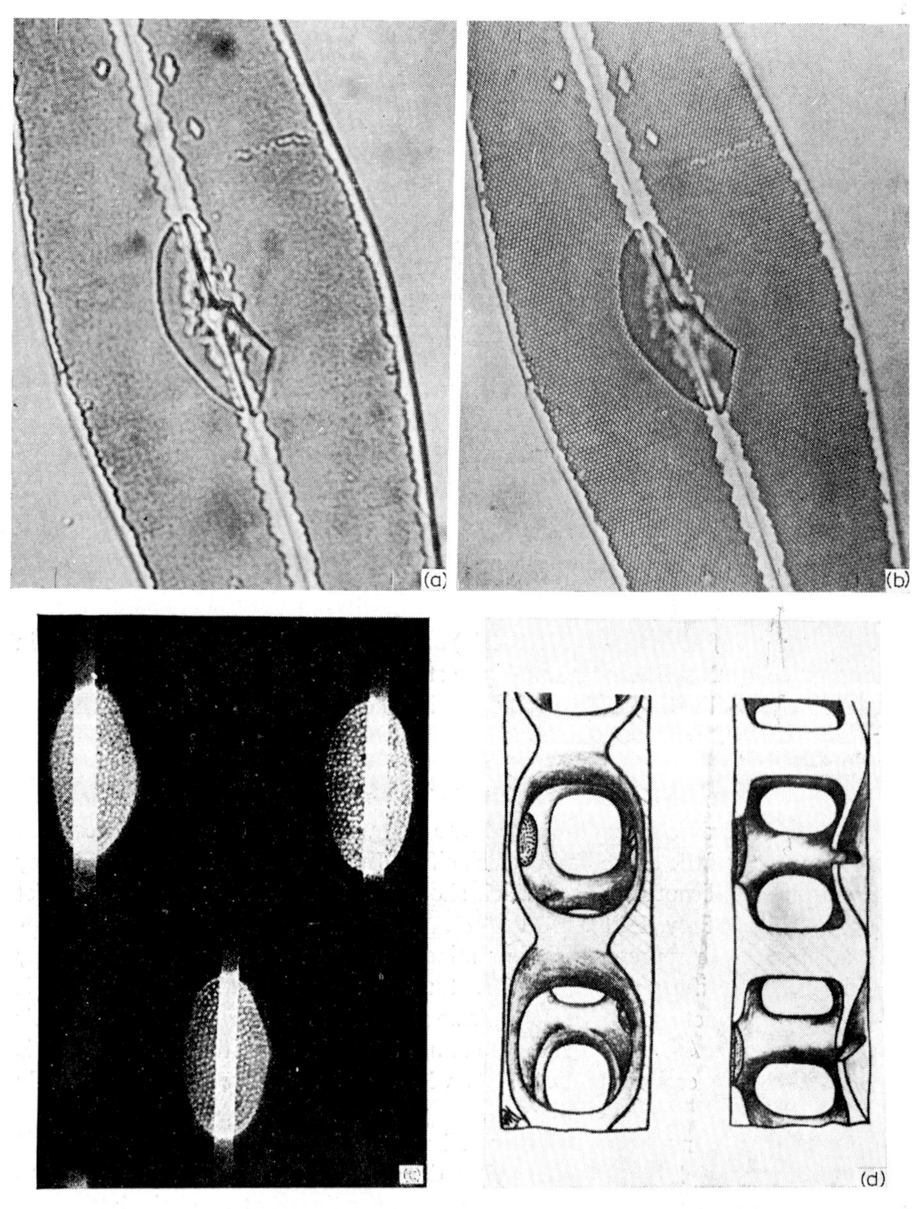

Fig. J–25. Fine structure of the *Pleurosigma* cell wall; (a) no resolution with light microscope of aperture 0.65; (b) pores visible with aperture 0.85; (c) sieve pores and slits visible in electron microscope, (d) space model (c and d after Müller and Pasewaldt, 1942).

without difficulty in the electron microscope, disclosing a remarkable diversity of sublight-microscopic pores and sieve plates (Kolbe and Gölz, 1943; Mühlethaler and Braun, 1946; Helmcke, 1953/54). Pfitzer (1882) had already established that the porous wall of the diatom shells of *Triceratium* is not solid but holds caverns whose outer and inner pores show different pore types. In the electron microscope the existence of void cavities in the diatomaceous membrane appears to be a general feature.

In *Pleurosigma angulatum*, the test object for light microscopes, (Fig. J–25a, b) the inner lamella exhibits slits; in contrast, the outer lamella is perforated by oval windows closed by sieve plates. The large depth of focus of the electron microscope permits both pore systems to be sharply represented in superposition (Fig. J–25c). With stereoscopic photographs, the interval of the two lamellae can be determined, so that a model of the fine structure of the wall can be constructed (Müller and Pasewaldt, 1942). This shows how the outer and the inner membranes with their different type of pore design are bound together by pillar-shaped struts which partition the ultrastructural space between the two lamellae into cavities (Fig. J–25d). Photogrammetry is used for the evaluation of such electron microscopic stereopictures (Helmcke and Richter, 1951), and this method has yielded remarkably precise models of the fine structure of various diatoms (Helmcke and Krieger, 1952). As a rule, the outer pores in the roof of the chamber system, which lie at the limit of the resolving power of the light microscope, are, as in *Pleurosigma*, closed by sieve plates with very fine pores.

How the ultrastructural cavities come to be left open during the deposition of the diatom wall is unknown. It is also not known whether cytoplasm flows into these cavities in the living state, or whether they are only accessible to aqueous solutions. Helmcke and Krieger (1952) assumed the sieve plates with their pores of only 150 Å diameter to be impermeable to the cytoplasm, so that the extracellular plasma characteristic of the diatoms could thus only pass out through the coarser openings of the raphes.

(*c*) *Suberisation*

The walls of periderm cells include a layer of suberin. This takes on a yellow colour with concentrated potassium hydroxide solution and is more resistant to strong oxidising agents than the remaining wall layers (v. Höhnel, 1877). Suberin shares its microchemical properties with cutin, although in contrast to the cuticle, the suberin lamellae can be dissolved in boiling aqueous alkali. Moreover suberisation takes place inside and cutinisation outside the primary wall.

The following is known concerning the ontogeny of suberised cell walls. The cells deposited by the cork cambium grow very quickly to their size without a suberin lamella becoming detectable (Mader, 1954). This lamella arises only when the cell is grown; inwards it is separated from the cell content by a carbohydrate lamella. After the formation of the suberin lamella the cell soon dies and, as in bottle cork, becomes filled with air, or excreted substances can be deposited in it, like the triterpene betulin in birch cork.

In the polarising microscope the suberin layer, like the cuticular layers, proves to be optically negatively birefringent with reference to the tangential direction of the cell. This birefringence disappears when suberin sections are heated in glycerol,

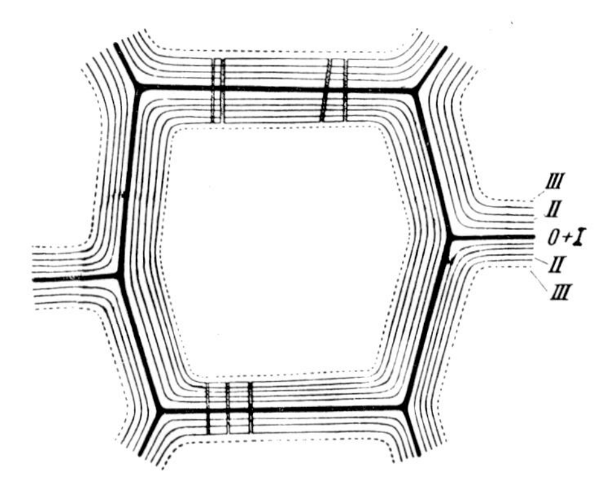

Fig. J–26. Suberised cell wall of cork; O–I middle lamella + primary wall, II secondary wall with suberin lamellae, III tertiary lamella with cellulose fibrils (Sitte, 1954).

and reappears again on cooling (Ambronn, 1888). It is due to embedded molecules of cork waxes oriented radially in the cell wall.

Using the electron microscope, Sitte (1954, 1955) found a primary wall containing cellulose microfibrils, which the suberin layer superimposes in the form of fine lamellae. It contains no framework substance, so that, in contrast to lignin, suberin is not an incrustation but an adcrustation. Ultrafine pores were found in the suberin layer of bottle cork, which were blocked by a substance alien to suberin. They represent obstructed plasmodesms, which functioned during the apposition growth of the laminated suberin layer and subsequently became closed. A tertiary lamella with cellulose microfibrils forms the closure from the cell interior. This lamella is absent in bottle cork, which must however be regarded as an exception which may be connected with the fact that the cork of *Quercus suber* used for technical purposes behaves somewhat irregularly owing to its very fast growth (Fig. J–26).

The investigations under discussion show that a middle layer consisting of middle lamella and primary wall is present in suberised cell walls. This is weakly lignified, as shown by the phloroglucinol reaction. Afterwards, the suberin layer forms as a secondary wall, by the apposition of very thin lamellae which contain, however, no framework substance, so that the suberin must be considered as an independent wall material. The final lamella of carbohydrates, on the other hand, again contains microfibrils; it is to be termed a tertiary wall. Since the tertiary lamella is explained as the atrophied membranogenic layer of the protoplast (p. 283), we should not be surprised that before the termination of its activity it still formed carbohydrates or, as in bottle cork, exhausted itself completely in the production of suberin. Mader (1954) overlooked the primary wall in the middle layer, and thus explained the tertiary lamella as the primary wall. There is something in this conception, insofar as suberin, like cutin, is then secreted outside the primary wall. Since, however, Sitte (1955) has detected a cellulose-containing disperse textured lamella in contact with the middle lamella, a distinction between the secretion of suberin and cutin appears to exist on the grounds of this ontogenetic evidence.

The significance of suberisation is seen generally in the formation of an efficient guard against transpiration. Actually, the impermeability of the suberin walls to water is so great that suberised cells die shortly after production of the cork lamella. Moreover, cork is an excellent thermal insulator, so that the vital phloem under the periderm appears to be very effectively protected against desiccation and thermal damage by the heat of the sun.

(*d*) *Cutinisation*

The cutin substances are found in pure form in the cuticles, and are mixed with other wall materials in the underlying cuticular layer of the epidermal cells.

With the formation of the cuticle, procutin is secreted through the outer wall of the differentiating cell, as a semi-solid material. The following chemical conception concerning the exudation of procutin may be made. According to Chibnall and Piper (1934) unsaturated fatty acids are the precursors of the hydroxy-fatty acids, saturated fatty acids, and paraffins of the plant waxes. If we imagine for example oleic acid or linoleic acid as the starting compounds for the C_{18}-chains, then these could pass through the cell wall as water-soluble soaps. Since the pasty secretion cannot be scraped off the leaf surface, the procutin is probably covered by a thin lamella of other wall substances. As a rule, it is produced in excess so that ridge-like folds arise, which then become passively straightened during the growth in area of the cell wall. Their procutin content can be interpreted as a reserve for the surface increase during the great growth period of the cell. After cessation of growth, the cuticular material gradually hardens to a solid film on atmospheric oxygen, by oxidation and polymerisation, so that the folds lose their plasticity and produce rigid cuticular ridges. The hardening of cutin can be compared with the varnish formation of drying oils.

The cuticle shows weak birefringence (Roelofsen, 1952) in the polarising microscope, but in contrast to the strongly birefringent wall layer lying beneath it, it often appears to be almost isotropic. In the electron microscope, it proves to be completely structureless, for no corpuscular or fibrillar elements can be established within it (Roelofsen and Houwink, 1951; Scott *et al.*, 1957). In this respect it behaves like the cork lamella in the periderm cells.

In many xerophytes the epidermis is coated not only by a cuticle but also by a more or less large continuous layer which jointly covers all epidermal cells. The cuticular layer of *Clivia* is represented as an example in Fig. J–27. It is distinguished from the cellulose epidermal outer wall lying beneath it by its lack of lamination, by its cutin reactions (*e.g.* yellow with zinc chloride-iodine), by strong UV absorption, and by its negative birefringence, referred to the tangential direction.

The cuticular layer consists of a mixture of polysaccharides and cutin substances. As no cellulose fibrils were found in the electron microscope, hemicelluloses must be present. In addition to the insoluble cutin substances, it contains extractable membrane waxes (see below). Between the cuticular and cellulose layers is an isotropic lamella which stains with ruthenium red and is thus termed a pectin lamella. It must be interpreted as a primary wall (I + O in Fig. J–27b). The cuticular layer can thus be compared with the polysaccharides deposited as a mucilage layer on the epiderm of aquatic plants, which are however strongly mixed with or replaced by cutin substances on terrestrial plants.

The outermost region of the cuticular layer can become regenerated after

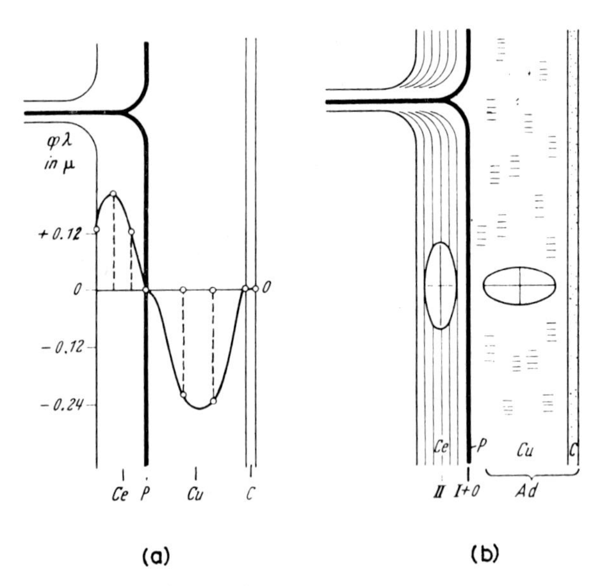

Fig. J–27. Cutinised cell wall, epidermal cell of a *Clivia* leaf; (a) birefringence (retardation $\varphi\lambda$) on longitudinal section in cellulose *Ce* and cutin *Cu* layer, *P* pectin lamella, *C* cuticula; (b) ontogeny; II secondary wall, I + O primary wall, *Ad* adcrustation (A. Frey, 1926).

artificial mechanical denudation but it then assumes cuticle-like qualities (Fritz, 1935). This points to the fact that the differentiated cells can only secrete the cutin substances, while evidently the remaining matrix-like wall substances of a normal cuticular layer can subsequently be supplied no more.

The cuticular layers are frequently anchored between the side walls of the epidermal cells (Fig. J–28b). These wedges sometimes push forward deeply between the epidermal cells. The cutinous substances can evidently be secreted through the entire surface of the epidermal cells, for cases have also been described in which these were cutinised on all sides, so that a so-called cuticular epithelium was found (Damm, 1901). If cutin is deposited in excess, swellings of the middle layer which show the cutin reaction, or even 'cutin cystoliths', are formed (Fritz, 1937).

Especially remarkable are the phenomena in the polarising microscope (Frey, 1926) between crossed nicols. The cellulose layer is positively birefringent with reference to the tangential direction of the epidermal cell, the pectic layer is isotropic, the cuticular layer negatively birefringent, and the cuticle almost isotropic. On suitable insertion of a first order red gypsum plate, the cellulose layer appears blue, the pectin layer red, the cuticular layer yellow, and the cuticle again red.

On heating the section in glycerol on a hot stage, the birefringence of the outer layer can be made to disappear (Ambronn, 1888). The cuticular layer then appears isotropic. When the section is cooled, the negative birefringence of the cutin layer again comes into force to its full extent. The liquefaction point of the fusible wall components between 50° and 70° C gives evidence of membrane waxes. They may be melted within the wall, and crystallise out again on cooling. If the phase differences are measured as a function of temperature, a hysteresis curve is obtained (Mad. Meyer, 1938) which indicates that a considerably higher temperature is

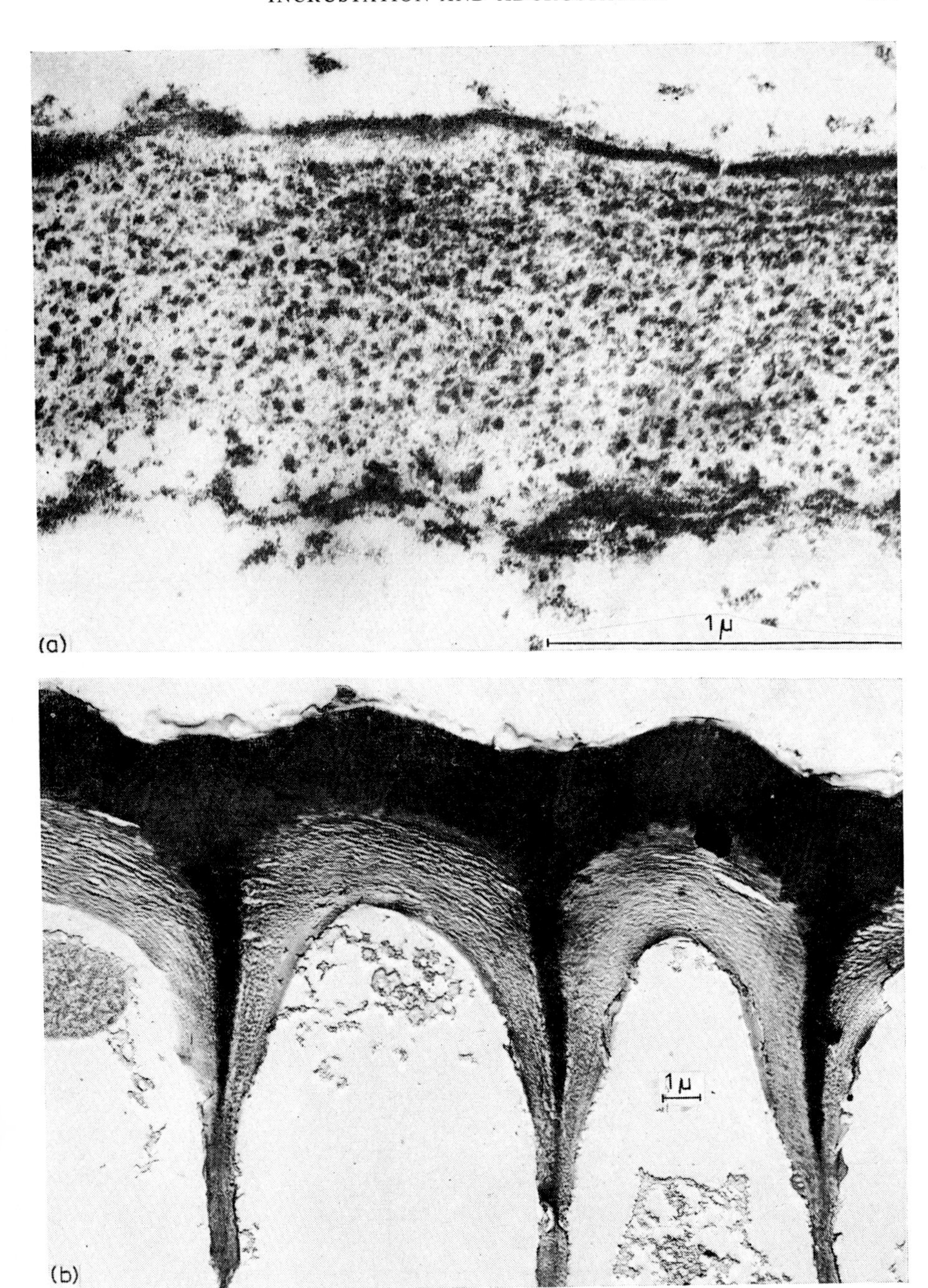

Fig. J–28. Cutinisation; (a) secretion of procutin across the epidermal cell wall, *Echeveria* leaf; (b) epidermal cells of *Clivia nobilis*; cellulose layer inside and cuticular layer outside, droplets of procutin visible at the boundary (Frey-Wyssling and Mühlethaler, 1959).

necessary for melting the waxes in the cell wall than for their crystallisation in the available ultrathin spaces.

The strength of the birefringence, measured by the retardation Γ (see p. 252), varies within the cuticular layer, being greatest in the central regions (Fig. J–27a). From this, we may conclude that the waxes are not distributed uniformly over the whole layer, but are concentrated in the middle. A radially directed orientation of the rod-shaped wax molecules follows from the negative birefringence (Fig. J–27b).

Ontogenetically, the translocation of the procutin across the young wall can be followed in the electron microscope (Frey-Wyssling and Mühlethaler, 1959). Fig. J–28a shows how the procutin droplets migrate through the matrix of the epidermal wall of *Echeveria secunda* to the outside. There is no cuticular layer in this object, and the cellulose layer is laminated as in *Clivia* (see Fig. J–27b). These droplets disappear in the grown cell, while the cuticle increases in thickness considerably, as we can see in Fig. J–28b, which shows the conclusion of the infiltration of cutin into the cuticular layer. It is seen that there are still some procutin particles between the outermost cellulose lamellae, which have not yet accomplished their migration. We do not know what force maintains this translocation of membrane substances, but it appears that a simple diffusion process is under consideration.

Similarly to the matrix, the cutin and the waxes are membrane substances which after their secretion behave passively (waxes crystallise, procutin becomes oxidised); thus their incorporation in the cell wall should not be compared with the formation and active textural organisation of the cellulose microfibrils, which we assume were synthesised in contact with the enzymes of the plasmalemma (see p. 306).

The strong ultraviolet absorption of the cuticular layer leads to the supposition that cutin absorbs short-wave light and in this way imparts to the leaves protection against UV radiation (Ursprung and Blum, 1917). Detailed examination showed however that because of a lack of any double bonds, cutin is transparent to ultraviolet (Wuhrmann-Meyer, 1941) so that many epiderms are unable to impart any UV protection. In most xerophytes, however, pale yellow flavone pigments having absorption maxima in the long-wave ultraviolet, are laid down in the cuticular layers (Bolliger, 1956). Only when the cuticular layer contains such UV dyestuffs as well as cutin is the striking impermeability to ultraviolet light observed.

More important than UV protection is perhaps the outstanding transpiration defence which is effected by the cuticular layer. As the cutin is not a pure lipophilic substance but contains small proportions of unesterified alcohol and carboxyl groups, its behaviour is of course not completely hydrophobic but more like that of a weakly polar material. It is therefore somewhat capable of swelling, so that small amounts of water are continuously lost through the cuticle, in the process of cuticular transpiration. In the cuticular layer however, this escape of water becomes, if not quite cancelled, considerably reduced, because of the wax deposit.

(*e*) *Sporoderm*

In pollen grains (microspores) and other spores, the cuticular layers are organised in a remarkably complicated way and show a surprising variety of forms. In the light microscope, the sporoderm is divided, like the outer epidermal wall of the xerophytes, into an inner layer of carbohydrates including callose called the intine, and a cutinised exine of sporopollenin deposited upon it. The exine of the pollen grains possesses a characteristic outer relief. There are warts, spines, areolations, etc., which characterise the pollen of certain plant families or genera. Thus for example verrucate pollens are characteristic of grasses, spiny pollens of Malvaceae and Compositae, and areolated pollens of Cruciferae.

On precise examination in the electron microscope, however, the exine frequently shows, in addition to its outer relief so important for pollen determination, a porous inner structure (Fernández-Morán and Dahl, 1952; Mühlethaler, 1953; Sitte, 1953) similar to that in the siliceous cell walls of diatoms (p. 310).

Differentiation of the exine into ectexine and endexine has been observed a long time ago. For the more recent theory of fine structure this classification is however too coarse. Thus Erdtman (1952) differentiates four different structural elements of the exine:

*S*culptured exine or *s*exine . . . . } ectosexine / endosexine

*N*on-sculptured exine or *n*exine . . . . } ectonexine / endonexine

The spatial relationships of these different lamellae are evident from Fig. J–29, taken from Erdtman. The continuous nexine is in general arranged in an often laminated (Afzelius, 1956b) basal layer (endonexine) and, laid above it, a somewhat thicker layer, the ectonexine. In contrast, the sexine shows the most varied

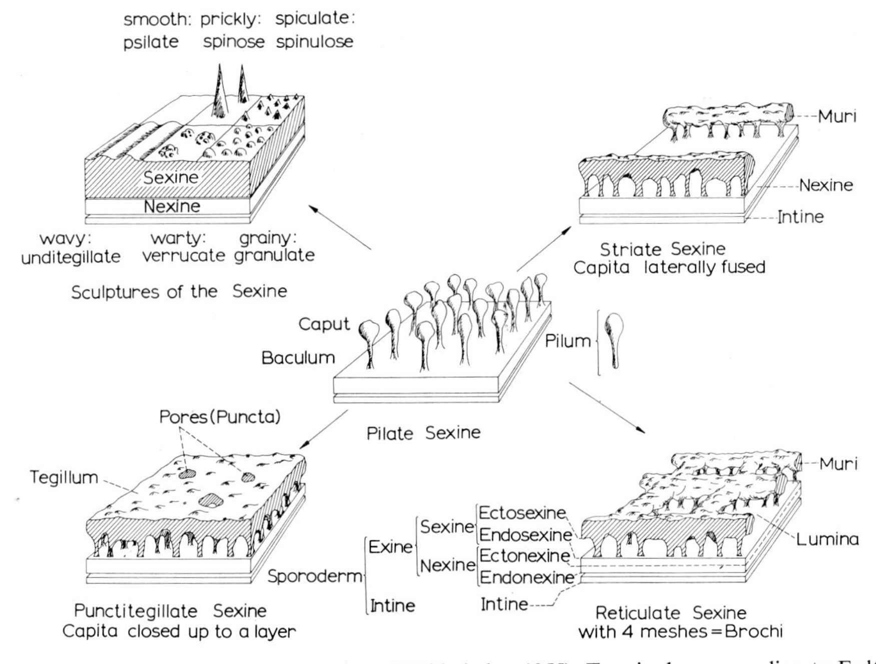

Fig. J–29. Fine structure of pollen exines (Mühlethaler, 1955). Terminology according to Erdtmann (1952).

relief. The diversity recorded by pollen analysts can best be understood if we correlate the different types, as in Fig. J–29, with one another (Mühlethaler, 1955); this is however by no means a phylogenetic but merely a morphological relationship. For this purpose, we can proceed from the so-called pilate type of sexine. In this case the sexine is resolved into a field of little spikes (pila), each of which consists of a shaft (baculum) and a small head (caput). The bacula correspond to the endosexine and the capita to the ectosexine. Complete fusion of all spikes with one another over their whole length gives a massive sexine, which by special development of the distal poles of the capita can be provided on its surface with various kinds of appendages; to this type belongs, for example, the spiny pollen of *Cucurbita*. If, however, only the heads fuse with one another, a baculated endosexine filled with air is formed, roofed over by ectosexine of litte columns (columellae) bearing a tegillum. Such tegillated exines are characteristic of the Fagaceae and the Betulaceae. If the warts and spines upon the tegillum do not lie in continuation of the bacula, as in the pollen of *Fagus*, they should not be considered as elongated heads of fused pila, but must be explained as additional structures of a special kind. The tegillum frequently shows pores; if these become large, a baculated endosexine bearing a reticulated ectosexine results, as in the honeycomb pollens of Solanaceae (like *Petunia*) and Gesneraceae (like *Aeschynanthes*). Finally, the pila may even be considered to be fused only in rows by their heads, giving rise to striped pollen with a striated sexine.

The pores of the ectosexine can be numerous and very fine (Fig. J–30a, c), and eventually even assume the character of plasmodesms which traverse not only the ectosexine but also the endosexine (bacula) and thus the tegillum and its columellae, and penetrate even into the ectonexine (Fig. J–30e). These small canals are not however true plasmodesms, for they terminate as blind ducts at the intine, and thus cannot be followed up to the cytoplasm of the microspore (Rowley *et al.*, 1959). On the same grounds, they can also have nothing to do with the so-called ectodesms of true epidermis, which advance from the inside of the cell up to the cuticle under which they stop (Schumacher, 1957). In fact, these pores and small canals of the sporoderm provide contact not with the intracellular plasma of the microspore but with the extracellular plasma of the tapetum cells which assumes the features of a plasmodium (Rowley, 1959). It is released after decomposition of the tapetal walls, then it surrounds the young microspores, and builds up from the outside the complex exine of the pollen grain. We can observe, for example, how spinules are set up on the ectosexine (Rowley, 1962). The plasmodium produces also interesting spheroids of sporopollenin, whose surface is furnished with the same type-specific fine structure as the ectosexine, although they represent only a functionless space-filling material around and between the tetraspores (Fig. J–30b and d).

⟶

Fig. J–30. Structure and ontogeny of the pollen exine; (a) coarse pores in the exine of *Phragmites communis*, black areas are raised portions (exinous spines); (b) fine pores between the spinules in the exine of *Zea mays*; (c) formation of exine by extracellular tapetal protoplasm *T*. Between pollen grains and tapetal plasmodium, spinose spheroids. Pollen grain to the right with sectioned germinal pore where exine absent. *L* loculi wall; (d) enlargement of (c), *S* spheroid, *I* intine; (e) exine of *Sorghum vulgare*; *c* channels through exine *E* and columella *co* along lacuna *la*, terminate short of intine *I* (Rowley *et al.*, 1959).

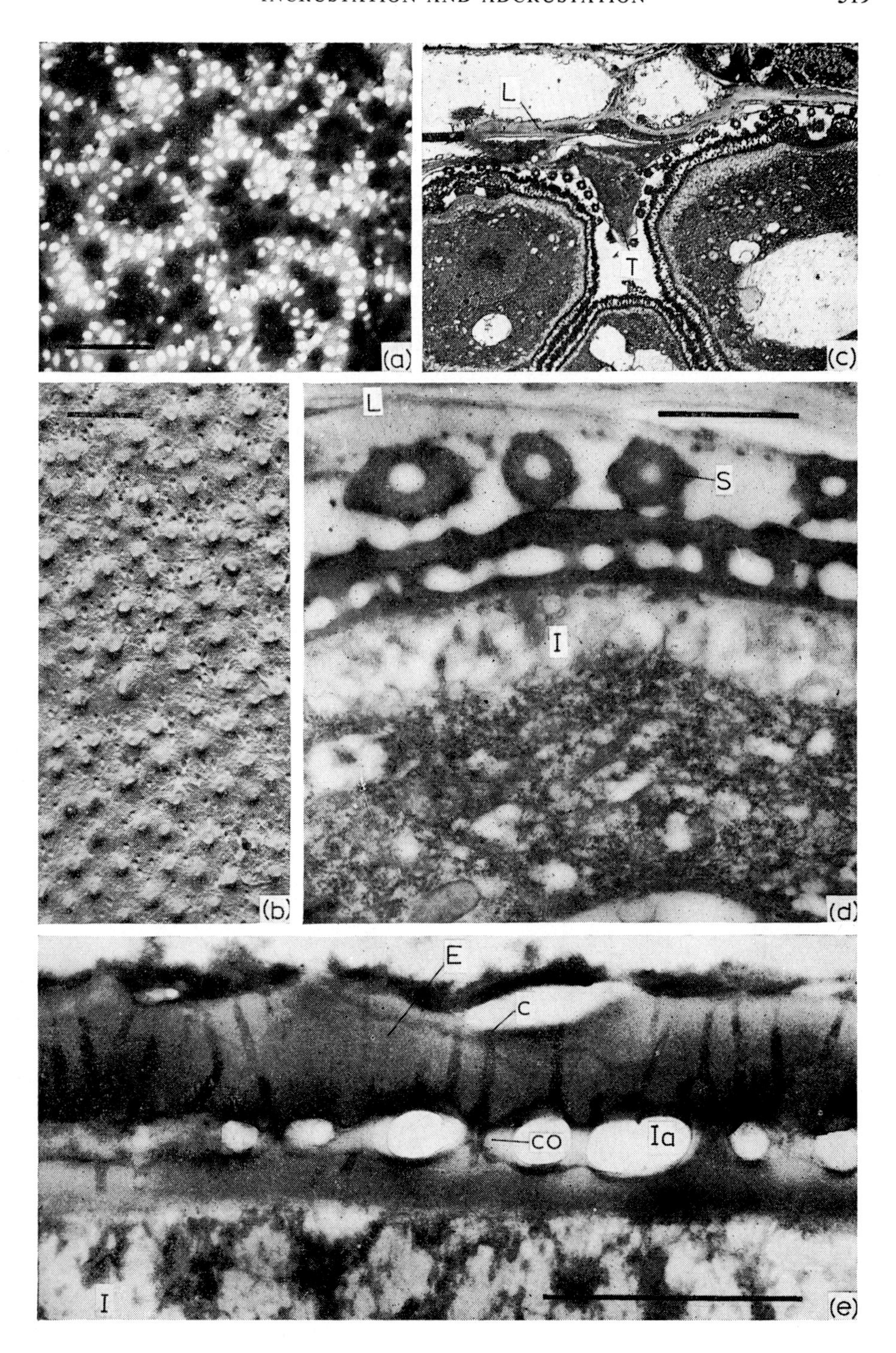
(a)
L
T
(c)
(b)
L
S
I
(d)
E
c
co
Ia
I
(e)

As the secretion of the exine occurs not by the haploid microspores themselves, but by the extracellular plasmodium, a normal fine structure of the sexine with all the type-specific details can be observed in pollen grains with induced male sterility which exhibit vacuolated plasma and a degenerated nucleus (Heslop-Harrison, 1958, 1962).

6. DIFFERENTIATION

(*a*) *Simple pits and perforations*

Adjacent protoplasts remain in mutual contact through the plasmodesms left open in the cell plate, as long as the primary wall is not covered by secondary apposition layers. There are in fact several cases, especially in thick-walled endosperm cells, where the plasmodesms permeate the entire thickness of the wall, which led to the discovery of the plasmodesms. Usually, however, the exchange of material between neighbouring cells through the secondary wall is made so difficult that thin regions of the wall must be left open. These are called pits.

In the light microscope, the cell communications known as simple pits consist of a canal divided into two pit cavities by a pit membrane.

It was a problem to the classical cytologists how the two adjacent pit cavities were left open in the secondary wall, although the two cells are separated by the pit membrane. This problem has now been solved with the help of the electron microscope. It is known that where pits will later arise, the primary wall is penetrated by particularly numerous clusters of plasmodesms, which are later still visible in the pit membrane (Fig. J–31a). The cellulose microfibrils are deposited around the future pit (Fig. J–32) and plugs of cytoplasm on both sides of the pit area prevent the apposition of the secondary wall. Thus in this place the cell wall remains confined to the primary wall and a pit is differentiated.

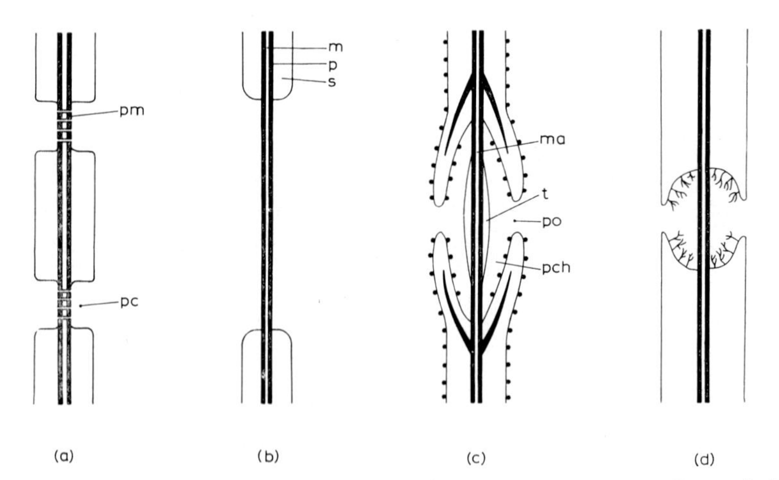

Fig. J–31. Diagrams of pits; (a) simple pit in metabolising cells, (b) fenestriform pit in ray parenchyma cells of *Pinus*, (c) bordered pit in tracheids and vessels, (d) vestured pit in xylem of certain broad leafed trees; *m* middle lamella, *p* primary wall, *s* secondary wall, *pm* pit membrane with plasmodesms, *pc* pit cavity, *pch* pit chamber, *po* porus, *t* torus, *ma* margo.

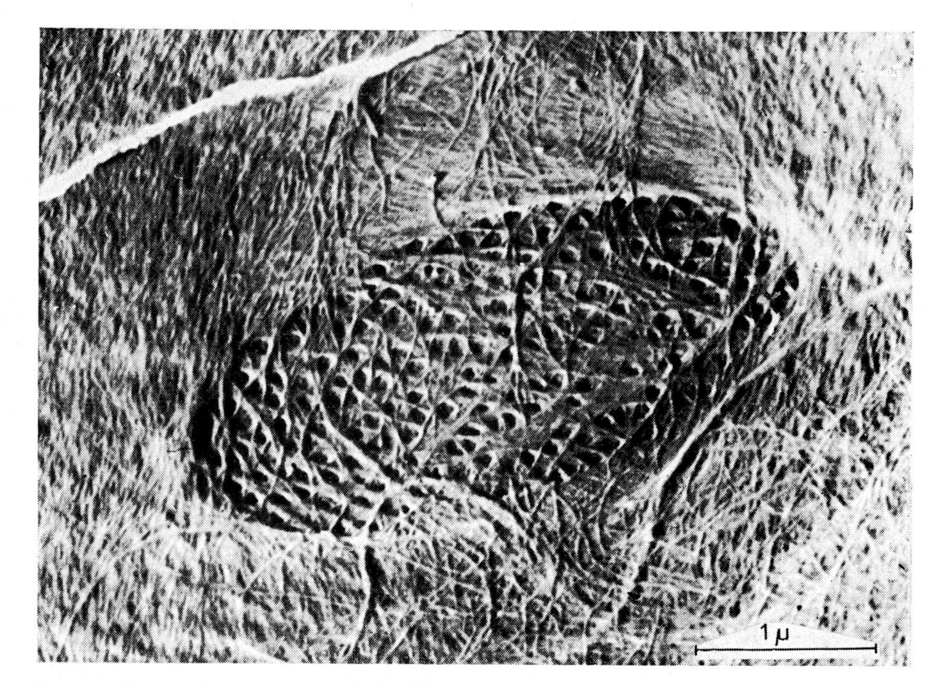

Fig. J–32. Pit membrane with plasmodesms in a meristematic cell of a young root of *Zea mays* (Mühlethaler, 1950a).

As the simple pits are predetermined by the clusters of plasmodesms, their number on the longitudinal walls cf extending cells is not increased during further cell differentiation (Wardrop, 1955; Scott *et al.*, 1956). On the transverse walls of parenchymatic cells which undergo no extension, the number of pits per unit area is therefore as a rule considerably larger than on longitudinal walls.

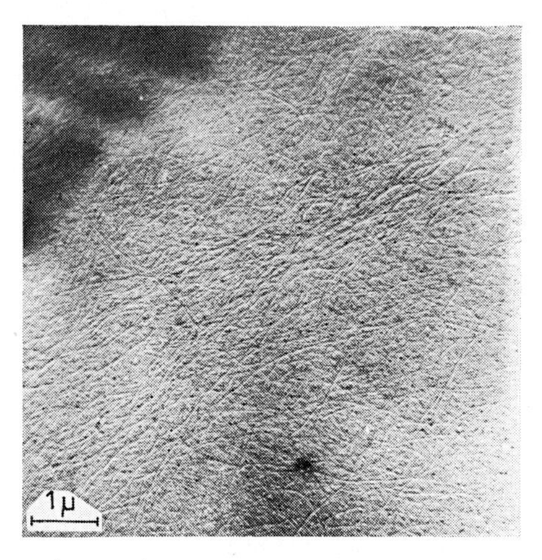

Fig. J–33. Fenestriform pit; membrane without plasmodesms (Frey-Wyssling *et al.*, 1956a).

In rare cases, as in the medullary ray cells of the *Pinus* species, the pit membrane coats the whole breadth of the cell; we then speak of fenestriform pits (Fig. J–31b). The same principle is followed here as in the angular collenchyma, where, too, the wall is only strengthened by apposition layers in the corner regions of the cell. As is evident from Fig. J–33, the pit membrane shows an ordinary disperse texture without plasmodesms.

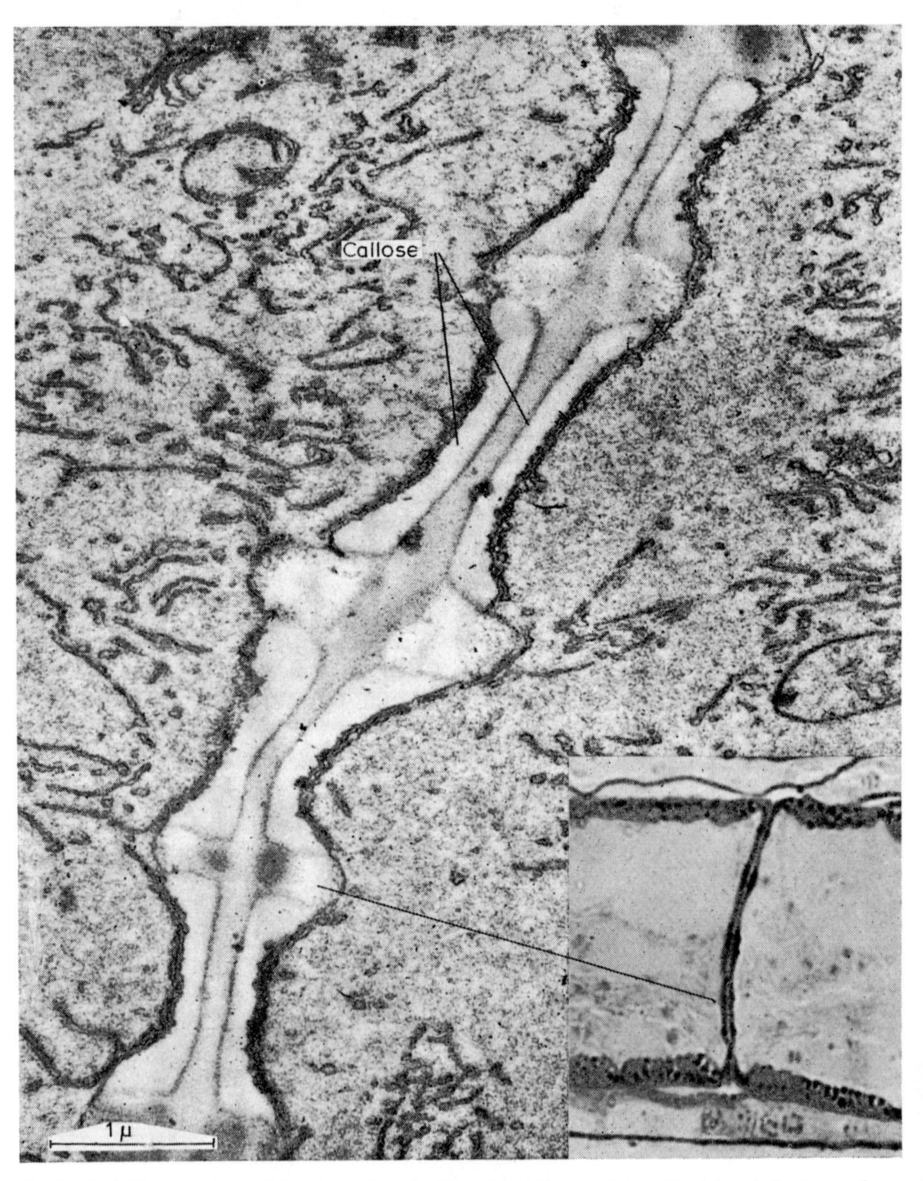

Fig. J–34. Differentiation of the sieve plate in *Cucurbita*. Comparison of light and electron micrographs. Formation of callose plugs instead of secondary walls at the place of future sieve pores. (Esau *et al.*, 1962).

When cells merge with one another (cell fusion), as is the case in the formation of vessels and articulated laticifers, the wall lying between the protoplasts, of the type of the pit membrane in Fig. J–33, become perforated or completely withdrawn. At this place the matrix of the wall is liquefied. It is not directly clear what happens in the meantime to the cellulose microfibrils for the enzyme cellulase is absent in higher plants. On this basis, already formed secondary walls remain, even when the cells concerned have become functionless (annular elements of protoxylem, heartwood). It may be assumed that the cellulose fibrils of the liquefying primary wall are mechanically pushed to the side and participate in the mechanical reinforcement of the circular marginal rim of the perforation. Fig. J–35 shows this process in the formation of sieve pores in sieve tubes.

(*b*) *Sieve pores*

In the formation of the large perforations in the transverse walls of the sieve tubes, the future pores are covered with callose platelets, and a close relationship is established between this adcrusting material and the endoplasmic reticulum (Esau *et al.*, 1962). Conspicuous bars of the secondary wall are visible between the callose plugs (Fig. J–34). Clearly the callose platelets hinder the apposition growth of the secondary wall at sites where a sieve pore is to be broken through the primary wall.

Pore formation begins with a funnel-shaped dissolution of the callose plugs from both sides and an initially narrow perforation of the primary wall where the funnels meet. Subsequently, the pore is widened to its definitive size. The microfibrils of the primary wall are combined with the bars of the secondary wall. As the middle lamella becomes disintegrated during maceration, in Fig. J–35 the cell walls of two adjacent sieve cells are displaced somewhat towards one another (Frey-Wyssling and Müller, 1957).

The communicating strands of plasma between the sieve cells of a sieve tube

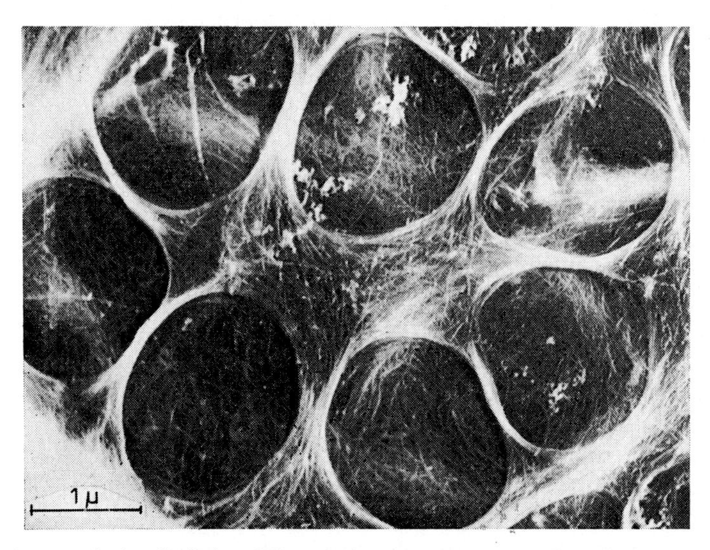

Fig. J–35. System of microfibrils in a differentiating sieve plate. The differentiation is less advanced in the adjacent wall seen below the pores (Frey-Wyssling and Müller, 1957).

are of great theoretical significance, for to some extent they obstruct the sieve pores through which the phloem sap must stream. The callose sheath coating the sieve pores, visible in the light microscope, also appears to be an obstruction to this flow, but it can be demonstrated that this narrowing of the pore canal is in fact a postmortal deposit of callose in injured phloem cells. Thus when the sieve tubes are fixed without sectioning the shoot, *e.g.* by injection of the fixative into the medullary cavity of *Cucurbita* stems, the sieve pores are seen to have no callose adcrustation (Eschrich, 1963). We can then see in the electron microscope that the axis-parallel ER strands which characterise the sieve tube plasma (Hohl, 1960) run into the plasma strand connecting the protoplasts of the tube members. According to Kollmann and Schumacher (1962), transverse sections, through a sieve pore, show that the plasma bridge is surrounded by the unit membrane of the plasmalemma, and in the interior of the strand one finds numerous small canals of the ER (see Fig. J–4d) comparable to plasmodesmata.

(*c*) *Bordered pits*

The recesses in the wall for the passage of sap from cell to cell are highly specialised not only in the phloem, but also in the xylem. In all cells which lose their living contents to serve as empty capillaries for water conduction, we find not simple but the so-called bordered pits. These possess a very large pit membrane, which is circularly overarched by a border (Fig. J–31c). The latter is originally laid down as a slender rim of the primary wall (Jutte and Spitt, 1963) from which the growth of the secondary wall starts. The border is separated from the pit membrane, and it can be shown that membranogenic plasma acts not only on the pit surface, but also under the growing pit border; for both the inner and the outer sides of the overarching border are covered with the warty tertiary layer (Fig. J–31c and J–36b) which we have explained as dried up plasma residues (see p. 283). The wart pattern of the bordered pits of the genus *Pinus* is specifically formed in different species, and a relationship exists not only between the warts and the striking dentiform wall sculptures of the ray tracheids (Frey-Wyssling *et al.*, 1956b) but also to the remarkable evaginations in the vestured pits of other types of wood (Côté and Day, 1962; Schmid and Machado, 1963). Hence it appears that in the wart formation of the tertiary wall a morphogenetic factor actively takes part in addition to the passive degeneration of the membranogenic plasma.

In bordered pits, the space known as the pit cavity in the simple pits is organised into a narrow entrance (porus) and a widened pit chamber (Fig. J–31c). The pit membrane is differentiated into a central cushion-like thickening (torus) and a thin marginal zone which is termed the margo. The aqueous flow translocated through the wall concerned is filtered through the margo, and is throttled if the torus is pressed against the porus. The bordered pits have, therefore, been explained as valve organelles. When however, the torus is pushed off from its middle position, as a rule, it covers the porus entirely (as observed in older tracheids) and then agglutinates so intimately with the pore margin that a permanent closure takes place. The main purpose of this arrangement is therefore to exclude tracheids of older year rings from the water conduction system.

The pit membrane in the bordered pits of deciduous trees possesses no torus (Fig. J–31d) and shows a disperse texture like that characteristic of the primary wall. In conifers, by comparison, the margo consists of radially directed, more or less fasciated microfibrils, which leave open slits between them (Fig. J–36a, b). This structure, which lies at the limit of the resolving power of the light microscope, had already been described by Russow (1883) and by Bailey (1913). The radial fibres holding the torus suspended in position, appear to be aggregated strands which are

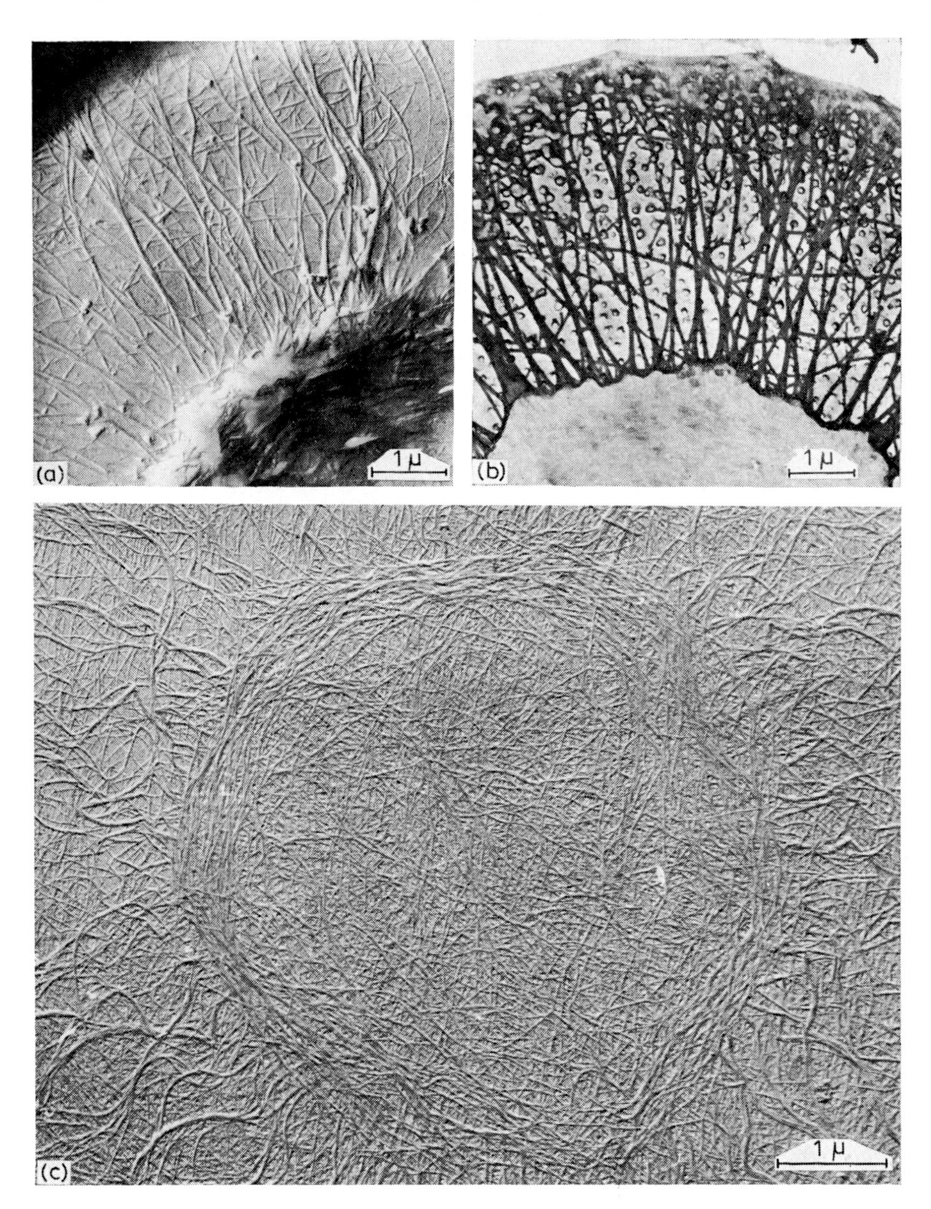

Fig. J–36. Bordered pits; (a) differentiated margo with fasciated microfibrils and circular texture of the torus; (b) view of the pit chamber; inner side of the border with warts, *Abies alba*; (c) first evidence of the torus and the radial strands of the margo in a differentiating pit of a young tracheid, *Pinus silvestris* (Frey-Wyssling *et al.*, 1956a).

anchored in the surrounding primary wall. Jayme *et al.* (1960) believe that the radial strands are artefacts which would result from a disperse textured margo subjected to tension forces when the preparation are dried. This argument is to be confronted by the fact that the bordered pits in green wood also are permeable to carbon and cinnabar particles (Bailey, 1913) and that the radial strands, on similar drying of different wood samples, show differing configurations. In sapwood, they are sublight-microscopically thin and yet widen in the course of the years into strands visible in the light microscope. Furthermore, ontogeny shows that radially directed microfibrils are laid down already during the differentiation of the cell wall (Fig. J–36c).

In the border we find, in contrast to the helical texture of the tracheids, a circular texture which is especially evident in the polarising microscope. The torus is also endowed with such a circular texture (Fig. J–36a), which is laid down very early, long before the pits start to function (Fig. J–36c). This texture of the borders was once explained as the result of circular stress during growth. As Fig. J–36c shows, however, during the differentiation of the bordered pits, microfibrils are produced simultaneously orientated tangentially in the torus and radially in the margo. From this it is clearly evident that the arrangement of the microfibrils is not induced by external forces (stress or strain) but by an active shaping force of a morphogenetic process (see p. 293).

(*d*) *Mucilaginous walls*

As has already been mentioned, the epidermis of aquatic plants secretes, as a rule, a hydrophilic layer of mucilage. The same is true of the root tips of terrestrial plants (Jenny and Grossenbacher, 1963), whose aerial organs are however protected by a hydrophobic cuticle. If a plant organ must be protected against the loss of water at a certain time but requires a water-accumulating environment for its development at a later time, as is the case with seeds, the secondary apposition layers in the inside of the epidermis can be developed as a mucilaginous wall. On moistening the epidermal cells, these are forced open by swelling of the slime, so that the seed appears to be coated with a macroscopic slime layer up to 0.5 mm thick. In certain seeds (*Cobaea*, Frey, 1927) the entire secondary wall swells up, while in other cases (as in linseed, Fauconnet, 1948) the mucilage originates chiefly from the extremely thick epidermal outer wall.

The histochemists distinguish between cellulose mucilage and pectin mucilage. The former is birefringent and gives the reaction with zinc chloride-iodine, whilst the second appears amorphous and, as uronides, can be stained with the basic pectin dyes (Ruthenium red, methylene blue). In the electron microscope the two types are easy to distinguish for the cellulose mucilages contain microfibrils (Mühlethaler, 1950b) while the pectin mucilages have no structure and are correspondingly softer than the cellulose slimes.

The cellulose threads can assume widths of 100 Å to 1 μ. Fig. J–37a shows the ultrafine fibrils from the mucilage of cress seeds (*Lepidium sativum*); in the mucilage of *Cobaea* seeds, on the other hand, they reach dimensions of macrofibrils (up to 5 cm long), so that they are visible in the light microscope. They show all the properties of strands having a parallel texture, *i.e.* birefringence with orthogonal extinction, pronounced dichroism when stained, slip planes, etc. In the epidermal

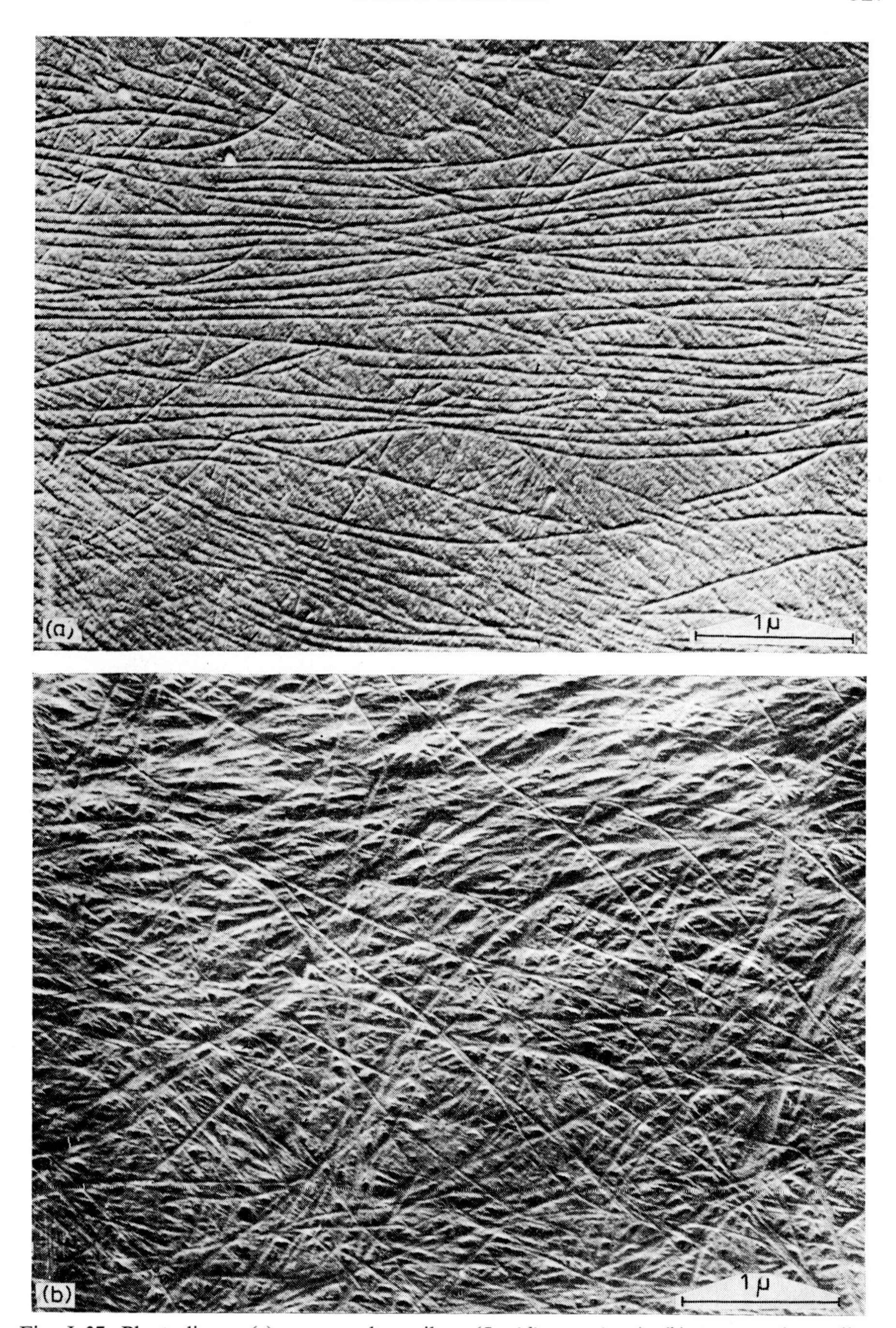

Fig. J–37. Plant slimes; (a) cress seed mucilage (*Lepidium sativum*), (b) tragacanth mucilage (Mühlethaler, 1950b).

cells these bands form closed secondary wall lamellae with a helical texture. The matrix laid down between them swells so forcibly in the presence of water that the whole cell lumen fills up and the cell is blown open. The mucilaginous content is then discharged to the outside, whence the helical strands of the secondary wall are torn out and unwound. In tragacanth mucilage (Fig. J–37b), which arises by an analogous choking with slime of the pith and primary ray cells, we can clearly recognise how the cellulose microfibrils of this mucilage were organised in a parallel textured layer in the now broken down secondary wall.

The ability of the matrix to swell so strikingly indicates that in such cell walls it consists of considerable amounts of unesterified galacturonic acid, with an especially large capacity for hydration (see Frey-Wyssling, 1959, p. 150 ff.).

The cellulose mucilages described show the great importance of the matrix for the physical properties of the cell wall. According to the quantity, capacity for hydration, and state of hydration of this ground substance, the system matrix + cellulose fibrils can produce soft slimes, firm mucilages, plastic primary walls, or elastic secondary walls.

K. Ectoplasmic Differentiations

Owing to the alloplasmic cell wall coating the plant protoplast, its ectoplasmic layer does not as a rule show special differentiations, such as pseudopodia (Gr. *pseudēs* = false, *pous* = foot), cilia (Lat. *cilium* = eyelash), flagella (Lat. *flagellum* = small whip) etc. Nevertheless, certain Protophyta, especially the Phytoflagellatae and the Spermatozoa of lower plant phyla, display undulipodia (Lat. *unda* = wave and additional organelles correlated to them. The ultrastructure of these differentiations has yielded important clues towards a better understanding of the evolutionary relationships of the plant and animal kingdoms.

1. UNDULIPODIA

There is no principal difference between the cilia and the flagella other than their length; a cilium measures 5–10 μ, a flagellum up to and over 150 μ. Moreover, the cilia are often more numerous than the number of flagella per cell, but their ultrastructure is alike. The collective name of undulipodia is therefore appropriate.

Undulipodia have a cylindrical shaft approximately 0.2 μ in width. They invariably hold 11 proteinaceous microfibrils, nine of which form a circle around a central pair. The nine peripheral fibrils are doublets, with two subfibrils, and the two central simple ones appear enclosed in a common sheath with a helical texture (Fig. K–5, *s*). They are embedded in a contrastless matrix which is enclosed by a membrane. Standard sizes of these structures according to Fawcett (1961) are given in Fig. K–1a. The figure refers to animal cilia, but Manton and Clarke (1952) had previously found the same ultrastructure in the spermatozoid tails of *Sphagnum*

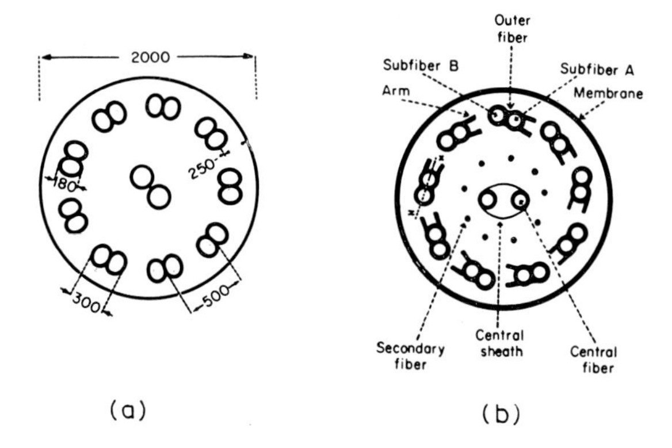

Fig. K–1. Ultrastructure of undulipodia; (a) diagram illustrating disposition and approximate size of fibrils and subfibrils in cross-section (Fawcett, 1961), (b) diagram of a flagellum with secondary fibrils (Gibbons and Grimstone, 1960).

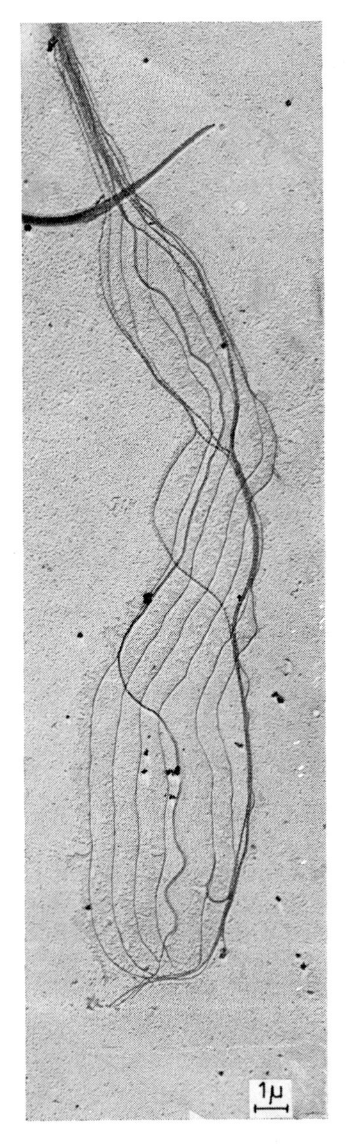

Fig. K–2. Flagellum of a *Sphagnum* spermatozoid decomposed in its 11 fibrils (Manton and Clarke 1952).

(Fig. K–2), Phaeophytes, and Pteridophytes (Manton, 1956). It is now established that all flagella, from the flagellates up to the mammalian sperms, belong to the 9 + 2 stranded type (Fawcett and Porter, 1952).

In 1960, Gibbons and Grimstone found that the doublets are furnished with arms on one side (Fig. K–1b) and that in the ground substance, which is poor in contrast, nine dark points appear in the cross-section; these were explained as secondary fibrils. The flagellar membrane passes over a deep fold in the plasmalemma (Fig. K–5, *fm* and *cm*); this is a unit membrane, whose dark stratum on the

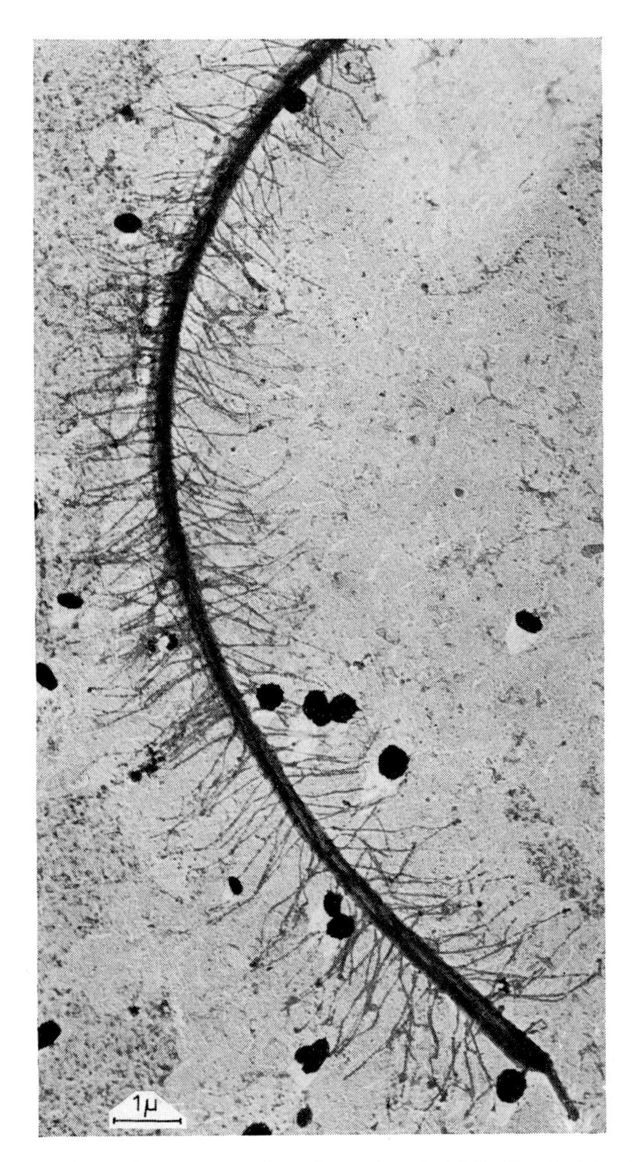

Fig. K–3. Hairy flagellum of a zoospore of the phaeophyte *Pylaiella litoralis* (Manton and Clarke, 1952).

plasma side is however, in *Pseudotrichonympha*, twice as thick as in the plasmalemma (40 Å instead of 20 Å). The dimensions of the three strata, measured from outside towards inside, are 20 + 30 + 40 = 90 Å.

Manton (1955) attempted to bring the symmetry of the eleven-stranded pattern into a relationship with the physical symmetry of the flagellated organism, and found, for example, in *Fucus*,that the frontal flagellum, related to the plane of symmetry of the bilateral spermatozoa, displays the orientation of the pattern of Fig. K–1b; *i.e.* the plane of symmetry of the body lies perpendicular to the line joining the central fibril pair and it passes through the odd median fibril situated outwards on the peripheral ring.

In certain algae the flagella are hairy (Fig. K–3) and are then called hispid or feathery. In cross-sections, it is seen that these outgrowths are branches of the strands of definite peripheral doublet fibrils, so that the hairs are arranged along the flagella in longitudinal rows. In the flagellates, which are related to *Euglena*, there is only one, and in the Phaeophyta and Heterkontae, on the other hand, there are two such rows. In *Fucus*, both third double strands in Fig. K–1b, counted from above to left and right, bear a row of such hairs.

The movement of the flagella is usually described as propeller-like, *i.e.* as three-dimensional helical. There are various theories on the contraction along the helical line necessary for this effect (Fawcett, 1961). Gray (1955), however, had made it plausible that only a two-dimensional planar wave motion takes place, and the screw movement comes about by rotation of the flagellum carrier. It is thus a case of two different motions superimposed on each other. According to Bishop (1958) the flagellum carries out the planar oscillation over the greatest part of its length, but the distal tail end seems involved in rotation. This author makes, for example, a rhythmic alternating contraction of microfibrils nos. 5 and 6 on the one hand, and nos. 1, 2, and 9 on the other hand, responsible for the wave vibration. It is, of course, not easy to give an obvious interpretation to the ninefold rotational symmetry of the microfibrils in relation to a function based on a planar bilateral symmetry, as the attribution of the impulses to two unequal groups with three and two contracting strands shows. The secondary fibrils could perhaps play a part in the rotational motion of the tail end; in any case, they widen in the flagellum root to septa having rotational symmetry, which in transverse section look remarkably like the radii of a cartwheel (Figs. K–4 and K–5).

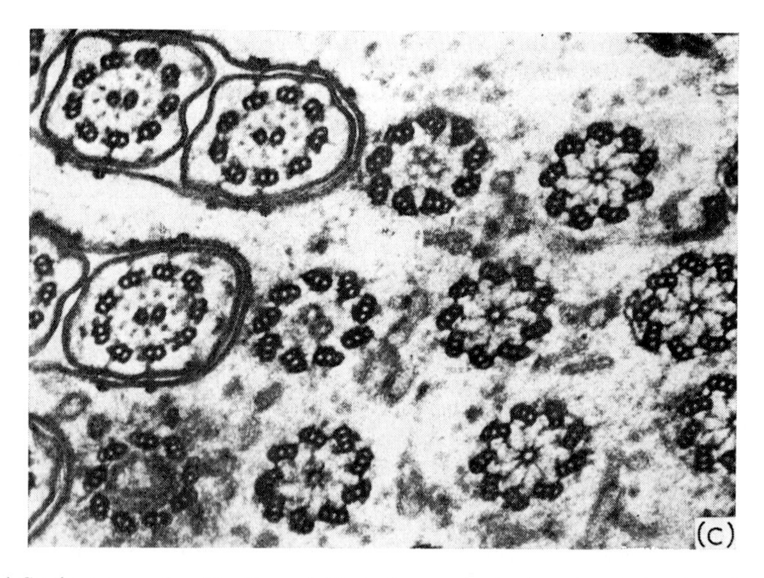

Fig. K–4. Sections across basal bodies and cilia in the anterior body-region of *Pseudotrichonympha* (see Fig. K–5) (Gibbons and Grimstone, 1960).

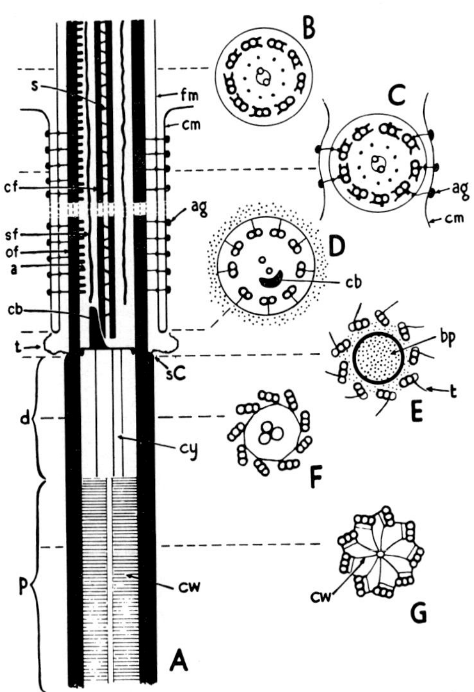

Fig. K–5. Diagram of cilium and basal body of *Pseudotrichonympha* (see Fig. K–4); (A) longitudinal section, (B–G) cross-sections at the levels indicated: *a* arms, *ag* anchor granule, *bp* basal plate, *cb* crescent body, *cf* central fibre, *cm* cell membrane, *cw* cartwheel structure, *cy* cylinders, *d* distal region of basal body, *fm* flagellar membrane of outer fibre *of*, *p* proximal region of basal body, *s* central sheath, *sC* distal end of subfibre *C*, *sf* secondary fibre, *t* transitional fibril (Gibbons and Grimstone, 1960).

The chemistry of the cilia has been best examined in the peritrichous bacterium *Proteus vulgaris*. In this object, they consist only of single 120 Å thick microfibrils (Astbury and Weibull, 1949) and may thus be compared rather with the individual fibrils than with the entire eleven-stranded flagella. The cilia can be separated from the bacterial body with a solution of sodium chloride, and obtained pure in such quantities that a chemical analysis is possible. They consist solely of protein. This therefore deserves attention for the bacterial membranes contain carbohydrates and glucosamine.

The nitrogen content of the flagellum material amounts to 16% and the usual amino acids were found as cleavage products; tryptophan, histidine, cysteine and hydroxyproline were however missing (Weibull, 1951). On the basis of its chemical composition the flagella protein comes close to the myosin of muscle. Characteristic differences still exist, in that the flagella lack cystine (Table K–I).

Tibbs (1957/58) has carried out similar analyses on the flagella of the Chlorophycean *Polytoma uvella*, in which cystine was found. The cystine content of various contractile fibre proteins is presented in Table K–I. It is low, especially when compared with keratin, so that in this respect the flagellum and the muscle protein come close to one another.

A fundamental distinction has, however, been established. After extraction of

TABLE K–I

THE CYSTINE CONTENT OF CONTRACTILE PROTEINS

Bacterial cilia	—	Weibull, 1951
Polytoma – flagella	0.81–0.87	Tibbs, 1957/58
Actin and myosin	1.5	Weibull, 1951
Keratin	10–15	Weibull, 1951

muscle fibres with glycerol, pure muscle protein remains which retains its contractability and which contracts vigorously and irreversibly on addition of ATP. If this experiment is performed with flagella, the contractile protein reacts to the influence of ATP by rhythmic activity, which can be followed for minutes or hours (Bishop, 1958).

2. BASAL BODIES

The undulipodia are rooted in a plasmatic particle which is commonly known as the basal granule, in the ciliates as the kinetosome (Gr. *kinein* = to move), and in the flagellates as the blepharoblast (Gr. *blepharon* = eyelid). As these structures are similar, Fawcett (1961) has proposed the simple topographical term basal body for these organelles. The name kinetosome presupposes that the rhythmic contractions of the undulipodia are controlled from this place.

The basal body is delimited from the extracellular flagellum by the basal plate (Fig. K–5 *bp*). In the case of *Pseudotrichonympha*, this plate carries a crescent-shaped body. The two central fibrils end somewhat above the basal plate. The nine doublets, on the other hand, continue into the basal body, but then appear as triplets of three subfibrils (Figs. K–4 and K–5). The added subfibril (labelled with *sC* in Fig. K–5A) ends on the basal plate in a thin radially-running tail, which was named the transitional fibril (*t*) by Gibbons and Grimstone (1960). The darker material of the matrix, which forms the secondary fibrils in the undulipodium, is organised in the form of three cylinders in the distal parts of the basal body and as bent septa in the proximal regions. A basal section shows a wheel (Gibbons 1960), which, on account of its curved spokes, remembers a pin wheel with pronounced rotational symmetry.

As a rule, the cylinder of the basal body runs openly into the groundplasm. There are however exceptions to this, in that the basal body can be closed or anchored in the groundplasm by a striated fibrous rootlet (Fawcett, 1961; Scholtyseck and Danneel, 1963). Here only the open basal body, such as that found in the Protobionta, can be considered.

The structure of the above-described basal body is reminiscent of the centriole. This cell organelle forms the central part of the centrosome from which, in animal cells, the spindle fibres radiate as an aster in mitosis. The centrosome contains usually two centrioles (diplosomes). Mitosis is initiated in cells which dispose of this apparatus, by division of the centrosome. We then find four centrioles together temporarily, in which case two daughter centrioles probably arise by budding from the existing pair of centrioles (Bernhard and De Harven, 1960).

The centriole is a cylinder about 0.15 μ in length and 0.3 to 0.5 μ in diameter. The centre of the cylinder appears 'empty' in contrast with the electron dense wall, which consists of 9 strands of triplet fibrils. In certain instances it was observed that the centrioles may carry a short cilium-like process. On the basis of this observation, an ontogenetic relationship appears to exist between the centriole and the basal bodies of the undulipodia. This connection had already been recognised by the classical cytologists (Henneguy, 1897; Lenhossek, 1898), who thought that ciliated cells lose their divisibility through enlargement and transformation of their centrioles into basal bodies. It seems, however, that an original diplosome can survive and induce further karyokinesis.

The basal body of the flagellum in the gastroderm of *Hydra* consists of two osmiophilic cylinders, one of which carries the undulipodium whilst the other lies oriented at right angles to it in the groundplasm, without a flagellum. Here also the kinetosome is formed from the two centrioles of a diplosome. Such observations are considered as evidence for the homology of the basal bodies with the centrioles,

How the chemical impulse for the rhythmic contraction is transmitted from the kinetosome to the flagellar fibrils is unknown. Since acetylcholine which plays a predominent role in nerve physiology, has been found in ciliated cells, comparisons have been made with neural transmission of stimuli. These are however unsatisfactory, because no ultrastructural conducting system can be established with the electron microscope.

3. EYE-SPOTS

Phytoflagellates like *Chromulina* (Rouiller and Fauré-Fremiet, 1958), *Euglena* (Wolken, 1956b), *Chlamydomonas* (Sager and Palade, 1957), etc., which are photosynthetically active, possess a spot of red pigment in the vicinity of the kinetosome of their locomotive apparatus, which has been termed the stigma (Gr. *stigma* = mark) or eye-spot. This is sensitive to light and permits the organisms in question to seek out places of optimal light intensity for the assimilatory activity of their chloroplasts.

The stigma consists of ultramicroscopic vesicles or platelets containing a carotenoid as the pigment. In some cases this is astacene, which was first found in Crustaceae (*Astacus*). The action spectrum of the phototactic motion of such flagellates corresponds to the absorption spectrum of the carotenoids, and for this reason the eye-spot is regarded as a photoreceptor.

The relations which the stigmata show to the undulipodia are interesting. In *Chromulina*, a chrysomonad with yellow-brown chromatophores, the pigment vesicles are formed in the end face of one of the two large plastids (Fig. K–6a). The plastid was termed a 'chromoplast' in the literature (Rouiller and Fauré-Fremiet, 1958), but as this is not a degenerated querontic form (see p. 231), but a photosynthetically active form of plastids with thylakoid lamellae, the classical term chromatophore is better if we want to avoid the term chloroplasts.

Near the plate with the pigment vesicles, a thickened cilium with 9 + 2 fibrils and a basal body is found in a plasma pouch (Fig. K–6a). This stands perpendicular

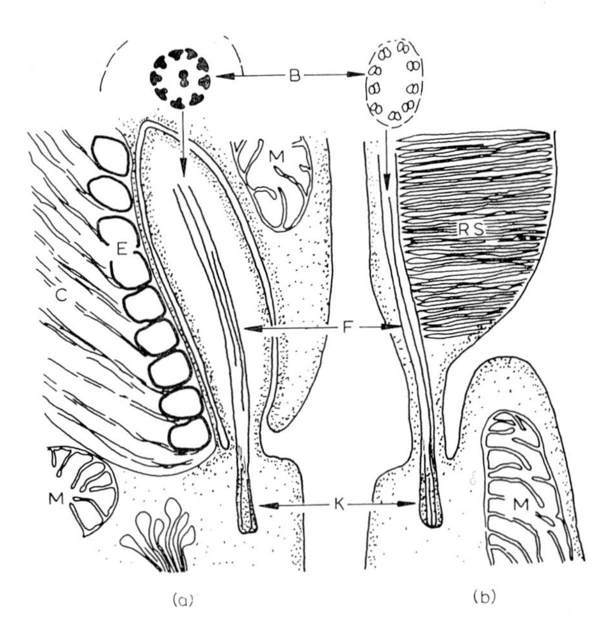

Fig. K–6. Comparison of the stigma with retinal rods; (a) stigma (eye-spot) of *Chromulina*, (b) junction of outer and inner segments of a retinal rod cell. *B* cross-section of a fibrillar bundle of *Chromulina* and a retinal rod cell, *C* chromatophore of *Chromulina*, *E* eye-spot chamber of *Chromulina*, *F* fibrillar bundle, *K* kinetosome, *M* mitochondria, *RS* flattened rod sacs (thylakoids) of the retinal cell (both *E* and *RS* contain carotene pigment) (Fauré-Fremiet, 1958).

to the basal body of the locomotively active flagellum and together they resemble a diplosome.

It appears that the eye-spot-cilium complex forms a primitive eye (Wolken, 1956b), for the retinal rods of the metazoa are also ciliary in nature. Their development takes place like the differentiation of a cilium (De Robertis, 1956). The outer segment of the retinal rod arises by swelling of the flagellum matrix, which, like the stroma of the proplastids, is full of small vesicles as precursors of a thylakoid stack. The inner segment consists of a bulge of protoplasm which contains the kinetosome and numerous mitochondria. The connecting strand between the two segments retains the original flagellum structure, although instead of 9 + 2 fibrils, only the 9 peripheral doublets appear. These statements are presented schematically in Fig. K–6b. As the light stimulus is registered in the layered composite body of the outer segment, the fibrils of the connecting strand must have a transmissive function.

Fauré-Fremiet (1958) is of the opinion that the conduction of the light stimulus in the phytoflagellates also takes place through ciliary systems. He has thus compared the structure of the eye-spots of *Chromulina*, together with their cilia, with the structure of a retinal rod (Fig. K–6). The homology is astonishing. There are of course differences which must not be overlooked: in the retinal rods the light-absorbing system is built into the cilium, whilst in *Chromulina* it lies in a neighbouring plastid, separated from the cilium by a slit; moreover, the cilium in the first case contains only 9, and in the second case the original 11 fibrils.

As eye-spots are found only in the phytoflagellates and clearly a phylogenetic connection exists between these organelles and the light-sensitive organs of the metazoa, we come to the surprising conclusion that animals were derived not from the heterotrophic, colourless, but rather from the autotrophic, green or yellow-green, flagellates!

Apart from such bold speculations, we must give meritorious credit to electron microscopy, which produced the proof of a close morphological relationship between the locomotive (undulipodia) and sensory cell organelles.

Retrospect

The cell is the smallest living unit equipped with an individual metabolism, independent energy cycle, and the tendency and capacity for autoreproduction. In the Protobionta, the cells have special organelles for all important vital functions (pseudopodia, undulipodia, sometimes an ostium, pulsating vacuoles, eye-spots, etc.). In the Metabionta a specialisation of the cells has occurred, resulting in division of labour with regard to specific functions (assimilatory cells, supporting cells, muscle cells, nerve cells, etc.). In the year 1849, the founders of the science of cells, Schleiden and Schwann, could only point to the protoplasm and the cell nucleus as general features common to all the remarkably differentiated cells, in which inventory the plasmalemma was added later. The striking organelles of the plant cells, such as the cell wall, the plastids, and the vacuoles with their tonoplasts, or the centrioles and cilia of animal cells, could not be termed obligatory organelles necessary for life, for there are certain types of cell which pursue their vital functions without them.

Here the electron microscope has now brought about an astonishing change. The mitochondria, whose universal presence was earlier disputed, are found in all types of cell, just like the ribosomes, the Golgi apparatus, and the endoplasmic reticulum with its relationship to the nuclear membrane. This extension of the general cell inventory speaks strongly in favour of a monophyletic development of all Eukaryonta, since the basic structure of the living substance is, inspite of its complexity and diversity, similar at all stages of the phylogenetic tree. Only the Prokaryonta form an exception in that they characteristically lack not only a typical nucleus, but frequently also typical mitochondria, Golgi apparatus, or a typical ER. It is difficult to say to what extent their cytological ultrastructures are precursors or impoverished forms of the normal cell structure, or whether they are polyphyletically arisen special structures with definite adaptations and a certain convergence to the general cell type. The remarkable consistency of the ultrastructure of all Eukaryonta is all the more surprising.

This statement leads to the legitimate question of why we restricted ourselves to plant cytology in the present book, rather than treat the field of cytology as a whole. The reason is that the specific cytological differentiations in the plant cells (chloroplasts, vacuoles, tonoplast, cell wall) and the animal tissues (blood cells, muscle, nerves, kidney, liver, pancreas, sensory organs) are so diverse, that their description would exceed the scope of a manageable textbook, especially if one would like to pursue the ontogenetic development of all the ultrastructures and textures.

New insight has been gained with the ontogeny of organelles in plant cells. The autonomous system of the cell nucleus, responsible for the identical replication of the hereditary material and providing for the correct sequence of the activation of morphogenetic informations stored in the genes, was supplemented by plant cytologists by a second autonomous system, the plastidome. The genetic independence of self-reproducing proplastids which do not arise *de novo* suggested itself by the pure maternal inheritance of certain plastid properties and finally was accepted to be established. By analogy, cytologists sought to ascribe such an autonomy to the mitochondria (chondriome), to the spherosomes (spherome), and even to the ribosomes (at that time called microsomes) so that all visible cell particles were explained as independent elementary replicators (Strugger, 1954). If we followed this concept to its logical conclusion, the cell would physiologically be not an entity but a mosaic or symbiosis of various autonomous systems characterised by self-replication and independent mutation.

Electron microscopy has now revealed that the cell organelles regarded as autoreplicators in ova, meristem cells, and de-differentiated parenchyma cells arise from ultrasmall particles, the so-called initials, which as precursors of proplastids and mitochondria can originate with the cooperation of the nucleus. Even though these questions are at present still incompletely clarified, they do open up the possibility of getting rid of a cell theory suggesting a symbiotic interaction of various autonomous organelle systems, and of considering the cell not only in a morphological, but also in physiological and genetic respects as an entity whose vital manifestation, development, and heredity are controlled only from one central organ, the nucleus.

In an interesting way, the pursuit of the ultrastructure has not only placed ontogenetic relationships under discussion, but has also brought problems of phylogeny closer to solution. The surprising fact that all undulipodia (flagella and cilia) from the flagellates to the sperm tails of all mobile male gametes (mosses, ferns, echinoderms, vertebrates) and up to the ciliated epithelia of mammals show always eleven fibrils in the 9 + 2 pattern, speaks again for a monophyletic relationship of all these taxonomic groups, which from a point of view of kinship lie so far apart. In the light of the ultrastructure of the sensory organs of phytoflagellates, a certain homology can be shown with the retinal rods of higher animals, and thus the thesis becomes involved in that the development of the animal kingdom appears, unexpectedly, to be derived not from the heterotrophic colourless flagellates, but from the autotrophic phytoflagellates. This is in agreement with other hypotheses according to which only plant cells with defined outlines (without pseudopodia) and absorbing food through the entire cell surface (without differentiation of an ostium) are suitable for the formation of facultative (colonies) and obligatory (Metabionta) cell associations. In this connection, it is also worth noting the interesting hypothesis according to which the strikingly analogous ultrastructures of chloroplasts and mitochondria, and the conformity of their active enzymes, are possibly due to homology. If these two organelles once arose phylogenetically from one another, the plastids, which are missing through a loss mutation in the heterotrophic organisms, must claim the seniority, because the chloroplasts could fulfil their metabolic function in an oxygen free atmosphere such as must have prevailed at the time of the origin of life on earth, whilst mitochondria require the oxygen liberated by the activity of the plastids.

The main object of ultrastructural investigations is to build a bridge between the morphological and the chemical sciences – an endeavour which forms an important part of molecular biology. As was pointed out in the introduction, biochemistry *in vitro* can only lead to an effective cell physiology, when the course of the biochemical reactions and the site of the enzyme systems have been successfully localised within the cell. In order to reach this goal, the chemical composition of the ultrastructural elements (particles, membranes etc.) and the mutual positions of the macro- and micromolecules present must first be clarified.

In this respect a wide field of investigation still lies before us, for results in such a direction have only been attained in the field of viruses and of the meta- and alloplasmic differentiation products of the cell, such as nerve sheaths, collagen fibres, silk, cellulose and chitin fibrils, starch granules, and the like. This means that the metabolisms of these objects are permanently or temporarily withdrawn so that their ultrastructure has a static formation which is occasionally even stabilized by processes of crystallisation. The problem may then be solved with the help of X-ray methods used in the elucidation of crystal structure. Thus we know, from stereochemical considerations, how the glucose rings are arranged in the chain molecule of cellulose, how the chains come together to a chain lattice, how the elementary fibrils visible in the electron microscope are formed, and how these associate into microfibrils and finally assemble to macrofibrils visible in the light microscope. To continue, we know how the fibre cells are constructed from such macrofibrils and how the macroscopic fibre strands used in the textile industry arise by aggregation and fasciation of such cells. Thus in the case of this object the structure is known through all orders of magnitude, from micromolecules up to the tangible fibre strand, which in the plant carries out its function as supporting tissue.

In the case of labile ultrastructures which owing to growth or metabolism are subject to constant change, and which must thus be termed dynamic, we stand at the very beginning of refined macromolecular analysis. Even for the chloroplasts and the retinal rods, where the revealed periodicity of the fine structure and the presence of pigments offer additional possibilities for investigation, a conclusive elucidation of the molecular structure is still outstanding. It is all the more difficult to solve the problem of the unit membranes of the plasmalemma, the endoplasmic reticulum, the Golgi apparatus, and so on, though progress will certainly be made in this respect in the next few years.

For the explanation of the molecular structure in membranes of a metabolising cell organelle, the principle of biochemical cycles is also a welcome help, for it requires that the enzymes involved in such a process be so arranged that the substrate can be passed along them in the direction of the chain reaction.

In this respect there is a fundamental distinction between the course of biochemical processes (metabolic physiology) and the ontogentic developmental processes (developmental physiology). While the biochemical cycles are actual cyclic processes, the so-called life cycles of cell organelles or whole organisms proceed one-sidedly, monotropically, *i.e.* the individual developmental stages cannot be reversed, in contrast to the individual steps of a chemical reaction. Thus the proplastids and the chloroplasts cannot change back into plastid initials, although the chloroplasts in de-differentiated cells can form new initials by budding, and thus give rise to the start of a new plastid generation. A similar pattern is observed as in the gamete formation of adult organisms and the succeeding formation of zygotes, from which a new generation arises. Such a zygote is not identical with the zygote from which the parent generation sprung forth, for mutations can occur in the course of development. Because of this the so-called life cycle of an organism does not follow a perfectly repetitive circle, but rather proceeds in a series of cycloids, (Fig. K–7), for the ontogenetic and the phylogenetic developments progress monotropically. Such a 'cycle' never returns to its point of departure, but it creates continually new starting points from which other individuals

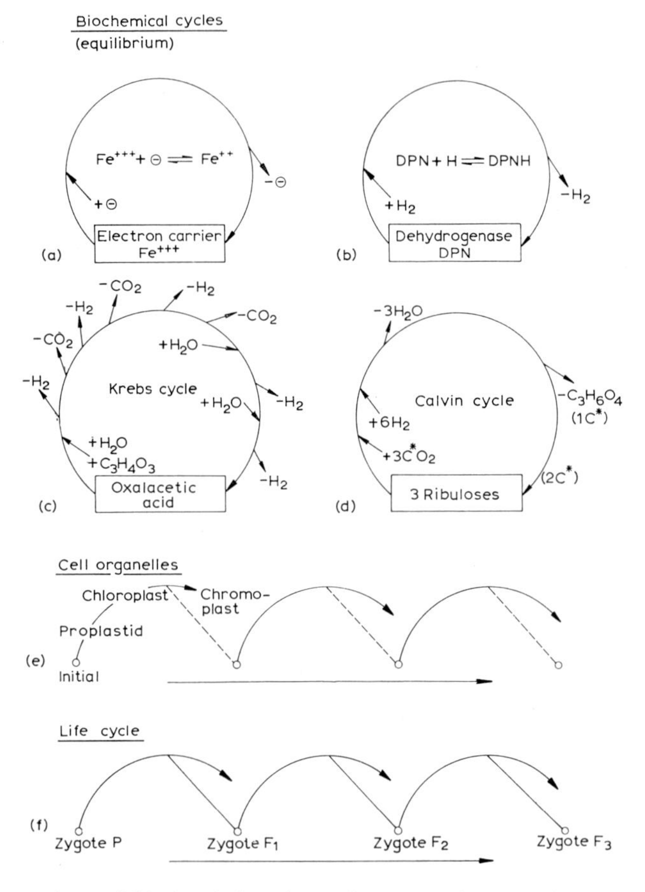

Fig. K–7. Comparison of biochemical cycles and monotropic cytological differentiation; (a) oxidation–reduction equilibrium; (b) hydrogenation–dehydrogenation equilibrium; (c) respiration cycle: oxalacetic acid is combined with pyruvic acid $C_3H_4O_3$ and is regenerated at the end of the cycle; (d) assimilation cycle: ribulose is combined with atmospheric CO_2 and is regenerated at the end of the cycle; (e) monotropic differentiation of plastids: initial → proplastid → chloroplast → chromoplast, new initials do not arise at the end of the 'cycle'; (f) so-called 'life cycles': gametes are not given off at the end but during the course of a life-time.

begin their ontogeny. The difference shown between biochemical cycles and life cycles manifests itself in that in the course of phylogeny the physiological-chemical cyclic processes have remained the same for millions of years, whilst, progressing morphologically, the organisms themselves have meanwhile altered phylogenetically out of recognition.

In the biochemical cyclic processes, the starting molecule is restored after a series of reactions to its original state (Fig. K–7 a/d). This is not the case in ontogenetic cycles, in which, after a large number of irreversible developmental steps, parts of the biological object which has passed through one cycle become dissociated as initials or gametes to start a new generation. These carry along the information for the new ontogenetic cycle based on the genetic code of their DNA.

The code is not however absolute; it is only constant to a first approximation, for the DNA chains are liable to mutation in the course of a sufficiently long period of time. Its approximate constancy and its variability depend on the chemical structure of the nucleic acid chains. The spectacular achievement of molecular biology is to have established how this structure becomes converted to the chemical structure of protein substances. Without such a detailed constructional plan which, at first approximation, is identically reproduced through the generations, the synthesis of chemical primary and secondary structures, as well as the formation of cytological ultrastructures and ultratextures, would be inconceivable. It is thus established that all biological structures arise from pre-existing structures:

Structura omnis e structura.

Bibliography

Accola, P. (1960) Dr. Thesis ETH, Zürich, 1960 and *Ber. Schweiz. Botan. Ges.*, **70**, 352.
Adam, N. K. (1941) *The Physics and Chemistry of Surfaces*, 3rd ed., University Press, Oxford.
Afzelius, B. A. (1955) *Exptl. Cell Res.*, **8**, 147.
Afzelius, B. A. (1956a) *Exptl. Cell Res.*, **11**, 67.
Afzelius, B. A. (1956b) *Grana Palynologica*, **1**, 22.
Ahearn, M. J. and J. J. Biesele (1965) *J. Cell Biol.*, in the press.
Albersheim, P., K. Mühlethaler and A. Frey-Wyssling (1960) *J. Biophys. Biochem. Cytol.*, **8**, 501.
Albertini, A. von (1947) *Schweiz. Z. Allgem. Pathol. Bakteriol.*, **7**, 4.
Altmann, R., *Die Elementarorganismen und ihre Beziehungen zu den Zellen* (1894) 2nd ed.; 1st ed. 1890, Veit, Leipzig.
Ambronn, H. (1888) *Ber. Deut. Botan. Ges.*, **6**, 226.
Ambronn, H. and A. Frey (1926) *Das Polarisationsmikroskop*, Akademische Verlagsgesellschaft, Leipzig.
Anderer, E. A. and D. Handschuh (1962) *Z. Naturforsch.*, **17b**, 536.
Anderson, D. B. (1927) *Sitz'ber. Akad. Wiss. Wien, Kl. 1*, **136**, 429.
Anderson, D. B. and Th. Kerr (1938) *Ind. Eng. Chem.*, **30**, 48.
Arndt, N. W. and D. P. Riley (1953) *Nature*, **172**, 803.
Arnold, C. G. (1963) *Ber. Deut. Botan. Ges.*, **76**, 3.
Arnon, D. I. (1955) *Science*, **122**, 9.
Arnon, D. I. (1961) *Federation Proc.*, **20**, 1012.
Arnon, D. I., M. B. Allan and F. R. Whatley (1956) *Biochim. Biophys. Acta*, **20**, 449.
Aronoff, S. (1960) *Handbuch Pflanzenphysiol.*, **5/1**, 234 Springer Berlin.
Aspinall, G. O. and C. T. Greenwood (1962) *J. Inst. Brewing*, **68**, 167.
Astbury, W. T. (1933) *Fundamentals of Fibre Structure*, Oxford University Press, H. Milford London.
Astbury, W. T. (1947) *Proc. Roy. Soc. London*, **B 134**, 303.
Astbury, W. T. (1949) *Exptl. Cell Res., Suppl.*, **1**, 234.
Astbury, W. T. (1961) *Nature*, **190**, 1124.
Astbury, W. T. and W. R. Atkin (1933) *Nature*, **132**, 348.
Astbury, W. T. and F. O. Bell (1938) *Nature*, **141**, 747.
Astbury, W. T. and T. C. Marwick (1932) *Trans. Faraday Soc.*, **29**, 206.
Astbury, W. T., T. C. Marwick and J. D. Bernal (1932) *Proc. Roy. Soc. London*, **B 109**, 443.
Astbury, W. T. and C. Weibull (1949) *Nature*, **163**, 280.
Avers, Ch. J. and E. E. King (1960) *Am. J. Botany*, **47**, 220.
Badenhuizen, N. P. (1961) *Proc. Koninkl. Ned. Akad. Wetenschap.*, **C 65**, 123.
Badenhuizen, N. P. (1959/61) *Recent Advances in Botany*, Univ. Press, Toronto, p. 1258.
Bahr, G. F. (1954) *Exptl. Cell Res.*, **7**, 457.
Bahr, G. F. (1955) *Exptl. Cell Res.*, **9**, 277.
Bailey, I. W. (1913) *Forestry Quart.*, **11**, 12.
Bailey, I. W. and M. R. Vestal (1937) *J. Arnold Arboretum*, **18**, 196.
Bailey, K., W. T. Astbury and K. M. Rudall (1943) *Nature*, **151**, 716.
Bairati, A. and F. E. Lehmann (1952) *Experientia*, **8**, 60.
Baker, J. R. (1957) *Symp. Soc. Exptl. Biol.*, **10**, 1.
Bakhuizen van de Sande, H. L. (1925) *Proc. Soc. Exptl. Biol. N.Y.*, **23**, 302.
Bakker, J. D. (1953) *Electrotechniek*, Nr. **15/16**. July 1953, Den Haag.
Balls, W. L. (1919) *Proc. Roy. Soc. London*, **B 90**, 542.
Bamford, C. H., L. Brown, A. Elliot, W. E. Hanby and I. F. Trotter (1954) *Nature*, **173**, 27.
Barnum, C. P. and R. A. Huseby (1948) *Arch. Biochem.*, **19**, 17.
Bates, F. L., D. French and R. E. Rundle (1943) *J. Am. Chem. Soc.*, **65**, 142.
Baud, Ch. A. (1949) *Bull. Histol. Appl.*, **26**, 99.

Bautz, E. (1955) *Naturwiss.*, **42**, 619.
Beams, H. W., T. C. Evans, W. W. Baker and V. van Breemen (1950) *Anat. Record*, **107**, 329.
Bear, R. S., K. J. Palmer and F. O. Schmitt (1941) *J. Cellular Comp. Physiol.*, **17**, 357.
Becker, W. A. (1934) *Acta Soc. Botan. Polon.*, **11**, 139.
Beer, M. and G. Setterfield (1958) *Am. J. Botany*, **45**, 571.
Beermann, W. (1961) *Chromosoma (Berl.)*, **12**, 1.
Beermann, W. (1962) Riesenchromosomen, *Protoplasmatologia*, VI D, Springer, Wien.
Bélar, A. (1929) *Arch. Entwicklungsmech. Organ.*, **118**, 359.
Bell, P. R. (1961) *Proc. Roy. Soc. London*, **153**, 421.
Bell, P. R. and K. Mühlethaler (1962a) *Nature*, **195**, 198.
Bell, P. R. and K. Mühlethaler (1962b) *J. Ultrastruct. Res.*, **7**, 452.
Bell, P. R. and K. Mühlethaler (1964a) *J. Cell Biol.*, **20**, 235.
Bell, P. R. and K. Mühlethaler (1964b) *J. Mol. Biol.*, **8**, 853.
Benda, C. (1897/98) *Verhandl. Physik. Ges. Berlin.*
Benda, C. (1901) *Verhandl. Anat. Ges. Jena.*
Bendet, I. J., D. A. Goldstein and M. A. Lauffer (1960) *Nature*, **187**, 781.
Bennett, H. St. (1956) *J. Biophys. Biochem. Cytol.*, **2** (Suppl.), 99.
Bergmann, M. and C. Niemann (1937) *J. Biol. Chem.*, **118**, 301.
Bernal, J. D. (1932) *Chem. Ind. (Berlin)*, **51**, 466.
Bernal, J. D. (1946) *Trans. Faraday Soc.*, **42 B**, 1.
Bernal, J. D. and D. Crowfoot (1934) *Nature*. **133**, 794.
Bernal, J. D. and I. Fankuchen (1941) *J. Gen. Physiol.*, **25**, 111.
Bernhard, W., A. Bauer, A. Gropp, F. Haguenau and Ch. Oberling (1955) *Exptl. Cell Res.*, **9**, 88.
Bernhard, W. and E. De Harven (1958/60) *4th Intern. Kongr. Elektronenmikroskopie Berlin, 1958, Verhandl.* **2**, 217, Springer, Berlin, 1960.
Berthold, G. (1886) *Studien über Protoplasmamechanik*. Verlag Arthur Felix, Leipzig.
Bessis, M. (1958) *Bull. Acad. Natl. Méd.*, **13**, 363.
Birnstiel, M. L., Margaret I. H. Chipchase and W. G. Flamm (1963) *Calif. Inst. Technol., Pasadena, Internal Rept.*
Birnstiel, M. L., J. H. Rho and Margaret I. H. Chipchase (1962) *Calif. Inst. Technol., Pasadena, Internal Rept.*
Bishop, D. W. (1958) *Nature*, **182**, 1638.
Bochsler, A. (1948) *Ber. Schweiz. Botan. Ges.*, **58**, 73.
Bolliger, R. (1956) Diploma Thesis ETH Zürich (unpublished).
Bonner, J. (1958) *Eng. Sci. Monthly, Calif. Inst. Technol., Pasadena*, Oct.
Bonner, J. (1959) *Am. J. Botany*, **46**, 58.
Bonner, J. (1960) *Symp. Protein Synthesis*, Wassenaar 1960, Acad. Press London.
Bopp-Hassenkamp, Gisela (1958) *Protoplasma*, **50**, 243.
Bosshard, Ursula (1964) Dr. Thesis ETH Zürich 1964 and *Z. Wiss. Mikroskopie*, **65**, 391.
Bot, G. M. (1939) Dr. Thesis, Leiden.
Boysen Jensen, P. (1954) *Dan. Biol. Medd.*, **22**, 3.
Brachet, J. (1949) *Acidi nucleici*. Rosenberg and Fellier, Torino.
Brachet, J. (1953) *Quart. J. Microscop. Sci.*, **94**, 1.
Brandt, P. W. (1958) *Exptl. Cell Res.*, **15**, 300.
Branton, D. and H. Moor (1964) *J. Ultrastruct. Res.*, **11**, 401.
Brenner, S. (1949) *S. Afr. J. Med. Sci.*, **14**, 13.
Brenner, S. (1953) *Biochim. Biophys. Acta*, **11**, 480.
Brenner, S. and R. W. Horne (1959) *Biochim. Biophys. Acta*, **34**, 103.
Brenner, S., G. Streisinger, R. W. Horne, S. P. Champe, L. Barnet, S. Benzer and M. W. Rees (1959) *J. Mol. Biol.*, **1**, 281.
Brown, D. E. S. (1934) *J. Cellular Comp. Physiol.*, **4**, 257 and **5**, 335.
Brown, G. L., H. G. Callan and G. Leaf (1950) *Nature*, **165**, 600.
Brown, W. V., H. Mollenhauer and C. Johnson (1962) *Am. J. Botany*, **49**, 57.
Bucher, H. (1957) *Holzforsch*, **11**, 1.
Burghardt, H. and J. Brandes (1957) *Naturwiss.*, **44**, 266.
Bütschli, O. (1892) *Untersuchungen über mikroskopische Schäume und das Protoplasma*. W. Engelmann, Leipzig.

Buttrose, M. S. (1960) Dr. Thesis ETH Zürich 1960 and *J. Ultrastruct. Res.* **4**, 231.
Buttrose, M. S. (1962a) *Naturwiss.*, **49**, 307.
Buttrose, M. S. (1962b) *J. Cell Biol.*, **14**, 159.
Buttrose, M. S. (1963a) *Australian J. Biol. Sci.*, **16**, 305.
Buttrose, M. S. (1963b) *Die Stärke*, **15**, 85.
Buttrose, M. S. (1963c) *Australian J. Biol. Sci.*, **16**, 768.
Buttrose, M. S., A. Frey-Wyssling and K. Mühlethaler (1960) *J. Ultrastruct. Res.*, **4**, 258.
Buvat, R. (1959) *Compt. Rend. Acad. Sci. Paris*, **248**, 1014.
Buvat, R. (1962) *5th Intern. Congr. Electron Microscopy, Philadelphia, 1962*, **2**, W-1 (Acad. Press, New York, 1962).
Buvat, R. and Anne Mousseau (1960) *Compt. Rend. Acad. Sci. Paris*, **251**, 3051.
Buvat, R. and A. Puissant (1958) *Compt. Rend. Sci. Paris*, **247**, 233.
Cairns, J. (1963a) *J. Mol. Biol.*, **6**, 208.
Cairns, J. (1963b) *Endeavour*, **22**, 141.
Callan, H. G. and L. Lloyd (1960) *Phil. Trans. Roy. Soc. London*, **B 243**, 135.
Callan, H. G., J. T. Randall and S. G. Tomlin (1949) *Nature*, **163**, 280.
Calvin, M. (1959) *Rev. Mod. Physics*, **31**, 147.
Calvin, M. (1962) *Science*, **135**, 879.
Camefort, H. (1962) *5th Intern. Congr. Electron Microscopy Philadelphia, 1962*, **2**, NN-7 (Acad. Press, New York, 1962).
Carlström, D. (1957) *J. Biophys. Biochem. Cytol.*, **3**, 669.
Caspersson, T. (1941) *Naturwiss.*, **29**, 33.
Caspersson, T. (1950) *Cell Growth and Cell Function.* Norton, New York.
Chance, B., R. W. Estabrook and Ch. P. Lee (1963) *Science*, **140**, 379.
Chance, B. and D. F. Parsons (1963) *Science*, **142**, 1176.
Chibnall, A. C. and S. H. Piper (1934) *Biochem. J.*, **28**, 2215.
Chipchase, Margaret I. H. and M. L. Birnstiel (1963) *Proc. Natl. Acad. Sci. (Wash.)*, **50**, 1101.
Chodat, R. (1922) *Bull. Soc. Bot. Genève*, **14**, 50.
Church, A. H. (1904) *The Relation of Phyllotaxis to Mechanical Laws.* Williams and Norgate, London.
Claude, A. (1943) *Science*, **97**, 451.
Claude, A. (1946) *J. Exptl. Med.*, **84**, 51.
Cohn, E. J. (1947) *Experientia*, **3**, 125.
Cohn, E. J. and J. T. Edsall (1943) *Proteins, Amino Acids and Peptides.* Reinhold Publ. Corp., New York.
Collander, R. (1937) *Ann. Rev. Biochem.*, **6**, 1.
Colvin, J. R. (1963) Oral communication, *2nd Intern. Cabot Symp., Harvard Forest.*
Colvin, J. R. (1964) *The biosynthesis of cellulose*, in M. H. Zimmermann (Ed.) *The Formation o Wood in Forest Trees.* Academic Press, New York, 1964, p. 189.
Colvin, J. R., S. T. Bayley and M. Beer (1957) *Biochim. Biophys. Acta*, **23**, 652.
Comar, C. L. (1942) *Bot. Gaz.*, **104**, 122.
Corey, R. B. and J. Donohue (1950) *J. Am. Chem. Soc.*, **72**, 2899.
Côté, W. A. and A. C. Day (1962) *Tappi*, **45**, 906.
Coxeter, H. S. M. (1961) *Introduction to Geometry.* J. Wiley and Sons, London.
Crick, F. H. C. and J. D. Watson (1956) *Nature*, **177**, 473.
Crowfoot, D. (1938) *Proc. Roy. Soc. London*, **A 164**, 580.
Damm, O. (1901) *Beih. Bot. Zentr.*, **11**, 219.
Dangeard, P. (1951) *Botaniste*, **35**, 35.
Dangeard, P. A. (1919) *Compt. Rend. Acad. Sci. Paris*, **169**, 1005.
Danielli, J. F. (1936) *J. Cellular Comp. Physiol.*, **7**, 393.
Danielli, J. F. (1938) *Cold Spring Harbor Symp. Quant. Biol.*, **6**, 190.
Danielli, J. F. and E. N. Harvey (1935) *J. Cellular Comp. Physiol.*, **5**, 483.
Danneel, R. (1959) *Arbeitsgemeinschaft Forsch. Nordrhein-Westfalen*, Heft **79**, 25.
Danneel, Silvia (1964) *Naturwiss.*, **51**, 368.
Das, N. K. (1963) *Science*, **140**, 1231.
Davenport, H. E. (1952) *Nature*, **170**, 1112.
Davison, P. F. (1960) *Nature*, **185**, 918.

Davison, P. F. and D. Freifelder (1962) *J. Mol. Biol.*, **5**, 635.
Davson, H. and J. F. Danielli (1943) *The Permeability of Natural Membranes*. University Press, Cambridge.
Dawson, I. M., J. Hossack and B. M. Wyburn (1955) *Proc. Roy. Soc. London*, **144**, 132.
De Duve, Chr. (1959) Lysosomes, a New Group of Cytoplasmic Particles, in T. Hayashi (Ed.), *Subcellular Particles*, Ronald Press, New York, 1959, pp. 128–159.
De Duve, Chr. (1963) *Sci. Am.*, **208**, 2.
De Duve, Chr. and J. Berthet (1954) *Inter. Rev. Cytol.*, **3**, 225.
De Robertis, E. (1956) *J. Biophys. Biochem. Cytol.*, **2** (Suppl.), 209.
De Robertis, E., C. M. Franchi and M. Podolsky (1953) *Biochim. Biophys. Acta*, **11**, 507.
Deuel, H. E. (1943) *Ber. Schweiz. Bot. Ges.*, **53**, 219.
De Vries, H. (1885) *Jahrb. Wiss. Botanik*, **16**, 465.
Diehl, J. M. and G. van Iterson (1935) *Kolloid-Z.*, **73**, 142.
Drawert, H. (1953) *Ber. Deut. Botan. Ges.*, **66**, 134.
Drawert, H. and Marianne Mix (1962a) *Planta*, **58**, 448.
Drawert, H. and Marianne Mix (1962b) *Ber. Deut. Botan. Ges.*, **75**, 128.
Drews, G. and W. Niklowitz (1956) *Arch. Mikrobiol.*, **24**, 147.
Drochmans, P. (1962) *J. Ultrastruct. Res.*, **6**, 141.
Du Buy, H. G., M. W. Woods and M. D. Lackey (1950) *Science*, **111**, 572.
Düvel, D. (1963) *Beitr. Biol. Pflanzen*, **39**, 83.
Düvel, D. and W. Mevius Jr. (1952) *Naturwiss.*, **39**, 23.
Eggmann, H. (1962) Diploma Thesis ETH. Zürich (unpublished).
Eichenberger, M. (1953) *Exptl. Cell Res.*, **4**, 275.
Eichhorn, E. L. and Caroline H. McGillavry (1959) *Acta Cryst.*, **12**, 872.
Engström, A. and F. Ruch (1951) *Proc. Natl. Acad. Sci. (Wash.)*, **37**, 459.
Erdtman, G. (1952) *Pollen Morphology and Plant Taxonomy of Angiosperms*. Almquist and Wiksell, Stockholm.
Erikson, G., A. Kahn, B. Walles and D. v. Wettstein (1961) *Ber. Deut. Botan. Ges.*, **74**, 221.
Esau, Katherine, V. I. Cheadle and E. B. Risley (1962) *Botan. Gaz.*, **123**, 233.
Eschrich, W. (1963) *Planta*, **59**, 243.
Estabrook, R. W. and A. Holowinsky (1961) *J. Biophys. Biochem. Cytol.*, **9**, 19.
Fajans, K. (1923) *Naturwiss.*, **11**, 165.
Fajans, K. (1925) *Z. Krist.*, **61**, 18.
Falk, H. (1961) *Protoplasma*, **54**, 594.
Falk, H. (1962a) *Protoplasma*, **54**, 432.
Falk, H. (1962b) *Protoplasma*, **55**, 237.
Falk, H. (1962c) *Z. Naturforsch.*, **17b**, 862.
Farrant, J. L. (1954) *Biochim. Biophys. Acta*, **13**, 569.
Fauconnet, L. (1948) *Pharm. Acta Helv.*, **23**, 101.
Fauré-Fremiet, E. (1958) *Quart. J. Microscop. Sci.*, **99**, 123.
Fauré-Fremiet, E., A. Mayer and G. Schaeffer (1909) *Compt. Rend. Soc. Biol. Paris*, **66**, 921.
Fawcett, D. (1961) Cilia and Flagella, in J. Brachet and A. E. Mirsky (Eds.) *The Cell*. Academic Press, New York, 1961, Vol. II, p. 217.
Fawcett, D. W. and K. R. Porter (1952) *Anat. Record*, **113**, 539; Abstract 33.
Feldherr, C. M. (1962) *J. Cell Biol.*, **12**, 159.
Fernández-Morán, H. (1962) *Circulation*, **26**, 1039.
Fernández-Morán, H. and A. O. Dahl (1952) *Science*, **116**, 465.
Ferry, J. D., S. Katz and I. Tinoco Jr. (1954) *J. Polymer Sci.*, **12**, 509.
Feulgen, R. and H. Rossenbeck (1924) *Z. Physiol. Chem.*, **135**, 203.
Finean, J. B. (1953) *Exptl. Cell Res.*, **5**, 202.
Finean, J. B. (1961) *Chemical Ultrastructure in Living Tissues*. Ch. Thomas, Springfield, Ill., 1961.
Finean J. B., F. S. Sjöstrand and E. Steinmann (1953) *Exptl. Cell Res.*, **5**, 557.
Fitz-James, Ph. C. (1960) *J. Biophys. Biochem. Cytol.*, **8**, 507.
Flemming, W. (1882) *Zellsubstanz, Kern- und Zellteilung*. Verlag Vogel, Leipzig.
Fraenkel-Conrat, H. (1956) *J. Am. Chem. Soc.*, **78**, 882.
Frank, H., M. L. Zarnitz and W. Weidel (1963) *Z. Naturforsch.*, **18b**, 281.
Franklin, R. E. (1955) *Nature*, **175**, 379.

Franklin, R. E. (1956) *Biochim. Biophys. Acta*, **19**, 203.
Franklin, R. E. and R. G. Gosling (1953) *Nature*, **172**, 156.
Franklin, R. E. and K. C. Holmes (1956) *Biochim. Biophys. Acta*, **21**, 405.
Frei, Eva and R. D. Preston (1961) *Proc. Roy. Soc. London*, **B 154**, 70.
Freudenberg, K. (1933) *Stereochemie*, Wien.
Freudenberg, K. (1954) *Fortschr. Chem. Org. Naturstoffe*, Springer, Wien. Bd. **11**, 43.
Freudenberg, K., H. Zocher and W. Dürr (1929) *Ber. Deut. Chem. Ges.*, **62**, 1814.
Frey, A. (1926) *Jahrb. Wiss. Botanik*, **65**, 195.
Frey, A. (1927) *Jahrb. Wiss. Botanik*, **67**, 597.
Frey-Wyssling, A. (1930) *Z. Wiss. Mikroskopie*, **47**, 1.
Frey-Wyssling, A. (1937) *Protoplasma*, **29**, 279.
Frey-Wyssling, A. (1938) *Submikroskopische Morphologie des Protoplasmas und seiner Derivate.* Borntraeger, Berlin.
Frey-Wyssling, A. (1940) *Kolloid-Z.*, **90**, 33.
Frey-Wyssling, A. (1942) *Jahrb. Wiss. Botanik*, **90**, 705.
Frey-Wyssling, A. (1945) *Arch. J. Klaus-Stift.*, Zürich, **20E**, 381.
Frey-Wyssling, A. (1948a) *Schweiz. Brauerei-Rundschau*, 1948, Nr. **1**, 3.
Frey-Wyssling, A. (1948b) *Growth Symp.*, **12**, 151.
Frey-Wyssling, A. (1949a) *Exptl. Cell Res., Suppl.* **1**, 33.
Frey-Wyssling, A. (1949b) *Research*, **2**, 300.
Frey-Wyssling, A. (1951) *Holz als Roh- und Werkstoff*, **9**, 333.
Frey-Wyssling, A. (1952) *Ber. Schweiz. Botan. Ges.*, **62**, 583.
Frey-Wyssling, A. (1953) *Submicroscopic Morphology of Protoplasm*, 2nd Eng. ed., Elsevier, Amsterdam.
Frey-Wyssling, A. (1954) *Nature*, **173**, 596.
Frey-Wyssling, A. (1955a) *Biochim. Biophys. Acta*, **17**, 155.
Frey-Wyssling, A. (1955b) *Biochim. Biophys. Acta*, **18**, 166.
Frey-Wyssling, A. (1957) *Macromolecules in Cell Structure*, Prather Lectures 1956, Harvard Univ. Press, Cambridge Mass.
Frey-Wyssling, A. (1959) *Die pflanzliche Zellwand.* Springer, Berlin.
Frey-Wyssling, A. (1959) *IX. Intern. Botan. Congr. Montreal, 1959*, in: *Recent Advances in Botany*, University Press Toronto, 1959/61, p. 737.
Frey-Wyssling, A. (1960) *Nova Acta Leopoldina*, **22**, 17.
Frey-Wyssling, A. (1962) *Symp. Intern. Soc. Cell Biol.*, **1**, 307.
Frey-Wyssling, A. (1962/63) *Lecture 6, Permeability Conf. Wageningen, 1962.* Publ. Willerik, Zwolle (Netherl.) 1963.
Frey-Wyssling, A. (1964a) in *Formation of Wood in Forest Trees*, M. H. Zimmermann (Ed.), Acad. Press, New York, p. 153.
Frey-Wyssling, A. (1964b) *Z. Wiss. Mikroskopie*, **66**, 45.
Frey-Wyssling, A. (1964c) *Arch. Julius Klaus Stift.* (Zürich), **39**, (107).
Frey-Wyssling, A., H. H. Bosshard and K. Mühlethaler (1956) *Planta*, **47**, 115.
Frey-Wyssling, A. and M. S. Buttrose (1961) *Makromol. Chem.*, **44–46**, 173.
Frey-Wyssling, A., E. Grieshaber and K. Mühlethaler (1963) *J. Ultrastruct. Res.*, **8**, 506.
Frey-Wyssling, A. and E. Kreutzer (1958a) *Planta*, **51**, 104.
Frey-Wyssling, A. and E. Kreutzer (1958b) *J. Ultrastruct. Res.*, **1**, 397.
Frey-Wyssling, A., J. F. López-Sáez and K. Mühlethaler (1964) *J. Ultrastruct. Res.*, **10**, 422.
Frey-Wyssling, A. and K. Mühlethaler (1949) *Vierteljahresschr. Naturforsch. Ges. Zürich*, **94**, 179.
Frey-Wyssling, A. and K. Mühlethaler (1950) *Vierteljahresschr. Naturforsch. Ges. Zürich*, **95**, 45.
Frey-Wyssling, A. and K. Mühlethaler (1951a) *Fortschr. Chem. Org. Naturstoffe*, **8**, 1.
Frey-Wyssling, A. and K. Mühlethaler (1951b) *Mikroskopie (Wien)*, **6**, 28.
Frey-Wyssling, A. and K. Mühlethaler (1959) *Vierteljahresschr. Naturforsch. Ges. Zürich*, **104**, 294.
Frey-Wyssling, A. and K. Mühlethaler (1960) *Schweiz. Z. Hydrobiol.*, **22**, 121.
Frey-Wyssling, A. and K. Mühlethaler (1963) *Makromol. Chem.*, **62**, 25.
Frey-Wyssling, A., K. Mühlethaler and H. H. Bosshard (1955/56) *Holz (Berlin)*, **13**, 245; **14**, 161.
Frey-Wyssling, A., K. Mühlethaler and H. H. Bosshard (1959) *Holzforsch. Holzverwert. (Wien)*, **11**, 107.
Frey-Wyssling, A., K. Mühlethaler and R. W. G. Wyckoff (1948) *Experientia*, **4**, 475.

Frey-Wyssling, A. and H. R. Müller (1957) *J. Ultrastruct. Res.*, **1**, 38.
Frey-Wyssling, A., F. Ruch and X. Berger (1955) *Protoplasma*, **45**, 97.
Frey-Wyssling, A. and E. Steinmann (1948) *Biochim. Biophys. Acta*, **2**, 254.
Frey-Wyssling, A. and E. Steinmann (1953) *Vierteljahresschr. Naturforsch. Ges. Zürich*. **98**, 20
Frey, H. P. (1959) Dr. Thesis ETH, Zürich, 1959 and *Holz als Roh- und Werkstoff*, **17**, 313.
Frey, R. (1950) Dr. Thesis ETH, Zürich, 1950 and *Ber. Schweiz. Botan. Ges.*, **60**, 199.
Fricke, H. (1926) *J. Gen. Physiol.*, **9**, 137.
Friedrich-Freksa, H., O. Kratky and A. Sekora (1944) *Naturwiss.*, **32**, 78.
Fritz, F. (1935) *Jahrb. Wiss. Botanik*, **81**, 718.
Fritz, F. (1937) *Planta*, **26**, 693.
Fujii, K. (1926) *Proc. Japan. Assoc. Advan. Sci.*, **2**, 1.
Gall, J. G. (1958) *John Hopkins Symp.* 1958, p. 103.
Gavaudan, P., H. Poussel and M. Guyot (1960) *Compt. Rend. Sci. Paris*, **250**, 429.
Gay, H. (1956) *Cold Spring Harbor Symp. Quant. Biol.*, **21**, 257.
Geitler, L. (1938) *Chromosomenbau*. Borntraeger, Berlin.
Geitler, L. (1940) *Ber. Deut. Botan. Ges.*, **58**, 131.
Gibbons, I. R. (1960) *Proc. Europ. Reg. Conf. Electron Microscopy, Delft, 1960*, Vol. **2**, 929, A. L. Houwink and B. J. Spit (Eds.). Ned. Ver. Electronenmicroscopie, Delft.
Gibbons, I. R. and A. V. Grimstone (1960) *J. Biophys. Biochem. Cytol.*, **7**, 697.
Gibor, A. and M. Izawa (1963) *Proc. Natl. Acad. Sci. (Wash.)*, **50**, 1164.
Gibson, K. D., M. Matthew, A. Neuberger and G. H. Tait (1961) *Nature*, **192**, 204.
Gicklhorn, J. (1932) *Protoplasma*, **15**, 90.
Gierer, A. and G. Schramm (1956) *Nature*, **177**, 702.
Giesbrecht, P. (1958) *4th Intern. Kongr. Elektronenmikroskopie Berlin, 1958, Verhandl.* Vol. **2**, 251 (Springer, Berlin, 1958/60).
Giesbrecht, P. (1960) *Zentr. Bakteriol. Hyg.*, **179**, 538.
Giesbrecht, P. (1961) *Zentr. Bakteriol. Hyg.*, **183**, 1.
Goedheer, J. L. (1955) *Biochim. Biophys. Acta*, **16**, 471.
Golgi, C. (1882/85) *Riv. Sper. Frenetria (Reggio-Emilia)*, **8**, 165, 361 (1882); **9**, 1, 161, 385 (1883); **11**, 11, 72 (1885).
Goodman, H. M. and A. Rich (1963) *Nature*, **199**, 318.
Goodwin, T. W. (1958) *Handbuch Pflanzenphysiol.*, Springer, Berlin, 1958, **10**, 186.
Goodwin, T. W. (1960) *Handbuch Pflanzenphysiol.*, Springer, Berlin, 1960, **5/1**, 394.
Gorter, E. and F. Grendel (1925) *J. Exptl. Med.*, **41**, 439.
Gray, J. (1955) *J. Exptl. Biol.*, **32**, 775.
Green, D. E. (1964) *Sci. Am.*, **210**, 63.
Green, D. E. and S. Fleischer (1963) *Biochim. Biophys. Acta*, **70**, 554.
Green, P. B. (1958) *Am. J. Botany*, **45**, 111.
Green, P. B. (1960) *J. Biophys. Biochem. Cytol.*, **7**, 289.
Green, P. B. (1962) *Science*, **138**, 1404.
Green, P. B. (1963) In *Cytodifferential and Macromolecular Synthesis*. Academic Press, New York, p. 203.
Green, P. B. (1964) *Am. J. Botany*, **51**, 334.
Green, P. B. and J. C. Chen (1960) *Z. Wiss. Mikroskopie*, **64**, 482.
Grieshaber, E. (1964) Dr. Thesis ETH, Zürich and *Vierteljahresschr. Naturforsch. Ges. Zürich*, **109**, 1.
Guénin, H. A. (1963) Oral communication.
Guilliermond, A. (1922) *Compt. Rend. Acad. Sci. Paris*, **175**, 283.
Guilliermond, A., G. Mangenot and L. Plantefol (1933) *Traité de Cytologie végétale*, Le François, Paris.
Guttes, E. and S. Guttes (1960) *Exptl. Cell Res.*, **20**, 239.
Haguenau, Françoise (1958) *Rev. Cytol.*, **7**, 423.
Håkansson, A. and A. Levan (1942) *Hereditas (Lund)*, **28**, 436.
Hall, C. E. (1960) *J. Biophys. Biochem. Cytol.*, **7**, 613.
Hall, C. E. and M. Litt (1958) *J. Biophys. Biochem. Cytol.*, **4**, 1.
Hall, C. E. and H. S. Slayter (1959) *J. Biophys. Biochem. Cytol.*, **5**, 11.
Hanstein, J. v. (1880) *Bot. Abhandl. Morphol. Physiol. (Bonn)*, **4**, Heft 2.

Harris, J. I. and P. Roos (1956) *Nature*, **178**, 90.
Harris, J. I., F. Sanger and M. A. Naughton (1956) *Arch. Biochem. Biophys.*, **65**, 427.
Harris, Patricia (1962) *J. Cell Biol.*, **14**, 475.
Harris, Patricia and D. Mazia (1962) *Symp. Intern. Soc. Cell Biol.*, **1**, 279.
Harvey, E. B. (1936) *Biol. Bull.*, **71**, 101.
Harvey, E. N. (1936) *J. Cellular Comp. Physiol.*, **8**, 251.
Harvey, E. N. (1937) *Trans. Faraday Soc.*, **33**, 943.
Hassenkamp, Gisela (1957) *Naturwiss.*, **44**, 334.
Haworth, W. N. (1925) *Nature*, **116**, 430.
Haworth, W., E. L. Hirst and J. J. Webb (1928) *J. Chem. Soc. London*, 1928, p. 2681.
Hegetschweiler, R. (1949) Dr. Thesis ETH, Zürich, 1949 and *Makromol. Chem.*, **4**, 156.
Heidenhain, M. (1907) *Plasma und Zelle*. Fischer, Jena, Bd. **1**.
Heierle, E. (1935) Dr. Thesis ETH, Zürich 1935 and *Ber. Schweiz. Botan. Ges.*, **44**, 17.
Heilbronn, A. (1914) *Jahrb. Wiss. Botanik*, **54**, 357.
Heinen, W. (1960) *Acta Botan. Neerl.*, **9**, 167.
Heinen, W. (1961) *Acta Botan. Neerl.*, **10**, 171.
Heitz, E. (1935) *Z. indukt. Abstamm.- Vererb. Lehre*, **70**, 402.
Heitz, E. (1954) *Exptl. Cell Res.*, **7**, 606.
Heitz, E. (1958) *Z. Naturforsch.*, **13b**, 663.
Heitz, E. (1959) *Z. Naturforsch.*, **14b**, 179.
Heitz, E. and R. Maly (1953) *Z. Naturforsch.*, **8b**, 243.
Helmcke, J. G. (1953/54) *Diatomeenschalen im elektronenmikroskopischen Bild.* Atlas in 2 Volumen, Transmare-Photo, Berlin.
Helmcke, J. G. (1954) *Naturwiss.*, **41**, 254.
Helmcke, J. G. and W. Krieger (1951) *Ber. Deut. Botan. Ges.*, **64**, (27).
Helmcke, J. G. and W. Krieger (1952) *Ber. Deut. Botan. Ges.*, **65**, 70.
Helmcke, J. G. and H. Richter (1951) *Z. Wiss. Mikroskopie*, **60**, 189.
Hengartner, Helen (1961) Dr. Thesis ETH, Zürich 1961 and *Holz als Roh- u. Werkstoff*, **19**, 303.
Hengstenberg, J. (1928) *Z. Krist.*, **67**, 583.
Henneguy, L. F. (1897) *Arch. Anat. Microscopie*, **1**, 481.
Herčík, F. (1955) *Biochim. Biophys. Acta*, **18**, 1.
Hersh, R. T. and H. K. Schachmann (1958) *Virology*, **6**, 234.
Hershey, A. D. and M. Chase (1952) *J. Genet. Physiol.*, **36**, 39.
Heslop-Harrison, J. (1962) *Nature*, **195**, 1069.
Heslop-Harrison, J. (1963) *Planta*, **60**, 243.
Heslop-Harrison, J. and Y. Heslop-Harrison (1958) *Portugaliae Acta Biologica*, **A5**, 79.
Heyn, A. N. J. (1936) *Nature*, **137**, 277.
Heyn, A. N. J. (1936) *Protoplasma*, **25**, 372.
Hirai, N., T. Yasui, S. Fujita and Y. Yamashita (1963) *Chem. High Polymers (Japan)*, **20**, 413.
Hirano, T. and C.C. Lindegren (1963) *J. Ultrastruct. Res.*, **8**, 322.
Hirsch, G. C. (1931) *Verhandl. Deut. Zool. Ges.*, 1931, p. 302.
Hirsch, G. C. (1958) *Naturwiss.*, **45**, 349.
Hirschle, J. (1942) *Naturwiss.*, **30**, 642.
Höber, R. (1922) *Physikalische Chemie der Zelle und der Gewebe*. Engelmann, Leipzig.
Hodge, A. J. (1959) *Rev. Mod. Physics*, **31**, 331.
Hodge, A. J., J. D. McLean and F. V. Mercer (1955) *J. Biophys. Biochem. Cytol.*, **1**, 605.
Hodge, A. J., J. D. McLean and F. V. Mercer (1956) *J. Biophys. Biochem. Cytol.*, **2**, 597.
Höfler, K. (1932) *Protoplasma*, **15**, 462.
Höfler, K. (1934) *Sitzungsber. Akad. Wiss. Wien, Math. Naturwiss. Kl., Abt.* I, **143**, 213.
Höfler, K. (1957) *Protoplasma*, **48**, 167.
Hofmeister, W. (1867) *Die Lehre von der Pflanzenzelle*. Leipzig.
Hogeboom, G. H., W. C. Schneider and G. E. Palade (1947) *Proc. Soc. Exptl. Biol. Med. N.Y.*, **65**, 320.
Hohl, H. R. (1960) *Ber. Schweiz. Botan. Ges.*, **70**, 395.
Höhnel, F. v. (1877) *Sitzungsber. Akad. Wiss. Wien, Abt. I*, **76**, 507.
Holter, H. (1959) *Intern. Rev. Cytol.*, **8**, 481.
Holter, H., M. Ottesen and R. Weber (1953) *Experientia*, **9**, 346.

Hooke, R. (1667) *Micrographia (or some physiological description of minute bodies made by magnifying glasses)*. London.
Horne, R. W. and A. P. Waterson (1959) *J. Mol. Biol.*, **2**, 75.
Horne, R. W. and P. Wildy (1961) *Virology*, **15**, 348.
Hoshino, M. (1961) *Exptl. Cell Res.*, **24**, 606.
Houwink, A. L. (1953) *Biochim. Biophys. Acta*, **10**, 360.
Houwink, A. L. and P. A. Roelofsen (1954) *Acta Botan. Neerl.*, **3**, 387.
Howatson, A. F. (1962) *Science*, **135**, 625.
Hubert, B. (1935) *Rec. Trav. Botan. Néerl.*, **32**, 323.
Huggins, M. L. (1957) *Proc. Natl. Acad. Sci. (Wash.)*, **43**, 209.
Husemann, E. and H. Ruska (1940) *Naturwiss.*, **28**, 534.
Huskins, C. L. and S. G. Smith (1935) *Ann. Botany*, **49**, 119.
Huxley, H. E. and G. Zubay (1960) *J. Mol. Biol.*, **2**, 10.
Hyde, B. B., A. J. Hodge and M. L. Birnstiel (1962) *5th Intern. Congr. Electron Microscopy Philadelphia, 1962*, Vol. **2**, T-1 (Acad. Press New York, 1962).
Ingen-Housz, J. (1779) *Experiments upon Vegetables.*
Ingram, V. M. (1957) *Nature*, **180**, 326.
Innamorati, M. (1963) *Giorn. Botan. Ital.*, **70**, 537.
Inoué, Sh. (1953) *Chromosoma*, **5**, 487.
Iwanowski, D. (1892) *St. Petersburg Acad. Imp. Sci. Bull.*, **35**, 67.
Jaccard, P. and A. Frey (1928) *Jahrb. Wiss. Botanik*, **68**, 844.
Jacob, F. and J. Monod (1961) *J. Mol. Biol.*, **3**, 318.
Jacquez, J. A. and J. J. Biesele (1954) *Exptl. Cell Res.*, **6**,17.
Jagendorf, A. T. and S. G. Wildman (1954) *Plant Physiol.*, **29**, 270.
Jakobi, G. (1963) *Z. Naturforsch.*, **18b**, 711.
Jarosch, R. (1961) *Protoplasma*, **53**, 34.
Jayme, G., G. Hunger and D. Fengel (1960) *Holzforsch.*, **14**, 97.
Jenny, A. and K. Grossenbacher (1963) *Proc. Soil Sci. Soc. Am.*, **27**, 273.
Johnston, F. B., G. Setterfield and H. Stern (1959) *J. Biophys. Biochem. Cytol.*, **6**, 53.
Jordan, P. (1947) *Eiweissmoleküle.* Wissenschaftliche Verlagsges., Stuttgart.
Jørgensen, E. G. (1953) *Physiol. Plantarum*, **6**, 301.
Jungers, V. (1930/33) *Cellule*, **40**, 5 (1930); **42**, 5 (1933).
Jutte, S. M. and B. J. Spit (1963) *Holzforsch.*, **17**, 168.
Kaesberg, P. (1956) *Science*, **124**, 626.
Kahler, H. and B. J. Lloyd (1953) *Biochim. Biophys. Acta*, **10**, 355.
Karlson, P. (1961) *Kurzes Lehrbuch der Biochemie*, 2. Auflage, Thieme Verlag, Stuttgart.
Karrer, P. (1935) *Schweiz. Med. Wochschr.*, **65**, 898.
Katz, J. R. and J. C. Derksen (1933) *Z. Physik. Chem.*, **A 165**, 228.
Kellenberger, E. (1957) *Nova Acta Leopoldina*, **19**, 55.
Kellenberger, E. (1961) *Advan. Virus Res.*, **8**, 1.
Kellenberger, E. (1963) Oral communication, Jan. 1963.
Kendrew, J. C., G. Bodo, H. M. Dintzis, R. G. Parrish, H. Wyckoff and D. C. Phillips (1958) *Nature*, **181**, 662.
Kendrew, J. C., R. E. Dickerson, B. E. Strandberg, R. G. Hart, D. R. Davies, D. C. Phillips and V. C. Shore (1960) *Nature*, **185**, 422.
Keosian, J. (1960) *Science*, **131**, 479.
Kerr, Th. and I. W. Bailey (1934) *J. Arnold Arboretum*, **15**, 327.
Kessler, G. (1958) Dr. Thesis ETH, Zürich 1957 and *Ber. Schweiz. Botan. Ges.*, **68**, 5.
Kessler, G., D. S. Feingold and W. Z. Hassid (1960) *Plant Physiol.*, **35**, 505.
Klein, J. (1882) *Jahrb. Wiss. Botanik*, **13**, 60.
Kleinschmidt, A. K. and D. Lang (1962) *5th Intern. Congr. Electron Microscopy Philadelphia, 1962*, Vol. **2**, 0-8 (Acad. Press, New York, 1962).
Kleinschmidt, A. K., D. Lang, D. Jacherts and R. K. Zahn (1962) *Biochim. Biophys. Acta*, **61**, 857.
Klug, A. and D. L. D. Caspar (1960) *Advan. Virus Res.*, **7**, 225.
Koehler, J. K. (1962) *Nature*, **194**, 757.
Kolbe, R. W. and E. Gölz (1943) *Ber. Deut. Botan. Ges.*, **61**, 91.

Kollmann, R. and W. Schumacher (1961) *Planta*, **57**, 583.
Kollmann, R. and W. Schumacher (1962) *Planta*, **58**, 366.
Kratky, O. and S. Kuriyama (1931) *Z. Physik. Chem.*, **B 11**, 363.
Kratky, O., W. Menke, A. Sekora, B. Paletta and M. Bischof (1959) *Z. Naturforsch.*, **14b**, 305.
Kreger, D. R. (1951) *Biochim. Biophys. Acta*, **6**, 406.
Kreger, D. R. (1954) *Biochim. Biophys. Acta*, **13**, 1.
Kreger, D. R. (1957) *Nature*, **180**, 914.
Kreger, D. R. (1960) *Proc. Koninkl. Ned. Akad. Wetenschap*, **C 63**, 613.
Kreutz, W. and W. Menke (1962) *Z. Naturforsch.*, **17b**, 675.
Kroeger, H. (1963) *J. Cellular Comp. Physiol.*, **62**, Suppl. 1, 45.
Küster, E. (1933) *Hundert Jahre Tradescantia*. G. Fischer, Jena.
Küster, E. (1935) *Die Pflanzenzelle*. G. Fischer, Jena. 1st ed., 1935; 2nd ed. 1951.
Labaw, L. W. and R. W. G. Wyckoff (1955) *Nature*, **176**, 455.
Labaw, L. W. and R. W. G. Wyckoff (1958) *J. Ultrastruct. Res.*, **2**, 8.
Lambertz, P. (1954) *Planta*, **44**, 147.
Lange, P. W. (1944/45) *Svensk Papp. Tidn.*, **47**, 263 (1944); **48**, 241 (1945).
Lange, P. W. (1954) *Svensk Papp. Tidn.*, **57**, 525.
Latta, H. (1962) *J. Ultrastruct. Res.*, **6**, 407.
Lauffer, M. A. (1938) *J. Physic. Chem.*, **42**, 935.
Lauffer, M. A. (1950) *Med. Physics*, **2**, 1142.
Lazarow, A. and S. J. Cooperstein (1953) *Exptl. Cell Res.*, **5**, 56.
Ledbetter, M. C. and K. R. Porter (1963) *J. Cell Biol.*, **19**, 239.
Lee, J. C. (1964) *J. Roy. Microscop. Soc.*, **83**, 229.
Lehmann, F. E. (1947) *Rev. Suisse Zool.*, **54**, 246.
Lenhossek, M. (1898) *Verhandl. Deut. Anat. Ges., Jena*, **12**, 106.
Leupold, U. (1958) *Cold Spring Harbor Symp. Quant. Biol.*, **23**, 161.
Leuthardt, F. (1949) *Vierteljahresschr. Naturforsch. Ges. Zürich*, **94**, 132.
Leuthardt, F. (1961) *Lehrbuch der Physiologischen Chemie*. 14. Auflage, de Gruyter, Berlin, 1961.
Leuthardt, F. and B. Exer (1953) *Helv. Chim. Acta*, **36**, 500.
Levinthal, C. and H. R. Crane (1956) *Proc. Natl. Acad. Sci. (Wash.)*, **42**, 436.
Lewis, W. H. (1931) *Bull. John Hopkins Hosp.*, **49**, 17.
Leyon, H. (1951) *Arkiv för Kemi*, **3**, 105.
Leyon, H. (1954) *Exptl. Cell Res.*, **7**, 609.
Liang, C. Y. and R. H. Marchessault (1959) *J. Polymer Sci.*, **37**, 385.
Lichtenthaler, H. K. and R. B. Park (1963) *Nature*, **198**, 1070.
Lindberg, O. and L. Ernster (1954) Chemistry and Physiology of Mitochondria and Microsomes. *Protoplasmatologia*, Band III A 4, Springer, Wien.
Linnane, A. W., E. Vitols and A. G. Nowland (1962) *J. Cell Biol.*, **13**, 345.
Loeser, Ch. N. and S. S. West (1962) *Ann. New York Acad. Sci.*, **97**, 346.
Lundegårdh, H. (1961) *Nature*, **192**, 243.
Lundegårdh, H. (1962) *Physiol. Plantarum*, **15**, 390.
Luyet, B. J. and R. Ernst (1934) *Proc. Soc. Exptl. Biol.*, **31**, 1225.
McAlear, J. H. and G. A. Edwards (1959) *Exptl. Cell Res.*, **16**, 689.
McArthur, I. (1943) *Nature*, **152**, 38.
Mader, H. (1954) *Planta*, **43**, 163.
Magnus, A. (1922) *Z. Anorg. Chem.*, **124**, 291.
Maltzahn, K. v. and K. Mühlethaler (1962a) *Experientia*, **18**, 315.
Maltzahn, K. v. and K. Mühlethaler (1962b) *Naturwiss.*, **49**, 308.
Mangin, L. (1890) *Compt. Rend. Acad. Sci. Paris*, **110**, 644.
Manton, I. (1950) *Biol. Rev.*, **25**, 486.
Manton, I. (1952) *Symp. Soc. Exptl. Biol.*, **6**, 306.
Manton, I. (1955) *Nature*, **176**, 123.
Manton, I. (1956) Plant cilia and associated organelles, in *Cellular Mechanisms in Differentiation and Growth*, D. Rudnick (Ed.), Princeton University Press, Princeton N.J., 1956, p. 61.
Manton, I. and B. Clarke (1952) *J. Exptl. Botany*, **3**, 265.
Marchessault, R. H. and C. Y. Liang (1962) *J. Polymer Sci.*, **59**, 357.
Marinos, N. G. (1960) *J. Ultrastruct. Res.*, **3**, 328.

Marinos, N. G. (1963) *J. Ultrastruct. Res.*, **9**, 177.
Marks, M. H., R. S. Bear and Ch. H. Blake (1949) *J. Exptl. Zool.*, **111**, 55.
Marmur, J. and P. Doty (1959) *Nature*, **183**, 1427.
Marquardt, H. (1957) Programm *7. Tagung Deut. Ges. Elektronenmikroskopie, Darmstadt, 23/25 Sept., 1957.*
Marsh, R. E., R. B. Corey and L. Pauling (1955) *Biochim. Biophys. Acta*, **16**, 1.
Marsland, D. A. (1942) Protoplasmic streaming in relation to gel structure in the cytoplasm, in: *The Structure of Protoplasm*, W. Seifriz (Ed.), Iowa State College Press, Ames Iowa, 1942, p. 127.
Marx, M. (1955) *Makromol. Chem.*, **16**, 157.
Marx-Figini, M. and G. V. Schulz (1963) *Makromol. Chem.*, **62**, 49.
Matile, Ph. (1964) Oral communication.
Matthey, R. (1945) *Experientia*, **1**, 50.
Matzke, E. B. (1946) *Am. J. Botany*, **33**, 58.
May, L. H. and M. S. Buttrose (1959) *Australian J. Biol. Sci.*, **12**, 146.
Mazia, D. (1959) The role of thiol groups in the structure and function of the mitotic apparatus, in *Sulphur in Proteins*. R. Benesch *et al.* (Eds.), Academic Press, New York, 1959, p. 367.
Mazia, D. and K. Dan (1952) *Biol. Bull.*, **103**, 283.
Mazia, D. and L. Jaeger (1939) *Proc. Natl. Acad. Sci. (Wash.)*, **25**, 456.
Meier, H. (1955) Dr. Thesis ETH, Zürich and *Holz als Roh- und Werkstoff*, **13**, 323.
Meier, H. (1956) *Proc. Stockholm Conf. Electron Microscopy, 1956*, p. 298.
Meier, H. (1961) *J. Polymer Sci.*, **51**, 11.
Menke, W. (1934) *Protoplasma*, **21**, 279; **22**, 56.
Menke, W. (1938) *Z. Physiol. Chem.*, **257**, 43.
Menke, W. (1943) *Biol. Zentr.*, **63**, 326.
Menke, W. (1958) *Z. Botanik*, **46**, 26.
Menke, W. (1961) *Z. Naturforsch.*, **17b**, 188.
Menke, W. (1962) *Ann. Rev. Plant Physiol.*, **13**, 27.
Mercer, E. H. (1957) *Nature*, **180**, 87.
Mercer, F. V. and N. Rathgeber (1962) *5th Intern. Congr. Electron Microscopy, Philadelphia 1962*, Vol. **2**, WW-11 (Acad. Press, New York, 1962).
Meves, F. (1904) *Ber. Deut. Botan. Ges.*, **22**, 284.
Meves, F. (1908) *Arch. Mikroskop. Anat. Entwicklungsgeschichte*, **72**, 816.
Meyer, A. (1883) *Das Chlorophyllkorn in chemischer, morphologischer und biologischer Beziehung.* Leipzig.
Meyer, K. H. (1940) *Die hochpolymeren Verbindungen.* Akad. Verlagsges., Leipzig.
Meyer, K. H. (1943) *Melliand Textilber.*, No. 3.
Meyer, K. H. and H. Mark (1928) *Ber. Deut. Chem. Ges.*, **61**, 1932.
Meyer, K. H. and H. Mark (1930) *Der Aufbau der hochpolymeren organischen Naturstoffe.* Akad. Verlagsges., Leipzig.
Meyer, K. H. and L. Misch (1937) *Helv. chim. Acta*, **20**, 232.
Meyer, Madeleine (1938) Dr. Thesis ETH, Zürich, 1938 and *Protoplasma*, **29**, 552.
Michel, K. (1943) *Zeiss Nachrichten*, **4**, 236.
Miller, S. L. and H. C. Urey (1959) *Science*, **130**, 245.
Mitchison, J. M. (1950) *Nature*, **166**, 313.
Moelwyn-Hughes, E. A. (1957) *Physical Chemistry.* Pergamon Press, London.
Mollenhauer, H. H., W. G. Whaley and J. H. Leech (1961) *J. Ultrastruct. Res.*, **5**, 193.
Mollenhauer, H. H. and W. G. Whaley (1962) *5th Intern. Congr. Electron Microscopy, Philadelphia 1962*, **2**, YY-3, Acad. Press, New York.
Monné, L. (1948) *Advan. Enzymol.*, **8**, 1.
Moor, H. (1959) Dr. Thesis ETH, Zürich, 1959 and *J. Ultrastruct. Res.*, **2**, 393.
Moor, H. (1964) *Z. Zellforsch.*, **62**, 546.
Moor, H. and K. Mühlethaler (1963) *J. Cell Biol.*, **17**, 609.
Moor, H., K. Mühlethaler, H. Waldner and A. Frey-Wyssling (1961) *J. Biophys. Biochem. Cytol.*, **10**, 1.
Moore, R. T. and J. H. McAlear (1963a) *J. Ultrastruct. Res.*, **8**, 144.
Moore, R. T. and J. H. McAlear (1963b) *J. Cell Biol*, **16**, 131.

Moses, M. J. (1958) *J. Biophys. Biochem. Cytol.*, **4**, 633.
Moses, M. J. (1958/60) *4. Intern. Kongr. Elektronenmikroskopie Berlin, 1958, Verhandl.*, **2**, 199, Springer, Berlin, 1958/60.
Mühlethaler, K. (1949) *Biochim. Biophys. Acta*, **3**, 15.
Mühlethaler, K. (1950a) *Biochim. Biophys. Acta*, **5**, 1.
Mühlethaler, K. (1950b) *Exptl. Cell Res.*, **1**, 341.
Mühlethaler, K. (1950c) *Ber. Schweiz. Botan. Ges.*, **60**, 614.
Mühlethaler, K. (1953) *Mikroskopie (Wien)*, **8**, 143.
Mühlethaler, K. (1955) *Planta*, **46**, 1.
Mühlethaler, K. (1958) *4. Intern. Kongr. Elektronenmikroskopie Berlin, 1958, Verhandl.*, **2**, 491, Springer, Berlin, 1958/60.
Mühlethaler, K. (1959) *Fortschr. Botanik*, **21**, 55.
Mühlethaler, K. (1959) *IX. Intern. Botan. Congr. Montreal, 1959*, in *Recent Advances in Botany*, University Press, Toronto, 1959/61, p. 732.
Mühlethaler, K. (1960a) *Beih. Z. Schweiz. Forstverein*, No. **30**, 55.
Mühlethaler, K. (1960b) *Z. Wiss. Mikroskopie*, **64**, 444.
Mühlethaler, K. (1961) Plant Cell Walls, in *The Cell*, J. Brachet and A. E. Mirsky (Eds.), Academic Press, New York, 1961, II, p. 86.
Mühlethaler, K. and P. R. Bell (1962) *Naturwiss.*, **49**, 63.
Mühlethaler K. and R. Braun (1946) *Ber. Schweiz. Botan. Ges.*, **56**, 360.
Mühlethaler, K. and A. Frey-Wyssling (1959) *J. Biophys. Biochem. Cytol.*, **6**, 507.
Mühlethaler, K. and H. Moor (1964) *10th Intern. Botan. Congr. Edinburgh, 1964*, Abstracts p. 214.
Müller, A. (1929) *Z. Krist.*, **70**, 386.
Müller, H. O. and C. W. A. Pasewaldt (1942) *Naturwiss.*, **30**, 55.
Murakami, S. and A. Takamiya (1962) *5th Intern. Congr. Electron Microscopy Philadelphia, 1962*, **2**, XX-12, Acad. Press, New York, 1962.
Myrbäck, K. and K. Ahlborg (1942) *Biochem. Z.*, **311**, 213.
Nägeli, C. (1858) *Die Stärkekörner*. Schulthess, Zürich.
Nägeli, C. (1862) *Sitzungsber. Akad. Wiss. München*, **4(2)**, 121.
Nägeli, C. and C. Cramer (1855) *Pflanzenphysiologische Untersuchungen*. 1. Heft, F. Schulthess, Zürich.
Nathanson, A. (1904) *Jahrb. Wiss. Botanik*, **39**, 607.
Nebel, B. R. and E. M. Coulon (1962) *Chromosoma*, **13**, 292.
Neurath, H. and K. Bailey (1953) *The Proteins*. Acad. Press, New York, Vol. **1 A**, p. 220.
Nicolai, E. and A. Frey-Wyssling (1938) *Protoplasma*, **30**, 401.
Nirenberg, M. W. (1963) *Sci. Am.*, **208**, 80.
Nirenberg, M. W. and J. H. Matthaei (1961) *Proc. Natl. Acad. Sci.*, **47**, 1588.
Noack, K. and E. Timm (1942) *Naturwiss.*, **30**, 453.
Novikoff, A. B. (1961) Lysosomes and related particles, in *The Cell*, J. Brachet and A. E. Mirsky (Eds.), Academic Press, New York, 1961, Vol. II, p. 423.
Novikoff, A. B., H. Beaufay and C. de Duve (1956) *J. Biophys. Biochem. Cytol.*, **2** (Suppl.), 179.
Ochoa, S. (1964a) *Experientia*, **20**, 57.
Ochoa, S. (1964b) *XXV Aniversario Consejo superior de investigaciones cientificas*. Coloquio sobre problemas actuales de Biologia. Madrid, Oct., 1964.
O'Kelley, J. C. (1953) *Plant Physiol.*, **28**, 281.
Oosawa, F. and M. Kasai (1962) *J. Mol. Biol.*, **4**, 10.
Oparin, A. (1957/58) *Publ. Acad. Sci. USSR Moskau*, 1957. Ref. in *Science*, **127**, 346 (1958) and *Nature*, **180**, 886 (1957).
Osborne, T. B. (1924) *The Vegetable Proteins*, 2nd ed., Longmans Green, London.
Oster, G. (1950) *J. Gen. Physiol.*, **33**, 445.
Oster, G. (1951) *Z. Elektrochem.*, **55**, 529.
Overton, E. (1899) *Vierteljahresschr. Naturforsch. Ges. Zürich*, **44**, 88.
Palade, G. E. (1953) *J. Histochem. Cytochem.*, **1**, 188.
Palade, G. E. (1955) *J. Biophys. Biochem. Cytol.*, **1**, 59.
Palade, G. E. (1956) *J. Biophys. Biochem. Cytol.*, **2** (Suppl.), 85.
Palade, G. E. (1958) Microsomes and ribunocleoprotein particles in *Microsomal Particles and*

Protein Synthesis, R. B. Roberts (Ed.), 1st. Symp. Biophys. Soc., Cambridge, Mass., 1958 Pergamon Press, New York, 1958, p. 36.
Palade, G. E. and K. R. Porter (1954) *J. Exptl. Med.*, **100**, 641.
Palmer, K. J. and M. B. Hartzog (1945) *J. Am. Chem. Soc.*, **67**, 2122.
Park, R. B. and J. Biggins (1964) *Science*, **144**, 1009.
Park, R. B. and N. G. Pon (1961) *J. Mol. Biol.*, **3**, 1.
Parsons, D. F. (1963) *Science*, **140**, 985.
Pätau, K. (1935) *Naturwiss.*, **23**, 537.
Pauling, L. (1953) *General Chemistry*. Freeman, San Francisco.
Pauling, L. (1960) *The Nature of the Chemical Bond*. 3rd. ed., Cornell University Press, Ithaca N.Y.
Pauling, L. and R. B. Corey (1953) *Nature*, **171**, 59.
Pauling, L. and R. B. Corey (1954) *Fortschr. Chem. Org. Naturstoffe*, **11**, 180.
Pearse, E. (1961) *Histochemistry*. Churchill, London.
Penso, G. (1955) *Protoplasma*, **45**, 251.
Perner, E. S. (1953) *Protoplasma*, **42**, 457.
Perner, E. S. (1958a) *Protoplasma*, **49**, 407.
Perner, E. S. (1958b) *Protoplasmatologia*, Bd. III **A 2**, Springer, Wien.
Perutz, M. F., M. G. Rossmann, A. F. Cullis, H. Muirhead and G. Will (1960) *Nature*, **185**, 416.
Peters, D. (1958) *4. Intern. Kongr. Elektronenmikroskopie Berlin, 1958, Verhandl.*, **2**, 552, Springer, Berlin, 1958/60.
Peveling, Elisabeth (1961) *Planta*, **56**, 530.
Peveling, Elisabeth (1962) *Protoplasma*, **55**, 429.
Pfeffer, W. (1886) *Botan. Ztg.*, **44**, 114.
Pfeffer, W. (1890) *Abhand. Sächs. Ges. Wiss., Math.-physik. Kl.*, **16**, 185.
Pfitzer, E. (1882) *Diatomeen*. In Schenks' *Handbuch der Botanik*. Bd. 2, Breslau.
Pischinger, A. (1937) *Z. Zellforsch. Mikroskop. Anat.*, **26**, 249.
Pischinger, A. (1950) *Protoplasma*, **39**, 567.
Plowe, J. Q. (1931) *Protoplasma*, **12**, 196.
Policard, A. (1958) *Compt. Rend. Acad. Sci. Paris*, **246**, 3194.
Pollister, A. W. and P. F. Pollister (1957) *Intern. Rev. Cytol.*, **6**, 85.
Polson, A. and R. W. G. Wyckoff (1947) *Nature*, **160**, 153.
Ponder, E. (1961) The cell membrane and its properties, in *The Cell*. J. Brachet and A. E. Mirsky (Eds.), Academic Press, New York, 1961, Vol. II, p. 1.
Porter, K. R. (1948) *Anat. Record*, **100**, 72.
Porter, K. R. (1961) *Biol. Struct. Function*, **1**, 127.
Porter, K. R. and J. B. Caulfield (1958) *4. Intern. Kongr. Elektronenmikroskopie Berlin, 1958, Verhandl.*, **2**, 503, Springer Berlin, 1958/60.
Porter, K. R., A. Claude and E. F. Fullam (1945) *J. Exptl. Med.*, **81**, 233.
Porter, K. R. and R. D. Machado (1960) *J. Biophys. Biochem. Cytol.*, **7**, 167.
Possingham, J. V., M. Vesk and F. V. Mercer (1964) *J. Ultrastruct. Res.*, **11**, 68.
Poux, N. (1962a) *5th Intern. Congr. Electron Microscopy Philadelphia, 1962*, **2**, W-2 (Acad. Press, New York).
Poux, N. (1962b) *J. Microscopie*, **1**, 55.
Preston, R. D. (1952) *The Molecular Architecture of Plant Cell Walls*. Wiley, New York.
Preston, R. D. (1962) *Symp. Intern. Soc. Cell Biol.*, **1**, 325.
Preston, R. D. and R. B. Duckworth (1946) *Proc. Leeds Phil. Soc. (Sci. Sect.)*, **4**, 343.
Preston, R. D., E. Nicolai, R. Reed and A. Millard (1948) *Nature*, **162**, 665.
Pringsheim, N. (1854) *Untersuchungen über den Bau und die Bildung der Pflanzenzelle*. Hirschwald, Berlin.
Regaud, Cl. (1910) *Arch. Anat. Microscopie*, **11**, 291.
Rich, A. (1963) *Sci. Am.*, **209**, 44.
Rich, A. and F. H. C. Crick (1958) Structure of collagen, in *Recent Advances in Gelatin and Glue Research*. G. Stainsby (Ed.), Pergamon, New York, 1958, p. 20.
Richards, F. M. (1958) *Proc. Natl. Acad. Sci.*, **44**, 162.
Riley, D. P. and G. Oster (1951) *Biochim. Biophys. Acta*, **7**, 527.
Ris, H. (1961) *Canadian J. Genet. Cytol.*, **3**, 95.

Ris, H. (1962) *5th Intern. Congr. Electron Microscopy Philadelphia, 1962*, **2**, XX-1, Acad. Press, New York, 1962.
Robertson, J. D. (1958) *4. Intern. Kongr. Elektronenmikroskopie Berlin, 1958, Verhandl.*, **2**, 159, Springer, Berlin, 1958/60.
Robertson, J. D. (1959) *Biochem. Soc. Symp. No. 16*, p. 3, Cambridge University Press, Cambridge, U.S.A., 1959.
Robinow, C. F. (1956) *Bacteriol. Rev.*, **20**, 207.
Roelofsen, P. A. (1952) *Acta Botan. Neerl.*, **1**, 99.
Roelofsen, P. A. (1959) *The Plant Cell Wall.* Borntraeger, Berlin.
Roelofsen, P. A. and A. L. Houwink (1951) *Protoplasma*, **40**, 1.
Rosselet, A. (1963) Diploma Thesis ETH, Zürich (unpublished).
Roth, L. E. and E. W. Daniels (1962) *J. Cell Biol.*, **12**, 57.
Roth, L. E. and R. A. Jenkins (1962) *5th Intern. Congr. Electron Microscopy Philadelphia, 1962*, **2**, NN-3, Acad. Press, New York, 1962.
Rouiller, Ch. and E. Fauré-Fremiet (1958) *Exptl. Cell Res.*, **14**, 47.
Rowley, J. R. (1959) *Grana Palynologica*, **2**, 3.
Rowley, J. R. (1962) *Science*, **137**, 526.
Rowley, J. R., K. Mühlethaler and A. Frey-Wyssling (1959) *J. Biophys. Biochem. Cytol.*, **6**, 537.
Rozsa, G. and R. W. G. Wyckoff (1950) *Biochim. Biophys. Acta*, **6**, 334.
Ruch, F. (1949) Dr. Thesis ETH, Zürich, 1949 and *Chromosoma*, **3**, 358 (1949).
Ruch, F. (1957) *Exptl. Cell Res.*, **4**, 58.
Ruch, F. (1960) *Z. Wiss. Mikroskopie*, **64**, 453.
Ruch, F. (1961) *Leitz-Mitt. Wiss. Techn.*, **1**, 250.
Ruch, F. and Helen Hengartner (1960) *Beih. Z. Schweiz. Forstver.*, No. **30**, 75.
Ruhland, W. (1912) *Jahrb. Wiss. Botanik*, **51**, 376.
Ruhland, W. and Ursula Heilmann (1951) *Planta*, **39**, 91.
Rundle, R. E., J. F. Foster and R. R. Baldwin (1944) *J. Am. Chem. Soc.*, **66**, 2116.
Runnström, J. (1929) *Protoplasma*, **5**, 201.
Ruppel, H. G. (1964) *Biochim. Biophys. Acta*, **80**, 63.
Ruska, H. (1962) *4. Int. Kongr. Neuropathol.*, Bd. **2**, 42, Thieme, Stuttgart.
Russow, E. (1883) *Botan. Zentr.*, **13**, 171.
Rustad, R. C. (1959) *Nature*, **183**, 1058.
Rybak, B. and M. Bricka (1952) *Experientia*, **8**, 265.
Ryle, A. P., F. Sanger, L. F. Smith and R. Kitai (1955) *Biochem. J.*, **60**, 541.
Ryter, A. and E. Kellenberger (1958) *Z. Naturforsch.* **13b**, 597.
Sager, R. and G. E. Palade (1957) *J. Biophys. Biochem. Cytol.*, **3**, 463.
Sanders, E. and C. T. Ashworth (1961) *Exptl. Cell Res.*, **22**, 137.
Sanger, F. and H. Tuppy (1951) *Biochem. J.*, **49**, 463, 481.
Satô, S. (1958) *Cytologia*, **23**, 383.
Satô, S. (1960) *Cytologia*, **25**, 119.
Sauer, K. and M. Calvin (1962) *J. Mol. Biol.*, **4**, 451.
Saurer, W. (1962) Diploma Thesis ETH, Zürich (unpublished).
Scarth, G. W. (1927) *Protoplasma*, **2**, 189.
Schaffer, F. L. and C. E. Schwerdt (1959) *Advan. Virus Res.*, **6**, 159.
Scheraga, H. A. (1961) *Protein Structure.* Academic Press, New York and London.
Scherrer, A. (1915) Dr. Thesis University, Zürich, 1914 and *Flora*, **7**, 1 (1915).
Schimper, A. F. W. (1885) *Jahrb. Wiss. Botanik*, **16**, 1.
Schlote, F. W. (1960) *Zool. Anz., Suppl.*, **23**, 478.
Schmid, R. and R. D. Machado (1963) *Holz als Roh- und Werkstoff*, **21**, 41.
Schmidt, W. J. (1932) *Naturwiss.*, **20**, 658.
Schmidt, W. J. (1936) *Z. Zellforsch. Mikroskop. Anat.*, **23**, 657.
Schmidt, W. J. (1937a) *Z. Wiss. Mikroskopie*, **54**, 159.
Schmidt, W. J. (1937b) *Die Doppelbrechung von Karyoplasma, Zytoplasma und Metaplasma* Borntraeger, Berlin.
Schmidt, W. J. (1939) *Protoplasma*, **32**, 193.
Schmitt, F. O. (1938) *J. Appl. Physics*, **9**, 109.
Schmitt, F. O. (1960) *Bull. New York Acad. Med.*, **36**, 725.

Schmitt, F. O., J. Gross and J. H. Highberger (1953) *Proc. Natl. Acad. Sci. (Wash.)*, **39**, 459.
Schmitt, F. O., J. Gross and J. H. Highberger (1955) *Symp. Soc. Exptl. Biol.*, **9**, 148.
Schmitt, F. O. and K. J. Palmer (1940) *Cold Spring Harbor Symp. Quant. Biol.*, **8**, 94.
Schmitz, F. (1884) *Jahrb. Wiss. Botanik*, **15**, 1.
Schneider, L. and K. E. Wohlfarth-Bottermann (1959) *Protoplasma*, **51**, 377.
Schnepf, E. (1961) *Z. Naturforsch.*, **16b**, 605.
Schnepf, E. (1963) *Flora*, **153**, 1.
Schnepf, E. (1964) *Protoplasma*, **63**, 137.
Scholtyseck, E. and R. Danneel (1963) *Protoplasma*, **56**, 99.
Schoute, J. C. (1913) *Rec. Trav. Botan. Néerl.*, **10**, 153.
Schramm, G., G. Schumacher and W. Zillig (1955) *Natura*, **175**, 549.
Schulz, G. V. (1936) *Z. Physik. Chem.*, **A 176**, 317.
Schumacher, W. (1957) *Ber. Deut. Botan. Ges.*, **70**, 335.
Schwegler, F. (1964) Diploma Thesis ETH, Zürich (unpublished).
Scott, F. M., K. C. Hamner, E. Baker and E. Bowler (1956) *Am. J. Botany*, **43**, 313.
Scott, F. M., K. C. Hamner and E. Baker (1957) *Science*, **125**, 399.
Seki, M. (1952) *Arch. Histol. Japon.*, **3**, 465.
Semmens, C. S. and P. N. Bhaduri (1939) *Stain Technol.*, **14**, 1.
Senn, G. (1908) *Gestaltveränderungen der Pflanzenchromatophoren.* Engelmann, Leipzig.
Setterfield, G. and S. T. Bayley (1957) *Canad. J. Botany*, **35**, 435.
Setterfield. G. and S. T. Bayley (1958) *Exptl. Cell Res.*, **14**, 622.
Setterfield, G. and S. T. Bayley (1961) *Ann. Rev. Plant Physiol.*, **12**, 35.
Seybold, A. (1941) *Botan. Arch.*, **42**, 254.
Shatkin, A. J. and E. L. Tatum (1959) *J. Biophys. Biochem. Cytol.*, **6**, 423.
Shinke, N. (1959) *The Nucleus*, **2**, 161.
Siekevitz, Ph. and G. E. Palade (1958) *J. Biophys. Biochem. Cytol.*, **4**, 309.
Sievers, A. (1963) *Protoplasma*, **56**, 188.
Simon, E. W. and J. A. Chapman (1960) *J. Exptl. Botany*, **12**, 414.
Sitte, P. (1953) *Mikroskopie (Wien)*, **8**, 290.
Sitte, P. (1954) *Rapport European Congress T.E.M., Brussels, 1954*, p. 83.
Sitte, P. (1955) *Mikroskopie (Wien)*, **10**, 178.
Sitte, P. (1958) *Protoplasma*, **49**, 447.
Sitte, P. (1963a) *Protoplasma*, **56**, 197.
Sitte, P. (1963b) *Protoplasma*, **57**, 304.
Sjöstrand, F. S. (1953) *Nature*, **171**, 30.
Sjöstrand, F. S. (1956) *Intern. Rev. Cytol.*, **5**, 496.
Sjöstrand, F. S. (1963) *J. Ultrastruct. Res.*, **9**, 561.
Smith, J. H. C. (1948) *Arch. Biochem.*, **19**, 449.
Smith, J. H. C. (1960) Protochlorophyll Transformations, in *Comparative Biochemistry of Photoreactive Systems*, M. B. Allen (Ed.), Academic Press, New York and London, 1960, p. 257.
Smith, J. H. C. and D. W. Kupke (1956) *Nature*, **178**, 751.
Smith, K. M. and R. C. Williams (1958) *Endeavour*, **17**, 12.
Smyth, D. G., S. Moore and W. H. Stein (1963) *J. Biol. Chem.*, **238**, 227.
Speich, H. (1942) Dr. Thesis ETH, Zürich 1942 and *Ber. Schweiz. Botan. Ges.*, **52**, 175.
Staehelin, T., C. C. Brinton, F. O. Wettstein and H. Noll (1963) *Nature*, **199**, 865.
Stafford, H. A. (1951) *Physiol. Plantarum*, **4**, 696.
Stanley, W. M. (1936) *J. Biol. Chem.*, **115**, 673.
Stäubli, W. (1957) Dr. Thesis ETH, Zürich.
Staudinger, H. (1936) *Z. Angew. Chem.*, **49**, 801.
Staudinger, H. (1937) *Svensk. Kem. Tidsskr.*, **49**, 3.
Steere, R. L. (1963) *Science*, **140**, 1089.
Steffen, K. and F. Walter (1955) *Naturwiss.*, **42**, 395.
Steffen, K. and F. Walter (1958) *Planta*, **50**, 640.
Steinert, M. and A. B. Novikoff (1960) *J. Biophys. Biochem. Cytol.*, **8**, 563.
Steinmann, E. (1952) *Exptl. Cell Res.*, **3**, 367.
Steinmann, E. and F. S. Sjöstrand (1955) *Exptl. Cell Res.*, **8**, 15.
Stent, G. S. (1964) *Science*, **144**, 816.

Sterling, C. and B. J. Spit (1957) *Am. J. Botany*, **44**, 851.
Steward, F. C. and K. Mühlethaler (1953) *Ann. Botany*, **17**, 295.
Stich, H. (1956) *Experientia*, **12**, 7.
Stoeckenius, W. (1958) *4. Intern. Kongr. Elektronenmikroskopie, Berlin, 1958, Verhandl.*, **2**, 174, Springer, Berlin, 1958/60.
Stoeckenius, W. (1962) *Symp. Intern. Soc. Cell Biol.*, **1**, 349.
Stoeckenius, W. (1963) *J. Cell Biol.*, **17**, 443.
Stoll, A. (1936) *Naturwiss.*, **14**, 53.
Stoll, A., E. Wiedemann and A. Rüegger (1941) *Verhandl. Schweiz. Naturforsch. Ges. Basel*, p. 125.
Strasburger, E. (1875) *Ueber Zellbildung und Zellteilung*. Fischer, Jena.
Strasburger, E. (1882) *Ueber den Bau und das Wachstum der Zellhäute*. Jena.
Strasburger, E. (1888) *Ueber Kern- und Zellteilung im Pflanzenreiche*. Fischer, Jena.
Straus, W. (1950) *Science*, **112**, 745.
Straus, W. (1954) *Exptl. Cell Res.*, **6**, 392.
Straus, W. (1961) *Protoplasma*, **53**, 405.
Strugger, S. (1950) *Naturwiss.*, **37**, 166.
Strugger, S. (1954) *Protoplasma*, **43**, 120.
Strugger, S. (1956/57) *Naturwiss.*, **43**, 451 (1956); **44**, 543 (1957).
Stryer, L., C. Cohen and R. Langridge (1963) *Nature*, **197**, 793.
Stuart, H. A. (1934) *Die Molekülstruktur*. Berlin.
Stubbe, W. (1962) *Z. Vererb. Lehre*, **93**, 175.
Svedberg, Th. (1938) *Kolloid-Z.*, **85**, 119.
Szarkowski, J. W., M. S. Buttrose, K. Mühlethaler and A. Frey-Wyssling (1960) *J. Ultrastruct. Res.*, **4**, 222.
Tagawa, K., H. Y. Tsujimoto and D. I. Arnon (1963) *Proc. Natl. Acad. Sci.*, **49**, 567.
Takashima, S. (1952) *Nature*, **169**, 182.
Tangl, E. (1879) *Jahrb. Wiss. Botanik*, **12**, 170.
Taylor, J. H. (1958) *Exptl. Cell Res.*, **15**, 350.
Taylor, N. W., H. T. Epstein and M. A. Lauffer (1955) *J. Am. Chem. Soc.*, **77**, 1270.
Thomas, O. L. (1960) *Nature*, **185**, 703.
Tibbs, J. (1957/58) *Biochim. Biophys. Acta*, **23**, 275 (1957); **28**, 636 (1958).
Timberlake, H. G. (1900) *Botan. Gaz.*, **30**, 73 and 154.
Tischler, G. (1934) *Allgemeine Karyologie*, 2. Auflage, **1**, 163 in *Handbuch Pflanzenanat.*, Borntraeger, Berlin.
Tissière, A. and J. D. Watson (1958) *Nature*, **182**, 778.
Todd, A. (1960) *Nature*, **187**, 819.
Törnävä, S. R. (1939) *Protoplasma*, 32, 329.
Tschermak-Woess, Elisabeth and Gertrud Hasitschka (1954) *Oesterr. Botan. Z.*, **101**, 79.
Ts'O, P. O. P. (1962) *Ann. Rev. Plant. Physiol.*, **13**, 45.
Tsugita, A., H. Fraenkel-Conrat, M. W. Nirenberg and H. J. Matthaei (1962) *Proc. Natl. Acad. Sci.*, **48**, 846.
Tuppy, H. (1959) *Naturwiss.*, **46**, 35.
Ursprung, A. and G. Blum (1917) *Ber. Deut. Botan. Ges.*, **35**, 385.
Van Bruggen, E. F. J., E. H. Wiebenga and M. Gruber (1962) *J. Mol. Biol.*, **4**, 1.
Van Iterson, G., K. H. Meyer and W. Lotmar (1936) *Rec. Trav. Chim. Pays-Bas*, **55**, 61.
Van Iterson, Woutera and C. F. Robinow (1961) *J. Biophys. Biochem. Cytol.*, **9**, 171.
Virgin, H. I., A. Kahn and D. v. Wettstein (1962) *Photochem. Photobiol.*, **2**, 83.
Vogel, A. (1953) Dr. Thesis ETH, Zürich, 1953 and *Makromol. Chem.*, **11**, 111.
Vogel, A. (1960) *Beih. Z. Schweiz. Forstver.*, No. **30**, 113.
Vogel, A. (1962) Oral communication.
Wada, B. (1955) *Caryologia*, **7**, 389.
Waldschmidt-Leitz, E. (1958) *Handbuch Pflanzenphysiol.*, **8**, 277, Springer, Berlin.
Walek-Czernecka, Anna (1962) *Acta Soc. Botan. Polon.*, **31**, 539.
Wardrop, A. B. (1955) *Australian J. Botany*, **3**, 137.
Wardrop, A. B. (1956) *Biochim. Biophys. Acta*, **21**, 200.
Wardrop, A. B. (1962) *Botan. Rev.*, **28**, 242.
Wardrop, A. B. and H. E. Dadswell (1953) *Holzforsch.*, **7**, 33.

Wardrop, A. B. and H. E. Dadswell (1955) *Australian J. Botany*, **3**, 177.
Wardrop, A. B., W. Liese and G. W. Davies (1959) *Holzforsch.*, **13**, 115.
Warner, J. R., A. Rich and C. E. Hall (1962) *Science*, **138**, 1399.
Wartiovaara, V. (1944) *Acta Botan. Fennica*, **34**, 1.
Watson, J. D. (1954) *Biochim. Biophys. Acta*, **13**, 10.
Watson, J. D. (1963) *Science*, **140**, 17.
Watson, J. D. and F. H. C. Crick (1953) *Nature*, **171**, 737.
Watson, M. L. (1955) *J. Biophys. Biochem. Cytol.*, **1**, 257.
Watson, M. L. (1959) *J. Biophys. Biochem. Cytol.*, **6**, 147.
Watson, M. L. (1962) *5th Intern. Congr. Electron Microscopy Philadelphia, 1962*, **2**, O-5, Acad. Press, New York, 1962.
Weber, F. (1933) *Protoplasma*, **19**, 455.
Weber, F. (1937) *Protoplasma*, **29**, 427.
Weber, P. (1962) *Z. Naturforsch.*, **17b**, 683.
Wehrmeyer, W. (1963) *Planta*, **59**, 280.
Wehrmeyer, W. and E. Perner (1962) *Protoplasma*, **54**, 573.
Weibull, C. (1951) *Nature*, **167**, 511.
Weidel, W. (1957) *Virus*. Springer, Berlin.
Weidel, W. and E. Kellenberger (1955) *Biochim. Biophys. Acta*, **17**, 1.
Weier, T. E. (1961) *Am. J. Botany*, **48**, 614.
Weier, T. E. (1963) *Am. J. Botany*, **50**, 604.
Weier, T. E., C. R. Stocking, W. W. Thomson and H. Drever (1963) *J. Ultrastruct. Res.*, **8**, 122.
Weier, T. E. and W. W. Thomson (1962) *J. Cell Biol.*, **13**, 89.
Weiling, F. (1961) *Naturwiss.*, **48**, 411.
Weiling, F. (1962) *Protoplasma*, **55**, 372 and 452.
Werner, A. (1904) *Lehrbuch der Stereochemie*. Jena.
Wettstein, D. v. (1957) *Exptl. Cell Res.*, **12**, 427.
Wettstein, D. v. (1959) *J. Ultrastruct. Res.*, **3**, 234.
Whaley, W. G., H. H. Mollenhauer and J. E. Kephart (1959) *J. Biophys. Biochem. Cytol.*, **5**, 501.
Whaley, W. G. and H. H. Mollenhauer (1963) *J. Cell Biol.*, **17**, 216.
Wiener, O. (1912) *Abhandl. Sächs. Ges. Wiss. Leipzig, math.-physik. Kl.*, **32**, 507.
Wildman, S. G. and M. Cohen (1955) *Handbuch Pflanzenphysiol.*, **1**, 243, Springer, Berlin.
Wildman, S. G., T. Hongladarom and S. I. Honda (1962) *Science*, **138**, 434.
Wildy, P. and R. W. Horne (1960) *Proc. Europ. Reg. Conf. Electron Microscopy, Delft, 1960*, **2**, 955. A. L. Houwink and B. J. Spit (Eds.). Ned. Ver. Electronenmicroscopie, Delft.
Wilkins, M. H. F. (1956) *Cold Spring Harbor Symp. Quant. Biol.*, **21**, 75.
Wilkins, M. H. F. and J. T. Randall (1953) *Biochim. Biophys. Acta*, **10**, 192.
Wilkins, M. H. F., W. E. Seeds, A. R. Stokes and H. R. Wilson (1953a) *Nature*, **172**, 759.
Wilkins, M. H. F., A. R. Stokes and H. R. Wilson (1953b) *Nature*, **171**, 738.
Wilkins, M. H. F., A. R. Stokes, W. E. Seeds and G. Oster (1950) *Nature*, **166**, 127.
Winkler, K. C. and H. G. Bungenberg de Jong (1940/41) *Arch. Néerl. Physiol.*, **25**, 431 and 467.
Wischnitzer, S. (1958) *J. Ultrastruct. Res.*, **1**, 201.
Wittmann, H. G. (1961) *Naturwiss.*, **48**, 729.
Wittmann, H. G. (1962) *Z. Vererb. Lehre*, **93**, 491.
Wittmann, H. G. (1963) *Naturwiss.*, **50**, 76.
Wohlfarth-Bottermann, K. E. (1960) *Protoplasma*, **52**, 58.
Wohlfarth-Bottermann, K. E. and F. Krüger (1954) *Z. Naturforsch.*, **9b**, 30.
Wohlfarth-Bottermann, K. E. and V. Moericke (1959) *Z. Naturforsch.*, **14b**, 446.
Wolken, J. J. (1956a) *J. Cellular Comp. Physiol.*, **48**, 349.
Wolken, J. J. (1956b) *J. Protozool.*, **3**, 211.
Wolken, J. J. (1959) *Am. Scientist*, **47**, 202.
Wollgiehn, R. and K. Mothes (1963) *Naturwiss.*, **50**, 95.
Wrischer, M. (1960) *Naturwiss.*, **47**, 522.
Wuhrmann-Meyer, K. and M. (1941) *Planta*, **32**, 43.
Wuhrmann, K. and W. Pilnik (1945) *Experientia*, **1**, 330.
Wyckoff, R. W. G. (1947) *Biochim. Biophys. Acta*, **2**, 139.
Wyckoff, R. W. G. (1948) *Biochim. Biophys. Acta*, **2**, 27.

Wyckoff, R. W. G. (1949) *Electron Microscopy*. Interscience, New York-London.
Yamamoto, T. (1963) *J. Cell Biol.*, **17**, 413.
Yanofsky, S. A. and S. Spiegelman (1962) *Proc. Natl. Acad. Sci. (Wash.)*, **48**, 1069 and 1466.
Yasuzumi, G. and K. Ito (1954) *Heredity*, **45**, 135.
Zeiger, K. (1949) *Z. Zellforsch.*, **34**, 230.
Zeiger, K. (1958) *4. Intern. Kongr. Elektronenmikroskopie Berlin, 1958, Verhandl.*, **2**, 17 (Springer Berlin, 1958/60).
Ziegler, H. (1953) *Z. Naturforsch.*, **8b**, 662.
Zsigmondy, R. (1925) *Kolloidchemie*. 5. Auflage, Leipzig.
Zubay, G. and P. Doty (1959) *J. Mol. Biol.*, **1**, 1.

Author index

Subject Index

Glossary of terms and abbreviations

Å. Abbreviation for Ångström, 10^{-8} cm.

Adenine. A purine base present in DNA, RNA, and nucleotides such as ADP and ATP.

ADP. Adenosine diphosphate.

Aerobiont. An organism unable to live without free oxygen.

Aleurone. Small proteinaceous grains, occuring in seeds.

Amino acids. Organic acids containing an amino group which form the manomeric units of protein molecules.

Amyloplast. Starch synthesising proplastid.

Anaerobiont. An organism able to live in the absence of free oxygen.

Anaphase. The stage in nuclear division in which half chromosomes move to opposite poles of the cell.

Anthocyanin. A blue, purple, or red vacuolar pigment.

Apoenzyme. The protein carrier of a coenzyme.

Archegonium. Female gametangium or egg bearing organ, in which the egg is protected by a jacket of sterile cells.

ATP. Adenosine triphosphate, a high energy organic phosphate of great importance in energy transfer in cellular reactions.

Autoradiograph. A photographic print made by a radioactive substance acting upon a sensitive film.

Bacteriophage. Literally, an eater of bacteria; a virus that infects specific bacteria, multiplies therein, and usually dissolves the bacterial cells.

Base pair. The nitrogen bases that pair in the DNA molecule, adenine ... thymine, and guanine ... cytosine.

Blackman's reaction. Dark reaction of photosynthesis leading to the formation of carbohydrates.

Blepharoblast. A deeply staining granule at the base of a flagellum.

Bordered pit. A pit in tracheids or other cells involved in water conduction having a distinct rim of the cell wall overarching the pit membrane.

Callose. A carbohydrate present as an adcrusting constituent of cell walls, common in sieve tubes.

Calvin cycle. Formation of carbohydrates by condensation of pentose-diphosphate with CO_2 and subsequent cyclic regeneration of the CO_2 acceptor.

Capsid. The proteinic coat of virus particles.

Capsomeres. Particulate knobs observed in the capsid.

Carotene. A reddish-orange terpenoid pigment of plastids.

Cell plate. The first formed wall structure in a dividing cell.

Cellulose. A carbohydrate of the cell wall consisting of chain molecules composed of glucose units.

Centrioles. Small bodies frequently in pairs, situated in the cytoplasm near the nucleus, and involved in the formation of the spindle. (Specially in animal cells.)

Centromere. The point of attachment of the spindle fibre to the chromatide.

Centrosome. Dense globular elongated zone which surrounds the centriole.

Chitin. A nitrogen-containing polysaccharide forming the outer coat of insects and crustaceans and also found in the cell walls of many fungi.

Chlorophyll. The green pigment found in the chloroplast.

Chloroplast. A plastid containing chlorophyll.

Chondriome. The mitochondria of a cell, collectively.

Chromatid. The half chromosome during prophase and metaphase of nuclear divisions.

Chromatin. Substance in the nucleus which contains DNA and therefore readily stains with basic dyes.

Chromocentres. Any accumulation of chromatin in the nucleus.

Chromomeres. Darkly staining granules sometimes seen in a series along prophase chromosomes during meiosis and forming bands in polytenic salivary gland chromosomes.

Chromonemata. Smallest light microscopic strands in chromosomes and chromatids.

Chromoplast. A plastid containing yellow or orange pigments, common in fruits and flowers.

Chromosomes. Rod-like bodies into which the chromatin of the nucleus becomes condensed during meiosis or mitosis.

Cilia. Protoplasmic hairs which, by a whip-like motion, propel certain types of unicellular organisms, gametes, and zoospores through water.

Cistron. A complex unit of the gene probably consisting of from a hundred to thousands of nucleotide pairs.

Code. The template containing the message for protein synthesis.

Codon. A triplet of nucleotides. Each codon selects a particular amino acid for the sequence in polypeptide synthesis.

Coenzyme. A substance, usually nonproteinic and of low molecular weight, necessary for the action of some enzymes.

Coleoptile. The sheath in germination of grass seedlings, which protects the succeeding leaves.

Collenchyma. A supporting tissue composed of cells with walls thickened along the cell edges.

Colloid. Matter composed of particles, ranging in sizes between 0,006 μ and 0,1 μ.

Configuration. The configuration of a molecule specifies the spatial arrangement of bonds without regard to the unicomplexity of spatial arrangements that may arise by rotation about single bonds.

Conformation. The conformations of a molecule are the different spatial arrangements of the atoms that may arise by rotation about single bonds.

Cristae. Crests or ridges, used here to designate invaginations of the inner mitochondrial membrane.

Crossing over. The exchange of corresponding segments between chromatids of homologous chromosomes.

Crystalloid. A crystal of protein which has the ability to swell and to stain (old term).

Cuticle. Lipophilic layer on outer wall of epidermal cells.

Cutin. An insoluble lipid substance which is but slightly permeable to water.

Cytochromes. Organic compounds, containing iron, involved in electron transfer in mitochondria.

Cytoplasm. All the protoplasm of a protoplast besides nucleus, plastids, and mitochondria.

Cytosine. A pyrimidine base found in DNA and RNA.

Diakinesis. Last stage in meiotic prophase where the nuclear membrane breaks down and the chromosomes become shorter by a closer packing of the helices.

Diastase. A complex of enzymes which catalysis the lysis of starch.

Dictyosome. Used as a synonym for Golgi apparatus.

Diploid. Referring to nuclei or cells with a double set of chromosomes.

Diplotene. Fourth stage of meiotic prophase where the paired chromatids begin to separate.

DNA. Deoxyribose nucleic acid; the DNA molecule carries the code of hereditary information.

DP. Degree of polymerisation, number of monomers in a polymer.

DPN. Diphospho-pyridine-nucleotide or more recently NAD (nicotinamide-adenine-dinucleotide).

DPNH. Reduced diphospho-pyridine-nucleotide or more recently NADH (reduced nicotinamide-adenine-dinucleotide).

Ectoplasm. A layer of clear non-granular cytoplasm at the periphery of a cell.

Elaeoplasm. Oil producing protoplasm.

Elementary fibrils. Smallest cellulose strands having a diameter of 30–40 Å.

Endomitosis. Increase of the number of chromatids in a nucleus without visible chromosome divisions.

Endoplasm. The granular central portion of the cytoplasm of a cell.

Enchylema. Serum-like content of the ER.

Enzyme. A protein of complex chemical constitution produced in living cells, which, even in very low concentration, speeds up certain chemical processes but is not used up in the reaction (s. holoenzyme).

ER. Endoplasmic reticulum, a system of capillaries and crypts within the cytoplasm.

Ergastoplasm. Basophilic areas in the cytoplasm.

Etiolation. A condition involving increased stem elongation, poor leaf development, and lack of chlorophyll found in plants growing with unsufficient light or in the dark.

Euchromatin. Parts of chromosomes which usually stain less deeply than heterochromatin.

Exine. Outer coat of pollen.

Fasciation. An arrangement in the form of a bundle.

Ferredoxin. An electron-transferring protein of high iron content involved in photosynthesis and in the biological production and consumption of hydrogen. It transfers electrons with the greatest reducing power in cellular metabolism.

Feulgen. A histochemical test for DNA first published in 1932 by R. Feulgen, a German histologist. In the test the purine deoxyribose linkages are hydrolised by warm, diluted HCl to produce an aldehyde group which rapidly regenerates colour in a decolourised solution of basic fuchsin (Schiff's reagent).

Flagellum. A long slender whip of protoplasm.

Gel. Jelly-like colloidal system with semi-solid properties.

Gene. A discrete substance in the chromosome which determines or conditions one or more hereditary characters.

Globoid. A rounded inclusion in an aleuron grain, containing phytin.

Globulins. Proteins insoluble in water, but soluble in dilute salt solutions.

Glucosidase. Hydrolytic enzyme splitting glucosidic bonds, linking glucose to other carbohydrates, or various compounds termed aglucons.

Glycogen. A carbohydrate related to starch but found generally in the liver of animals and in fungi.

Glycolysis. Decomposition of sugar to pyruvic acid.

Golgi apparatus. Cell organelle named after the Italian cytologist Camillo Golgi (1844–1926) who first described this cell structure.

Grana. Structure within chloroplasts seen as green granules with the light microscope and as a series of parallel lamellae with the electron microscope.

Groundplasm. Homogeneous plasma remaining after cell organelles and particles have been excluded, also termed matrix.

Guanine. A purine base found in DNA and RNA.

Haem. Ferro-porphyrin complex of haemoglobin and various Fe holding enzymes.

Haploid. Referring to nuclei or cells with a single set of chromosomes.

Helix. A structure coiled along the surface of a cylindre.

Hemicelluloses. Polysaccharides (pentosans and hexosans) resembling cellulose.

Heterochromatin. Deeply staining segments of chromosomes.

Hexose. A six carbon sugar.

Hill reaction. Photochemical reaction of photosynthesis involving photolysis of water and formation of reducing agent.

Hilum. Centre of starch grain.

Histones. Proteins rich in basic amino acids found in the nucleus and associated with the chromosomes and the nucleolus.

Holoenzyme. An enzyme composed of apo- and coenzyme.

Hormone. A specific organic substance, produced in one part of a plant or animal body, and transported to another part where it controls or stimulates a specific process.

Hyaloplasm. The portion of the cytoplasm, which appears homogeneous in the light microscope.

Hydrogen acceptor. A substance capable of accepting hydrogen atoms in the oxidation-reduction reactions of metabolism.

Hydrolysis. Union of a compound with water, attended by decomposition into less complex compounds; usually controlled by enzymes.

Imbibition. The absorption of liquids or vapors into the ultrastructural spaces or pores found in such materials as cellulose.

Initials. Youngest stages of cytoplasmic organelles.

Interchromomeres. Regions between chromomeres in the chromosomes.

Intine. The innermost coat of a pollen grain.

Karyolymph. The ground substance of a nucleus in a liquid state (sol).

Karyoplasm. The ground substance of a nucleus in a semi-solid state (gel).

Keratin. Fibrous protein; it constitutes the horny part of hoofs, nails, and hair and contains a high percentage of sulphur.

Kinetosome. Basal granule of a cilium.

Krebs cycle. The breakdown of pyruvic acid to CO_2 and water with the release of energy in respiration.

Leptonema. The first stage of the meiotic prophase in which the unpaired chromosomes appear as very long and slender threads in diploid number.

Leucoplast. A colourless plastid.

Lignin. An organic polymer of phenolic substances incrusting the cellulose framework of certain plant cell walls.

Lipid. A chemical component incompatible with water.

Lipophanerosis. Abnormal formation of numerous lipid globules in ageing protoplasts or during the disintegration of cell organelles.

Lysosome. A cell organelle containing enzymes believed to aid in the digestion of intracellular or extracellular substrates.

Macromolecule. A large molecule with colloidal properties, MW $>$ 1000.

Matrix. Homogeneous ground substance.

Meiosis. Two special cell divisions occuring once in the life cycle of every sexually reproducing plant and animal, reducing the diploid to the haploid chromosome number.

Messenger RNA. The RNA that is produced in the nucleus and that moves into the cytoplasm. It determines the sequence of amino acids in the protein molecule.

Metaphase. The stage of nuclear division during which the chromosomes lie in the equatorial plane of the spindle, called metaphase plate.

Micelle. A term created by Nägeli (1817–1891) for elongated crystalline sublight-microscopic particles.

Microfibrils. Small cellulose strands having a diameter between 100–200 Å, composed of elementary fibrils.

Microtubules. Small tubules 200–300 Å in diameter present in the cytoplasm.

Middle lamella. First membrane separating two adjacent protoplasts and remaining as a distinct cementing layer between adjacent cell walls.

Mitochondrion. A cell organelle associated with intracellular respiration.

Mitosis. Nuclear division involving the appearance of chromosomes.

Mucilage. A substance containing polysaccharides and uronides having slimy properties.

Mutation. The alteration of a hereditable property.

Myelin. A lipid substance which forms the medullary sheath of nerve fibres.

NAD. See DPN.

NADH. See DPNH.

Nadi reaction. Reaction used for the histochemical demonstration of cytochrome oxidase.

NADP. See TPN.

Nexine. Nonsculptured exine.

Nitrogen bases. Adenine, guanine, cytosine, thymine and uracil.

nm. Nanometer (formerly millimicron mμ) = 10^{-9} m.

Nuclear envelope. The double membrane surrounding the nucleus. It is part of the ER.

Nucleolus. Dense protoplasmic body in the nucleus.

Nucleoplasm. The dense groundplasm of the nucleus (see karyoplasm and karyolymph).

Nucleosides. Components of nucleic acids consisting of a nitrogen base and a sugar.

Nucleotide. Monomer of nucleic acids: nucleoside (nitrogen base+sugar)+phosphoric acid.

Nucleus. A dense protoplasmic body essential in cellular synthetic and developmental activities.

Ontogeny. The course of development of an individual organism, an organ or an organelle.

Operon. A set of cistrons, all closely linked on the genetic map and hence residing in contiguous sectors of the DNA.

Osmosis. Diffusion of water through a semipermeable membrane.

Oxysomes. Small particles attached to the mitochondrial cristae.

Pachytene. Third meiotic stage during which the paired chromosomes contract.

Pectin. Compounds found in the cell wall composed primarily of hexuronic acids.

Pentose. A five carbon sugar.

Periderm. A protective layer which develops on those plant organs which last for some time; it consists of cork, the cork cambium, and usually some phelloderm.

Peristromium. A contractile, amoeboid sheath around the chloroplast.

Phage. Virus infecting bacteria.

Phosphorylation. A reaction in which phosphate is added to a compound, *e.g.*, the formation of ATP from ADP and inorganic phosphate.

Photophosphorylation. A reaction in which light energy is converted into chemical energy in the form of ATP.

Photon. A quantum of light; the energy of a photon is proportional to the frequency, *i.e.* reciprocal to the wave length of its radiation.

Photosynthesis. An endergonic process in which carbon dioxide is transformed into carbohydrate, the energy involved being light.

Phragmoplast. The modified spindle region at the equatorial plate after the chromosomes have moved to the poles.

Phylogeny. The course of evolution.

Phytin. Ca, Mg-salt of phytic acid which is inositol hexaphosphoric acid.

Pinocytosis. Active intake of water by invagination of the plasmalemma.

Pit. A recess in the secondary cell wall closed by the primary wall.

Plasmalemma. Plasma membrane surrounding the protoplast.

Plasmodesmata. Fine protoplasmic connections between cells which extend through the wall, common in primary walls and pit membranes.

Plasmolysis. The separation of the cytoplasm from the cell wall, due to the removal of water from the protoplast.

Plastidome. The plastids of a cell, collectively.

Plastids. A class of protoplasmic organelles involved principally in the formation and storage of carbohydrates.

Polymerisation. The chemical union of monomers such as glucose, or nucleotides to form polymers such as starch or nucleic acids.

Polypeptide. Polymeric chain of amino acids.

Polyploid. Referring to nuclei or cells with more than two complete sets of chromosomes.

Polysaccharide. A macromolecule composed of sugar monomers.

Polysomes. Groups of ribosomes associated

with a strand of messenger RNA involved in the production of protein.

Prolamellar body. Accumulation of vesicles formed by the invagination of the proplastid membrane during etiolation.

Promitochondria. Precursors of mitochondria.

Prophase. The early stage in nuclear division, characterised by the appearance of the chromosomes and their movement to the metaphase plate.

Proplastid. The poorly differentiated precursor of a plastid.

Protamins. Basic polypeptides with a molecular weight between 1000–5000, isolated from fish sperms.

Protease. An enzyme breaking down a protein.

Protein. A macromolecule composed of polypeptides.

Proton. The nucleus of a hydrogen atom which is a positively charged particle (H^+).

Protoplasm. Living substance.

Protoplast. The organised living unit of a single cell.

Pyrenoid. A dense starch-containing body occurring within the chloroplasts of certain algae and liverworts.

Quantasomes. Smallest units for photosynthesis in chloroplasts.

Ribosomes. Small particles 100–200 Å in diameter containing RNA. Centres for protein synthesis.

RNA. Ribonucleic acid. m-RNA = messenger, r-RNA = ribosomal, s-RNA = soluble, t-RNA = transfer ribonucleic acid.

s. Sedimentation constant corresponding to one Svedberg unit.

Sexine. Sculptured exine.

Sieve pores. Perforations in the transverse walls of sieve tubes, through which strands of cytoplasm pass. These perforated walls are known as sieve plates.

Sol. A liquid colloidal system.

Spherosomes. Cytoplasmic particles involved in fat synthesis and storage.

Spindle. The framework of achromatin fibres which is formed between the centrosomes during nuclear division.

Sporopollenin. Resistant cell wall substance in the pollen exine.

Starch. Reserve carbohydrate containing amylose and amylopectin.

Stroma. The ground substance of chloroplasts and mitochondria.

Suberin. An insoluble lipophilic material found in the cell walls of cork tissue.

Svedberg unit. Protein units with a molecular weight of 17 500.

Telophase. The last stage of nuclear division, in which daughter nuclei are reorganised.

Thylakoid. Lamellar unit in chloroplasts.

Thymidine-^{3}H. Tritiated or radioactive thymidine (a nucleoside incorporated in DNA; but not in RNA).

Thymine. A pyrimidine base present in DNA.

TMV. Tobacco mosaic virus.

Tonoplast. The cytoplasmic membrane bordering the vacuole.

TPN. Triphospho-pyridine-nucleotide, or more recently termed nicotinamide-adenine-dinucleotide-phosphate (NADP).

TPNH. Reduced triphospho-pyridine-nucleotide or more recently written NADPH as reduced from of NADP.

Tracheids. Water conducting, elongated cells with bordered pits in wood.

UDP. Uridine-diphosphate.

UMP. Uridine-monophosphate.

Unit cell. The smallest group of atoms, ions, or molecules, whose repetition at regular intervals, in three dimensions, produces the lattice of a given crystal.

Unit membrane. The basic unit of the membrane system of the cell; it is characterised by a transparent central stratum lined on both sides by a darker stratum.

Uracil. A pyrimidine base present in RNA.

Uridine. Nucleoside with the base uracil.

Uronides. Carbohydrates with carboxyl groups.

UTP. Uridine-triphosphate.

Vacuole. The centrally located usually large cavity of the plant cell limited by the tonoplast and filled with a dilute salt and sugar solution, called cell sap.

Van der Waal's bonds. Intermolecular forces (auxiliary valencies) caused by a lack of compensation of electric charge in the molecule.

Viruses. A group of disease-producing agents, parasitic in plants and animals, unable to multiply outside the host cells. Invisible in the light microscope.

Watson–Crick model. Referring to the twin-threaded model of the molecular DNA helixstructure.

Waxes. Ester of fatty acids with higher alcohols.

Zygotene. Second stage of meiotic prophase where chromosome pairing takes place.

Zymogene granules. Prosecret granules containing digestive enzymes formed by exocrine pancreas or salivary gland cells.